D1283553

# WCDMA FOR UMTS

# WCDMA FOR UMTS

## HSPA Evolution and LTE

## Fifth Edition

**Edited by**

**Harri Holma and Antti Toskala**

*Nokia Siemens Networks, Finland*

**WILEY**

A John Wiley and Sons, Ltd., Publication

*Library of Congress Cataloging-in-Publication Data*
WCDMA for UMTS: HSPA evolution and LTE / edited by Harri Holma, Antti Toskala. –
5th ed.
    p. cm.
Includes bibliographical references and index.
ISBN 978-0-470-68646-1 (cloth)
1. Code division multiple access. 2. Wireless communication systems – Standards.
3. Mobile communication systems – Standards. 4. Global system for mobile communications.
I. Holma, Harri, 1970- II. Toskala, Antti.
TK5103.452.W39 2010
621.3845 – dc22                                        2010013154

A catalogue record for this book is available from the British Library.
ISBN 978-0-470-68646-1 (H/B)

Typeset in 9/11 Times by Laserwords Private Limited, Chennai, India.
Printed and bound in the United Kingdom by Antony Rowe Ltd, Chippenham, Wiltshire.

# Contents

# Preface

Second generation telecommunication systems, such as GSM, enabled voice traffic to go wireless: the number of mobile phones exceeds the number of landline phones and the mobile phone penetration is approaching 100% in several markets. The data handling capabilities of second generation systems are limited, however, and third generation systems are needed to provide the high bit rate services that enable high quality images and video to be transmitted and received, and to provide access to the web with higher data rates. These third generation mobile communication systems are referred to in this book as UMTS (Universal Mobile Telecommunication System). WCDMA (Wideband Code Division Multiple Access) and its evolution HSPA (High Speed Packet Access) is the main third generation air interface globally. During the publication of the 5th edition, the number of WCDMA/HSPA subscribers has exceeded 500 million. It is expected that the 1 billion landmark will be passed in less than two years. There are over 300 commercial HSPA networks globally supporting peak data rates up to 42 Mbps. HSPA has grown to be the preferred radio network for providing wireless broadband access, for supporting an increasing number of smart phones and for offering high capacity and high quality voice service in an efficient way. This book gives a detailed description of the WCDMA/HSPA air interface and its utilization. The contents are summarized in Figure 1.

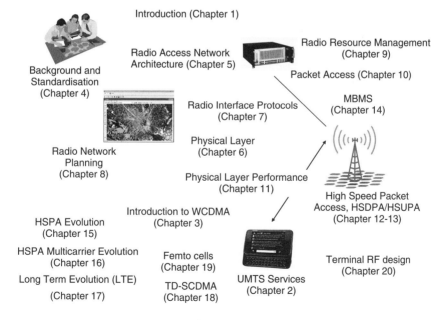

**Figure 1.** Contents of this book

The book is structured as follows. Chapters 1–4 provide an introduction to the technology and its standardization. Chapters 5–7 give a detailed presentation of the WCDMA standard, while Chapters 8–11 cover the utilization of the standard and its performance. Chapters 12–16 present HSPA and its evolution. TD-SCDMA is described in Chapter 18. The home base stations, also called femtocells, are explained in Chapter 19. Chapter 20 covers terminal RF design challenges.

Chapter 1 briefly introduces the background, development, status and future of WCDMA/HSPA radio. Chapter 2 presents examples of the current UMTS applications and the main uses cases. Chapter 3 introduces the principles of the WCDMA air interface, including spreading, Rake receiver, power control and handovers. Chapter 4 presents the background to WCDMA, the global harmonization process and the standardization. Chapter 5 describes the architecture of the radio access network, interfaces within the radio access network between base stations and radio network controllers (RNC), and the interface between the radio access network and the core network. Chapter 6 covers the physical layer (Layer 1), including spreading, modulation, user data and signalling transmission, and the main physical layer procedures of power control, paging, transmission diversity and handover measurements. Chapter 7 introduces the radio interface protocols, consisting of the data link layer (Layer 2) and the network layer (Layer 3). Chapter 8 presents the guidelines for radio network dimensioning, gives an example of detailed capacity and coverage planning, and covers GSM co-planning. Chapter 9 covers the radio resource management algorithms that guarantee the efficient utilization of the air interface resources and the quality of service. These algorithms are power control, handovers, admission and load control. Chapter 10 depicts packet access and presents the performance of packet protocols of WCDMA. Chapter 11 analyses the coverage and capacity of the WCDMA air interface. Chapter 12 presents the significant Release 5 feature, High Speed Downlink Packet Access, HSDPA, and Chapter 13 the corresponding uplink counterpart High Speed Uplink Packet Access, HSUPA in Release 6. Chapter 14 presents Multimedia Broadcast Multicast System, MBMS. Chapter 15 introduces HSPA evolution in Releases 7, 8 and 9. Chapter 16 describes HSPA multicarrier evolution up to four carriers. Long Term Evolution (LTE) in Releases 8 and 9 is presented in Chapter 17. The time division duplex (TDD) based TD-SCDMA (Time Division Synchronous Code Division Multiple Access) is illustrated in Chapter 18. The femtocells are presented in Chapter 19 and the challenges in the terminal RF design in Chapter 20.

The 2nd edition contained coverage of the recently introduced key features of 3GPP Release 5 specifications, such as High Speed Downlink Packet Access, HSDPA and IP Multimedia Subsystem (IMS). The 3rd edition of the book continued to deepen the coverage of several existing topics both based on the field experiences and based on more detailed simulation studies. The 3rd edition covered the main updates in 3GPP standard Release 6. The 4th edition added in detail 3GPP Release 6 features including High Speed Uplink Packet Access (HSUPA) Multimedia Broadcast Multicast System (MBMS), HSPA evolution and terminal RF design challenges.

The 5th edition of the book introduces new material in the areas of HSPA evolution including Releases 8 and 9, HSPA multicarrier solutions, GSM band refarming for HSPA, Integrated Mobile Broadcast (IMB), TD-SCDMA description, femtocells, terminal power consumption estimates, services and LTE.

This book is aimed at operators, network and terminal manufacturers, service providers, university students and frequency regulators. A deep understanding of the WCDMA/HSPA air interface, its capabilities and its optimal usage is the key to success in the UMTS business.

This book represents the views and opinions of the authors, and does not necessarily represent the views of their employers.

# Acknowledgements

The editors would like to acknowledge the time and effort put in by their colleagues in contributing to this book. Besides the editors, the contributors were Dominique Brunel, Leo Chan, Renaud Cuny, Karol Drazynski, Frank Frederiksen, Jacek Gora, Zhi-Chun Honkasalo, Seppo Hämäläinen, Kari Horneman, Markku Juntti, Jorma Kaikkonen, Troels Kolding, Martin Kristensson, Janne Laakso, Jaana Laiho, Fabio Longoni, Atte Länsisalmi, Nina Madsen, Preben Mogensen, Peter Muszynski, Laurent Noël, Maciej Pakulski, Klaus Pedersen, Johanna Pekonen, Patryk Pisowacki, Karri Ranta-aho, Jussi Reunanen, Oscar Salonaho, Jouni Salonen, Hanns-Jürgen Schwarzbauer, Kari Sipilä, Tommi Uitto, Jukka Vialén, Jaakko Vihriälä, Achim Wacker and Jeroen Wigard.

While we were developing this book, many of our colleagues from Nokia and Nokia Siemens Networks offered their help in suggesting improvements and finding errors. Also, a number of colleagues from other companies have helped us in improving the quality of the book. The editors are grateful for the comments received from Heikki Ahava, Erkka Ala-Tauriala, David Astely, Erkki Autio, Matthew Baker, Luis Barreto, Johan Bergman, Angelo Centonza, Kai Heikkinen, Kari Heiska, Kimmo Hiltunen, Klaus Hugl, Alberg Höglund, Kaisu Iisakkila, Ann-Louise Johansson, Kalle Jokio, Susanna Kallio, Istvan Kovacs, Ilkka Keskitalo, Pasi Kinnunen, Tero Kola, Petri Komulainen, Mika Laasonen, Lauri Laitinen, Olivier Claude Lebreton, Anne Leino, Arto Leppisaari, Pertti Lukander, Esko Luttinen, Peter Merz, Wolf-Dietrich Moeller, Risto Mononen, Jonathan Moss, Jari Mäkinen, Magdalena Duniewicz Noël, Olli Nurminen, Tero Ojanperä, Lauri Oksanen, Kari Pajukoski, Kari Pehkonen, Eetu Prieur, Mika Rinne, Sabine Roessel, Rauno Ruismäki, David Soldani, Agnieszka Szufarska, Pekka Talmola, Kimmo Terävä, Mitch Tseng, Antti Tölli, Veli Voipio, Helen Waite and Dong Zhao.

The team at John Wiley & Sons participating in the production of this book provided excellent support and worked hard to keep the demanding schedule. The editors especially would like to thank Sarah Tilley and Mark Hammond for assistance with practical issues in the production process, and especially the copy-editor, for her efforts in smoothing out the engineering approach to the English language expressions.

We are extremely grateful to our families, as well as the families of all the authors, for their patience and support, especially during the late night and weekend editing sessions near different production milestones.

Special thanks are due to our employer, Nokia Siemens Networks, for supporting and encouraging such an effort and for providing some of the illustrations in this book.

Finally, we would like to acknowledge the efforts of our colleagues in the wireless industry for the great work done within the 3rd Generation Partnership Project (3GPP) to produce the global WCDMA standard in merely a year and thus to create the framework for this book. Without such an initiative this book would never have been possible.

The editors and authors welcome any comments and suggestions for improvements or changes that could be implemented in forthcoming editions of this book. The feedback is welcome to editors' email addresses harri.holma@nsn.com and antti.toskala@nsn.com.

# Abbreviations

| | |
|---|---|
| 3GPP | 3rd Generation partnership project (produces WCDMA standard) |
| 3GPP2 | 3rd Generation partnership project 2 (produced cdma2000 standard) |
| AAL2 | ATM Adaptation Layer type 2 |
| AAL5 | ATM Adaptation Layer type 5 |
| ABB | Analog baseband |
| ACELP | Algebraic code excitation linear prediction |
| ACIR | Adjacent channel interference ratio, caused by the transmitter non-idealities and imperfect receiver filtering |
| ACK | Acknowledgement |
| ACL | Access control list |
| ACLR | Adjacent channel leakage ratio, caused by the transmitter non-idealities, the effect of receiver filtering is not included |
| ACTS | Advanced communication technologies and systems, EU research projects framework |
| ADC | Analog to digital conversion |
| AGC | Automatic gain control |
| A-GW | Access gateway |
| AICH | Acquisition indication channel |
| ALCAP | Access link control application part |
| AM | Acknowledged mode |
| AM | Amplitude modulation |
| AMD | Acknowledged mode data |
| AMR | Adaptive multirate (speech codec) |
| AMR-NB | Narrowband AMR |
| AMR-WB | Wideband AMR |
| ARIB | Association of radio industries and businesses (Japan) |
| AOL | America on-line |
| AP | Access point |
| ARP | Allocation and retention priority |
| ARQ | Automatic repeat request |
| ASC | Access service class |
| ASN.1 | Abstract syntax notation one |
| ATM | Asynchronous transfer mode |
| AWGN | Additive white Gaussian noise |
| AWS | Advanced wireless services |

| | |
|---|---|
| BB | Baseband |
| BB SS7 | Broadband signalling system #7 |
| BCCH | Broadcast channel (logical channel) |
| BCFE | Broadcast control functional entity |
| BCH | Broadcast channel (transport channel) |
| BER | Bit error rate |
| BLER | Block error rate |
| BMC | Broadcast/multicast control protocol |
| BM-SC | Broadcast multicast service center |
| BO | Backoff |
| BoD | Bandwidth on demand |
| BOM | Bill of material |
| BPSK | Binary phase shift keying |
| BS | Base station |
| BSC | Base station controller |
| BSS | Base station subsystem |
| CA-ICH | Channel assignment indication channel |
| CB | Cell broadcast |
| CBC | Cell broadcast center |
| CBS | Cell broadcast service |
| CCCH | Common control channel (logical channel) |
| CCH | Common transport channel |
| CCH | Control channel |
| CDD | Cyclic Delay Diversity |
| CDF | Cumulative distribution function |
| CD-ICH | Collision detection indication channel |
| CDMA | Code division multiple access |
| CFN | Connection frame number |
| CIF | Common intermediate format |
| CIR | Carrier to interference ratio |
| CM | Connection management or Cubic metric |
| CMOS | Complementary metal oxide semiconductor |
| CN | Core network |
| C-NBAP | Common NBAP |
| CODIT | Code division test bed, EU research project |
| CPC | Continuous packet connectivity |
| CPCH | Common packet channel |
| CPE | Customer premises equipment |
| CPICH | Common pilot channel |
| CQI | Channel quality indicator |
| CRC | Cyclic redundancy check |
| CRNC | Controlling RNC |
| C-RNTI | Cell-RNTI, radio network temporary identity |
| CS | Circuit Switched |
| CSCF | Call state control function |
| CSG | Closed subscriber group |
| CSICH | CPCH status indication channel |
| CTCH | Common traffic channel |
| CW | Continuous wave |
| CWTS | China wireless telecommunications standard group |

| | |
|---|---|
| DAC | Digital to audio conversion |
| DARP | Downlink advanced receiver performance |
| DBB | Digital baseband |
| DC | Direct current |
| DCA | Dynamic channel allocation |
| DCCH | Dedicated control channel (logical channel) |
| DCFE | Dedicated control functional entity |
| DCH | Dedicated channel (transport channel) |
| DC-HSDPA | Dual cell HSDPA |
| DC-HSPA | Dual cell HSPA |
| DC-HSUPA | Dual cell HSUPA |
| DCR | Direct conversion receiver |
| DDR | Direct digital receiver |
| DECT | Digital enhanced cordless telephone |
| DF | Decision feedback |
| DFCA | Dynamic frequency and channel allocation |
| DL | Downlink |
| D-NBAP | Dedicated NBAP |
| DNS | Domain name system |
| DPCCH | Dedicated physical control channel |
| DPDCH | Dedicated physical data channel |
| DPI | Deep packet inspection |
| DRNC | Drift RNC |
| DRX | Discontinuous reception |
| DS-CDMA | Direct spread code division multiple access |
| DSCH | Downlink shared channel |
| DSL | Digital subscriber line |
| DTCH | Dedicated traffic channel |
| DTX | Discontinuous transmission |
| DVB-T/H | Digital video broadcast terrestrial / handheld |
| DwPTS | Downlink pilot time slot |
| E-AGCH | E-DCH absolute grant channel |
| E-DCH | Enhanced uplink DCH |
| EDGE | Enhanced data rates for GSM evolution |
| E-DPCCH | E-DCH dedicated physical control channel |
| E-DPDCH | E-DCH dedicated physical data channel |
| EFR | Enhance full rate |
| EGSM | Extended GSM |
| E-HICH | E-DCH acknowledgement indicator channel |
| EIRP | Equivalent isotropic radiated power |
| EP | Elementary Procedure |
| EPC | Evolved Packet Core |
| E-PUCH | E-DCH physical uplink channel |
| E-RGCH | E-DCH relative grant channel |
| E-RUCCH | E-DCH random access uplink control channel |
| ETSI | European Telecommunications Standards Institute |
| E-UCCH | The E-DCH uplink control channel |
| E-UTRAN | Evolved UTRAN |
| EVM | Error vector magnitude |

| FACH | Forward access channel |
|------|------------------------|
| FBI | Feedback information |
| FCC | Federal communication commission |
| FCS | Fast cell selection |
| FDD | Frequency division duplex |
| FDMA | Frequency division multiple access |
| FER | Frame error ratio |
| FFT | Fast Fourier transform |
| FP | Frame protocol |
| FPACH | Fast physical access channel |
| FRAMES | Future radio wideband multiple access system, EU research project |
| FTP | File transfer protocol |
| GERAN | GSM/EDGE Radio Access Network |
| GGSN | Gateway GPRS support node |
| GMSC | Gateway MSC |
| GNSS | Global navigation satellite system |
| GP | Guard Period |
| GPRS | General packet radio system |
| GPS | Global positioning system |
| GSIC | Groupwise serial interference cancellation |
| GSM | Global system for mobile communications |
| GTP-U | User plane part of GPRS tunnelling protocol |
| GW | Gateway |
| HARQ | Hybrid automatic repeat request |
| HB | High band |
| HLR | Home location register |
| HNB | Home node B |
| HNBAP | Home node B application part |
| HP | High power |
| HPF | High pass filter |
| HSDPA | High speed downlink packet access |
| HS-DPCCH | Uplink high speed dedicated physical control channel |
| HS-DSCH | High speed downlink shared channel |
| HSS | Home subscriber server |
| HS-SCCH | High speed shared control channel |
| HSUPA | High speed uplink packet access |
| HTML | Hypertext markup language |
| HTTP | Hypertext transfer protocol |
| HUE | Home Node B UE |
| IC | Interference cancellation or Integrated circuit |
| ID | Identity |
| IETF | Internet engineering task force |
| IFFT | Inverse Fast Fourier Transform |
| IMB | Integrated mobile broadcast |
| IMD | Intermodulation |
| IMEISV | International Mobile Station Equipment Identity and Software Version |
| IMS | IP multimedia sub-system |
| IMSI | International mobile subscriber identity |
| IMT-2000 | International mobile telephony, 3rd generation networks are referred as IMT-2000 within ITU |
| IN | Intelligent network |

| | |
|---|---|
| IP | Internet protocol |
| IPDL | Idle periods in downlink |
| IPI | Inter-path interference |
| IPSec | IP security |
| IRC | Interference rejection combining |
| IS-95 | cdmaOne, one of the 2nd generation systems, mainly in Americas and in Korea |
| IS-136 | US-TDMA, one of the 2nd generation systems, mainly in Americas |
| IS-2000 | IS-95 evolution standard, (cdma2000) |
| ISDN | Integrated services digital network |
| ISI | Inter-symbol interference |
| ITU | International telecommunications union |
| ITUN | SS7 ISUP Tunnelling |
| Iu BC | Iu broadcast |
| L2 | Layer 2 |
| LAI | Location area identity |
| LAN | Local area network |
| LB | Low band |
| LCD | Liquid crystal display |
| LCS | Location services |
| LNA | Lower noise amplifier |
| LO | Local oscillator |
| LP | Low pass |
| LTE | Long term evolution |
| MAC | Medium access control |
| MAI | Multiple access interference |
| MAP | Maximum a posteriori |
| MBMS | Multimedia broadcast multicast service |
| MBSFN | Mobile broadcast single frequency network |
| MCCH | MBMS point-to-multipoint control channel |
| MCS | Modulation and coding scheme |
| MCU | Multipoint control unit |
| MDT | Minimization of drive test |
| ME | Mobile equipment |
| MF | Matched filter |
| MGCF | Media gateway control function |
| MGW | Media gateway |
| MHA | Mast head amplifier |
| MIMO | Multiple input multiple output |
| MLSD | Maximum likelihood sequence detection |
| MM | Mobility management |
| MME | Mobility management entity |
| MMS | Multimedia message |
| MMSE | Minimum mean square error |
| MNB | Macro Node B |
| MOS | Mean opinion score |
| MPEG | Motion picture experts group |
| MR-ACELP | Multirate ACELP |
| MRF | Media resource function |
| MS | Mobile station |
| MSCH | MBMS scheduling channel |

| | |
|---|---|
| MSC/VLR | Mobile services switching centre/visitor location register |
| MSN | Microsoft network |
| MT | Mobile termination |
| MTCH | MBMS point-to-multipoint control channel |
| MTP3b | Message transfer part (broadband) |
| MUD | Multiuser detection |
| MUE | Macro UE |
| NAS | Non access stratum |
| NBAP | Node B application part |
| NF | Noise figure |
| NITZ | Network identity and time zone |
| NRT | Non-real time |
| O&M | Operation and maintenance |
| OCNS | Orthogonal channel noise simulator |
| ODMA | Opportunity driven multiple access |
| OFDMA | Orthogonal frequency division multiple access |
| OSS | Operations support system |
| OTDOA | Observed time difference of arrival |
| OVSF | Orthogonal variable spreading factor |
| PA | Power amplifier |
| PAD | Padding |
| PAR | Peak to average ratio |
| PC | Power control |
| PCB | Printed circuit board |
| PCCC | Parallel concatenated convolutional coder |
| PCCCH | Physical common control channel |
| PCCH | Paging channel (logical channel) |
| PCCPCH | Primary common control physical channel |
| PCFICH | Physical control format indicator channel |
| PCH | Paging channel (transport channel) |
| PCI | Precoding information |
| PCMCIA | Personal computer memory card international association |
| PCPCH | Physical common packet channel |
| PCRF | Policy and Charging Rules Function |
| PCS | Personal communication systems, 2nd generation cellular systems mainly in Americas, operating partly on IMT-2000 band |
| PDC | Personal digital cellular, 2nd generation system in Japan |
| PDCP | Packet data convergence protocol |
| PDN | Public data network |
| PDP | Packet data protocol |
| PDSCH | Physical downlink shared channel |
| PDU | Protocol data unit |
| PEP | Performance enhancement proxy |
| PER | Packed encoding rules |
| PF | Proportional fair |
| P-GW | Packet Data Network Gateway |
| PHY | Physical layer |
| PI | Page indicator |
| PIC | Parallel interference cancellation |
| PICH | Paging indicator channel |

| PLL | Phase locked loop |
| PLMN | Public land mobile network |
| PM | Phase modulation |
| PNFE | Paging and notification control function entity |
| POC | Push-to-talk over cellular |
| PRACH | Physical random access channel |
| PS | Packet switched |
| PSC | Physical scrambling code |
| PSCH | Physical shared channel |
| PSTN | Public switched telephone network |
| P-TMSI | Packet-TMSI |
| PU | Payload unit |
| PUCCH | Physical uplink control channel |
| PUSCH | Physical uplink shared channel |
| PDCCH | Physical downlink control channel |
| PLCCH | Physical layer common control channel |
| PSD | Power spectral density |
| PVC | Pre-defined Virtual Connection |
| QAM | Quadrature amplitude modulation |
| QCIF | Quarter common intermediate format |
| QoS | Quality of service |
| QPSK | Quadrature phase shift keying |
| QVGA | Quarter video graphics array |
| RAB | Radio access bearer |
| RACH | Random access channel |
| RAI | Routing area identity |
| RAN | Radio access network |
| RANAP | RAN application part |
| RB | Radio bearer |
| RF | Radio frequency |
| RLC | Radio link control |
| RMC | Reference measurement channel |
| RN | Relay node |
| RNC | Radio network controller |
| RNS | Radio network sub-system |
| RNSAP | RNS application part |
| RNTI | Radio network temporary identity |
| ROHC | Robust header compression |
| RR | Round robin |
| RRC | Radio resource control |
| RRM | Radio resource management |
| RSS | Really Simple Syndication |
| RSSI | Received signal strength indicator |
| RSVP | Resource reservation protocol |
| RT | Real time |
| RTCP | Real-time transport control protocol |
| RTP | Real-time protocol |
| RTSP | Real-time streaming protocol |
| RU | Resource unit |
| RUA | RANAP user adaptation |

| SAAL-NNI | Signalling ATM adaptation layer for network to network interfaces |
|----------|-------------------------------------------------------------------|
| SAAL-UNI | Signalling ATM adaptation layer for user to network interfaces |
| SABP | Service Area Broadcast Protocol |
| SAE | System architecture evolution |
| SAIC | Single antenna interference cancellation |
| SAP | Service access point |
| SAP | Session announcement protocol |
| SAS | Stand alone SMLC |
| SAW | Surface acoustic wave |
| SCCP | Signalling connection control part |
| SCCPCH | Secondary common control physical channel |
| SC-FDMA | Single carrier frequency division multiple access |
| SCH | Synchronization channel |
| SCRI | Signaling connection release indication |
| SCTP | Simple control transmission protocol |
| SDD | Space division duplex |
| SDP | Session description protocol |
| SDQNR | Signal to distortion quantization noise ratio |
| SDU | Service data unit |
| SeGW | Sequrity gateway |
| SEQ | Sequence |
| SF | Spreading Factor |
| SFN | System frame number |
| SFN | Single frequency network |
| SGSN | Serving GPRS support node |
| S-GW | Serving Gateway |
| SHO | Soft handover |
| SIB | System information block |
| SIC | Successive interference cancellation |
| SID | Silence indicator |
| SINR | Signal-to-noise ratio where noise includes both thermal noise and interference |
| SIP | Session initiation protocol |
| SIR | Signal to interference ratio |
| SM | Session management |
| SMLC | Serving mobile location centre |
| SMS | Short message service |
| SN | Sequence number |
| SNR | Signal to noise ratio |
| SoC | System on chip |
| SON | Self optimized networks |
| SQ-PIC | Soft quantized parallel interference cancellation |
| SRB | Signalling radio bearer |
| SRNC | Serving RNC |
| SRNS | Serving RNS |
| SRS | Sounding reference symbol |
| SS7 | Signalling System #7 |
| SSCF | Service specific co-ordination function |
| SSCOP | Service specific connection oriented protocol |
| SSDT | Site selection diversity transmission |

| | |
|---|---|
| STD | Switched transmit diversity |
| STTD | Space time transmit diversity |
| SVOPC | Sinusoidal voice over packet coder |
| TCH | Traffic channel |
| TCP | Transport control protocol |
| TCTF | Target channel type field |
| TD/CDMA | Time division CDMA, combined TDMA and CDMA |
| TDD | Time division duplex |
| TDMA | Time division multiple access |
| TD-SCDMA | Time division synchronous CDMA, 1.28 Mcps TDD |
| TE | Terminal equipment |
| TF | Transport format |
| TFCI | Transport format combination indicator |
| TFCS | Transport format combination set |
| TFI | Transport format indicator |
| TFRC | Transport format and resource combination |
| THP | Traffic handling priority |
| TM | Transparent mode |
| TMGI | Temporary mobile group identity |
| TMSI | Temporary mobile subscriber identity |
| TPC | Transmission power control |
| TR | Transparent mode |
| TS | Technical specification |
| TSTD | Time switched transmit diversity |
| TTA | Telecommunications Technology Association (Korea) |
| TTC | Telecommunication Technology Commission (Japan) |
| TTI | Transmission time interval |
| TxAA | Transmit adaptive antennas |
| UDP | User datagram protocol |
| UE | User equipment |
| UL | Uplink |
| UM | Unacknowledged mode |
| UMD | Unacknowledged mode data |
| UMTS | Universal mobile telecommunication services |
| UpPTS | Uplink pilot time slot |
| URA | UTRAN registration area |
| URL | Universal resource locator |
| U-RNTI | UTRAN RNTI |
| USB | Universal serial bus |
| USCH | Uplink shared channel |
| USIM | UMTS subscriber identity module |
| US-TDMA | IS-136, one of the 2nd generation systems mainly in USA |
| UTRA | UMTS Terrestrial radio access (ETSI) |
| UTRA | Universal Terrestrial radio access (3GPP) |
| UTRAN | UMTS Terrestrial radio access network |
| VAD | Voice activation detection |
| VoIP | Voice over IP |
| VPN | Virtual private network |
| WAP | Wireless application protocol |

| | |
|---|---|
| WARC | World administrative radio conference |
| WCDMA | Wideband CDMA, Code division multiple access |
| WiMAX | Worldwide interoperability for microwave access |
| WLL | Wireless local loop |
| WML | Wireless markup language |
| WWW | World wide web |
| XHTML | Extensible hypertext markup language |
| ZF | Zero forcing |

# 1

# Introduction

Harri Holma and Antti Toskala

## 1.1 WCDMA Early Phase

The research work towards third generation (3G) mobile systems started in the early 1990s. The aim was to develop a radio system capable of supporting up to 2 Mbps data rates. The WCDMA air interface was selected in Japan in 1997 and in Europe in January 1998. The global WCDMA specification activities were combined into a third generation partnership project (3GPP) that aimed to create the first set of specifications by the end of 1999, called Release 99. The first WCDMA network was opened by NTT DoCoMo in Japan 2001, using a proprietary version of the 3GPP specifications. The first 3GPP-compliant network opened in Japan by the end of 2002 and in Europe in 2003 (3 April 2003).

The operators had paid extraordinary prices for the UMTS spectrum in the auctions in the early 2000s and expectations for 3G systems were high. Unfortunately, the take-up of 3G devices and services turned out to be very slow. The global number of WCDMA subscribers was less than 20 million by the end of 2004 and more than 50% of them were located in Japan. The slow take-up can be attributed to many factors: it took time to get the system working in a stable way – the protocol specifications in particular caused a lot of headaches. The terminal suffered from high power consumption and from short talk time. The terminal prices also remained high due to low volumes. The packet-based mobile services had not yet been developed and the terminal displays were not good enough for attractive applications. Also the coverage areas of 3G networks were limited partly due to the high frequency at 2100 MHz.

The early WCDMA networks still offered some benefits for the end users including data rate up to 384 kbps in uplink and in downlink and simultaneous voice and data. WCDMA was also a useful platform for debugging the UMTS protocol layers and the development of wideband RF implementation solutions in the terminals and in the base stations.

WCDMA/HSPA subscriber growth is shown in Figure 1.1. After the slow take-up, the growth accelerated, starting in 2006 and the total number of subscribers was 450 million by the end of 2009, that is seven years (2002–2009) after the launch of the first 3GPP compliant network.

---

*WCDMA for UMTS: HSPA Evolution and LTE, Fifth Edition*   Edited by Harri Holma and Antti Toskala
© 2010 John Wiley & Sons, Ltd

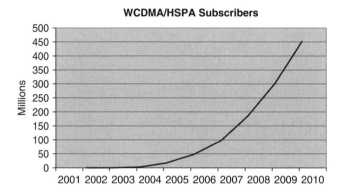

**Figure 1.1**   The growth of WCDMA/HSPA subscribers

## 1.2   HSPA Introduction and Data Growth

The early WCDMA deployments turned out to be important in preparing for the introduction of mobile broadband. 3GPP Release 5 included High Speed Downlink Packet Access (HSDPA) that changed the mobile broadband world. HSDPA brought a few major changes to the radio networks: the architecture became flatter with packet scheduling and retransmissions moving from RNC to the base station, the peak bit rates increased from 0.384 Mbps initially to 1.8–3.6 Mbps and later to 7.2–14.4 Mbps, the spectral efficiency and network efficiency increased considerably and the latency decreased from 200 ms to below 100 ms. The commercial HSDPA networks started at the end of 2005 and more launches took place during 2006. Suddenly, wide area networks were able to offer data rates similar to low end ADSL (Asymmetric Digital Subscriber Line) and were also able to push the cost per bit down so that offering hundreds of megabytes, or even gigabytes of data per month became feasible. The high efficiency also allowed changes to the pricing model, either to be flat rate or gapped flat rate. The HSDPA upgrade to the existing WCDMA network was a software upgrade in the best case without any site visits. The corresponding uplink enhancement, the High Speed Uplink Packet Access (HSUPA), was introduced in 3GPP Release 6. The combination of HSDPA and HSUPA is referred to as HSPA.

HSPA mobile broadband emerged as a highly successful service. The first use cases were PCMCIA (Personal Computer Memory Card International Association) and USB (Universal Serial Bus) modem connected to a laptop and using HSPA as the high data rate bit pipe similar to ADSL. Later also integrated HSPA modems were available in laptops. The typical modems are shown in Figure 1.2. The penetration of HSPA subscriptions exceeded 10% of the population in advanced markets in less than two years from the service launch which made HSPA connectivity one of the fastest growing mobile services.

The flat rate pricing together with high data rates allowed users to consume large data volumes. The average usage per subscriber is typically more than 1 gigabyte per month and it keeps increasing. The combination of more subscribers each using more data caused the total data volume to explode in HSPA networks. An example case from a West European country is shown in Figure 1.3. The growth of the total data volume is compared to the total voice traffic. The voice traffic has been converted to data volume by assuming 16 kbps data rate: 10 minutes of voice converts into 1.2 megabytes of data. The data volumes include both downlink and uplink transmission. Total voice traffic has been growing slowly from 2007 to 2009 from 4.4 terabytes per day to 5.0 terabytes per day. At the same time the data has grown from practically zero to 50 terabytes per day. In other words, 90% of the bits in the radio network are related to the data connections and only 10% to the voice connection in 2009. The wide area networks shifted

Integrated HSPA modem          USB HSPA modem

**Figure 1.2**   Examples of HSPA modems

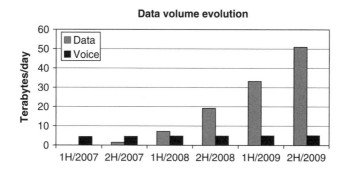

**Figure 1.3**   The growth of HSPA data usage – example European market

from being voice-dominated to being data-dominated in just two years. Note that the data is primarily carried by HSPA networks and the voice traffic by both GSM and WCDMA/HSPA networks. Therefore, if we only look at WCDMA/HSPA networks, the share of data traffic is even larger.

Fast data growth brings the challenge of cost efficiency. More voice traffic brings more revenue with minute-based charging while more data traffic brings no extra revenue due to flat rate pricing – more data just creates more expenses. The HSPA network efficiency has improved considerably especially with Ethernet-based Iub transport and compact new base stations with simple installation, low power consumption and fast capacity expansion. HSPA evolution also includes a number of features that can enhance the spectral efficiency. Quality of Service (QoS) differentiation is utilized to control excessive network usage to keep users happy also during the busy hours.

It is not only USB modems but also the increasingly popular smart phones that have created more traffic in HSPA networks. Example smart phones are shown in Figure 1.4. The smart phones enable a number of new applications including community access, push mail, navigation and widgets in addition to browsing and streaming applications. Those applications create relatively low data volumes but fairly frequent flow of small packets which created a few new challenges for end-to-end performance and for the system capacity. The first challenge was terminal power consumption. The frequent transmission of small packets keeps terminal RF parts running and increases the power consumption. Another challenge is the high signaling load in the networks caused by the frequent packet transmissions. HSPA evolution includes features that cut down the power consumption considerably and also improve the efficiency of small packet transmission in the HSPA radio networks.

**Figure 1.4**   Example 3G/HSPA smart phones

## 1.3   HSPA Deployments Globally

Globally, there are 341 HSPA networks running in 143 countries with a total of over 380 commitments for HSPA launches in May 2010 [1]. HSPA has been launched in all European countries, in practice, in all countries in the Americas, in most Asian countries and in many African countries. The largest HSPA network is run by China Unicom with the first year deployment during 2009 of approximately 150,000 base stations. Another large market – India – is also moving towards large-scale HSPA network rollouts during 2010 when the spectrum auctions are completed. The total number of HSPA base stations globally is expected to exceed 1 million during 2010.

Many governments have recognized that broadband access can boost the economy. If there is insufficient wireline infrastructure, the wireless solution may be the only practical broadband solution. HSPA has developed into a truly global area mobile broadband solution serving as the first broadband access for end users in many new growth markets.

The WCDMA networks started at 2100 MHz band in Asia and in Europe and at 1900 MHz in USA. The high frequency makes the cell size small which limits the coverage area. Therefore, the WCDMA/HSPA networks have recently been deployed increasingly at low frequencies of 850 and at 900 MHz. The lower frequency gives approximately three times larger coverage area than 1900 or 2100 MHz. The first commercial UMTS900 network was opened in 2007 and widespread UMTS900 rollouts started in 2009 when the European Union (EU) changed the regulation to allow UMTS technology in the 900 MHz band. UMTS850 and UMTS900 have clearly boosted the availability of HSPA networks in less densely populated areas.

The bands 850, 900 and 1900 previously were used mainly for GSM. WCDMA/HSPA specifications have been designed for co-existence with GSM on the same band. The commercial networks have shown that WCDMA/HSPA can be operated together with GSM on the same frequency band while sharing even the same base station. The minimum spectrum requirement for WCDMA is 4.2 MHz.

In addition to these four bands, also the AWS (Advanced Wireless Services) band (1700/2100) is used for HSPA in the USA, in Canada and in some Latin American countries, starting in Chile. Japanese networks additionally use two further frequency variants: 1700 by Docomo and 1500 by Softbank. The frequency variants are summarized in Figure 1.5.

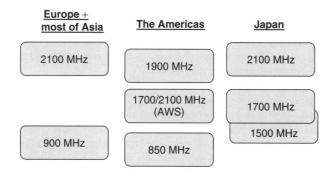

**Figure 1.5**   WCDMA/HSPA frequency variants

The typical HSPA terminals support two or three frequency variants with two upper bands (2100 and 1900) and one lower band (900 or 850). The wide support of 900 and 850 in the terminals makes the low band reframing a feasible option for the operators. Some high end terminals even support five frequency bands 850/900/1700/1900/2100. The number of global frequency variants in HSPA is still small and easier to manage compared to 3GPP LTE where more than 10 different frequency variants are required globally.

## 1.4   HSPA Evolution

3GPP Releases 5 and 6 defined the baseline for mobile broadband access. HSPA evolution in Releases 7, 8 and 9 has further boosted the HSPA capability. Development continues in Release 10 during 2010. The peak bit rate in Release 6 was 14 Mbps downlink and 5.76 Mbps in uplink. The downlink and uplink data rates improve with dual cell HSPA (DC-HSPA), with 3-carrier and 4-carrier HSPA and with higher-order modulation 64QAM downlink and 16QAM uplink. The multicarrier HSPA permits full benefit of 10–20 MHz bandwidth similar to LTE. The downlink data rate can also be increased by a multi-antenna solution (MIMO, Multiple Input Multiple Output). The peak bit rate in Release 9 is 84 Mbps downlink and 23 Mbps uplink. The downlink data rate is expected to double in Release 10 to 168 Mbps by aggregating four carriers together over 20 MHz bandwidth. The data rate evolution is illustrated in Figure 1.6. We can note that the HSPA peak rates are even higher than the best ADSL peak rates in the fixed copper lines, especially in uplink.

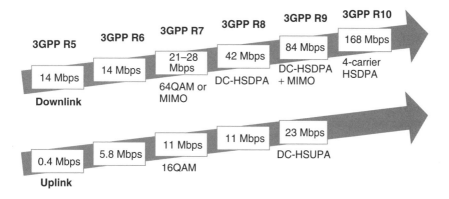

**Figure 1.6**   Evolution of HSPA maximum peak bit rate

End-to-end latency is another part of optimized end user performance. The commercial HSPA networks show that the average round trip time can be pushed to below 30 milliseconds with HSPA evolution offering faster response times for the applications. The radio latency in many cases is no longer the limiting factor. The latency development has been considerable since the early WCDMA networks had a latency of approximately 200 ms.

The terminal power consumption is reduced considerably with HSPA evolution by using discontinuous transmission and reception (DTX/DRX). The voice talk time can be extended to 10–15 hours. The usage time with data applications and always-on services can be pushed relatively even more by using new common channel structures in addition to DTX/DRX.

Voice service has traditionally been by circuit switched (CS) voice. HSPA evolution allows the traditional CS voice on top of HSPA packet radio to be run. The solution is a CS voice from the core network and from a roaming or charging point of view, but it is similar to Voice over IP (VoIP) in the HSPA radio network. The HSPA radio gives clear benefits also for the voice service: better talk time with discontinuous transmission and reception, higher spectral efficiency with HSPA-related performance enhancements and faster call setup time with less than 2 second mobile-to-mobile call setup time.

In short, the 3G network capability has improved enormously from Release 99 to Release 9. The simple reason is that radio has changed completely from the WCDMA circuit connection type operation to HSPA fully packet-based operation. It is possible to run all the service on top of HSPA in Release 9, including packet services, CS voice service, VoIP, common channels, signaling and paging. There are in practice only a few physical layer channels left from the early Release 99 specification in Layer 1 – everything else has been rewritten in 3GPP specifications.

Self Optimized Network (SON) features have been included in 3GPP specifications and in radio network products. SON features allow easier network configuration and optimization, leading to lower operation expenditures and better end user performance. SON features are related, for example, to plug-and-play installation, automatic neighborlist management or antenna optimization. The complexity of the network management increases when the operators use three different radio standards in parallel: GSM, HSPA and LTE. The SON algorithms can help reduce the complexity especially in these multi-radio networks.

## 1.5  HSPA Network Product

The performance and size of the radio network products have seen concentrated development lately. The first phase 3G base station weighed hundreds of kilograms, required more than 1 kW of power and supported less than 10 Mbps of total data capacity when HSDPA was not available. The latest base stations weigh less than 50 kg, consume less than 500 W and support over 100 Mbps data capacity. The fast product development drives down the cost per bit in terms of base station prices and also in terms of installation costs, electricity and transmission costs with the support of IP transport. The way of installing the base stations has also changed. The RF (Radio Frequency) parts of the base station can be installed close to the antenna to minimize losses in the RF cables and to maximize the radio performance. When installed this way, the RF parts are called Remote Radio Units and the signal is transferred to the baseband unit via optical fiber. The length of the fiber can be even up to several kilometers, making also the so-called base station hotel a possible option. The next step in the evolution could be the integration of the antenna and the RF parts. Such a solution is called an active antenna. Development has been even faster in the radio network controller (RNC) where the capacity has increased by a factor of 100 to tens of Gbps while the physical size of the product has become smaller.

Another trend in the radio network products is multi-radio capability where the single product is able to support multiple radio standards simultaneously. The multi-radio is also called Single RAN or Software Defined Radio (SDR) and it is one factor reducing the cost of radio networks. Running just one base station with up to three radio standards costs less than running three separate base stations.

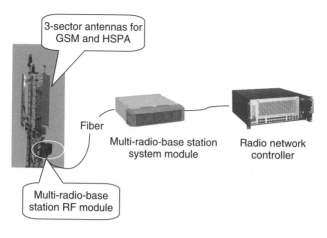

**Figure 1.7** HSPA radio network installation and product evolution

The cost savings come from lower site rental costs, less electricity consumption, smaller operation and maintenance costs, and also transmission costs.

The new base station RF units have much higher output power level capabilities compared to the RF units of the early WCDMA base stations. Originally the typical output power of the carrier was 20 watts while today it has increased to 60 watts and is likely to increase even more. The higher output power has increased the base station coverage area and increased the HSDPA capacity and data rates. Also the sensitivities of the RF receiver have improved which, together with the remote radio unit solution, has improved the overall radio performance significantly.

The typical site installation and the products are shown in Figure 1.7. The size of the base station modules and RNC modules are 20–30 kg which makes it possible for a single person to carry the products during installation.

## 1.6 HSPA Future Outlook

The power of HSPA lies in the capability to support simple CS voice service, high data rate broadband data and smart phone always-on applications all with a single network in an efficient way. There is no other radio technology with similar capabilities. Global HSPA market size has grown tremendously and it will keep the HSPA ecosystem running and evolving for many years. WCDMA/HSPA terminal sales exceeded CDMA sales in volume in 2008. WCDMA/HSPA has become the largest radio technology in terms of radio network sales and it is expected to become the largest technology in terminal sale volume by 2011. HSPA evolution continues in 3GPP in Releases 10 and beyond. Some of the work items in 3GPP are common between HSPA and LTE (Long-Term Evolution), such as femto cells. Some LTE-Advanced items are also being considered for introduction into HSPA specifications. The expected number of subscribers for different wide area radio technologies is shown in Figure 1.8. Figure 1.8 shows that HSPA is considered to be the main growth technology for the next five years.

The long-term data rate and capacity evolution utilize LTE technology. LTE will be the technology choice for the new frequency bands such as digital dividend 700/800, 1800 and 2600. LTE has been designed for the smooth co-existence with HSPA in terms of multimode terminals and base stations, inter-system handovers and common network management systems. The evolution from HSPA to LTE can take place smoothly and those two radios can co-exist for long time. LTE serves also as the long-term platform towards LTE-Advanced targeting for data rates up to 1 Gbps.

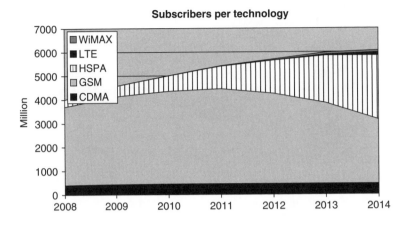

**Figure 1.8**   Expected growth of subscribers for wide area radio technologies [2]

# References

[1] Global Mobile Suppliers Association (GSA) Network survey, May 2010.
[2] Informa Telecoms & Media, WCIS+, June 2009.

# 2

# UMTS Services

Harri Holma, Martin Kristensson, Jouni Salonen, Antti Toskala and
Tommi Uitto

## 2.1 Introduction

This chapter will elaborate on UMTS services from the user's perspective. Successful practitioners of UMTS and WCDMA/HSPA technology need to understand the value of services to consumers and businesses, as well as the business models and value proposition options that operators have. The approach will be rather non-technical. Indeed, we can see that greatest successes in the market place are created when the underlying technology and complexity are hidden from the eventual user of the service. Readers are encouraged to reflect on the impact, requirements and trade-offs that various services have or imply to 3GPP-defined functionalities and network elements delivering the service.

In the early days of UMTS and WCDMA, the promise of the industry was that 'we will put the internet in every pocket'. It was envisioned that this would be accomplished by 'delivering up to 2 Mbps data rates'. This was very appealing since typical premium fixed broadband connections enabled similar data rates. Moreover, the improvement over 2G (second generation) cellular systems was very substantial: EDGE networks, for example, delivered tens of kbps or later at best close to 200 kbps data rates. Due to the legacy of the business models of 2G network operators, their market power and control points, the typical view was that operators should tightly control access of their UMTS subscribers to content and the Internet. Many operators wanted to avoid becoming 'bit pipes' and repeating the flat rate price competition experienced in fixed Internet access when it developed from narrowband to broadband. Some operators were keen to develop their portals to function as the control point and gateway to the Internet, to make sure that they could charge enough for various services in this 'walled garden'. The portals and operator offerings were to become '3G service kiosks' where there would not be just one 'killer application' but a wide variety of different services. Some people even envisioned a domain approach of having a parallel 'mmm' web separated from the world-wide web or 'www', in order to protect the operator control points. The challenge became even more relevant when incumbent and greenfield operators paid significant sums of money in auctions for their UMTS licenses in some parts of the world, such as the UK and Germany.

At the time of writing, we can now look back and conclude that UMTS business has been quite different from what was originally envisioned. As has often happened in the mobile industry, it takes

*WCDMA for UMTS: HSPA Evolution and LTE, Fifth Edition*   Edited by Harri Holma and Antti Toskala
© 2010 John Wiley & Sons, Ltd

a long time for successes to take place but when they do, they happen on a much bigger scale. 3GPP Release 99-based systems deployed during the first years certainly did not deliver 'up to 2 Mbps data rates'. The effective data rates of some 300 kbps enabled by the PS384 downlink bearer represented only a marginal improvement over GSM/EDGE. However, in early 2010, at the time of compiling this edition, peak data rates of up to 21–28 Mbps (64 QAM downlink and MIMO 2 × 2 downlink) are being deployed to networks, enabling mobile broadband operators to compete against some of the best ADSL networks in the fixed broadband domain. Due to the site density required by 2.1 GHz frequency, UMTS/WCDMA coverage remained very patchy in many European countries for several years from the network roll-out. However, in many countries WCDMA coverage has now reached well over 90% of the population. By deploying WCDMA900, some operators are now effectively matching the coverage of their GSM networks operating at 900 MHz. Subscribers with a UMTS device have gradually started to take the availability of UMTS service for granted, only to be disappointed when dropping to GSM/EDGE layer outside of WCDMA coverage. UMTS has become an everyday technology from the user perspective in most countries where licenses have been issued.

Let us take a look at the expectations, targets and promise of UMTS starting from the legacy of 2G systems. 2G systems, such as GSM, were originally designed for efficient delivery of voice services. Functionalities supporting circuit-switched or packet-switched data services were added only later. UMTS networks, on the contrary, were designed from the beginning for flexible delivery of any type of service, where each new service does not require particular network optimization. UMTS networks were designed from the outset for both circuit-switched and packet-switched services and for a number of simultaneous connections per terminal, called Multi-RABs. Compared with 2G systems, the UMTS and WCDMA/HSPA radio solution brings advanced capabilities that enable new types of services. Such capabilities are, for instance:

- High bit rates of up to 14.4 Mbps enabled in 3GPP Release 5, with the Release 5 and Release 6 terminals in the market supporting up to 10 Mbps, and with a further added capability enabled up to 28.8 Mbps in Release 7 specifications. Release 8 enabled 42 Mbps peak data rate and Release 9 specifications have enabled peak data rates up to 84 Mbps, and Release 10 is going to further increase the downlink peak data rates up to 168 Mbps with the technologies as discussed in Chapter 15. The practical bit rates were around 1–2 Mbps with the first Release 5 deployments while the development with the latest networks and devices can reach data rates up to 10 Mbps or even beyond. Such data rates, as shown in the example speedtest measurement plot in Figure 2.1, with 18 Mbps downlink and over 3 Mbps uplink, were totally unthinkable based on the experiences with first 3G networks in 2004.

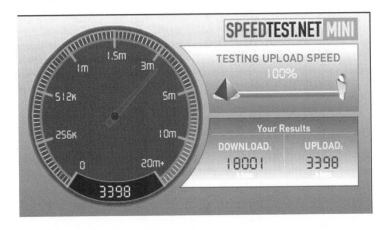

**Figure 2.1**   Example speedtest results from a release 7 capable HSPA network

- Low delays with packet round trip times below 100 ms with Release 5 and even below 50 ms with Release 6.
- Short connection set-up times and 'always-on' modes.
- Seamless mobility also for packet data applications.
- Quality of Service differentiation for high efficiency and segmentation of service delivery.
- Simultaneous voice and data capability.
- High bandwidth broadcasting.
- Interworking with other systems such as GSM/EDGE currently and LTE once LTE networks and terminals with 2G/3G inter-working are launched towards the end of 2010.

Based upon these capabilities, it is possible to cater for various types of services through UMTS systems. This chapter divides UMTS services into the following categories, many or all of which are obviously enabled by other cellular technologies as well but with different experience, implementation and cost:

- Voice
- Video telephony
- Messaging
- Mobile email
- Browsing
- Downloading (of applications)
- Streaming
- Gaming
- Mobile broadband for laptop and netbook connectivity
- Social networking
- Mobile TV
- Location-based services
- Machine-to-machine communications.

Examples in each category are provided in the sub-sections below. The categories are somewhat arbitrary and partly overlapping but they provide a way to present concrete examples. This chapter looks in addition at the above-mentioned service categories as well as service quality issues, the necessary network capacity from a service perspective, tariffs and the types of WCDMA devices currently available.

## 2.2 Voice

If there is one 'killer application' enabled by UMTS systems, it is still the voice service. In terms of the amount of traffic in bytes, laptop and netbook connectivity service, i.e. mobile broadband has surpassed voice traffic in many networks. However, in terms of service penetration, in other words the percentage of subscribers using a particular type of service, voice is still the dominant service in UMTS systems. The same can be said about the revenue share although many operators package voice service together with data services in their offering. However, operators are fiercely protecting their voice revenue, and even fighting against Voice-over IP delivered over Release 99 or HSPA bearers under a flat rate data package. Usually a maximum number of voice minutes are included in a flat rate package, after which a special price per minute or call will apply. The power of voice telephony combined with full mobility in a wide area indoors and outdoors is enormous in modern life.

We can list specific technical enablers and technical solutions for voice service in UMTS:

- circuit-switched narrowband AMR calls, including lower codec AMR;
- circuit-switched wideband AMR calls (WB-AMR);
- circuit-switched over HSPA, supporting both wideband and narrowband AMR;

- Push-to-Talk over Cellular (PoC);
- Voice-over IP (VoIP).

The first three are provided through the circuit switched core network connected to WCDMA/HSPA radio access network, whereas VoIP is switched through Packet Core and, for example, IMS (IP Multimedia System). The first two use Release 99 WCDMA bearers in air interface, whereas CS over HSPA uses an obviously high speed shared channel and HSPA transport between Node B and Radio Network Controller (RNC). The last two use PS bearers, either Release 99 or HSPA.

## 2.2.1 Narrowband AMR and Wideband AMR Voice Services

### 2.2.1.1 General

Today, voice calls in UMTS are typically carried as 3GPP Release 99 based circuit-switched calls. After the Node B and Radio Network Controller (RNC), calls are routed over the Iu-Cs interface to Mobile Switching Centers (MSC, Release 99) or Mobile Softswitches (MSS, Release 4). As described later in the chapter, CS calls can also benefit from HSPA air interface and transport up to RNC but they are still switched in CS Core. Voice-over-IP calls are packet-switched calls that can be carried over both Release 99 and HSPA data bearers and routed over the Iu-PS interface to Packet Core and switched in IP Multimedia Subsystem (IMS) or dedicated VoIP server.

The circuit-switched voice calls in UMTS employ the Adaptive Multi-Rate (AMR) technique. The multi-rate speech coder is a single integrated speech codec with eight source rates: 12.2 (GSM-EFR), 10.2, 7.95, 7.40 (IS-641), 6.70 (PDC-EFR), 5.90, 5.15 and 4.75 kbps. The AMR bit rates can be controlled by the radio access network. To facilitate interoperability with existing cellular networks, some of the modes are the same as in existing cellular networks. The 12.2 kbps AMR speech codec is equal to the GSM EFR codec, 7.4 kbps is equal to the US-TDMA speech codec, and 6.7 kbps is equal to the Japanese PDC codec. The AMR speech coder is capable of switching its bit rate every 20 ms speech frame upon command. For the AMR mode, switching in-band signaling is used.

The narrowband AMR coder operates on speech frames of 20 ms corresponding to 160 samples at the sampling frequency of 8000 samples per second, whereas wideband AMR (WB-AMR) is based on the 16,000 Hz sampling frequency, thus extending the audio bandwidth to 50–7000 Hz. The coding scheme for the multi-rate coding modes is the so-called Algebraic Code Excited Linear Prediction Coder (ACELP). The multi-rate ACELP coder is referred to as MR-ACELP. Every 160 speech samples, the speech signal is analysed to extract the parameters of the CELP model (LP filter coefficients, adaptive and fixed codebooks' indices and gains). The speech parameter bits delivered by the speech encoder are rearranged according to their subjective importance before they are sent to the network. The rearranged bits are further sorted based on their sensitivity to errors and are divided into three classes of importance: A, B and C. Class A is the most sensitive, and the strongest channel coding is used for class A bits in the air interface.

During a normal telephone conversation, the participants alternate so that, on average, each direction of transmission is occupied about 50% of the time. The AMR has three basic functions to effectively utilize discontinuous activity:

- Voice Activity Detector (VAD) on the TX side.
- Evaluation of the background acoustic noise on the TX side, in order to transmit characteristic parameters to the RX side.
- The transmission of comfort noise information to the RX side is achieved by means of a Silence descriptor (SID) frame, which is sent at regular intervals.
- Generation of comfort noise on the RX side during periods when no normal speech frames are received.

Discontinuous transmission (DTX) has some obvious positive implications: in the user terminal, talk time (time between recharging the battery) is prolonged or a smaller battery could be used for a given operational duration. From the network point of view, the average required bit rate is reduced, leading to a lower interference level and hence increased capacity.

The AMR specification also contains error concealment. The purpose of frame substitution is to conceal the effect of lost AMR speech frames. The purpose of muting the output in the case of several lost frames is to indicate the breakdown of the channel to the user and to avoid generating possibly annoying sounds as a result of the frame substitution procedure [1, 2]. The AMR speech codec can tolerate about a 1% frame error rate (FER) of class A bits without any deterioration of speech quality. For class B and C bits, a higher FER is allowed. The corresponding bit error rate (BER) of class A bits will be about $10^{-4}$.

The bit rate of the AMR speech connection can be controlled by the radio access network depending on the air interface loading and the quality of the speech connections. During high loading, such as during busy hours, it is possible to use lower AMR bit rates to offer higher capacity while providing slightly lower speech quality. This is often referred to as Lower Codec AMR. Also, if the mobile is running out of the cell coverage area and using its maximum transmission power, a lower AMR bit rate can be used to extend the cell coverage area. The capacity and coverage of the AMR speech codec are discussed in Chapter 12. With the AMR speech codec it is possible to achieve a trade-off between network capacity, coverage and speech quality according to the operator's requirements.

After Node B and Radio Network Controller (RNC), 3GPP Release 99 compatible circuit-switched AMR voice calls are switched in Mobile Switching Centers (3GPP Release 99) or Mobile Softswitches (3GPP Release 4 and up).

### 2.2.1.2   AMR Source-Based Rate Adaptation – Higher Voice Capacity [3]

Currently, AMR codec uses source-based rate adaptation with voice activity detection (VAD) driven discontinuous transmission (DTX) to optimize network capacity and power consumption of the mobile terminal. In AMR speech codec, voice activity detection (VAD) is used to lower the bit rate only during silence periods. However, active speech is coded by fixed bit rate that is selected by the radio network according to network capacity and radio channel conditions. Although the network capacity is optimized during silence periods using VAD/DTX, it can be further optimized during active speech with source-controlled rate adaptation. Thus, AMR codec mode is selected for each speech frame depending on the source signal characteristics, see Figure 2.2. The speech codec mode can be updated in every 20-ms frame in WCDMA.

AMR source adaptation allows the same voice quality with lower average bit rate. The bit rate reduction is typically 20–25% and is illustrated in Figure 2.3. The reduced AMR bit rate can be utilized to lower the required transmission power of the radio link, and thus further enhance AMR voice capacity. The WCDMA flexible layer 1 allows adaptation of the bit rate and the transmission power for each 20-ms frame. The estimated capacity gain is 15–20%. The bit stream format of source-adapted AMR is fully compatible with the existing fixed-rate AMR speech codec format, therefore, the decoding part is independent of source-based adaptation. The AMR source-based adaptation can be added as a software upgrade to the networks to enhance the WCDMA downlink capacity without any changes to the mobiles. The AMR source-controlled adaptation can also be implemented with wideband AMR speech codec.

### 2.2.1.3   Wideband AMR – Better Voice Quality [4]

3GPP Release 5 introduces wideband AMR (WB-AMR) speech codec, which gives substantial voice quality enhancements compared to narrowband AMR codec or a standard fixed telephone line. In case of packet switched streaming, WB-AMR is already part of Release 4. The WB-AMR codec has also been selected by the ITU-T in the standardization activity for a wideband codec around 16 kbps. This is of significant importance since this is the first time that the same codec has been adopted for wireless

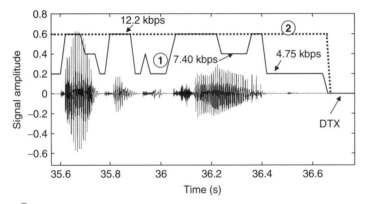

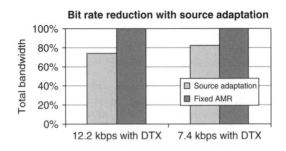

① = AMR with source adaptation changes its bit rate according to the input signal
② = AMR today uses fixed bit rate (+DTX)

**Figure 2.2** AMR source-based mode selection as a function of time and speech content

**Figure 2.3** Reduction of required bit rate with equal voice quality

as well as wireline services. This will eliminate the need for transcoding, and ease the implementation of wideband voice applications and services across a wide range of communications systems. The WB-AMR codec operates on nine speech coding bit-rates between 6.6 and 23.85 kbps. The term wideband comes from the sampling rate, which has been increased from 8 kHz to 16 kHz. This allows coverage of twice the audio bandwidth compared to the classic telephone voice bandwidth of 4 kHz. While all the previous codecs in mobile communication systems operate on narrow audio bandwidth limited to 200–3400 Hz, WB-AMR extends the audio bandwidth to 50–7000 Hz. Figure 2.4 shows the listening test result, where WB-AMR is compared to narrowband AMR. The results are presented as subjective mean opinion score (MOS) where a higher number indicates better experienced voice quality. The MOS results show that WB-AMR is able to improve the voice quality without increasing the required radio bandwidth. For example, WB-AMR 12.65 kbps clearly provides higher MOS than narrowband AMR at 12.2 kbps. The improved voice quality can be achieved because of higher sampling frequency.

## 2.2.2  Circuit-Switched over HSPA

With the introduction of High Speed Downlink Packet Access (HSDPA) in 3GPP Release 5, High Speed Uplink Packet Access (HSUPA) in 3GPP Release 6, it also became possible to prepare to improve the performance of voice calls and the associated capacity of the WCDMA/HSPA system compared

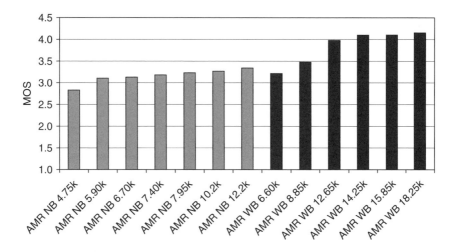

**Figure 2.4**   Mean opinion score (MOS) example with wideband and narrowband AMR

to Release 99 AMR calls. More capacity can be extracted from the system by adjusting the bit rate in air interface with the high speed shared channel in HSPA, compared with the dedicated Release 99 channel. Calls can be set up faster by using the faster control channels of HSPA compared to Release 99 WCDMA. It should be noted that exactly the same narrowband and wideband AMR codecs are in use here, but the benefits of air interface performance of HSPA are introduced. Circuit-Switched over HSPA (CS over HSPA or CSoHSPA) was standardized in 3GPP Release 7 to enable CS calls to be carried with a HSPA air interface and to materialize these benefits without any impact to core networks. In other words, CS over HSPA uses the same MSC/MSS circuit core as Release 99 AMR calls. In conjunction with CS over HSPA, talk time can be extended by utilizing Discontinuous Transmission (DTX) and Discontinuous Receiving (DRX) introduced in 3GPP Release 7, together sometimes referred to as Continuous Packet Connectivity (CPC).

Test results show that by deploying CS over HSPA with CPC, the following improvements can be achieved:

- Talk time can be extended by up to 50%, mainly thanks to DTX/DRX.
- Mobile-to-mobile call setup times can be improved from approximately 3.5 s down to 1.5 s, thanks to faster control channels.
- Voice capacity of the system can be improved by up to 100%, i.e. the number of voice calls per MHz of bandwidth can be doubled, thanks to the overall efficiency improvement in the air interface.

Talk time extension is of particular relevance, because in the first years of UMTS services being available in the market, one of biggest causes of dissatisfaction among users, according to market research by terminal manufacturers, was the short time between recharging. Users had become used to the long idle and talk times in GSM/EDGE and did not expect to have to recharge every day or even several times per day, which was the case in the early years of UMTS. Battery technologies and power consumption of UMTS terminals have developed, but especially data services are still relatively heavy on power consuming.

Interestingly enough, the capacity benefit of CS over HSPA is not limited to voice service only. The other way of presenting the same phenomenon is that for constant voice traffic, the capacity available for data traffic is increased by packing voice in smaller bandwidth.

## 2.2.3  Push-to-Talk over Cellular (PoC)

Push-to-talk over cellular (PoC) service is a niche service which has not gained wide adoption among UMTS services. It is interesting to make this observation when proprietary iDEN mobile service with Push-to-Talk has been able to capture some market share and price premium in some Latin American countries.

PoC is one-way communication from one-to-one or one-to-many: when one person speaks, the others listen. The call is normally established by simply pushing a single button and the receiving user(s) hears the speech without the need to press a button to answer. While ordinary voice is bi-directional (full-duplex), a PoC service is a one-directional (half-duplex) service. The basic PoC application may hence be described as a walkie-talkie application over the packet switched domain of a cellular network. In addition to the basic voice communication functionality, the PoC application provides the end user with complementary features such as:

- ad-hoc and pre-defined communication groups;
- access control so that a user may define who is allowed to make calls to him/her;
- 'do-not-disturb' if immediate reception of audio is not desirable.

With ordinary voice calls a bi-directional communication channel is reserved between the end users throughout the duration of the call, which typically lasts in the order of minutes. In PoC, the channel is, on the other hand, only set up to transfer a short speech burst from one to possibly multiple users. Once this speech burst has been transferred, the one-way packet switched communication channel is released. This difference is highlighted in Figure 2.5.

The speech packets in a PoC solution are carried from the sending mobile station to the server by PS bearers (Release 99 or HSPA) and the packet core network. The server then forwards the packets to the receiving mobile stations. However, the characteristics of any PoC service set tight requirements on the performance of the radio access network.

In order for a PoC service to be well perceived by the end users, it must fulfil some fundamental requirements:

- simple user interface, for example, a dedicated push-to-talk button;
- high voice quality and enough sound pressure in the speaker to work also in noisy environments;
- low delay from pressing the push to talk button until it is possible to start talking, called 'start-to-talk time';
- low delay to receive an answer from the peer end, called 'speech-round trip time'.

The end user is expected to be satisfied with the interactivity of the PoC service if the start-to-talk delay is around or below one second while the speech round trip time should be kept lower than or around four seconds. A radio network that hosts PoC connections must, for example, be capable of:

- providing always-on packet data connections;
- reserving and releasing radio access resources fast in order to keep start-to-talk and speech-round trip times low;
- delivering a constant bit rate with low packet jitter during the duration of one speech burst.

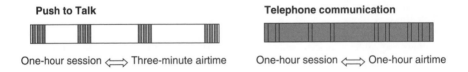

Figure 2.5   Push to talk versus ordinary telephone communication

## 2.2.4   Voice-over IP

The main driver for Voice-over-IP (VoIP) in fixed networks has been the rapid increase of affordable broadband connections (xDSL, WLAN, cable, etc.). WCDMA networks can also offer an adequate level of quality for VoIP services. A number of features have been included in 3GPP Releases 5, 6 and 7 specifications that improve the end-to-end performance and the capacity of VoIP service. Many device manufacturers have introduced VoIP-capable devices on the market which can make CS calls through GPRS/WCDMA networks as well as VoIP calls through both WLAN and WCDMA networks.

VoIP calls in UMTS are carried over Release 99 or HSPA data bearers, routed through Packet Core and 'switched' in IP Multimedia Subsystem (IMS) or dedicated VoIP server. However, carrying VoIP packets over Release 99 (e.g. PS64) is less efficient than circuit-switched Release 99 calls (e.g. 12.2 kbps plus overheads). In general, carrying short voice sample packets with long IP headers over air interface, typically the most expensive asset of an operator, is not particularly efficient without some optimization. The overhead is 40 bytes in IPv4 and 60 bytes in IPv6, representing some 60% overhead in VoIP application. The real-time nature of voice communication also sets real-time, short delay requirements to the connection. It can be said that the following are key enablers of a cost-efficient, high-quality, mass-market offering for VoIP in UMTS:

- HSPA;
- IP header compression, specifically Robust Header Compression (ROHC) algorithms;
- Conversational Quality of Service (QoS);
- IMS or dedicated VoIP server.

As discussed, VoIP can be carried over Release 99 packet-switched data connection, but this is not particularly efficient. ROHC shortens the IP header significantly before sending the packets over the air interface. When a UMTS system is not loaded, VoIP works quite well from the user perspective without any Quality of Service (QoS) mechanisms, but Conversational QoS is recommended to ensure the real-time experience especially in loaded environments. With Deep Packet Inspection (DPI) algorithms, e.g. in Packet Core, the end-to-end network can detect that a service is VoIP and can assign an optimized target bit rate and QoS class to it. Operators wishing to prevent such VoIP traffic and protect (CS) voice revenue, e.g. when a flat rate data tariff is offered, can use DPI for their purposes, although there are ways to bypass such attempts. IMS systems provide for SIP-based peer-to-peer IP connectivity, enabling VoIP as one of the possible services. However, VoIP service can be delivered with dedicated VoIP servers as well.

Any speech codec can obviously be carried over WCDMA/HSPA. In addition to narrowband and wideband AMR codecs such as 7.95 kbps or 12.2 kbps, VoIP calls can use, for example, G.729 or Sinusoidal Voice Over Packet Coder (SVOPC). At the time of writing, Skype uses at least these two codecs. Since narrowband and wideband AMR are the most efficient codecs when measured with quality and bit rate, they are the obvious choices for operators using IMS for VoIP.

Operators have generally not been keen to offer VoIP service over UMTS systems at the time of writing. Typically the reason is that operators want to protect voice revenue, and as long as voice is carried as a circuit-switched service, protecting it has been easy. This is probably the reason why operators have not been happy about terminal manufacturers including VoIP clients in their devices. Operators often consider robust and carrier-grade IMS systems of their own to be the tool to control the traffic and like being able to charge possibly more than the pure flat rate data tariff for it. Still, it is somewhat difficult for operators to prevent their subscribers from downloading a VoIP client to their terminal and calling through a third party VoIP server or system, such as Skype.

## 2.2.5   Key Performance Indicators for Voice

Operators of UMTS systems compete in various ways, but one typical competitive factor is the network and service quality. The quality can be measured objectively with key performance indicators (KPI) measuring certain characteristics of the service utilizing formulas and data collected from the network, as well as subjectively with users providing their feedback and scoring. A good example of the latter are the subjective Mean Opinion Scores (MOS) for voice quality as described above.

Typical key performance indicators for voice services include, for example:

- call set-up success rate (CSSR), i.e. the percentage of successful call set-up attempts, out of all call attempts;
- call completion success rate (CCSR), i.e. the percentage of calls ended intentionally, out of all calls;
- call set-up time, which obviously varies depending on where the call terminates.

The formulas used for these KPIs vary from operator and network, and the above is intended to be only an introduction to the topic.

## 2.3  Video Telephony

In the early UMTS era, some operators wanted to differentiate their 3G/UMTS service from 2G by launching video telephony as one of the services. The service was implemented with the circuit-switched (CS) 64 kbps transparent data bearer in 3GPP Release 99. Such service was not possible in 2G, and operators expected to be able to attract 2G subscribers to migrate to 3G and buy WCDMA devices once such a differentiating and novel service was available. The experience of NTT DoCoMo in Japan, using their early WCDMA FOMA (Freedom of Mobile Multimedia Access) system, suggested that there is market demand for such a service. Video telephony was used widely with high service penetration in Japan.

However, video telephony did not become a widely accepted UMTS service in the rest of the world. The resolution, size and quality of the screens in early UMTS terminals, the low bit rate of CS 64 kbps, and end-to-end network optimization challenge, among other things, resulted in relatively poor user experience. New WCDMA subscribers would typically try the service a few times or when a campaign was launched, but would not use it continuously and repeatedly. Video telephony remained a niche service with low service penetration and very low share of the overall traffic. In typical networks in Europe, the amount of video calls is well below 1% of all calls made.

The next step in video telephony was an attempt to move to the direction of a 'rich call', where a packet switched video connection (e.g. with Release 99 PS bearer) was combined with a voice call as a Multi-RAB. The service is called video sharing, referring to the use case of starting a video, e.g. in the middle of a voice call to 'share' something with the other calling party. Such service has not attracted a lot of usage either.

There are a couple of specific reasons why we may conclude that mobile video telephony as described above has not been a major success:

- The video picture is relatively small and not of particularly high quality.
- The audio quality is not as high as when the terminal is held close to the ear.
- Adding video to mobile communication simply does not add as much relative value as, e.g. voice.
- A truly mobile user may have to watch where he or she is walking, driving or generally moving, and simply cannot watch the screen.
- In order to position oneself well in the video picture, the user has to raise the arm to the same level with the face or tilt the face significantly down. Holding the terminal in one's hand with arms down, the other calling party will see the person's face from below. All this is not very practical.

At the time of writing, video telephony over UMTS is picking up to some extent through the laptop or netbook connectivity use case, where the user is stationary or nomadic and launches a video, and VoIP application through a client in a laptop or netbook connected to a server over a WCDMA/HSPA radio access network using a USB stick, dongle or datacard. The use case is very similar to that of the computer being connected with a LAN cable or WLAN (WiFi), but the value of UMTS comes from the ubiquitous coverage.

Let us elaborate, however, on the technical implementation of video telephony in UMTS networks. 3GPP has specified that ITU-T Recommendation H.324M should be used for video telephony in circuit-switched connections and Session Initiation Protocol (SIP) to support IP multimedia applications, including video telephony in 3GPP Release 5 core network environment.

### 2.3.1 Multimedia Architecture for Circuit Switched Connections

Originally ITU-T Rec. H.324 was intended for multimedia communication over a fixed telephone network (PSTN). 3GPP modified the H.324 Recommendation to make the system more suitable for digital domain and more robust against transmission errors. The overall picture of the H.324M system is shown in Figure 2.6 [5, 6].

H.324M consists of the following mandatory elements: H.223 for multiplexing and H.245 for control. Elements that are optional but are typically employed are H.263/MPEG-4 video codec and G.723.1/AMR speech codec. The recommendation defines the seven phases of a call: set-up, speech only, modem training, initialization, message, end, and clearing. Level 0 of H.223 multiplexing is exactly the same as that of H.324, thus providing backward compatibility with older H.324 terminals. With a standardized in-band negotiation procedure the terminal can adapt to the prevailing radio link conditions by selecting the appropriate error resiliency level.

One of the recent developments of H.324 is an operating mode that makes it possible to use an H.324 terminal over ISDN links. This mode of operation is defined in Annex D of the H.324 recommendation and is also referred to as H.324/I. H.324/I terminals use the I.400 series ISDN user-network interface in place of the V.34 modem. The output of the H.223 multiplex is applied directly to each bit of the digital channel, in the order defined by H.223. Operating modes are defined bit rates ranging from 56 kbps to 1920 kbps, so that H.324/I allows the use of several 56 or 64 kbps links at the same time, thus providing direct interoperability with H.320 ISDN terminals.

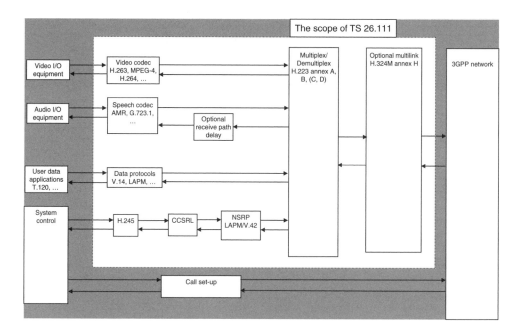

**Figure 2.6**   Scope of ITU Rec. H.324M

### 2.3.2    Video Codec

It is recommended that all H.324M terminals support both H.263 and MPEG-4 video codecs. Error resiliency and high efficiency make the MPEG-4 video codec particularly well suited for mobile video telephony. MPEG-4 Visual is organized into Profiles. Within a Profile, various Levels are defined. Profiles define subsets of tool sets. Levels are related to computational complexity. Of these Profiles, Simple Visual Profile provides error resilience (through data partitioning, RVLC, resynchronization marker and header extension code) and low complexity. It is recommended that the Simple Visual Profile @ Level 0 is supported to achieve adequate error resilience for transmission error and low complexity simultaneously. No other Profiles are recommended to be supported. Higher Levels for the Simple Visual Profile may be supported depending on the terminal capabilities [7].

MPEG-4 allows various input formats, including general formats such as CIF (Common Intermediate Format) and QCIF (Quarter CIF). H.324M encoders and decoders are recommended to support the 1:1 pixel format (square format). Encoders should signal this capability using H.245 capability exchange and the appropriate header fields in video codecs so that unnecessary pixel shape conversions can be avoided. It is also baseline compatible with H.263.

Regardless of which specific video codec standard is used, all video decoder implementations should in practice include basic error concealment techniques. These techniques may include replacing erroneous parts of the decoded video frame with interpolated picture material from previous decoded frames or from spatially different locations of the erroneous frame. The decoder should aim to prevent the display of substantially corrupted parts of the picture. In any case, it is recommended that the terminal could tolerate every possible bit stream without catastrophic behavior (such as the need for a user-initiated reset of the terminal). The picture size of QCIF for Level 1 should be used for the sake of interoperability.

Video telephony has roughly similar delay requirements to the speech service. However, due to the nature of video compression, the higher compression factor, the BER requirement is more demanding than that of speech. Figure 2.7 shows examples of 3G video telephony application.

**Figure 2.7**    Nokia N73 video telephony application

## 2.4 Messaging

In this section, we elaborate on messaging services such as:

- Short Messaging Service (SMS)
- Multimedia Messaging Service (MMS)
- Voice mail and audio messaging
- Instant Messaging.

### 2.4.1 Short Messaging Service (SMS)

In addition to basic voice service, the Short Messaging Service (SMS) can be called a 'killer application' in UMTS (or GSM/EDGE for that matter) both in terms of service penetration and share of operator revenue. The term SMS is typically used for both the service and the text message itself. Many users find it particularly convenient and effective to type short, up to 160 character text messages (or in case of longer text, chains of such 160 character messages parsed to one message at the receiving terminal). Such a message is less intrusive to the receiving party than a voice call, but more immediate and instant than e.g. an email. After voice, SMS is in all likelihood the second most successful mobile service to date, followed recently by the success of laptop connectivity, i.e. mobile broadband and application downloads with smartphones.

In addition to person-to-person messaging, SMS can be used for, e.g. person-to-content orders or queries. For example, the user can send a standard format message such as 'Find John Doe' or 'Order Hit Tune 6' to a specific number, in order to receive by SMS all phone numbers and addresses where the name matches 'John Doe' or a ring tone matching with 'Hit Tune 6'.

The production cost of an SMS is very low due to the small amount of data sent. In fact, the SMS is sent as part of the control signaling due to low data volume. Yet at the same time, the high value of SMS has enabled operators to charge a premium for the service. This has traditionally made SMS a very profitable service for operators. In the early days of 3G, operators often boasted about the share of data representing e.g. 15−20% of their revenue, while most of the revenue was actually brought in by SMS. The share of SMS as a percentage of data revenue has since decreased and due to package tariffs, it is nowadays often impossible to calculate the exact share.

### 2.4.2 Multimedia Messaging Service (MMS)

Picture messaging was developed on top of SMS to convey simple grey scale bitmap pictures along with text. Multimedia messaging service (MMS) enabling colour photographs with better resolution to be sent was then a natural development step towards richer person-to-person messaging. MMS is an example of a store and forward type of service, where a message is composed on a mobile device, consisting typically of a still image taken with an in-built digital camera and a short descriptive text. An MMS is sent to a server where it is stored until fetched by the recipient's device. The fetching of the message is triggered by WAP Push message, which is fundamentally an SMS, including details such as the sender's MSISDN, the subject field and the location of the message on the server. Most handsets nowadays also support so-called Synchronized Multimedia Integration Language (SMIL), allowing users to create rich timed multimedia presentations with multiple images or videos as well as text and speech. MMS messages are typically delivered over PS bearers in WCDMA, either Release 99 or HSPA.

As the MMS service is of a store-and-forward type, it does not inherently impose rules on delivery time, thereby suggesting timewise a loose 3GPP quality of service class. What is more important for the users is that the content is delivered with a high probability and that the delivered message is as close to the original one as possible. There are thousands of MMS-capable device models as well as a few WAP gateways and Multimedia Messaging Service Centers (MMSC) available, and all of them should work seamlessly together. In order to facilitate interoperability, Open Mobile Alliance (OMA), that works closely with 3GPP, has created MMS specifications. OMA MMS version 1.2 specification defines a minimum set of requirements and conformance to enable end-to-end interoperability of MMS applications, MMS-capable handsets and servers, and content provisioning. For instance, it limits the maximum MMS size to 300 kB. The OMA specification MMS 1.3 has raised the maximum size to 600 kB [8, 9]. Another important requirement from the end user's perspective is that it should be possible to send MMS messages simultaneously while in a telephone call. This in turn calls for support for multiple radio access bearers (Multi-RABs).

### 2.4.3   Voice Mail and Audio Messaging

Classic voice mail is one of the most popular services in UMTS in terms of service penetration. The vast majority of users have activated their voice mail box, allowing incoming calls to be diverted to play a greeting message and allow the calling party to record a voice message if the call is not answered within a specified time. The voice mail messages can be listened to at the discretion of the receiving party. An SMS can be sent to the receiving party notifying him or her of voice mails in the voice mail box. Voice mails are stored in Voice Mail Systems (VMS) in the network, rather than, e.g. in the handset.

Audio messaging is a special type of MMS service that consists only of the audio component in a SMIL presentation. Unlike voice mail, audio messages can be stored on the handset to listen to later. They can also be forwarded to other users with MMS-capable handsets. Audio messaging is easy to implement in any MMS network as it needs no additional hardware or software. Subscribers can also start using it on any MMS-enabled handset as soon as they have MMS service. For operators, this is a new way of generating revenue from their existing MMSC investments. The service is very cost-effective and can be priced to differentiate it from other MMS-based services. It is a service that holds great potential to boost messaging in new growth regions where written skills may be less common than in mature markets. A one-minute audio message takes up only about 35 kB.

### 2.4.4   Instant Messaging

Instant messaging is a very popular internet service that is now also available for mobile users. All major internet service providers, such as Yahoo, MSN, AOL, ICQ, Jabber and Google, allow mobile clients to access the service with the smartphone. There are also mobile messaging solutions that combine access to all the aforementioned services in a single application. The main features of the IM service are: real-time chatting, sending/receiving images or other media elements and sending/receiving documents.

## 2.5   Mobile Email

Most of us use email almost every day, but how many of us are able to use mobile email to connect either to a business email or a private email? It was estimated back in 2007 that there were more than 650 million corporate email boxes in the world, but only about 10 million of them had mobile connectivity, and only half of that deployed push email. However, Research In Motion Ltd (RIM), the provider of one of the most successful mobile email services, disclosed in conjunction of their quarterly reporting on June 18, 2009, that the number of BlackBerry accounts had grown by 3.8 million net in the previous quarter to a total of approximately 28.5 million [10]. Nokia, the market leader in handheld devices, recently joined forces with Microsoft to challenge RIM's position in corporate mobile email, following the 2005 acquisition of Intellisync to improve Nokia's mobile email capabilities, among other things. The battle here is essentially between RIM's more proprietary client/server system relying on user experience and the loyalty of BlackBerry users and the alliance of Nokia and Microsoft relying on the Nokia brand, user experience and customer loyalty, together with the ubiquity of Microsoft Outlook for email in corporate world. It can be concluded that mobile email is one of the fastest growing segments in mobile communication and UMTS.

There are different ways to connect to an email server and deliver email messages. Now, as mobile browsers have evolved to full HTML browsers, it is possible to connect directly with a browser to an email service on the internet, for instance, MSN/Hotmail, Gmail, Yahoo. Correspondingly it is possible to connect email services possessing an XHTML interface.

Often mobile devices have an in-built email client, that makes connection to the service straightforward, provided that the user can configure the device or request the configuration settings over the air. The most recent evolution is push email, where either the entire email message or its header is pushed silently (without user action) to the device. This may be based on the proprietary protocols or standardized OMA Email notification 1.0 standard. Push email is in fact very similar to MMS. A message is sent via the packet data network to the server, which will send a push notification to a recipient device, that will in turn fetch the message. The main differences between MMS and push email are: MMS supports SMIL presentation, but is limited in size, email does not support SMIL, but can cope with the content of several megabytes. From the network point of view, email is a typical service that clearly belongs to a background QoS class – delay is not as important, instead, error-free delivery is.

## 2.6   Browsing

Browsing the Internet is an obvious use case and service for UMTS. There is significant value in being able to browse the Internet with a mobile device as the location, presence or context of the user may make browsing with a PC in a fixed position, for instance, cumbersome or even impossible. The browsing experience in the early 3G era with WAP protocol and small low-resolution screens has since improved with the introduction of HTML browsing and larger, high-resolution screens. Even today, however, some websites do not work particularly well or at all with mobile handset browsers. Some website owners have made mobile-optimized versions of their sites to improve the user experience, often with automatic detection of the device and suggestion to divert to the optimized site.

From a user perspective it is crucial that browsing is easily accessible, fast and reactive. Rough performance requirements for browsing are such that the first page download time should be lower than 10 s and the second page download lower than 4 to 7 s [11]. Another user requirement is that it should be possible to browse smoothly also when moving or traveling by e.g. car, train, subway or bus. This requires efficient handling of cell reselections in order to prevent connection breaks at cell reselections. Since WCDMA utilizes soft handover for packet switched data, there are no breaks at cell reselection.

Progressive download is a relatively new technique used to play media content while the media is still being downloaded to the player. Before the implementation of the progressive download, the entire media file was loaded into the memory before it was played. The benefit of progressive download is obvious when playing large media files and especially when having play list functionality. Progressive download can be achieved using a regular web (http) server. The client handles the buffering and playing during the download process. If the playback rate exceeds the download rate, playback is delayed (buffering occurs) until more data is downloaded. After the entire media file has been downloaded it can normally be saved in a device memory or in a memory card.

WAP/XHTML browsing is losing ground to full HTML browsing in mobile use case. There are a number of reasons for this: display sizes of mobile devices are growing, access speed is getting higher, and the browsers are supporting various scripting languages, RSS/Atom feeds and other features previously found only in the PC world.

Podcasting is fairly new service type, where internet users upload their audio or video content to the internet for free downloading by other web users. Frequently updated audio or video content is also referred to as a feed, which typically conforms to either Atom or RSS formats. The feed contains the descriptive text about the content as well as enclosures, which refer to the binary media content within the feed. State-of-the-art browsers support both RSS and Atom feeds, making it easy to follow certain news feeds, even automatic downloading of the feeds is possible nowadays. Reading podcasts tends to create a lot of data traffic, therefore flat rate tariffing is needed in order to make this service appealing to the end user.

Another service that commonly utilizes feed technology is blogging. A typical blog combines text, images, and links to other blogs, web pages, and other media related to its topic. The ability for readers to leave comments in an interactive format is an important part of blogs. Most blogs are primarily textual although some focus on photographs, video or audio (podcasting), and are part of a wider network of social media.

## 2.7  Application and Content Downloading

Application downloading and the use of such downloaded applications with a handheld device have become very popular recently. In the early days of 3G/UMTS, the portfolio of applications that could be downloaded was more limited than today, consisting of simple applications such as ringing tones, wall papers, games, dictionaries, etc. Applications could typically be loaded from operators' own portals or third party websites. The business model worked fairly well and application and content downloading proliferated especially among the segments of young people, early adopters and technology enthusiasts (it should be noted that such segmentation is fairly arbitrary).

Application and content downloading for business swiftly accelerated with the introduction of new types of handsets and business models. Apple, having made iPod music and video player together with their iTunes music and video store a big success, entered the mobile domain with the iPhone. Nokia had earlier launched similar services and platform as a complementary service to mobile phones, but following the success of Apple, decided to make application and content downloading a strategic growth area for their business.

At the time of writing, Apple reported that more than two billion applications had been downloaded from its 'App Store' website, some with a price of e.g. 80 US cents, some even free of charge. The quarterly rate of downloads had reached more than half a billion. Apple had stated that there were more than 85,000 applications available in App Store. The applications could be categorized in two groups: those which deliver a serious value and those which are simply for fun. In parallel with Apple, Nokia opened its own application store called 'Ovi' (door in Finnish) with example visual layout as shown in Figure 2.8. The business model is similar to the one of Apple's App Store. Players like Apple and Nokia have created an opportunity for an eco-system to form, where various SW developers and HW or accessory providers in mobile or adjacent industries develop products that can be delivered through or benefit from the existence of the application stores. For example, these handheld device suppliers attract a lot of small and agile SW developers to create their own applications and offer them for distribution through the application stores. For every application downloaded, the developer and application store owner can share the revenue. Other examples include accessories that are compatible with the core HW/SW of the likes of Apple and Nokia, enhancing the user experience in one way or another.

It is interesting to observe the fight over customers and associated control points. In the early days of 3G/UMTS, many operators planned and attempted to capture the customers, tying them to their own portals providing the access to content and applications, with some success. When Nokia introduced similar services for Nokia mobile phone users over their own website, some operators considered it an offensive move and tried to push Nokia back into its role as only a device supplier, while other operators accepted the offering as complementary to their own proposition. With the superior appeal of some new handheld devices and application stores, operators have had to accept that device suppliers will take their share of the value in the applications space. Some operators have concluded that they have to get distribution rights, even exclusive ones, for certain smartphones simply to prevent the device supplier from giving exclusivity to the operators' competitors. In such situations operators are obviously concerned with their subscribers changing their service provider to get the handheld device they want.

**Figure 2.8**   Apple's App Store (left) and Nokia's Ovi Store (right)

While application downloads and use of such applications are fairly transparent to the underlying radio bearers and use of UMTS technology – as they can be downloaded or used typically with whatever radio bearers are available – it is very interesting to analyse the impact on the network traffic profile and judge the implications to the network and network element designs. Indeed, proliferation of smartphones like iPhone, the Nokia N-series or the BlackBerry, just to name a few, have changed the traffic profile of UMTS networks significantly, exposing network bottlenecks and even causing network quality problems. Regardless of whether downloading such applications or using them requires Release 99 PS bearers or HSPA, the overall traffic profile has an important impact on the network design. Operators and network equipment suppliers have in some instances been caught by surprise by the emerging requirements created by the traffic generated by smartphones. Some of the smartphone applications, such as push email solutions or social networks, together with some specific terminal implementations, have been creating a lot of signaling traffic for the network due to the battery-saving solutions that aim to release the connection always when possible (depending on the application, this may end up setting up the connection in a matter of seconds again), this has been addressed in 3GPP through the Release 8 fast dormancy solution to reduce the resulting signaling load to the network from these solutions in early smartphones using non-3GPP based solutions. The fast dormancy development is addressed further in Chapter 15.

It can be generalized that some smartphone applications, including but certainly not limited to mobile email and various widgets for e.g. news services, weather services or social networks, establish connections to the network and servers very frequently, or even use continuous connection or stand-by connection states, or change often between these states. By the stand-by states we mean the connection states described later in this book, such as Cell_FACH, Cell_PCH, URA_PCH, as opposed to idle state or connected state Cell_DCH. Such stand-by states have been standardized to provide for and enable different trade-offs between network capacity, call set-up times, battery time and data speeds. For example, some email applications benefit from the 'always-on' functionality when connection set-up times are minimized, compromising slightly on battery times. Establishing connections to the network and changing the connection state between Idle, Cell_DCH and the stand-by states creates signaling traffic in the control plane abstraction layer of the network. At the same time, various states consume network element capacity in different ways. For example, Radio Network Controllers (RNC) may have limited and different capacities for simultaneous users in different connection states. If there are applications that keep the user in a connected or stand-by state for prolonged times without release, the connected state capacity of an RNC can run out, blocking access for new calls and connections. A similar phenomenon could be seen in the radio connection set-up capacity of UMTS NodeBs. On the other hand, not utilizing the stand-by states effectively means that establishing calls from the idle state requires much more signaling and creates a load in the control plane (as opposed to the user plane). Some early smartphones used a so-called 'fast dormancy' feature pushing the phones to idle to save the battery, which then caused a significant increase in signaling and control plane traffic when connection was established.

Successful operators, network suppliers, network planners and practitioners of UMTS and WCDMA/HSPA technology will take such anticipated, planned or materialized changes in the traffic profile into account when designing, deploying and optimizing networks or designing and developing products for UMTS networks.

## 2.8   Streaming

Multimedia streaming is a technique for transferring data so that it can be processed as a steady and continuous stream. Just as with a progressive download, a browser plug-in or a special streaming player starts displaying the data after a few seconds of buffering. Streaming technologies are becoming increasingly important with the growth of the Internet, and due to somewhat tricky digital content rights issues. Since the content cannot be stored and therefore cannot be forwarded and reused, it is easier to agree on the content delivery as streaming media than as downloadable media. Also, mobile station memory may limit the size of the downloads.

For streaming to work, the client side receiving the data must be able to collect the data and send it as a steady stream to the application that is processing the data and converting it to sound or pictures. Streaming applications are very asymmetric and therefore typically withstand more delay than more symmetric conversational services. This also means that they tolerate more jitter in transmission, which can easily be smoothed out by buffering.

A streaming server works with the client to send audio and/or video over the Internet and play it almost immediately without saving the media content. This allows real-time 'broadcasting' of live events, and the ability to control the playback of on-demand content. Playback begins as soon as sufficient data has downloaded.

## 2.9   Gaming

We first characterize the existing multiplayer games into key categories based on their end user requirements. Three reasonable categories are according to the study in [12, 13]: real-time action games, real-time strategy games and turn-based strategy games, see Figure 2.9.

For real-time action games, the end-to-end network delay requirements can be as low as 50 ms to satisfy the most demanding users. The end-to-end network delays are particularly noticeable to the users if some users have low delays, such as 30 ms, while others suffer higher delays, such as 200 ms. Bearing in mind that HSPA networks can provide end-to-end network delays of below 50 ms, it is possible to provide real-time strategy and turn-based strategy games, and even real-time action games over HSPA.

The real-time action games are constantly transmitting and receiving packets with typical bit rates of 10–20 kbps. Such bit rates can easily be delivered over cellular networks. However, these packets must be delivered with a very low delay which sets high requirements for the network performance. For real-time strategy and turn-based strategy games, both the requirements on the bit rate and the end-to-end network delays are looser, and there is more freedom on how to map these services to radio channels. This mapping is discussed later in Chapter 10.

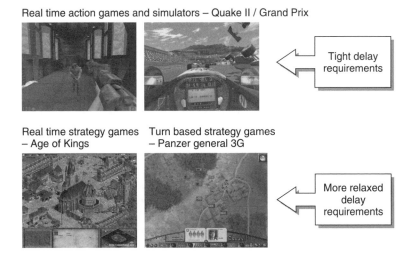

**Figure 2.9**   Multiplayer game classification

## 2.10   Mobile Broadband for Laptop and Netbook Connectivity

In the GSM/EDGE era, before the launch of WCDMA/HSPA, it was foreseen that 'data will go wireless' with GPRS, and that this trend would be a success similar to that of 'voice going wireless'. GPRS certainly enabled data to be sent and consumed over a mobile GSM/EDGE network, but actually a large share of the targeted wireless data was actually captured and delivered by another technology: Wireless LAN (WLAN) or WiFi. The data rates of WLAN/WiFi were far superior to those of GSM/EDGE/GPRS, and WLAN/WiFi hot spots and easy access and pricing models proliferated. Moreover, the added value of full wide area mobility and the ubiquity of a GSM/EDGE system in terms of data delivery were not as important as the more nomadic user experience of WLAN/WiFi, as long as a sufficient number of hot spots were available. It can be said that GSM/EDGE/GPRS remained a somewhat marginal vehicle in delivering significant amounts of mobile data. UMTS systems especially with High Speed Packet Access (HSPA) were going to change the balance.

It could be seen in early to mid-2007 that something remarkable was happening in UMTS networks and mobile broadband business. The portfolio of High Speed Downlink Packet Access (HSDPA)-capable devices had become richer since the introduction of the first PCMCIA data cards in late 2005. Their performance, together with the networks updated with latest HSPA functionalities, had improved significantly since the first QPSK-modulating devices enabling peak data rates of 1.8 Mbps. With peak data rates of 3.6–7.2 Mbps in downlink and 0.384–2.0 Mbps in uplink (and with the latest generation of network and devices even close to 20 Mbps as shown in Figure 2.1), with corresponding improvement in effective and typical data rates, the mobile broadband service started to be 'good enough' and somewhat comparable to that of WLAN/WiFi networks. Perhaps more importantly, operators started to launch so-called flat-rate tariff packages, where a user could consume any amount of data (typically up to a relatively high threshold of e.g. a few GB) per month.

In addition to the mobile broadband solution, some operators have even competed against fixed broadband ADSL proposition with HSPA in home PC use case. There is, for instance, a Kuwaiti operator who sells or rents Customer Premises Equipment (CPE) with HSPA capability, where HSPA is used for backhauling the traffic from a home PC!

At the Mobile World Congress (MWC) in Barcelona in February 2008, there was significant excitement among mobile operators with regards to traffic finally growing rapidly in their networks, together with mobile broadband subscriptions and sales of data packages. By the MWC of February 2009, the atmosphere had become a bit more concerned: many operators were facing challenges with their Base Station backhauling capacity and a small share of users consuming huge amounts of bandwidth with peer-to-peer traffic of sharing movies, among other things. Despite the challenges, mobile broadband remains one of the most exciting growth segments in UMTS.

Figure 2.10 shows an example of the rapid growth in data traffic, with each curve representing the amount of traffic flowing through one real sample network in a day. As we can see, in this sample from Europe, APAC, the Middle East and Latin America, there are networks approaching the level of 30–40 TB (TeraBytes) of data delivered per day. Many phenomena in nature and business develop according to an S-shaped curve. If this is the case here, these networks have not even met the point where the second derivative is zero. In other words, the growth continues to accelerate. In practice, this means that operators need to add a second, third, even fourth carrier to the Node Bs depending on the limit in their operating license, together with channel capacity, RNC, Packet Core and transport network capacity expansions. Additionally, it is becoming more important to control the bandwidth in smarter and more segmented ways, by e.g. introducing Quality of Service algorithms and differentiation by customer segments. Subscribers 'abusing' the system need to be controlled by e.g. deep packet inspection and reducing data rates based upon detected application. In some countries regulation makes such control difficult if not even impossible.

In some advanced mobile broadband markets, operators have even seen a decline in fixed broadband ADSL subscriptions. If a user wants to have mobile broadband enabled in the laptop anyway, why pay for an additional subscription for fixed broadband? Obviously the quality and coverage of good indoor penetration are key to such a trend.

Let us now consider two aspects of Mobile Broadband: end-to-end security and the effect of radio latency on the application performance.

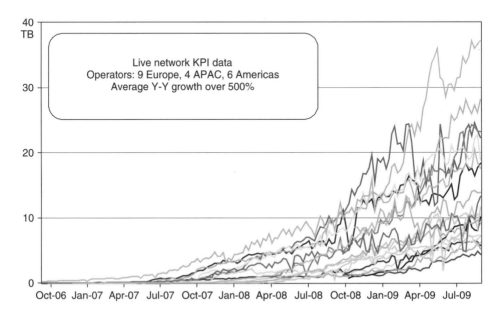

**Figure 2.10**  Data growth in selected Mobile Broadband networks: total HSDPA data volume (MAC-d) at Iub per network. *Source:* Nokia Siemens Networks.

## 2.10.1   End-to-End Security

End-to-end security can be obtained using Virtual Private Networks (VPN) for the encryption of the data. One option is to have VPN client located in the laptop and the VPN gateway in the corporate premises. Such approach is often used by large corporates whose responsibility is to obtain and maintain required equipment for the remote access service. Another approach uses VPN connection between the mobile operator core site and the company intranet. The mobile network uses standard UMTS security procedures. In this case the company only needs to subscribe to the operator's VPN service and obtain a VPN gateway. These two approaches are illustrated in Figure 2.11.

## 2.10.2   Impact of Latency on Application Performance

Application performance in mobile broadband should preferably be similar or superior to the performance of DSL or WLAN. The application performance depends on the available bit rate but also on the network latency. The network latency is here measured as the round trip time. The round trip time is the delay taken for the small IP packet to travel from the mobile to a server and back. The effect of the latency is illustrated below using a file download over Transmission Control Protocol, TCP, in Figure 2.12. The

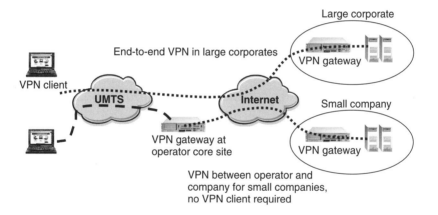

**Figure 2.11**   Virtual private network architectures

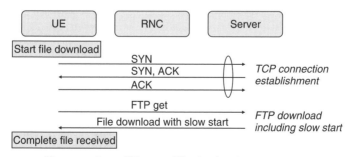

**Figure 2.12**   Example signallng flow in a file download using TCP

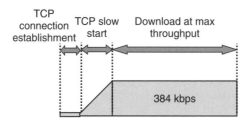

**Figure 2.13**   Example file download using TCP

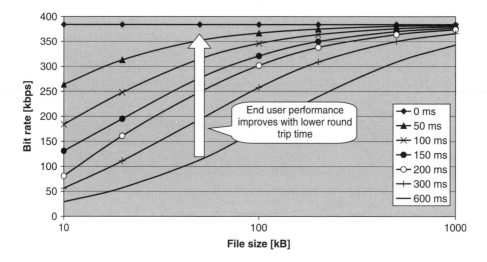

**Figure 2.14**   Effect of round-trip time on the user's bit rate with layer 1 bit rate of 384 kbps

download process includes TCP connection establishment and file download including TCP slow start. The bit rate experienced by the end user is defined here as the download file size divided by the total time. The delay components are illustrated in Figure 2.13.

The bit rates experienced by the user with round trip times between 0 and 600 ms are shown in Figure 2.14. Figure 2.14 assumes that a dedicated channel with 384 kbps already exists and no channel allocation is required. The curves show that a low round trip time is beneficial especially for small file sizes due to the TCP slow start. WCDMA round trip time is analysed in detail later and it is typically 150–200 ms. Figure 2.15 shows the download time with different round trip times. The download time of below 100 kB file is below 3 s as long as the round trip time is below 300 ms. Low round trip time will be more relevant if we need to download several small files using separate TCP sessions.

## 2.11   Social Networking

A seasoned technology executive in California's Silicon Valley told one of the editors in 2009: 'My teenage daughter and her friends say they only use email with the old people. The way they communicate with each other is through Twitter using their mobile devices, rather than using fixed or mobile email.' Social networking applications like Twitter, together with instant messaging, can be seen as either eating into or complementing traditional email use.

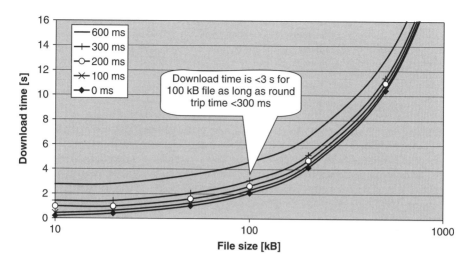

**Figure 2.15**    Effect of round-trip time on the download times with layer 1 bit rate of 384 kbps

While it can be said that social networking applications such as Facebook, Twitter, or their more professional peer of LinkedIn are only some downloadable applications or websites usable over UMTS networks, it is interesting to elaborate on them as a service category of their own. Social networking applications are by nature adding value based upon sharing information, sharing content, sharing and updating presence information, linking people with each other, communicating as a network of people, etc. Simply stated, there is typically more value in doing so quickly, in real time, online, regardless of the location and presence of the users, or even precisely because of the location and presence! We can see that social networking applications benefit greatly from mobility or nomadicity, high data speeds, short call and connection set-up times, messaging, location determination capabilities, and other characteristics and attributes of UMTS systems. From the UMTS technology viewpoint, they generate an interesting cocktail of requirements to maximize the end user quality experience. Just as with smartphones, we can see a lot of signaling traffic and low average data rates as opposed to the use case of e.g. peer-to-peer traffic of sharing movies over mobile broadband.

## 2.12   Mobile TV

Mobile TV can be delivered over regular Release 99 PS or HSDPA bearers. With small screens, even a PS128 bearer is sufficient for reasonably good user experience, although use of at least PS384 or HSDPA is advised. In order to cater for a larger capacity of mobile TV delivery, Multimedia Broadcast and Multicast Service (MBMS) was standardized as part of 3GPP Release 6. MBMS benefits of sending the same 'broadcast' channel to a number of users at the same time, using data rate of 256 kbps. It has been estimated that the number of users watching the same content (with the same timing also) needs to exceed three or four per cell, on average, before MBMS makes commercial sense over regular Release 99 PS or HSDPA delivery.

To date, MBMS has not been deployed on a wide scale anywhere in the world. There has not been enough demand to justify the investment in the full end-to-end capability including the user equipment (UEs). Operators still consider regular Release 99 or HSPA data service adequate for mobile TV in 2009.

One possible reason for the slow or no adoption of MBMS at all lies in the nature of the delivery. Whereas MBMS is by nature a broadcast system, users often want to watch content as 'video on demand', which makes it by nature a unicast delivery. In some countries other technologies than MBMB that are only providing the mobile TV application (but typically with some linking with the cellular subcription) are in use, such as Digital Video Broadcasting for Handheld devices (DVB-H) which is optimized only for the purpose of providing mobile TV service.

## 2.13   Location-Based Services

It is easy to understand why location-based services were expected to become one of the killer application areas in the early days of UMTS: mobility enhanced with high wireless data speeds, together with the existence of various techniques to determine the user's location, could open significant avenues to providing value leveraging location information. However, the early visions of location-based services did not materialize until digital maps and industry players with access to and willingness to use digital maps emerged. The proliferation of GPS receivers in devices has accelerated the adoption of location-based services.

A survey released by Strategy Analytics in April 2006 predicted that the share of smartphone-based GPS devices will grow in the coming years, while the share of dedicated devices will start to decline from 2007 onwards. This is exactly what has happened. It is clear that combining location data with the communication capabilities and extensibility of a smartphone provides novel user experiences and business opportunities. Maps and city guides including points of interest data can be kept always up-to-date, which also means increased data traffic. In fact, there is a growing list of value-added services that can be updated over the air: new maps, city guides, new voice commands, weather, traffic information, safety cameras, traffic cameras, and so forth.

A location-based service is provided either by an operator or by a third party service provider that utilizes available information on the terminal location. The service is either push type (e.g. automatic distribution of local information) or pull type (e.g. localization of emergency calls). Other possible location-based services are discount calls in a certain area, broadcasting of a service over a limited number of sites (broadcasting video on demand), and retrieval and display of location-based information, such as the location of the nearest gas stations, hotels, restaurants, and so on. Figure 2.16 shows an example. Depending on the service, the data may be retrieved interactively or in the background. For instance, before traveling to an unknown city abroad one may request night-time download of certain points of interest in the city. The downloaded information typically contains a map and other data to be displayed on top of the map. By clicking the icon on the map, one gets information from the point. Information to be downloaded as background or interactively can be limited by certain criteria and personal interest.

Mobile search is a special type of location service that can be characterized by three verbs: search, find, and connect. Typically it combines web search, for instance Yahoo! or Windows Live, with local search (Yellow pages), such as Eniro in Nordic countries. Typically search results are shown on the map

**Figure 2.16**   Nokia 6110 Navigator with integrated GPS receiver

on the display of a smartphone and the user can easily check which of the results is best for the purpose and then connect to the service either by launching a web browser or making a call. If the location of the user is known, it can be used to prioritize the search results.

The location information can be input by the user or detected by the network or mobile station. The network architecture of the location services is discussed in Chapter 5. In the following, we consider cell coverage-based and GPS-based positioning methods. These methods are complementary rather than competing, and are suited for different purposes. These approaches are introduced in the following sections.

## 2.13.1   Cell Coverage-Based Location Calculation

Cell coverage-based location method is a network-based approach, i.e., it does not require any new functionalities in the mobile. The radio network has the location information with a cell level accuracy when the mobile has been allocated a dedicated channel or when the mobile is in cell_FACH or cell_PCH states. These states are introduced in Chapter 7. If the mobile is in idle state, its location with cell accuracy can be obtained by forcing the mobile to cell_FACH state with a location update as illustrated in Figure 2.17.

The accuracy of the cell coverage-based method depends heavily on the cell size. The typical cell ranges in the urban area are below 1 km and in the dense urban a few hundred meters, providing fairly accurate location information.

The accuracy of the cell coverage based-approach can be improved by using the round trip time measurement that can be obtained from the base station. That information is available in the cell_DCH state and it gives the distance between the base and the mobile station.

## 2.13.2   Assisted GPS (A-GPS)

Most accurate location measurements can be obtained with an integrated GPS receiver in the mobile. The network can provide additional information, such as visible GPS satellites, reference time and Doppler, to assist the mobile GPS measurements. The assistance data improves the GPS receiver sensitivity for

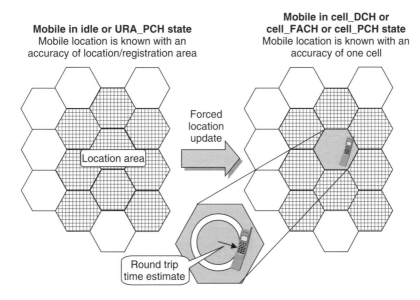

**Figure 2.17**   Location calculation with cell coverage combined with round-trip time

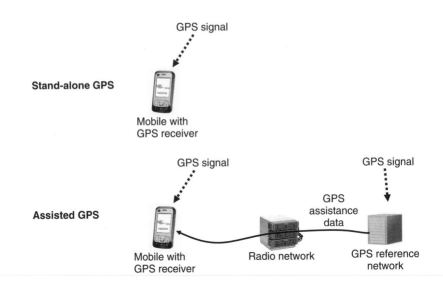

**Figure 2.18**  Stand-alone and assisted GPS

indoor measurements, makes the acquisition times faster and reduces the GPS power consumption. The principle of assisted GPS is shown in Figure 2.18.

A reference GPS receiver in every base station provides most accurate assistance data and most accurate GPS measurements by the mobile. The assisted GPS measurements can achieve accuracy of 10 meters outdoors and a few tens of meters indoors. That accuracy meets also the FCC requirements in the USA. If the most stringent measurement probabilities and accuracies are not required, the reference GPS receiver is not needed in every base station, but only a few reference GPS receivers are needed in the radio network. It is also possible to let the mobile GPS make the measurements without any additional assistance data.

Standalone GPS devices and services can be further divided into three categories: on-board location service, off-board location service and a hybrid service. The first holds all the location data – maps, points of interest, addresses – on the memory card of the device, the second, on the other hand, fetches the data over the air from the server, creating a lot of data traffic. The hybrid system falls in between these two: it fetches new maps from the server, but stores them locally on the memory card for future use. The hybrid system can also retrieve more frequently changing data, such as information about road construction, local weather, traffic jams, etc.

## 2.14   Machine-to-Machine Communications

By machine-to-machine communication in UMTS context we mean any service or application connecting two machines over a UMTS radio network, as opposed to an immediate human role or intervention. It can be said that the key driver behind various machine-to-machine communication applications and services is to improve the efficiency, cost, speed, quality and ease of use of applications and communication that traditionally required human intervention or activity. Obviously there has to be value over a fixed, cabled connection. While it is easy to come up with various examples of such use, it is correct to note that machine-to-machine communications represents a fairly small share of operator revenues today. However, it is expected to be one of the areas to experience rapid growth in the future. The growth is to a great extent driven by the decreasing cost of UMTS chipsets and device technologies and the willingness of UMTS chipset and device makers and industry players to include small and cost-effective UMTS radio in more and more devices within the converging digital world. Also the improved UMTS coverage makes it easier for the UMTS radio to enter the market so far dominated by GSM-based solutions.

Let us provide a very short list of simple examples:

- Connecting private home alarm systems with the monitoring room of a guard company over a UMTS network, providing a connection to send the alarm message or even photographs of the event even if cabled network connections are cut or not available.
- Sending information or alarms from domestic appliances, such as refrigerators, heat pumps or saunas, to a subscriber's UMTS device.
- Enabling electricity, gas or water companies to 'read' meters of their customers over a UMTS network for billing purposes in the absence of a cabled network.

3GPP is investigating further improvements for machine-to-machine communications as part of the Release 10 studies. The intention is to see whether there could be some measures, for example, to reduce signaling (e.g. paging or random access) from the potential high number of machine-to-machine installations.

## 2.15 Quality of Service (QoS) Differentiation

Chapter 10 covers application end-to-end performance assuming that the system load is reasonably low. When the system load gets higher, it becomes important to prioritize the different services according to their requirements. This prioritization is called QoS differentiation. 3GPP QoS architecture is designed to provide this differentiation [14]. The terminology is shown in Figure 2.19.

The most relevant parameters of UMTS QoS classes are summarized in Table 2.1. The main distinguishing factor between the four traffic classes is how delay-sensitive the traffic is: the conversational class is meant for most delay-sensitive traffic, while the background class is the most delay-insensitive. There are a further three different priority categories, called allocation/retention priority, within each QoS class. Interactive has also three traffic handling priorities. Conversational and streaming class parameters include the guaranteed bit rate and the transfer delay parameters. The guaranteed bit rate defines the minimum bearer bit rate that UTRAN must provide and it can be used in admission control and in resource allocations. The transfer delay defines the required 95th percentile of the delay. It can be used to define the RLC operation mode (acknowledged, non-acknowledged mode) and the number of retransmissions.

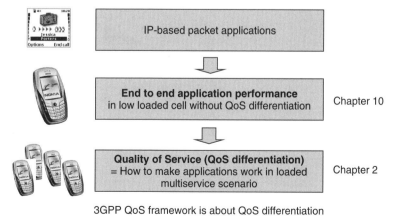

**Figure 2.19**   Definition of quality of service differentiation

**Table 2.1** UMTS QoS classes and their main parameters

|  | Conversational class | Streaming class | Interactive class | Background class |
|---|---|---|---|---|
| Transfer delay | 80 ms – | 250 ms – |  |  |
| Guaranteed bit rate (kbps) | Up to 2 Mbps | Up to 2 Mbps |  |  |
| Traffic handling priority |  |  | 1, 2, 3 |  |
| Allocation/Retention priority | 1, 2, 3 | 1, 2, 3 | 1, 2, 3 | 1, 2, 3 |

The conversational class is characterized by low end-to-end delay and symmetric or nearly symmetric traffic between uplink and downlink in person-to-person communications. The maximum end-to-end delay is given by the human perception of video and audio conversation: subjective evaluations have shown that the end-to-end delay has to be less than 400 ms. The streaming class requires bandwidth to be maintained like conversational class but streaming class tolerates some delay variations that are hidden by the dejitter buffer in the receiver. The interactive class is characterized by the request response pattern of the end-user. At the message destination there is an entity expecting the message (response) within a certain time. The background class assumes that the destination is not expecting the data within a certain time.

UMTS QoS classes are not mandatory for the introduction of any low delay service and have not been widely deployed in the networks so far. It is possible to support streaming video or conversational Voice over IP from the end-to-end performance point of view by using just background QoS class. QoS differentiation becomes useful for the network efficiency during high load when there are services with different delay requirements. If the radio network has knowledge about the delay requirements of the different services, it would be able to prioritize the services accordingly and improve the efficiency of the network utilization. The qualitative gain of the QoS differentiation is illustrated in Figure 2.20. Considerable efficiency gains can be obtained in Step 2 just by introducing a few prioritization classes within the interactive or background class by using allocation and retention parameters (ARP). The pure prioritization in packet scheduling alone is not enough to provide full QoS differentiation gains. Users within the same QoS and ARP class will share the available capacity. If the number of users is simply too high, they will all suffer from bad quality. In that case it would be better to block a few users to guarantee the quality of the existing connections, like streaming videos. That is provided in Step 3 in Figure 2.20

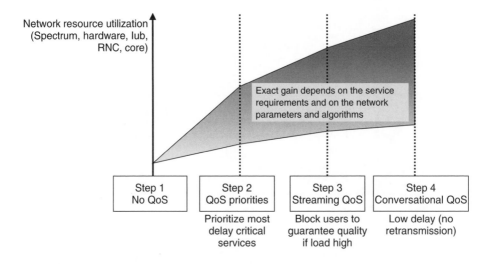

**Figure 2.20** Qualitative gain illustration for QoS differentiation

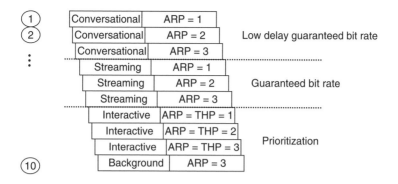

**Figure 2.21**    Example of 10 categories by taking a subset of UMTS QoS classes

with guaranteed bit rate streaming. The radio network can estimate the available radio capacity and block an incoming user if there is no room to provide the required bandwidth without sacrificing the quality of the existing connections. The final Step 4 further differentiated between guaranteed bit rate services with different delay requirements. If the delay requirements are known, the WCDMA RAN can allocate suitable radio parameters – such as retransmission parameters – for the new bearer.

An example QoS differentiation scheme is shown in Figure 2.21 with 10 different QoS categories: 6 guaranteed bit rate categories and 4 non-real time categories. It is assumed in this case that traffic handling priority is equal to allocation and retention priority, and there is no prioritization within background class.

The following two figures illustrate Steps 1 and 2. Figure 2.22 shows an example where all the services have the same QoS parameters and the same treatment. In this case all services share the network

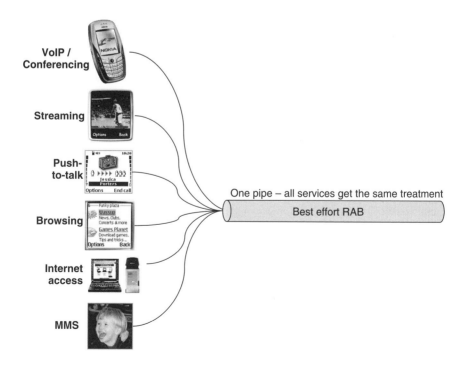

**Figure 2.22**    No QoS differentiation – all services use the same QoS parameters

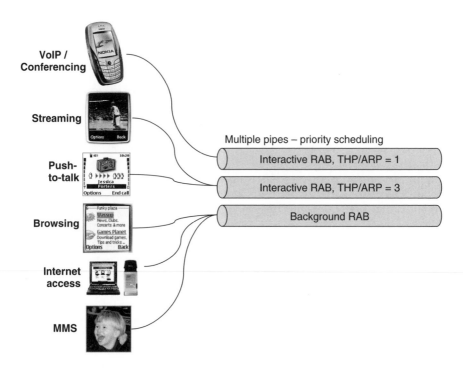

**Figure 2.23**   QoS prioritization used with three classes

resources equally: they get the same bit rate and experience the same delay. The network dimensioning must be done so that this bit rate or delay fulfils the most stringent requirements of the services provided in the network. The background type of services, like sending of MMS, will get the same quality, which is unnecessarily good and wastes network resources. Figure 2.23 shows the case where there are three different pipes with QoS prioritization in packet scheduling. This approach already provides QoS differentiation and makes the network dimensioning requirements less stringent.

The layered architecture of a UMTS bearer service is depicted in Figure 2.24; each bearer service on a specific layer offers its individual services using those provided by the layers below. The QoS parameters are given by the core network to the radio network in the radio access bearer set-up.

Figure 2.25 illustrates the mechanisms to define the QoS parameters in a radio access bearer set-up:

1. UE can request QoS parameters. In particular, if the application requires guaranteed bit rate streaming or conversational class, it has to be requested by UE, otherwise, it cannot be given by the network.
2. Access point node (APN) in GGSN can give QoS parameters according to operator settings. Some services may be accessed via certain APNs. That allows the operator to control the QoS parameters for different services and makes it also possible to prioritize operator-hosted services compared to accessing other services.
3. Home location register (HLR) may contain subscriber-specific limitations for the QoS parameters.
4. WCDMA radio network must be able to provide the QoS differentiation in packet handling.

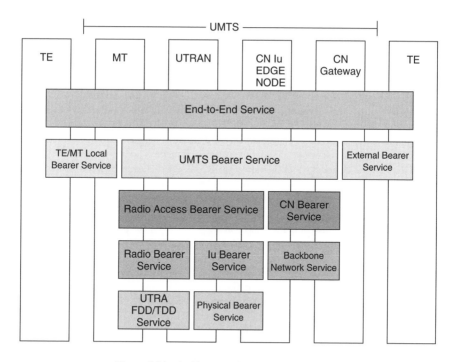

**Figure 2.24**    Architecture of a UMTS bearer service

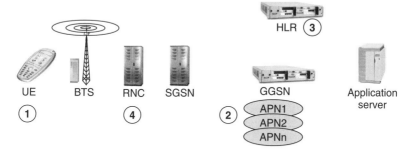

(1) = UE must request QoS class if it wants streaming or conversational QoS

(2) = Different APNs in GGSN can be used to provide different QoS

(3) = User specific QoS limitations can be defined in HLR

(4) = Radio access network must be able to provide QoS differentiation

**Figure 2.25**    The role of UE, GGSN and HLR in defining QoS class

## 2.16   Maximum Air Interface Capacity

Until now, we have elaborated on various service categories that are enabled by UMTS technology. In practice, service strategy definition does not happen in isolation of network capacity. Any service becomes useless if the network runs out of capacity to cater for the service. A classic dilemma that operators are facing is to create and price a value proposition in such a way that the services will not congest the network and provide an upside for revenue growth, while capitalizing on network investments efficiently. Let us estimate the maximum network level capacity with HSPA technology by using typical traffic distribution assumptions from live networks. The target is to obtain an approximate value for the maximum number of subscribers that can be supported with the given base station site density and with the given spectrum resources. The overview of the estimation process is shown in Figure 2.26.

HSDPA UEs send a Channel Quality Indication (CQI) every few milliseconds when UE is in Cell_DCH state. CQI indicates the maximum possible data rate that UE can receive with error rate less than 10%, as elaborated in Chapter 12. CQI reports are mainly used for the link adaptation and for the packet scheduling algorithms but CQI reports can also be used to estimate the maximum air interface capacity for the network dimensioning purposes. An example CQI distribution over the whole network is shown in Figure 2.27. The median CQI value in this case is 21 which corresponds to approximately 5.5 Mbps throughput. The cell throughput can be slightly higher due to system-level packet scheduling gains with a proportional fair scheduler. We assume in this example 6 Mbps average cell throughput. The CQI values are affected by the network RF planning, by the amount of other-cell interference and by the performance of the UE receivers. Better UE receivers and network optimization may improve the CQI values while higher other-cell loading may lower the CQI values.

We also need to reserve some margin for the network dimensioning in the busy hour in order to guarantee low delays and reasonably good data rates. We assume in this example maximum 50% load factor over the busy hour, which leads to average busy hour throughput of 6 Mbps × 50% = 3 Mbps per cell. The maximum load level depends on the operator target settings in terms of end user data rates. It also depends on QoS differentiation algorithms. If QoS is used, the network could be loaded higher by the less important background traffic.

The traffic is never equally distributed between the sites. There are several sites in the network that are needed to provide coverage and they will not be fully loaded. An example traffic distribution is shown in Figure 2.28 where 50% of the traffic is carried by 15% of the cells. When we estimate the maximum network capacity, it is those 15% of the cells that get congested and limit the total capacity. The network capacity could be improved by adding cell sites to the congested areas. The traffic distribution depends on the network type and on the traffic profiles.

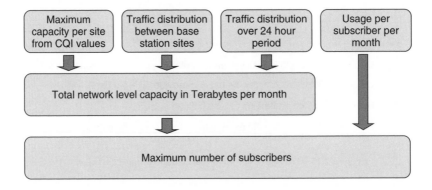

**Figure 2.26**   Estimation of the maximum number of supported subscribers in the network

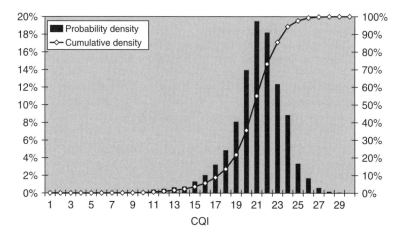

**Figure 2.27**  Example CQI distribution

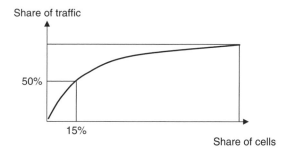

**Figure 2.28**  Example traffic distribution between cells

The traffic is not equally distributed over a 24-hour period. The busy hour in data networks is typically in the evening but there is some data traffic also during the night hours. An example traffic distribution is shown in Figure 2.29. The busy hour carries 7% of the daily traffic on the network level in this example.

We can do an example capacity calculation with the following assumptions:

- 10,000 or 20,000 base station sites each with three sectors. That corresponds to a large European country.

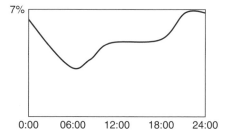

**Figure 2.29**  Example traffic distribution over a daily 24-hour period

- Total 15 MHz or 20 MHz spectrum equal to three or four HSPA carriers. Most operators have 15 MHz at 2100 MHz and potentially 5 MHz at 900 MHz.
- Cell throughput 6 Mbps and no voice traffic assumed in this example. In real networks part of the capacity would be taken by voice usage. Busy hour average cell throughput assumed to be 50% of 6 Mbps.
- Busy hour carries 7% of the daily traffic.
- 15% of sites carry 50% of the traffic.
- Each subscriber consumes 2 GB of data per month in downlink.

The results are shown in Figure 2.30 and Figure 2.31. If there are 10,000 sites and 15 MHz spectrum, the maximum carried data volume in downlink will be approximately 500 Terabytes per day. With 20,000 sites and 20 MHz spectrum, the capacity would be nearly 1400 Terabytes per day. The capacity corresponds to 8 million and 21 million data subscribers. These calculations indicate that HSPA radio with current spectrum allocation can provide 2 GB of data for all or for most existing voice subscribers.

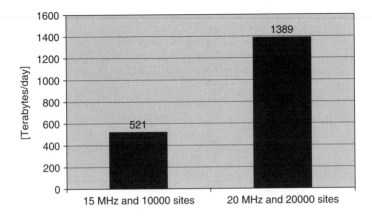

**Figure 2.30**   Maximum downlink data volume for the whole network

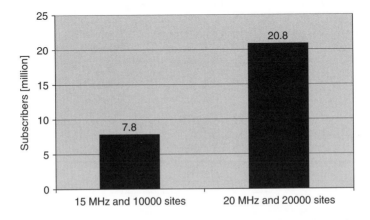

**Figure 2.31**   Maximum number of broadband subscribers each 2 GB/month

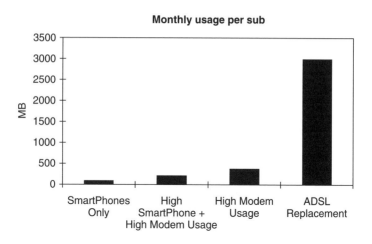

**Figure 2.32** HSDPA data usage per subscriber per month

The most sensitive parameter in this calculation is the data usage per subscriber. If each subscriber uses 10 GB instead of 2 GB, the number of supported subscribers would be substantially lower. The monthly usage per subscriber depends greatly on the type of users in the network. If the majority of the users are using so-called smartphones, the monthly usage tends to be just 100–200 MB. If most subscribers use HSPA for DSL replacement, the monthly usage is substantially higher, typically several GBs. Some examples of typical HSDPA data usage per subscriber are shown in Figure 2.32 and Figure 2.33 with an indication of dominating terminal type. Similar trends can also be seen if we analyze the average data volume for each HS-DSCH allocation. In case of smartphones, the average data volume is below 100 kB in each allocation while with DSL use, the data volume is 500 kB per allocation. The daily HSDPA data volume can be extremely high even if there are plenty of smartphones in the network. In those cases the

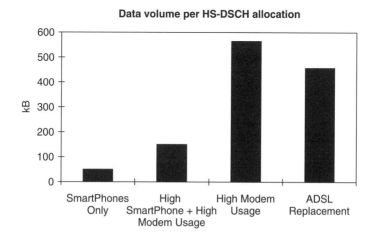

**Figure 2.33** HSDPA data volume per HS-DSCH allocation

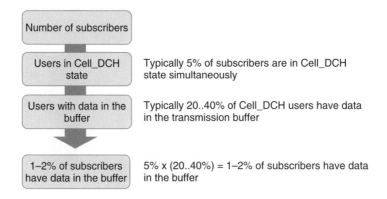

**Figure 2.34** Simultaneous users in HSPA networks

subscriber amount is significantly higher compared to the cases where HSPA network is mainly used as an ADSL replacement.

The next step is to estimate the user data rates during the high load. From Figure 2.31 we take 7.8 million subscribers and 10,000 sites. The busy sectors have then 7.8e6/10e3/3*50%/15% = 867 subscribers assuming 15% of cells carry 50% of traffic. That equals 289 subscribers per 5 MHz cell. Now we need to estimate what is the percentage of those subscribers that are downloading data simultaneously in the busy areas during the busy hour. The network statistics shows that typically 5% of the subscribers are in the Cell_DCH state during the busy hour. The statistics further shows that 20–40% of those users in Cell_DCH have data in the buffer. These values are from those networks where HSPA is mainly used as the wireless modem with laptops. Therefore, we can estimate that 1–2% of subscribers are downloading data simultaneously. The calculation is illustrated in Figure 2.34. That corresponds to 3–6 simultaneous users downloading data in the high capacity case. If the cell throughput is on average 6 Mbps, then the high load situation gives 1–2 Mbps user throughputs.

There are solutions to provide even more capacity and support higher traffic volumes with the following solutions:

- 6-sector sites can offer up to 80% more capacity per site compared to 3-sector sites. The 6-sector solution may not be feasible in all sites if the antenna installation increases wind load excessively.
- QoS differentiation can be applied to lower the priority of the heavy users when they have exceeded their monthly quota. Typically, a small percentage (<20%) of the users take most of the capacity (>80%).
- Offloading the traffic from macro network to small cells like micro or femto (home Node B) cells. 3GPP specifications include more relaxed RF requirements for micro base stations, called medium area base stations in [15]. 3GPP Releases 8 and 9 includes a number of enhancements for femto cells including optimized architecture and incoming handovers. Femtocells are considered in detail in Chapter 19.
- More macro sites can potentially be added to the congested areas. The latest base station products are small and compact which can make the site acquisition simpler compared to the earlier products.

## 2.17 Terminals

In the very early days of UMTS, terminal or user equipment (UE) availability was a clear bottleneck to the operator's business. There was a race between network maturity and UE availability, with some evidence of a 'chicken and egg' problem: without networks being mature enough, it did not make sense for UE manufacturers to invest aggressively in devices. Without UEs being mature enough and available in volumes, it did not make sense for network manufacturers to introduce features and functionalities. Lack of availability of devices even complicates network functionality testing. The UE portfolio was very limited, resembling very much the portfolio of 2G or GSM/EDGE devices, albeit with some futuristic designs but inferior battery time performance. Operators were struggling to secure enough volume of attractive enough handsets to migrate their subscribers to UMTS systems and launch UMTS services.

Today, we can say that there is an abundance of UMTS devices and device categories in the market, and the portfolio and functionality of UEs are certainly not an issue. There is still a price premium in multimode devices supporting 2G and 3G (and in many cases GPS, WiFi and Bluetooth), but UMTS devices have become more affordable and accessible to the masses. To date, more than 1000 different UMTS devices have been launched. UE categories are no longer limited to handheld devices. The following UE categories can be defined as available in large volumes:

- Classic handheld devices in various form-factors (monoblocks, clamshells, etc.)
- Smartphones
- PCMCIA data cards
- CPE using HSPA for backhauling PC traffic
- USB sticks and dongles
- Laptop, netbook and PC-embedded modems.

## 2.18   Tariff Schemes

The legacy of 2G network technology operators was to base their tariff schemes based upon price per minute, either post-paid or pre-paid depending on the payment reliability of the user segment. In the transition to 3G and UMTS, it was the intention of many operators to continue tariffing of both voice and data based upon similar logic, i.e. price per volume or time unit. Many operators wanted to avoid becoming bit pipes and attempted to build 'walled gardens' for mobile data. Unfortunately, it can be said that this delayed the adoption and proliferation of mobile data usage for several years.

In general, pricing schemes can be categorized as follows:

- price per minute of use, for both voice and data;
- price per message in all messaging services;
- price per kb of data transferred;
- fixed price per month, i.e. 'flat rate' tariffs;
- combinations of the above.

Initially, many operators attempted to price data per kB transferred. Users found this structure nontransparent and even threatening, as there is no way for the user to understand before using a service how much data would be transferred. Pricing data per minute of connection was not successful either.

Today, we can see that flat rate tariffing schemes with upper limits for data limiting excessive loading of the system have become the most popular among operators. The initial concerns of flat rate schemes congesting the networks and removing any upside potential have been alleviated, even if some mobile broadband operators are indeed seeing their networks fill up with traffic rapidly, driving capacity expansions.

## References

[1] 3GPP, Mandatory Speech Codec Speech Processing Functions, AMR Speech Codec: General Description (3G TS 26.071).
[2] 3GPP, Mandatory Speech Codec Speech Processing Functions, AMR Speech Codec: Frame Structure General Description (3G TS 26.101 version 1.4.0), 1999.

[3] Holma, H., Melero, J., Vainio, J., Halonen, T. and Mäkinen, J., 'Performance of Adaptive Multirate (AMR) Voice in GSM and WCDMA', *Proceedings of IEEE Vehicular Technology Conference* (VTC 2003, Spring), April 2003, pp. 2177–2181.

[4] 3GPP, Technical Specification Group Services and System Aspects, Speech Coded Speech Processing Functions, AMR Wideband Speech Codec, General Description, (3G TS 26.171 version 5.0.0), 2001.

[5] 3GPP, Technical Specification Group Services and System Aspects, Codec for Circuit Switched Multimedia Telephony Service: General Description (3G TS 26.110 version 6.0.0 Release 6), 2004.

[6] ITU-T H.324, Terminal for Low Bitrate Multimedia Communication, 1998.

[7] Stockhammer, T., Hannuksela, M. and Wiegand, T., 'H.264/AVC in Wireless Environments', July, 2003.

[8] 3GPP, Multimedia Messaging Service (3GPP TS 22.140), 2001.

[9] 3GPP, MMS Architecture and Functionality (3GPP TS 23.140), 2001.

[10] Research In Motion Ltd Press Release, 'Research In Motion Reports First Quarter Results', June 18, 2009.

[11] 3GPP, 'Service Aspects; Services and Service Capabilities (Release 6)', 3GPP TS 22.105, version 6.2.0, June 2003.

[12] Nokia, 'Multiplayer Game Performance over Cellular Networks', Forum Nokia, version 1.0, January 20, 2004.

[13] Anttila, J. and Lakkakorpi, J., 'On the Effect of Reduced Quality of Service on Multiplayer Online Games', *IJIGS*, vol. 2, no. 2, 2003.

[14] 3GPP, Technical Specification Group Services and System Aspects, QoS Concept and Architecture (3G TR 23.107).

[15] 3GPP Technical Specification, TS 25.104, Base Station (BS) radio transmission and reception (FDD), version 9.2.0, December 2009.

# 3

# Introduction to WCDMA

Peter Muszynski and Harri Holma

## 3.1 Introduction

This chapter introduces the principles of the WCDMA air interface. Special attention is drawn to those features by which WCDMA differs from GSM and IS-95. The main parameters of the WCDMA physical layer are introduced in Section 3.2. The concept of spreading and despreading is described in Section 3.3, followed by a presentation of the multipath radio channel and Rake receiver in Section 3.4. Other key elements of the WCDMA air interface discussed in this chapter are power control and soft and softer handovers. The need for power control and its implementation are described in Section 3.5, and soft and softer handover in Section 3.6.

## 3.2 Summary of the Main Parameters in WCDMA

We present the main system design parameters of WCDMA in this section and give brief explanations for most of them. Table 3.1 summarizes the main parameters related to the WCDMA air interface. Here we highlight some of the items that characterize WCDMA:

- WCDMA is a wideband Direct-Sequence Code Division Multiple Access (DS-CDMA) system, i.e. user information bits are spread over a wide bandwidth by multiplying the user data with quasi-random bits (called chips) derived from CDMA spreading codes. In order to support very high bit rates (up to 2 Mbps), the use of a variable spreading factor and multicode connections is supported. An example of this arrangement is shown in Figure 3.1.
- The chip rate of 3.84 Mcps leads to a carrier bandwidth of approximately 5 MHz. DS-CDMA systems with a bandwidth of about 1 MHz, such as IS-95, are commonly referred to as narrowband CDMA systems. The inherently wide carrier bandwidth of WCDMA supports high user data rates and also has certain performance benefits, such as increased multipath diversity. Subject to his operating license, the network operator can deploy multiple 5 MHz carriers to increase capacity, possibly in the form of hierarchical cell layers. Figure 3.1 also shows this feature. The actual carrier spacing can be selected on a 200 kHz grid between approximately 4.4 and 5 MHz, depending on interference between the carriers.

*WCDMA for UMTS: HSPA Evolution and LTE, Fifth Edition*   Edited by Harri Holma and Antti Toskala
© 2010 John Wiley & Sons, Ltd

**Table 3.1**  Main WCDMA parameters

| | |
|---|---|
| Multiple access method | DS-CDMA |
| Duplexing method | Frequency division duplex/time division duplex |
| Base station synchronization | Asynchronous operation |
| Chip rate | 3.84 Mcps |
| Frame length | 10 ms |
| Service multiplexing | Multiple services with different quality of service requirements multiplexed on one connection |
| Multirate concept | Variable spreading factor and multicode |
| Detection | Coherent using pilot symbols or common pilot |
| Multiuser detection, smart antennas | Supported by the standard, optional in the implementation |

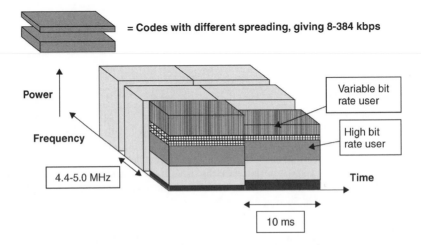

= **Codes with different spreading, giving 8-384 kbps**

Power

Frequency

4.4-5.0 MHz

Variable bit rate user

High bit rate user

Time

10 ms

**Figure 3.1**  Allocation of bandwidth in WCDMA in the time–frequency–code space

- WCDMA supports highly variable user data rates, in other words, the concept of obtaining Bandwidth on Demand (BoD) is well supported. The user data rate is kept constant during each 10 ms frame. However, the data capacity among the users can change from frame to frame. Figure 3.1 also shows an example of this feature. This fast radio capacity allocation will typically be controlled by the network to achieve optimum throughput for packet data services.
- WCDMA supports two basic modes of operation: Frequency Division Duplex (FDD) and Time Division Duplex (TDD). In the FDD mode, separate 5 MHz carrier frequencies are used for the uplink and downlink respectively, whereas in TDD only one 5 MHz is time-shared between the uplink and downlink. Uplink is the connection from the mobile to the base station, and downlink is that from the base station to the mobile.
- The TDD mode is based heavily on FDD mode concepts and was added in order to leverage the basic WCDMA system also for the unpaired spectrum allocations of the ITU for the IMT-2000 systems. The TDD mode is described in detail in Chapter 18.
- WCDMA supports the operation of asynchronous base stations, so that, unlike in the synchronous IS-95 system, there is no need for a global time reference such as a GPS. Deployment of indoor and micro-base stations is easier when no GPS signal needs to be received.

- WCDMA employs coherent detection on uplink and downlink based on the use of pilot symbols or common pilot. While already used on the downlink in IS-95, the use of coherent detection on the uplink is new for public CDMA systems and will result in an overall increase of coverage and capacity on the uplink.
- The WCDMA air interface has been crafted in such a way that advanced CDMA receiver concepts, such as multiuser detection and smart adaptive antennas, can be deployed by the network operator as a system option to increase capacity and/or coverage. In most second generation systems no provision was made for such receiver concepts and as a result they are either not applicable or can be applied only under severe constraints with limited increases in performance.
- WCDMA is designed to be deployed in conjunction with GSM. Therefore, handovers between GSM and WCDMA are supported in order to be able to leverage the GSM coverage for the introduction of WCDMA.

In the following sections of this chapter we will briefly review the generic principles of CDMA operation. In the subsequent chapters, the above-mentioned aspects specific to the WCDMA standard will be presented and explained in more detail. The basic CDMA principles are also described in [1, 2, 3 and 4].

## 3.3   Spreading and Despreading

Figure 3.2 depicts the basic operations of spreading and despreading for a DS-CDMA system. User data is here assumed to be a BPSK-modulated bit sequence of rate $R$, the user data bits assuming the values of $\pm 1$. The spreading operation, in this example, is the multiplication of each user data bit with a sequence of 8 code bits, called chips. We assume this also for the BPSK spreading modulation. We see that the resulting spread data is at a rate of $8 \times R$ and has the same random (pseudo-noise-like) appearance as the spreading code. In this case we would say that we used a spreading factor of 8. This wideband signal would then be transmitted across a wireless channel to the receiving end.

During despreading, we multiply the spread user data/chip sequence, bit duration by bit duration, with the very same 8 code chips as we used during the spreading of these bits. As shown, the original

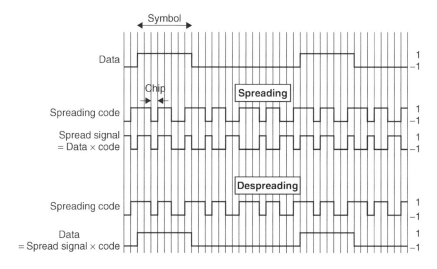

**Figure 3.2**   Spreading and despreading in DS-CDMA

user bit sequence has been recovered perfectly, provided we have (as shown in Figure 3.2) also perfect synchronization between the spread user signal and the (de)spreading code.

The increase of the signaling rate by a factor of 8 corresponds to a widening (by a factor of 8) of the occupied spectrum of the spread user data signal. Due to this virtue, CDMA systems are more generally called spread spectrum systems. Despreading restores a bandwidth proportional to $R$ for the signal.

The basic operation of the correlation receiver for CDMA is shown in Figure 3.3. The upper half of the figure shows the reception of the desired own signal. As in Figure 3.2, we see the despreading operation with a perfectly synchronized code. Then, the correlation receiver integrates (i.e. sums) the resulting products (data × code) for each user bit.

The lower half of Figure 3.3 shows the effect of the despreading operation when applied to the CDMA signal of another user whose signal is assumed to have been spread with a different spreading code. The result of multiplying the interfering signal with the own code and integrating the resulting products leads to interfering signal values lingering around 0.

As can be seen, the amplitude of the own signal increases on average by a factor of 8 relative to that of the user of the other interfering system, i.e. the correlation detection has raised the desired user signal by the spreading factor, here 8, from the interference present in the CDMA system. This effect is termed 'processing gain' and is a fundamental aspect of all CDMA systems, and in general of all spread spectrum systems. Processing gain is what gives CDMA systems the robustness against self-interference that is necessary in order to reuse the available 5 MHz carrier frequencies over geographically close distances. Let's take an example with real WCDMA parameters. Speech service with a bit rate of 12.2 kbps has a processing gain of 25 dB = $10 \times \log_{10}$ (3.84e6/12.2e3). After despreading, the signal power needs to be typically a few decibels above the interference and noise power. The required power density over the interference power density after despreading is designated as $E_b/N_0$ in this book, where $E_b$ is the energy, or power density, per user bit and $N_0$ is the interference and noise power density. For speech service $E_b/N_0$ is typically in the order of 5.0 dB, and the required wideband signal-to-interference ratio is therefore 5.0 dB minus the processing gain = −20.0 dB. In other words, the signal power can be 20 dB under the interference or thermal noise power, and the WCDMA receiver can still detect the signal. The wideband signal-to-interference ratio

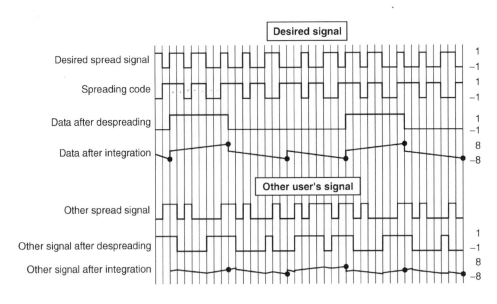

**Figure 3.3**  Principle of the CDMA correlation receiver

is also called the carrier-to-interference ratio $C/I$. Due to spreading and despreading, $C/I$ can be lower in WCDMA than, for example, in GSM. A good quality speech connection in GSM requires $C/I = 9 - 12\,\text{dB}$.

Since the wideband signal can be below the thermal noise level, its detection is difficult without knowledge of the spreading sequence. For this reason, spread spectrum systems originated in military applications where the wideband nature of the signal allowed it to be hidden below the omnipresent thermal noise.

Note that within any given channel bandwidth (chip rate) we will have a higher processing gain for lower user data bit rates than for high bit rates. In particular, for user data bit rates of 2 Mbps, the processing gain is less than 2 ($= 3.84\,\text{Mcps}/2\,\text{Mbps} = 1.92$ which corresponds to 2.8 dB) and some of the robustness of the WCDMA waveform against interference is clearly compromised.

Both base stations as well as mobiles for WCDMA essentially use this type of correlation receiver. However, due to multipath propagation (and possibly multiple receive antennas), it is necessary to use multiple correlation receivers in order to recover the energy from all paths and/or antennas. Such a collection of correlation receivers, termed 'fingers', is what comprises the CDMA Rake receiver. We will describe the operation of the CDMA Rake receiver in further detail in the following section, but before doing so, we make some final remarks regarding the transformation of spreading/despreading when used for wireless systems.

It is important to understand that spreading/despreading by itself does not provide any signal enhancement for wireless applications. Indeed, the processing gain comes at the price of an increased transmission bandwidth (by the amount of the processing gain).

All the WCDMA benefits come rather 'through the back door' by the wideband properties of the signals when examined at the system level, rather than the level of an individual radio link:

1. The processing gain together with the wideband nature suggests a frequency reuse of 1 between different cells of a wireless system (i.e. a frequency is reused in every cell/sector). This feature can be used to obtain high spectral efficiency.
2. Having many users share the same wideband carrier for their communications provides inter-ferer diversity, i.e. the multiple access interference from many system users is averaged out, and this again will boost capacity compared to systems where one has to plan for the worst-case interference.
3. However, both the above benefits require the use of tight power control and soft handover to avoid one user's signal blocking the others' communications. Power control and soft handover will be explained later in this chapter.
4. With a wideband signal, the different propagation paths of a wireless radio signal can be resolved at higher accuracy than with signals at a lower bandwidth. This results in a higher diversity content against fading, and thus improved performance.

## 3.4   Multipath Radio Channels and Rake Reception

Radio propagation in the land mobile channel is characterized by multiple reflections, diffractions and attenuation of the signal energy. These are caused by natural obstacles such as buildings, hills, and so on, resulting in so-called multipath propagation. There are two effects resulting from multipath propagation that we are concerned with in this section:

1. The signal energy (pertaining, for example, to a single chip of a CDMA waveform) may arrive at the receiver across clearly distinguishable time instants. The arriving energy is 'smeared' into a certain multipath delay profile, see Figure 3.4, for example. The delay profile extends typically from 1 to 2 μs in urban and suburban areas, although in some cases delays as long as 20 μs or more with significant signal energy have been observed in hilly areas. The chip duration at 3.84 Mcps

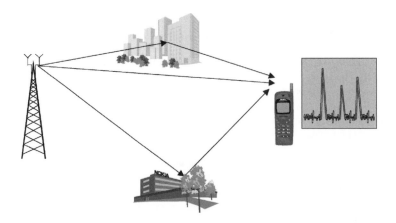

**Figure 3.4**   Multipath propagation leads to a multipath delay profile

is 0.26 μs. If the time difference of the multipath components is at least 0.26 μs, the WCDMA receiver can separate those multipath components and combine them coherently to obtain multipath diversity. The 0.26 μs delay can be obtained if the difference in path lengths is at least 78 m (= speed of light ÷ chip rate = $3.0 \cdot 10^8 \text{ ms}^{-1} \div 3.84 \text{ Mcps}$). With a chip rate of about 1 Mcps, the difference in the path lengths of the multipath components must be about 300 m, which cannot be obtained in small cells. Therefore, it is easy to see that the 5 MHz WCDMA can provide multipath diversity in small cells, which is not possible with IS-95.

2. Also, for a certain time delay position there are usually many paths nearly equal in length along which the radio signal travels. For example, paths with a length difference of half a wavelength (at 2 GHz this is approximately 7 cm) arrive at virtually the same instant when compared to the duration of a single chip, which is 78 m at 3.84 Mcps. As a result, signal cancellation, called fast fading, takes place as the receiver moves across even short distances. Signal cancellation is best understood as a summation of several weighted phasors that describe the phase shift (usually modulo radio wavelength) and attenuation along a certain path at a certain time instant.

Figure 3.5 shows an exemplary fast fading pattern as would be discerned for the arriving signal energy at a particular delay position as the receiver moves. We see that the received signal power can drop considerably (by 20–30 dB) when phase cancellation of multipath reflections occurs. Because of the underlying geometry causing the fading and dispersion phenomena, signal variations due to fast fading occur several orders of magnitude more frequently than changes in the average multipath delay profile. The statistics of the received signal energy for a short-term average are usually well described by the Rayleigh distribution, see, e.g. [5] and [6]. These fading dips make error-free reception of data bits very difficult, and countermeasures are needed in WCDMA. The countermeasures against fading in WCDMA are shown below.

1. The delay dispersive energy is combined by utilizing multiple Rake fingers (correlation receivers) allocated to those delay positions on which significant energy arrives.
2. Fast power control and the inherent diversity reception of the Rake receiver are used to mitigate the problem of fading signal power.
3. Strong coding and interleaving and retransmission protocols are used to add redundancy and time diversity to the signal and thus help the receiver in recovering the user bits across fades.

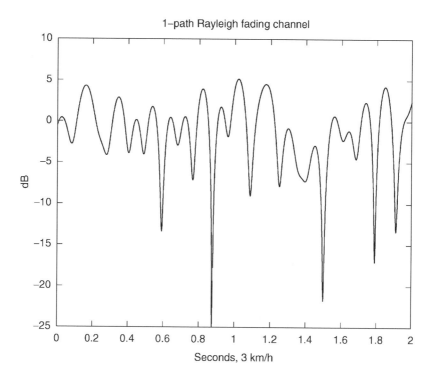

**Figure 3.5** Fast Rayleigh fading as caused by multipath propagation

The dynamics of the radio propagation suggest the following operating principle for the CDMA signal reception:

1. Identify the time delay positions at which significant energy arrives and allocate correlation receivers, i.e. Rake fingers, to those peaks. The granularity for acquiring the multipath delay profile is in the order of one chip duration (typically within the range of $\frac{1}{4} - \frac{1}{2}$ chip duration) with an update rate in the order of some tens of milliseconds.
2. Within each correlation receiver, track the fast-changing phase and amplitude values originating from the fast fading process and remove them. This tracking process has to be very fast, with an update rate in the order of 1 ms or less.
3. Combine the demodulated and phase-adjusted symbols across all active fingers and present them to the decoder for further processing.

Figure 3.6 illustrates points 2 and 3 by depicting modulation symbols (BPSK or QPSK) as well as the instantaneous channel state as weighted complex phasors. To facilitate point 2, WCDMA uses known pilot symbols that are used to sound the channel and provide an estimate of the momentary channel state (value of the weighted phasor) for a particular finger. Then the received symbol is rotated back, so as to undo the phase rotation caused by the channel. Such channel-compensated symbols can then be simply summed together to recover the energy across all delay positions. This processing is also called Maximal Ratio Combining (MRC).

Figure 3.7 shows a block diagram of a Rake receiver with three fingers according to these principles. Digitized input samples are received from the RF front-end circuitry in the form of I and Q branches (i.e. in complex low-pass number format). Code generators and a correlator perform the despreading

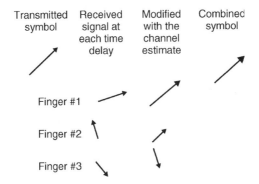

**Figure 3.6**   The principle of maximal ratio combining within the CDMA Rake receiver

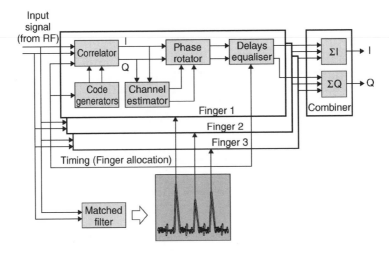

**Figure 3.7**   Block diagram of the CDMA Rake receiver

and integration to user data symbols. The channel estimator uses the pilot symbols to estimate the channel state which will then be removed by the phase rotator from the received symbols. The delay is compensated for the difference in the arrival times of the symbols in each finger. The Rake combiner then sums the channel-compensated symbols, thereby providing multipath diversity against fading. Also shown is a matched filter used for determining and updating the current multipath delay profile of the channel. This measured and possibly averaged multipath delay profile is then used to assign the Rake fingers to the largest peaks.

In typical implementations of the Rake receiver, processing at the chip rate (correlator, code generator, matched filter) is done in ASICs, whereas symbol-level processing (channel estimator, phase rotator, combiner) is implemented by a DSP. Although there are several differences between the WCDMA Rake receiver in the mobile and the base station, all the basic principles presented here are the same.

Finally, we note that multiple receive antennas can be accommodated in the same way as multiple paths received from a single antenna: by just adding additional Rake fingers to the antennas, we can then receive all the energy from multiple paths *and* antennas. From the Rake receiver's perspective, there is essentially no difference between these two forms of diversity reception.

## 3.5  Power Control

Tight and fast power control is perhaps the most important aspect in WCDMA, in particular on the uplink. Without it, a single overpowered mobile could block a whole cell. Figure 3.8 depicts the problem and the solution in the form of closed loop transmission power control.

Mobile stations MS1 and MS2 operate within the same frequency, separable at the base station only by their respective spreading codes. It may happen that MS1 at the cell edge suffers a path loss, say, 70 dB above that of MS2 which is near the base station BS. If there were no mechanism for MS1 and MS2 to be power-controlled to the same level at the base station, MS2 could easily overshoot MS1 and thus block a large part of the cell, giving rise to the so-called near–far problem of CDMA. The optimum strategy in the sense of maximizing capacity is to equalize the received power per bit of all mobile stations at all times.

While one can conceive open loop power control mechanisms that attempt to make a rough estimate of path loss by means of a downlink beacon signal, such a method would be far too inaccurate. The prime reason for this is that the fast fading is essentially uncorrelated between uplink and downlink, due to the large frequency separation of the uplink and downlink bands of the WCDMA FDD mode. Open loop power control is, however, used in WCDMA, but only to provide a coarse initial power setting of the mobile station at the beginning of a connection.

The solution to power control in WCDMA is fast closed loop power control, also shown in Figure 3.8. In closed loop power control in the uplink, the base station performs frequent estimates of the received Signal-to-Interference Ratio (SIR) and compares it to a target SIR. If the measured SIR is higher than the target SIR, the base station will command the mobile station to lower the power; if it is too low, it will command the mobile station to increase its power. This measure–command–react cycle is executed at a rate of 1500 times per second (1.5 kHz) for each mobile station and thus operates faster than any significant change of path loss could possibly happen and, indeed, even faster than the speed of fast Rayleigh fading for low to moderate mobile speeds. Thus, closed loop power control will prevent any power imbalance among all the uplink signals received at the base station.

The same closed loop power control technique is also used on the downlink, though here the motivation is different: on the downlink there is no near–far problem due to the one-to-many scenario. All the signals within one cell originate from the one base station to all mobiles. It is, however, desirable to provide a marginal amount of additional power to mobile stations at the cell edge, as they suffer from increased other-cell interference. Also on the downlink a method of enhancing weak signals caused by Rayleigh fading with additional power is needed at low speeds when other error-correcting methods based on interleaving and error correcting codes do not yet work effectively.

Figure 3.9 shows how uplink closed loop power control works on a fading channel at low speed. Closed loop power control commands the mobile station to use a transmit power proportional to the inverse of the received power (or SIR). Provided the mobile station has enough headroom to ramp the

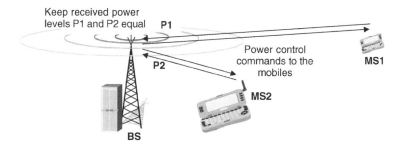

**Figure 3.8**  Closed loop power control in CDMA

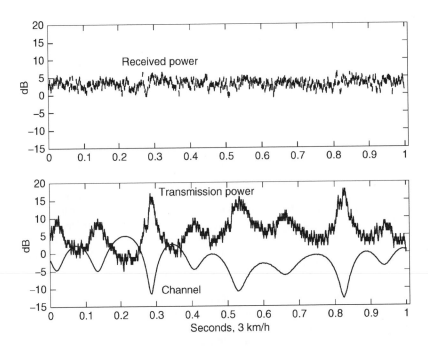

**Figure 3.9** Closed-loop power control compensates a fading channel

power up, only very little residual fading is left and the channel becomes an essentially non-fading channel as seen from the base station receiver.

While this fading removal is highly desirable from the receiver point of view, it comes at the expense of increased average transmit power at the transmitting end. This means that a mobile station in a deep fade, i.e. using a large transmission power, will cause increased interference to other cells. Figure 3.9 illustrates this point. The gain from the fast power control is discussed in more detail in Section 9.2.1.1.

Before leaving the area of closed loop power control, we mention one more related control loop connected with it: outer loop power control. Outer loop power control adjusts the target SIR set point in the base station according to the needs of the individual radio link and aims at a constant quality, usually defined as a certain target bit error rate (BER) or block error rate (BLER). Why should there be a need for changing the target SIR set point? The required SIR (there exists a proportional $E_b/N_0$ requirement) for, say, BLER = 1% depends on the mobile speed and the multipath profile. Now, if one were to set the target SIR set point for the worst case, i.e. high mobile speeds, one would waste much capacity for those connections at low speeds. Thus, the best strategy is to let the target SIR set point float around the minimum value that just fulfils the required target quality. The target SIR set point will change over time, as shown in the graph in Figure 3.10, as the speed and propagation environment changes. The gain of outer loop power control is discussed in detail in Section 9.2.2.1.

Outer loop control is typically implemented by having the base station tag each uplink user data frame with a frame reliability indicator, such as a CRC check result obtained during decoding of that particular user data frame. Should the frame quality indicator indicate to the Radio Network Controller (RNC) that the transmission quality is decreasing, the RNC in turn will command the base station to increase the target SIR set point by a certain amount. The reason for having outer loop control reside in the RNC is that this function should be performed after a possible soft handover combining. Soft handover will be presented in the next section.

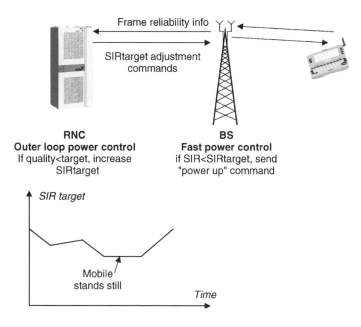

**Figure 3.10**   Outer loop power control

## 3.6   Softer and Soft Handovers

During softer handover, a mobile station is in the overlapping cell coverage area of two adjacent sectors of a base station. The communications between mobile station and base station take place concurrently via *two* air interface channels, one for each sector separately. This requires the use of two separate codes in the downlink direction, so that the mobile station can distinguish the signals. The two signals are received in the mobile station by means of Rake processing, very similar to multipath reception, except that the fingers need to generate the respective code for each sector for the appropriate despreading operation. Figure 3.11 shows the softer handover scenario.

In the uplink direction a similar process takes place at the base station: the code channel of the mobile station is received in each sector, then routed to the same baseband Rake receiver and the maximal ratio combined there in the usual way. During softer handover only one power control loop per connection is active. Softer handover typically occurs in about 5–15% of connections.

Figure 3.12 shows the soft handover. During soft handover, a mobile station is in the overlapping cell coverage area of two sectors belonging to different base stations. As in softer handover, the communications between mobile station and base station take place concurrently via two air interface channels from each base station separately. As in softer handover, both channels (signals) are received at the mobile station by maximal ratio combining Rake processing. Seen from the mobile station, there are very few differences between softer and soft handover.

However, in the uplink direction soft handover differs significantly from softer handover: the code channel of the mobile station is received from both base stations, but the received data is then routed to the RNC for combining. This is typically done so that the same frame reliability indicator as provided for outer loop power control is used to select the better frame between the two possible candidates within the RNC. This selection takes place after each interleaving period, i.e. every 10–80 ms. Note that during soft handover two power control loops per connection are active, one for each base station. Power control in soft handover is discussed in Section 9.2.1.3.

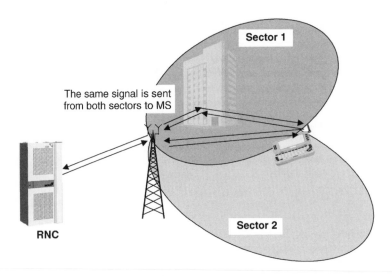

**Figure 3.11**  Softer handover

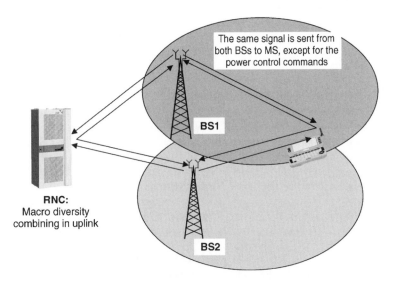

**Figure 3.12**  Soft handover

Soft handover occurs in about 20–40% of connections. To cater for soft handover connections, the following additional resources need to be provided by the system and must be considered in the planning phase:

- additional Rake receiver channels in the base stations;
- additional transmission links between base station and RNC;
- additional Rake fingers in the mobile stations.

We also note that soft and softer handover can take place in combination with each other.

Why are these CDMA-specific handover types needed? They are needed for similar reasons as closed loop power control: without soft/softer handover there would be near–far scenarios of a mobile station penetrating from one cell deeply into an adjacent cell without being power-controlled by the latter. Very fast and frequent hard handovers could largely avoid this problem; however, they can be executed only with certain delays during which the near–far problem could develop. So, as with fast power control, soft/softer handovers are an essential interference-mitigating tool in WCDMA. Soft and softer handovers are described in more detail in Section 9.3.

In addition to soft/softer handover, WCDMA provides other handover types:

- Inter-frequency hard handovers that can be used, for example, to hand a mobile over from one WCDMA frequency carrier to another. One application for this is high capacity base stations with several carriers.
- Inter-system hard handovers that take place between the WCDMA FDD system and another system, such as WCDMA TDD or GSM.

# References

[1] Ojanperä, T. and Prasad, R., *Wideband CDMA for Third Generation Mobile Communications*, New York: Artech House, 1998.

[2] Viterbi, A., *Principles of Spread Spectrum Communication*, Reading, MA: Addison-Wesley, 1997.

[3] Cooper, G. and McGillem, C., *Modern Communications and Spread Spectrum*, New York: McGraw-Hill, 1998.

[4] Dixon, R., *Spread Spectrum Systems with Commercial Applications*, Chichester: John Wiley & Sons, 1994.

[5] Jakes, W., *Microwave Mobile Communications*, Chichester: John Wiley & Sons, 1974.

[6] Saunders, S., *Antennas and Propagation for Wireless Communication Systems*, Chichester: John Wiley & Sons, 1999.

# 4

# Background and Standardization of WCDMA

Antti Toskala

## 4.1 Introduction

In the first phase of third-generation standardization, the basic process of selecting the best technology for multiple radio access was conducted in several regions. This chapter describes the selection process during 1997, as well as the decisions made by the regional standardization organizations in early 1998. Different standardization bodies that carried out WCDMA-related work are introduced, and then 3rd Generation Partnership Project (3GPP), the common standardization effort to create a global standard for WCDMA, is described, including a description of how a proposal becomes the specifications in the WCDMA standardization process. Then the developments in ITU for work on the IMT-2000 are presented. Then follows a look beyond the first WCDMA release, Release 99, in terms of content of Releases 4, 5, 6, 7, 8, 9 and 10. This chapter concludes with a look at the IMT-Advanced and resulting eco-system convergence for the evolution towards the 4th generation systems with 3GPP Long Term Evolution (LTE) system specification work.

## 4.2 Background in Europe

In Europe, a long period of research preceded the selection of third-generation technology. The RACE I (Research of Advanced Communication Technologies in Europe) programme started the basic third-generation research work in 1988. This programme was followed by RACE II, with the development of the CDMA-based Code Division Testbed (CODIT) and TDMA-based Advanced TDMA Mobile Access (ATDMA) air interfaces in 1992–1995. In addition, wideband air interface proposals were studied in a number of industrial projects in Europe, e.g. see [1].

The European research programme Advanced Communication Technologies and Services (ACTS) was launched at the end of 1995 in order to support mobile communications research and

*WCDMA for UMTS: HSPA Evolution and LTE, Fifth Edition*   Edited by Harri Holma and Antti Toskala
© 2010 John Wiley & Sons, Ltd

development. Within ACTS, the Future Radio Wideband Multiple Access System (FRAMES) project [2] was set up with the aim of defining a proposal for a UMTS radio access system. The main industrial partners in FRAMES were Nokia, Siemens, Ericsson, France Télécom and CSEM/Pro Telecom, with participation also from several European universities. Based on an initial proposal evaluation phase in FRAMES, a harmonized multiple access platform was defined, consisting of two modes: FMA1, a wideband TDMA [3], and FMA2, a wideband CDMA [4]. The FRAMES wideband CDMA and wideband TDMA proposals were submitted to ETSI as candidates for UMTS air interface and ITU IMT-2000 submission.

The proposals for the UMTS Terrestrial Radio Access (UTRA) air interface received by the milestone were grouped into five concept groups in ETSI in June 1997, after their submission and presentation during 1996 and early 1997.

The following groups were formed:

- Wideband CDMA (WCDMA)
- Wideband TDMA (WTDMA)
- TDMA/CDMA
- OFDMA
- ODMA.

The concept groups formed in ETSI are introduced briefly in the following section. The evaluation of the proposals was based on the requirements defined in the ITU-R IMT-2000 framework (and in ETSI defined specifically in UMTS 21.01 [5], as well as on the evaluation principles and conditions covered in UMTS 30.03 [6]). The results of the evaluation were collated in UMTS 30.06 [7].

## 4.2.1   Wideband CDMA

The WCDMA concept group was formed based on the WCDMA proposals from FRAMES/FMA2, Fujitsu, NEC and Panasonic. Several European, Japanese and US companies contributed to the development of the WCDMA concept. The physical layer of the WCDMA uplink was adopted mainly from FRAMES/FMA2, while the downlink solution was modified following the principles of the other proposals made to the WCDMA concept group.

The basic system features consisted of:

- wideband CDMA operation with 5 MHz;
- physical layer flexibility for integration of all data rates on a single carrier;
- reuse 1 operation.

The enhancements covered included:

- transmit diversity;
- adaptive antenna operation;
- support for advanced receiver structures.

The WCDMA concept received most support, one of the technical motivating issues being the flexibility of the physical layer to accommodate different service types simultaneously. This was considered to be an advantage, especially with respect to low and medium bit rates. Among the drawbacks of WCDMA, it was recognized that in an unlicensed system in the Time Division Duplexing (TDD) band, with the continuous transmit and receive operation, pure WCDMA technology does not facilitate interference avoidance techniques in cordless-like operating environments.

## 4.2.2 Wideband TDMA

The WTDMA concept group was formed by taking the non-spread option from the FRAMES/FMA1 proposal. FRAMES/FMA1 was basically a TDMA-based system concept with 1.6 MHz carrier spacing for wideband service implementation. The concept aimed at high capacity with the aid of interference averaging over the operator bandwidth, with fractional loading and frequency hopping.

The basic system features consisted of:

- equalization with training sequences in TDMA bursts;
- interference averaging with frequency hopping;
- link adaptation;
- two basic burst types, 1/16th and 1/64th burst lengths for high and low data rates respectively;
- low reuse sizes.

The enhancements covered included:

- inter-cell interference suppression;
- support of adaptive antennas;
- TDD operation;
- less complex equalizers for large delay-spread environments.

The main limitation associated with the system was considered to be the range with respect to low bit-rate services. This is due to the fact that, in TDMA-based operation, the slot duration is, at a minimum, only 1/64th of the frame timing, which results in either very high peak power or a low average output power level. This means that for large ranges with, for example, speech, the WTDMA concept would not have been competitive on its own, but would have required a narrowband option as a companion.

## 4.2.3 Wideband TDMA/CDMA

The WTDMA/CDMA group was based on the spreading option in the FRAMES/FMA1 proposal, resulting in the hybrid CDMA/TDMA concept with 1.6 MHz carrier spacing.

The basic system features consisted of:

- a TDMA burst structure with midamble for channel estimation;
- the CDMA concept applied on top of the TDMA structure for additional flexibility;
- a reduction of intra-cell interference by multi-user detection for users within a timeslot on the same carrier;
- low reuse sizes, down to 3.

Enhancements covered included:

- frequency hopping;
- inter-cell interference cancellation;
- support of adaptive antennas;
- operation in TDD mode;
- dynamic channel allocation (DCA).

This proposal, especially the issues related to receiver complexity, led to lively discussions during the selection process.

## 4.2.4   OFDMA

The OFDMA group was based on OFDMA technology with inputs mainly from Telia, Sony and Lucent. The system concept was shaped by discussions about OFDMA in other forums, such as the Japanese standardization forum, ARIB.

The basic concept features included:

- operation with slow frequency hopping with TDMA and OFDM multiplexing;
- a 100 kHz wide bandslot from the OFDM signal as the basic resource unit;
- higher rates built by allocating several bandslots, creating a wideband signal;
- diversity provided by dividing the information among several bandslots over the carrier.

The enhancement techniques covered were:

- transmit diversity;
- multi-user detection for interference cancellation;
- adaptive antenna solutions.

One main technical weakness of the system concept was the uplink transmission direction, where the resulting envelope variations caused concern about power amplifier design.

## 4.2.5   ODMA

Vodafone proposed Opportunity Driven Multiple Access (ODMA), basically a relaying protocol, not a pure multiple access as such. ODMA was later integrated in the WCDMA and WCDMA/TDMA concept groups and was not considered in the selection process as a concept on its own. ODMA was later considered for a while in 3GPP standardization, but was eventually not included in the specifications.

## 4.2.6   ETSI Selection

All the proposed technologies were basically able to fulfil the UMTS requirements, although it was difficult to reach a consensus on issues such as system capacity, since the results of simulations can vary greatly depending on the assumptions. However, it soon became evident in the selection process that WCDMA and TDMA/CDMA were the main candidates. Also, issues such as the global potential of a technology naturally had an impact in cases where obvious technical conclusions were very limited; in this respect, the outcome of the ARIB technology selection in Japan garnered support for WCDMA.

ETSI decided between the technologies in January 1998 [8], selecting WCDMA as the standard for the UTRA air interface on the paired frequency bands, i.e. for Frequency Division Duplexing (FDD) operation, and WTDMA/CDMA for operation with unpaired spectrum allocation, i.e. for TDD operation. As illustrated in Figure 4.1, it took 10 years from the initiation of the European research programs to reach a decision on the UTRA technology. The detailed standardization of UTRA proceeded within ETSI until the work was handed over to the 3GPP. The technical work was transferred to 3GPP with the contribution of UTRA in early 1999.

## 4.3   Background in Japan

In Japan, the Association for Radio Industries and Businesses (ARIB) evaluated possible third-generation systems based on three different main technologies based on WCDMA, WTDMA and OFDMA.

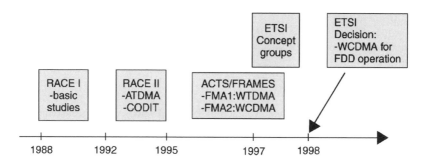

**Figure 4.1**  European research programmes towards third-generation systems and the ETSI decision

The WCDMA technology in Japan was very similar to that being considered in Europe in ETSI; indeed, the members of ARIB contributed their technology to ETSI's WCDMA concept group. Details of FRAMES/FMA2 were provided from Europe for consideration in the ARIB process. The other technologies considered, WTDMA and OFDMA, also had many similarities to the candidates in the ETSI selection process.

The result of the ARIB selection process in 1997 was WCDMA, with both FDD and TDD modes of operation. Since WCDMA had been chosen in ARIB before the process was completed in ETSI, it carried more weight in the ETSI selection as the global technological alternative. Since the creation of 3GPP for the third-generation standardization framework, ARIB have contributed their WCDMA to 3GPP, in the same way as ETSI have contributed UTRA. In Japan, work on higher layer specifications is the responsibility of the Telecommunication Technology Committee (TTC), which has also changed the focus to 3GPP.

## 4.4   Background in Korea

In Korea, the Telecommunications Technology Association (TTA) adopted a two-track approach to the development of third-generation CDMA technology. The TTA1 and TTA2 air interface proposals (later renamed Global CDMA 1 and 2, respectively) were based on synchronous and asynchronous wideband CDMA technologies respectively. TTA1 WCDMA was similar to WCDMA technology in ETSI, ARIB and T1P1, while TTA2 was similar to cdma2000 in TR45.5.

Several technical details of the Korean technology that differed from the ETSI and ARIB solutions were submitted to the ETSI and ARIB standardization processes, leading to a high degree of commonality between the ETSI, ARIB and TTA WCDMA solutions. The Korean standardization efforts were later moved to 3GPP and 3GPP2 to contribute to WCDMA and cdma2000 standardization respectively.

## 4.5   Background in the United States

In the United States there are several second-generation technologies, the most widely distributed digital systems being those based on either GSM-1900, US-TDMA (D-AMPS) or US-CDMA (IS-95) standards. For all those technologies, a natural path of evolution towards the third generation had been defined. In addition, a third-generation CDMA proposal that had no direct relation to second-generation systems, namely WIMS W-CDMA, came from the TR46.1 standardization committee.

### 4.5.1   W-CDMA N/A

Work on GSM-1900-related standardization was carried out within T1P1, as with GSM standardization in ETSI, with similar discussions concerning technology selection. As a result, W-CDMA N/A (N/A for North America) was submitted to the ITU-R IMT-2000 process. The proposals had much in common with the ETSI and ARIB WCDMA technologies, since the contributing companies had also been active in the ETSI and ARIB selection processes.

## 4.5.2 UWC-136

In TR45.3, discussions concerning the evolution of IS-136 (Digital AMPS) technology towards the third generation took place. The resulting selection was a combination of narrowband and wideband TDMA technologies, with the narrowband component identical to the EDGE concept, part of GSM evolution in ETSI and T1P1. The wideband part for indoor service provision up to 2 Mbps was based on the same WTDMA concept as was considered in ETSI. The selection of EDGE technology in TR45.3 will create a clear connection between TDMA technologies within TR45.3, T1P1 and ETSI. The development of TDMA-based technology has accelerated around the common component in the air interface in the form of the EDGE component and, thus, several existing IS-136 operators have announced they will roll-out GSM/EDGE as part of their radio access evolution.

## 4.5.3 cdma2000

The cdma2000 air interface proposal to ITU was the result of work in TR45.5 on the evolution of IS-95 towards the third generation. The cdma2000 proposal is based partly on IS-95 principles with respect to synchronous network operation, common pilot channels, and so on, but it is a wideband version with three times the bandwidth of IS-95. The ITU proposal contains further bandwidth options as well as the multi-carrier option for downlink. The cdma2000 proposal had a high degree of commonality with the Global CDMA 1 ITU proposal from TTA, Korea.

The cdma2000 multi-carrier option, as standardized by 3GPP2, was covered in earlier versions of this book but is no longer considered, as there was no market demand (or actual impelementations) for that and thus the chapter on it has been removed from the 4th edition onwards.

Following the industry convergence, 3GPP2 is doing some maintenance work for different cdma2000 revisions but is not developing any more next generation systems (or major new versions of cdma2000) as discussed in Section 4.12.

## 4.5.4 TR46.1

The WIMS W-CDMA was not based on work derived from an existing second-generation technology, but was a new third-generation technology proposal with no direct link to any second-generation standardization. It was based on the constant processing gain principle with a high number of multicodes in use; thus, it shows some fundamental differences, but also a level of commonality with WCDMA technology in other forums.

## 4.5.5 WP-CDMA

Wideband Packet CDMA (WP-CDMA) resulted from the convergence between W-CDMA N/A of T1P1 and WIMS W-CDMA of TR46.1 in the United States. The main features of the WIMS W-CDMA proposal were merged with the principles of W-CDMA N/A. The merged proposal was submitted to the ITU-R IMT-2000 process towards the end of 1998, and to the 3GPP process at the beginning of 1999. Its most characteristic feature, compared with the other WCDMA-based proposals, was a common packet mode channel operation for the uplink direction, but there were also a few smaller differences.

## 4.6  Creation of 3GPP

As similar technologies were being standardized in several regions around the world, it became evident that achieving identical specifications to ensure equipment compatibility globally would be very difficult with work going on in parallel. Also, having to discuss similar issues in several places was naturally a waste of resources for the participating companies. Therefore, initiatives were made to create a single forum for WCDMA standardization for a common WCDMA specification.

The standardization organizations involved in the creation of the 3GPP [9] were ARIB (Japan), ETSI (Europe), TTA (Korea), TTC (Japan) and T1P1 (USA). T1P1 later became ATIS. The partners agreed on joint efforts for the standardization of UTRA, now standing for Universal Terrestrial Radio Access, as distinct from UTRA (UMTS Terrestrial Radio Access) from ETSI, also submitted to 3GPP. Companies such as manufacturers and operators are members of 3GPP through the respective standardization organization to which they belong. as illustrated in Figure 4.2.

Later during 1999, the China Wireless Telecommunication Standard Group (CWTS) also joined 3GPP and contributed technology from TD-SCDMA, a TDD-based CDMA third-generation technology already submitted to ITU-R earlier. The name was later changed to CCSA (China Communication Standards Association).

3GPP also includes market representation partners: the GSM Association, the UMTS Forum, the Global Mobile Suppliers Association, the IPv6 Forum and the Universal Wireless Communications Consortium (UWCC).

The work was formally initiated at the end of 1998 and the detailed technical work was started in early 1999, with the aim of having the first version of the common specification, called Release-99, ready by the end of 1999.

Within 3GPP, four different technical specification groups (TSGs) were set up as follows:

1. Radio Access Network TSG
2. Core Network TSG
3. Service and System Aspects TSG
4. Terminals TSG.

Of these groups, the one most relevant to the WCDMA technology is the Radio Access Network TSG (RAN TSG), which was originally divided into four different working groups, but later (2005), with restructuring of 3GPP, a fifth group was added to cover terminal testing activities, as illustrated in Figure 4.3. From the Terminals TSG the activities were split to other TSGs and the Core Network TSG was then renamed the Core and Terminals TSG.

The RAN TSG will produce Release-99 of the UTRA air interface specification. The work done within the 3GPP RAN TSG working groups has been the basis of the technical description of the UTRA air interface covered in this book. Without such a global initiative, this book would have been forced to focus on a single regional specification, though with many similarities to those of other regions. Thus, the references throughout this book are to the specification volumes from 3GPP.

**Figure 4.2**  3GPP organizational partners

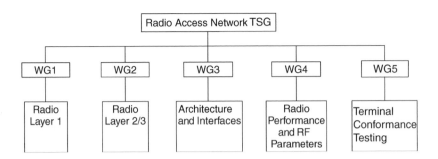

**Figure 4.3**   3GPP RAN TSG working groups

During the first half of 1999 the inputs from the various participating organizations were merged into a single standard, leaving the rest of the year to finalize the detailed parameters for the first full release, Release 99, of UTRA from 3GPP. The member organizations have undertaken individually to produce standard publications based on the 3GPP specification. Thus, for example, the Release-99 UMTS specifications from ETSI are identical to the Release 99 specifications produced by 3GPP. The latest specifications can be obtained from 3GPP [9].

During 2000, further work on GSM evolution was moved from ETSI and other forums to 3GPP, including work on GPRS and EDGE. A new TSG, TSG GERAN, was set up for this purpose.

## 4.7   How Does 3GPP Operate?

In 3GPP, the work is organized around work items, which basically define the justification and objective for a new feature. For a smaller topic there can be only a single work item in one working group if the impacts are limited to that group or at least are mainly for the specifications under the responsibility of that group. For bigger items, such as High Speed Downlink Packet Access (HSDPA), work tasks were undertaken for each of the four RAN working groups and these work tasks were under a common work item, a building block, named HSDPA.

The work items sheets also usually contain the specifications to be implemented and the expected schedule of the work, although the latter is usually rather optimistic. Work items need to have four supporting companies, but they also need to have some justification that can be agreed at the respective TSG RAN level. (Note that some variations of the way of working exist between TSGs.) For the larger topic, quite often a feasibility study (or study item in TSG RAN) is needed before the decision of actually creating a work item is taken. A feasibility study will simply focus on the pain versus gain ratio of the new feature, comparing what are the advantages and what are the resulting impacts to the equipment and existing features (known as backwards compatibility).

For each work item a rapporteur is nominated, who has the responsibility of coordinating the work and reporting the progress of the working groups to the TSG level. At the TSG level every TSG-level meeting (called a plenary) monitors the progress every three months and establishes the necessary coordination between the working groups and TSGs if needed. Sometimes a work item is determined as not reaching the expected target and it may be altered or removed from the work programme. Once the work item is completed in all working groups, Change Requests (CRs) are brought to the plenary for approval. CRs contain the changes needed in each particular specification, and once the plenary level approval is obtained, the specification will be updated to a new version with the changes resulting from the new feature. A simplified illustration of the process from feasibility study to specification finalization is shown in Figure 4.4. The latest work item descriptions can be found in [9].

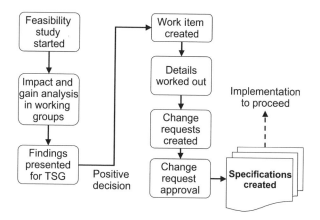

**Figure 4.4** Example of 3GPP standardization process

Creation of the specifications does not necessarily mean that everything is 100% completed yet; typically, some meeting rounds then take place regarding potential corrections, which typically emerge as the implementations proceed and as the details are being verified in implementation and testing. The CRs are also used to introduce the corrections; they are agreed in working groups and, once approved by the following TSG plenary meeting, the CRs are then included in the specification.

## 4.8   Creation of 3GPP2

Work done in TR45.5 and TTA was merged to form 3GPP2, focused on the development of cdma2000 Direct-Sequence (DS) and Multi-Carrier (MC) mode for the cdma2000 third-generation component. This activity has been running in parallel with the 3GPP project, with participation from ARIB, TTC and CWTS as member organizations. The focus shifted to MC mode after global harmonization efforts, but then later work started to focus more on the narrowband IS-95 evolution, reflected in the IS-2000 standards series. As mentioned ealier, now the activity in 3GPP2 has been decreasing as the effort of 3GPP2 towards next generation systems has been contributing to LTE development in 3GPP instead.

## 4.9   Harmonization Phase

During 1999, attempts were made to further harmonize the CDMA-based third-generation solutions. For the 3GPP framework the ETSI, ARIB, TTA and T1P1 concepts had already been merged into a single WCDMA specification, while cdma2000 was still on its own in TR45.5. Eventually the manufacturers and operators agreed to adopt a harmonized global third-generation CDMA standard consisting of three modes: MC, Direct Spread (DS) and TDD. The MC mode was based on the cdma2000 MC option, the DS mode on WCDMA (UTRA FDD), and the TDD mode on UTRA TDD.

The main technical impacts of these harmonization activities were the change of UTRA FDD- and TDD-mode chip rate from 4.096 Mcps to 3.84 Mcps and the inclusion of a common pilot for UTRA FDD. The work in 3GPP2 focused on the MC mode, and the DS mode from cdma2000 was abandoned. Eventually the work in 3GPP2 resumed on the 1.28 Mcps evolution and development for the MC mode has been stopped. The result is that, globally, there is only one DS wideband CDMA standard, namely WCDMA.

## 4.10   IMT-2000 Process in ITU

In the ITU, recommendations have been developed for third-generation mobile communications systems, the ITU terminology being called IMT-2000 [10], formerly FPLMTS. In the ITU-R, ITU-R TG8/1 has worked on the radio-dependent aspects, while the radio-independent aspects have been covered in ITU-T SG11.

In the radio aspects, ITU-R TG8/1 received a number of different proposals during the IMT-2000 candidate submission process. In the second phase of the process, evaluation results were received from the proponent organizations as well as from the other evaluation groups that studied the technologies. During the first half of 1999 the recommendation IMT.RKEY was created, which describes the IMT-2000 multimode concept.

The ITU-R IMT-2000 process was finalized at the end of 1999, when the detailed specification (IMT-RSCP) was created and the radio interface specifications were approved by ITU-R [11]. The detailed implementation of IMT-2000 will continue in the regional standards bodies. The ITU-R process has been an important external motivation and timing source for IMT-2000 activities in regional standards bodies. The requirements set by ITU for an IMT-2000 technology have been reflected in the requirements in the regional standards bodies, e.g. in ETSI UMTS 21.01 [5], in order for the ETSI submission to fulfil the IMT-2000 requirements. The ITU-R interaction between regional standardization bodies in the IMT-2000 process is reflected in Figure 4.5.

The ITU-R IMT-2000 grouping, with TDMA- and CDMA-based groups, is illustrated in Figure 4.6. The UTRA FDD (WCDMA) and cdma2000 are part of the CDMA interface, as CDMA Direct Spread and CDMA Multi-Carrier respectively. UWC-136 and DECT are part of the TDMA-based interface in the concept, as TDMA Single Carrier and TDMA Multi-Carrier respectively. The TDD part in CDMA consists of UTRA TDD from 3GPP and TD-SCDMA from CWTS. Harmonization has been completed for the FDD part in the CDMA interface, and the harmonization process for the CDMA TDD modes within 3GPP resulted in the 1.28 Mcps TDD being included in the 3GPP Release 4 specifications, completed in March 2001.

## 4.11   Beyond 3GPP Release 99 WCDMA

Upon completion of the Release 99 specifications, work has concentrated on specifying new features for following releases. Once a release has been finalized, there is also typically some period when corrections are done to the earlier release in parallel while making the following release. Typically, such corrections arise as the implementation proceeds and as the test systems are updated to include the latest changes in the specifications. As experience in various forums has shown, a major step forward

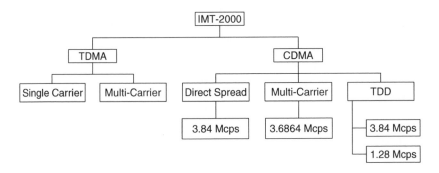

**Figure 4.5**   ITU-R IMT-2000 grouping

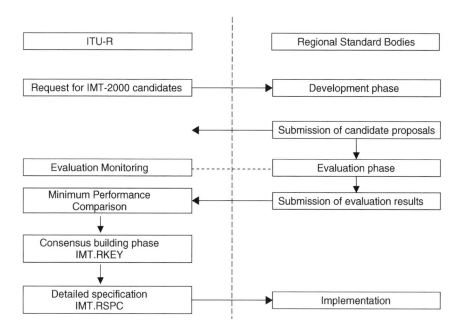

**Figure 4.6**  Relationship of ITU-R to the regional standard bodies

in system capabilities with many new features requires a phasing-in period for the specifications. Fortunately, the main functions have been verified in the various test systems in operation since 1995, but only the actual implementation will reveal any errors and inconsistencies in the fine detail of the specifications. Once a particular release has been progressed far enough in the implementation and devices deployed (and working) in the field, possible problems are then worth fixing only in the next release, where devices have not yet been rolled out.

In 3GPP, the next version of the specifications was originally considered as Release 2000, but since then the release naming was adjusted so that the next release was called Release 4, due in March 2001. Release 4 contained only minor adjustments with respect to Release 99. Bigger items that were included in Release 5 were HSDPA and an IP-based transport layer; see Chapters 12 and 5 respectively. Release 5 was completed in March 2002 for the WCDMA radio aspects. Release 99 specifications have a version number starting with 3, whereas Release 4 and 5 specifications have version numbers starting logically with 4 and 5 respectively.

On the TDD side, the narrowband (1.28 Mcps) TDD mode originally from CWTS (China) was included in 3GPP Release 4. The 1.28 Mcps UTRA TDD mode, or TD-SCDMA, is covered in Chapter 18. Release 7 contains a third chip rate for TDD, i.e. 7.68 Mcps.

Besides the IP-based transport option in Release 5, the protocols developed by the Internet Engineering Task Force have also influenced the WCDMA specifications in other ways. The Release 4 specifications contain a robust IP header compression suitable for cellular transmission to enable efficient Voice-over-IP service.

Inclusion of bigger items was continued in Release 6, which included High Speed Uplink Packet Access (HSUPA; covered in Chapter 13), Multimedia Broadcast Multicast System (MBMS; Chapter 14), as well as further HSDPA enhancements (discussed in Chapter 12). While the first Release 6 specifications were available at the end of 2003, the HSUPA was completed with specification availability in December 2004.

Release 7, finalized for June 2007, includes support for flat radio access network architecture in large deployments, higher-order modulation enhancements and Multiple Input Multiple Output antenna technology along with other inprovements to boost capacity and device battery life as discussed in Chapter 15 on HSPA evolution. Also, for MBMS, Release 7 added MBMS-dedicated carrier support, with further additions in Release 8 also to enable a MBMS-dedicated carrier in TDD bands based on FDD structures as covered in Chapter 14. Also in Release 8 several important features were included such as support for CS voice over HSPA, improved uplink data rates in Cell_FACH state and dual-carrier (dual-cell) HSDPA, and Release 9 continued with Dual-carrier HSUPA and dual-band HSDPA. Now 3GPP is working with Release 10 items, which include topics such as support for 4-carrier HSDPA and studies for improved network energy efficiency. The Release 10 protocols specifications are expected to be frozen during the first half of 2011.

Parallel to Release 7, work was carried out of the UTRAN Long Term Evolution (LTE), known also as Evolved UTRAN for Release 8, as discussed in Chapter 17. The full set of specifications was available in December 2007 as part of Release 8, with the RRC specification for backwards compatibility achieved in March 2009 and for the protocol on the interfaces between radio and core in May 2009. Further features were then added for Release 9 LTE for December 2009, such as MBMS for LTE.

Besides adding new features, 3GPP has also carried out the removal of unused features from the WCDMA specifications, and during the latest round, completed in June 2005, several features from Release 99 mainly related to the physical layer, including Downlink Shared Channel and Common Packet Channel, were removed from the specifications as they were not implemented in the market, as covered in Chapter 6.

Having been in operation now more than 10 years, 3GPP has proven its ability to create systems that have global acceptance and the ability to reach wide deployment and are supported by a healthy ecosystem with multiple operator, chip set and device vendors as well as infrastructure providers. With the creation of LTE and the global convergence of technology (as discussed in the next section), the 3GPP ecosystem is also attracting such players that traditionally have had presence only in other standardization forums e.g. 3GPP2.

## 4.12   Industry Convergence with LTE and LTE-Advanced

Following the parallel work on different paths between 3GPP and 3GPP2, it was truly remarkable to see the industry come together with the Long Term Evolution (LTE) [12] work. Some time after the start of the LTE work in 2005, when approaching the specification availability, 3GPP2 key players moved their efforts to work in 3GPP instead of developing further the planned 3GPP2 solution for the 4th generation radio access technology. By the milestone of approaching Release 8, LTE specification availability was frozen, as shown in Figure 4.7, and the key operators from 3GPP2 technology had joined together and started to contribute to 3GPP LTE development.

The approach to migrating from 3GPP2 technology to 3GPP varies slightly between operators. In some cases an existing cdma2000 operator has rolled out HSPA network (for example, SK Telecom in Korea or Telus in Canada), while in other cases an operator is going directly to LTE from cdma2000 (as has been announced by, for example, Verizon Wireless in the USA or by KDDI in Japan). All this is working towards ensuring a fully functional ecosystem around 3GPP technologies.

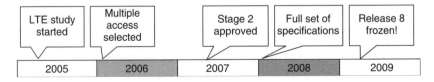

**Figure 4.7**   LTE Release 8 standardization milestones

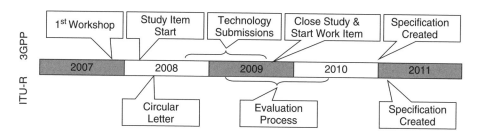

**Figure 4.8** LTE-Advanced schedule in 3GPP and in ITU-R

Convergence is also visible in the next phase of the ITU-R process, with the IMT-Advanced solution. 3GPP is developing as part of Release 10 specifications the 3GPP proposals for the IMT-Advanced solution and is working with the timeline as shown in Figure 4.8. Compared to the multiple proposals in IMT-2000 process, only two submissions were done for the IMT-Advanced process with one being LTE-Advanced and other coming from WiMAX direction (802.16 m). The technology components of LTE-Advanced are further addressed in Chapter 17 and also given in [13]. While the work also continues on new WCDMA/HSPA features, only LTE-Advanced has been submitted as part of the IMT-Advanced process while WCDMA/HSPA developments have been provided to ITU-R as updates to the existing IMT-2000 technology.

# References

[1] Pajukoski, K. and Savusalo, J., 'Wideband CDMA Test System', in *Proceedings of IEEE International Conference on Personal Indoor and Mobile Radio Communications*, PIMRC'97, Helsinki, Finland, 1–4 September 1997, pp. 669–672.

[2] Nikula, E., Toskala, A., Dahlman, E., Girard, L. and Klein, A., 'FRAMES Multiple Access for UMTS and IMT-2000', *IEEE Personal Communications Magazine*, April 1998, pp. 16–24.

[3] Klein, A., Pirhonen, R., Sköld, J. and Suoranta, R., 'FRAMES Multiple Access Mode 1 – Wideband TDMA with and without Spreading', *Proceedings of IEEE International Conference on Personal Indoor and Mobile Radio Communications*, PIMRC'97, Helsinki, Finland, 1–4 September 1997, pp. 37–41.

[4] Ovesjö, F., Dahlman, E., Ojanperä, T., Toskala, A. and Klein, A., 'FRAMES Multiple Access Mode 2 – Wideband CDMA', *Proceedings of IEEE International Conference on Personal Indoor and Mobile Radio Communications*, PIMRC'97, Helsinki, Finland, 1–4 September 1997, pp. 42–46.

[5] Universal Mobile Telecommunications System (UMTS), Requirements for the UMTS Terrestrial Radio Access System (UTRA), ETSI Technical Report, UMTS 21.01 version 3.0.1, November 1997.

[6] Universal Mobile Telecommunications System (UMTS), Selection Procedures for the Choice of Radio Transmission Technologies of the UMTS, ETSI Technical Report, UMTS 30.03 version 3.1.0, November 1997.

[7] Universal Mobile Telecommunications System (UMTS), UMTS Terrestrial Radio Access System (UTRA) Concept Evaluation, ETSI Technical Report, UMTS 30.06 version 3.0.0, December 1997.

[8] ETSI Press Release, SMG Tdoc 40/98, 'Agreement Reached on Radio Interface for Third Generation Mobile System, UMTS', Paris, France, January 1998.

[9] http://www.3GPP.org.

[10] http://www.itu.int/imt/.

[11] ITU Press Release, ITU/99-22, 'IMT-2000 Radio Interface Specifications Approved in ITU Meeting in Helsinki', 5 November 1999, Helsinki, Finland.

[12] Holma, H. and Toskala, A., *LTE for UMTS*, Chichester: John Wiley & Sons, Ltd, 2009.

[13] 3GPP Technical Report TR 36.912, version 9.1.0, Feasibility Study for Further Advancements for E-UTRA (LTE-Advanced), December 2009.

# 5

# Radio Access Network Architecture

Fabio Longoni, Atte Länsisalmi and Antti Toskala

## 5.1 Introduction

This chapter gives a wide overview of the Universal Mobile Telephone System (UMTS) architecture, including an introduction to the logical network elements and the interfaces. The UMTS utilizes the same well-known architecture that has been used by all main second-generation systems and even by some first-generation systems. The reference list contains the related 3GPP specifications [1–24].

The UMTS consists of a number of logical network elements that each has a defined functionality. In the standards, network elements are defined at the logical level, but this quite often results in a similar physical implementation, especially since there are a number of open interfaces (for an interface to be 'open', the requirement is that it has been defined to such a detailed level that the equipment at the endpoints can be from two different manufacturers). The network elements can be grouped based on similar functionality, or based on which sub-network they belong to.

Functionally, the network elements are grouped into the Radio Access Network (RAN; UMTS Terrestrial RAN (UTRAN)) that handles all radio-related functionality, and the Core Network (CN), which is responsible for switching and routing calls and data connections to external networks. To complete the system, the User Equipment (UE) that interfaces with the user and the radio interface is defined. The high-level system architecture is shown in Figure 5.1.

From a specification and standardization point of view, both UE and UTRAN consist of completely new protocols, the designs of which are based on the needs of the new WCDMA radio technology. On the contrary, the definition of CN is adopted from GSM. This gives the system with new radio technology a global base of known and rugged CN technology that accelerates and facilitates its introduction, and enables such competitive advantages as global roaming.

Another way to group UMTS network elements is to divide them into sub-networks. The UMTS is modular in the sense that it is possible to have several network elements of the same type. In principle, the minimum requirement for a fully featured and operational network is to have at least one logical network element of each type (note that some features and consequently some network elements are

*WCDMA for UMTS: HSPA Evolution and LTE, Fifth Edition*   Edited by Harri Holma and Antti Toskala
© 2010 John Wiley & Sons, Ltd

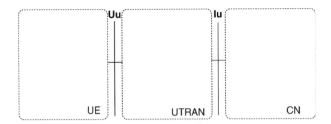

**Figure 5.1** UMTS high-level system architecture

optional). The possibility of having several entities of the same type allows the division of the UMTS into sub-networks that are operational either on their own or together with other sub-networks, and that are distinguished from each other with unique identities. Such a sub-network is called a UMTS Public Land Mobile Network (PLMN). Typically, one PLMN is operated by a single operator, and is connected to other PLMNs as well as to other types of network, such as ISDN, PSTN, the internet, and so on. Figure 5.2 shows elements in a PLMN and, in order to illustrate the connections, also external networks.

The UTRAN architecture is presented in Section 5.2. A short introduction to all the elements is given below. The UE consists of two parts:

1. The Mobile Equipment (ME) is the radio terminal used for radio communication over the Uu interface.
2. The UMTS Subscriber Identity Module (USIM) is a smartcard that holds the subscriber identity, performs authentication algorithms, and stores authentication and encryption keys and some subscription information that is needed at the terminal.

UTRAN also consists of two distinct elements:

1. The Node B converts the data flow between the Iub and Uu interfaces. It also participates in radio resource management. (*Note*: the term 'Node B' from the corresponding 3GPP specifications is used throughout this chapter. The more generic term 'base station' used elsewhere in this book means exactly the same thing.)
2. The Radio Network Controller (RNC) owns and controls the radio resources in its domain (the Node Bs connected to it). The RNC is the service access point for all services that UTRAN provides the CN, e.g. management of connections to the UE.

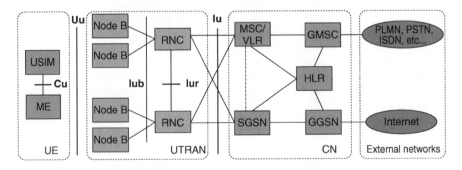

**Figure 5.2** Network elements in a PLMN

The main elements of the GSM CN (there are other entities not shown in Figure 5.2, such as those used to provide IN services) are as follows:

- The Home Location Register (HLR) is a database located in the user's home system that stores the master copy of the user's service profile. The service profile consists of, for example, information on permitted services, forbidden roaming areas, and Supplementary Service information such as the status of call forwarding and the call forwarding number. It is created when a new user subscribes to the system, and remains stored as long as the subscription is active. For the purpose of routing incoming transactions to the UE (e.g. calls or short messages), the HLR also stores the UE location on the level of MSC/VLR and/or SGSN, i.e. on the level of serving system.
- The Mobile Services Switching Center/Visitor Location Register (MSC/VLR) is the switch (MSC) and database (VLR) that serves the UE in its current location for Circuit-Switched (CS) services. The MSC function is used to switch the CS transactions, and the VLR function holds a copy of the visiting user's service profile, as well as more precise information on the UE's location within the serving system. The part of the network that is accessed via the MSC/VLR is often referred to as the CS domain. MSC also has a role in the early UE handling, as discussed in Chapter 7.
- The Gateway MSC (GMSC) is the switch at the point where UMTS PLMN is connected to external CS networks. All incoming and outgoing CS connections go through GMSC.
- The Serving General Packet Radio Service (GPRS) Support Node (SGSN) functionality is similar to that of MSC/VLR but is typically used for Packet-Switched (PS) services. The part of the network that is accessed via the SGSN is often referred to as the PS domain. Similar to MSC, SGSN support is needed for the early UE handling operation, as covered in Chapter 7.
- Gateway GPRS Support Node (GGSN) functionality is close to that of GMSC but is in relation to PS services.

The external networks can be divided into two groups:

1. *CS networks*. These provide circuit-switched connections, like the existing telephony service. ISDN and PSTN are examples of CS networks.
2. *PS networks*. These provide connections for packet data services. The Internet is one example of a PS network.

The UMTS standards are structured so that the internal functionality of the network elements is not specified in detail. Instead, the interfaces between the logical network elements have been defined. The following main open interfaces are specified:

- *Cu interface*. This is the electrical interface between the USIM smartcard and the ME. The interface follows a standard format for smartcards.
- *Uu interface*. This is the WCDMA radio interface, which is the subject of the main part of this book. The Uu is the interface through which the UE accesses the fixed part of the system and, therefore, is probably the most important open interface in UMTS. There are likely to be many more UE manufacturers than manufacturers of fixed network elements.
- *Iu interface*. This connects UTRAN to the CN and is introduced in detail in Section 5.4. Similar to the corresponding interfaces in GSM, A (CS) and Gb (PS), the open Iu interface gives UMTS operators the possibility of acquiring UTRAN and CN from different manufacturers. The permitted competition in this area has been one of the success factors of GSM.
- *Iur interface*. The open Iur interface allows a soft handover between RNCs from different manufacturers and, therefore, complements the open Iu interface. Iur is described in more detail in Section 5.5.1.

- *Iub interface*. The Iub connects a Node B and an RNC. UMTS is the first commercial mobile telephony system where the Controller–Base Station interface is standardized as a fully open interface. Like the other open interfaces, open Iub is expected to motivate further competition between manufacturers in this area. It is likely that new manufacturers concentrating exclusively on Node Bs will enter the market.

## 5.2   UTRAN Architecture

The UTRAN architecture is highlighted in Figure 5.3. UTRAN consists of one or more Radio Network Sub-systems (RNSs). An RNS is a sub-network within UTRAN and consists of one RNC and one or more Node Bs. RNCs may be connected to each other via an Iur interface. RNCs and Node Bs are connected with an Iub Interface. During Release 7, work study on the support of small RNSs was done, meaning use of co-located RNC and Node B functionalities in a flat architecture, and that was found feasible without mandatory specification changes. The flat architecture is further discussed in Chapter 15.

Before entering into a brief description of the UTRAN network elements (in this section) and a more extensive description of UTRAN interfaces (in the following sections), we present the main characteristics of UTRAN that have also been the main requirements for the design of the UTRAN architecture, functions and protocols. These can be summarized in the following points:

- *Support of UTRA* and all the related functionality. In particular, the major impact on the design of UTRAN has been the requirement to support *soft handover* (one terminal connected to the network via two or more active cells) and the WCDMA-specific *Radio Resource Management* algorithms.
- Maximization of the *commonalities in the handling of PS and CS data*, with a unique air interface protocol stack and with use of the same interface for the connection from UTRAN to both the PS and CS domains of the CN.
- Maximization of the *commonalities with GSM*, when possible.
- Use of the *Asynchronous Transfer Mode (ATM) transport* as the main transport mechanism in UTRAN.
- Use of the Internet Protocol (IP)-based transport as the alternative transport mechanism in UTRAN from Release 5 onwards.

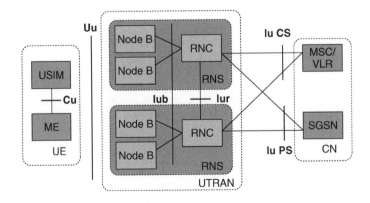

**Figure 5.3**   UTRAN architecture

## 5.2.1   The Radio Network Controller (RNC)

The RNC is the network element responsible for the control of the radio resources of UTRAN. It interfaces the CN (normally to one MSC and one SGSN) and also terminates the Radio Resource Control (RRC) protocol that defines the messages and procedures between the mobile and UTRAN. It logically corresponds to the GSM BSC.

### 5.2.1.1   Logical Role of the RNC

The RNC controlling one Node B (i.e. terminating the Iub interface towards the Node B) is indicated as the *Controlling RNC* (CRNC) of the Node B. The CRNC is responsible for the load and congestion control of its own cells, and also executes the admission control and code allocation for new radio links to be established in those cells.

If one mobile–UTRAN connection uses resources from more than one RNS (see Figure 5.4), the RNCs involved have two separate logical roles (with respect to this mobile–UTRAN connection):

- *Serving RNC (SRNC)*. The SRNC for one mobile is the RNC that terminates both the Iu link for the transport of user data and the corresponding RAN application part (RANAP) signaling to/from the CN (this connection is referred to as the RANAP connection). The SRNC also terminates the RRCl Signaling, i.e. the signaling protocol between the UE and UTRAN. It performs the L2 processing of the data to/from the radio interface. Basic Radio Resource Management operations, such as the mapping of Radio Access Bearer (RAB) parameters into air interface transport channel parameters, the handover decision, and outer loop power control, are executed in the SRNC. The SRNC may also (but not always) be the CRNC of some Node B used by the mobile for connection with UTRAN. One UE connected to UTRAN has one and only one SRNC.
- *Drift RNC (DRNC)*. The DRNC is any RNC, other than the SRNC, that controls cells used by the mobile. If needed, the DRNC may perform macrodiversity combining and splitting. The DRNC does not perform L2 processing of the user plane data, but routes the data transparently between the Iub and Iur interfaces, except when the UE is using a common or shared transport channel. One UE may have zero, one or more DRNCs.

Note that one physical RNC normally contains all the CRNC, SRNC and DRNC functionality.

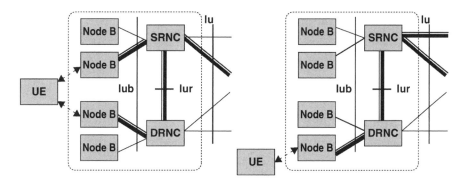

**Figure 5.4**   Logical role of the RNC for one UE UTRAN connection. The left-hand scenario shows one UE in inter-RNC soft handover (combining is performed in the SRNC). The right-hand scenario represents one UE using resources from one Node B only, controlled by the DRNC

## 5.2.2   The Node B (Base Station)

The main function of the Node B is to perform the air interface L1 processing (channel coding and interleaving, rate adaptation, spreading, etc.). It also performs some basic Radio Resource Management operation as the inner loop power control. It logically corresponds to the GSM Base Station. The enigmatic term 'Node B' was initially adopted as a temporary term during the standardization process, but then never changed. The logical model of the Node B is described in Section 5.5.2.

# 5.3   General Protocol Model for UTRAN Terrestrial Interfaces

## 5.3.1   General

Protocol structures in UTRAN terrestrial interfaces are designed according to the same general protocol model. This model is shown in Figure 5.5. The structure is based on the principle that the layers and planes are logically independent of each other and, if needed, parts of the protocol structure may be changed in the future while other parts remain intact.

## 5.3.2   Horizontal Layers

The protocol structure consists of two main layers: the Radio Network Layer and the Transport Network Layer. All UTRAN-related issues are visible only in the Radio Network Layer, and the Transport Network Layer represents standard transport technology that is selected to be used for UTRAN but without any UTRAN-specific changes.

## 5.3.3   Vertical Planes

### 5.3.3.1   Control Plane

The Control Plane is used for all UMTS-specific control signaling. It includes the Application Protocol (i.e. RANAP in Iu, Radio Network System Application Part (RNSAP) in Iur and Node B Application Part (NBAP) in Iub), and the Signaling Bearer to transport the Application Protocol messages.

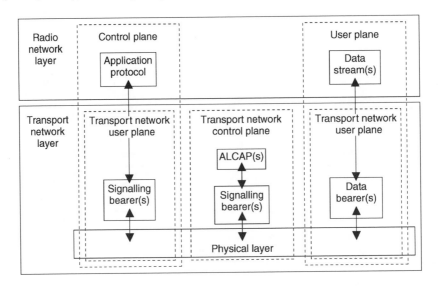

**Figure 5.5**   General protocol model for UTRAN terrestrial interfaces

The Application Protocol is used, among other things, for setting up bearers to the UE (i.e. the RAB in Iu and subsequently the Radio Link in Iur and Iub). In the three-plane structure the bearer parameters in the Application Protocol are not directly tied to the User Plane technology, but rather are general bearer parameters.

The Signaling Bearer for the Application Protocol may or may not be of the same type as the Signaling Bearer for the ALCAP. It is always set up by operation and maintenance (O&M) actions.

### 5.3.3.2  User Plane

All information sent and received by the user, such as the coded voice in a voice call or the packets in an Internet connection, are transported via the User Plane. The User Plane includes the Data Stream(s), and the Data Bearer(s) for the Data Stream(s). Each Data Stream is characterized by one or more frame protocols specified for that interface.

### 5.3.3.3  Transport Network Control Plane

The Transport Network Control Plane is used for all control signaling within the Transport Layer. It does not include any Radio Network Layer information. It includes the ALCAP protocol that is needed to set up the transport bearers (Data Bearer) for the User Plane. It also includes the Signaling Bearer needed for the ALCAP.

The Transport Network Control Plane is a plane that acts between the Control Plane and the User Plane. The introduction of the Transport Network Control Plane makes it possible for the Application Protocol in the Radio Network Control Plane to be completely independent of the technology selected for the Data Bearer in the User Plane.

When the Transport Network Control Plane is used, the transport bearers for the Data Bearer in the User Plane are set up in the following fashion. First, there is a signaling transaction by the Application Protocol in the Control Plane, which triggers the se-up of the Data Bearer by the ALCAP protocol that is specific for the User Plane technology.

The independence of the Control Plane and the User Plane assumes that an ALCAP signaling transaction takes place. It should be noted that ALCAP might not be used for all types of Data Bearers. If there is no ALCAP signaling transaction, then the Transport Network Control Plane is not needed at all. This is the case when it is enough simply to select the user plane resources, e.g. selecting end-point addresses for IP transport or selecting a preconfigured Data Bearer. It should also be noted that the ALCAP protocols in the Transport Network Control Plane are not used for setting up the Signaling Bearer for the Application Protocol or for the ALCAP during real-time operation.

The Signaling Bearer for the ALCAP may or may not be of the same type as that for the Application Protocol. The UMTS specifications assume that the Signaling Bearer for ALCAP is always set up by O&M actions, and do not specify this in detail.

### 5.3.3.4  Transport Network User Plane

The Data Bearers in the User Plane and the Signaling Bearers for the Application Protocol also belong to the Transport Network User Plane. As described in the previous section, the Data Bearers in the Transport Network User Plane are directly controlled by the Transport Network Control Plane during real-time operation, but the control actions required to set up the Signaling Bearer(s) for the Application Protocol are considered O&M actions.

## 5.4  Iu, the UTRAN–CN Interface

The Iu interface connects UTRAN to CN. Iu is an open interface that divides the system into radio-specific UTRAN and CN, which handles switching, routing and service control. As can be seen from Figure 5.3, the Iu can have two main different instances, which are Iu CS to connect UTRAN to CS

CN, and Iu PS to connect UTRAN to PS CN. The additional third instance of Iu, the Iu Broadcast (Iu BC, not shown in Figure 5.3), has been defined to support Cell Broadcast Services (see Section 5.4.5). The original design goal in the standardization was to develop only one Iu interface, but then it was realized that fully optimized User Plane transport for CS and PS services can only be achieved if different transport technologies are permitted. Consequently, the Transport Network Control Plane is different. One of the main design guidelines has still been that the Control Plane should be the same for Iu CS and Iu PS, and the differences are minor.

A third instance of the Iu interface, Iu BC, is used to connect UTRAN to the Broadcast domain of the CN. The Iu BC interface is not shown in Figure 5.3.

## 5.4.1 Protocol Structure for Iu CS

The Iu CS overall protocol structure is depicted in Figure 5.6. The three planes in the Iu interface share a common ATM transport which is used for all planes. The physical layer is the interface to the physical medium: optical fibre, radio link or copper cable. The physical layer implementation can be selected from a variety of standard off-the-shelf transmission technologies, such as SONET, STM1, or E1.

### 5.4.1.1   Iu CS Control Plane Protocol Stack

The Control Plane protocol stack consists of RANAP, on top of Broad Band (BB) Signaling System #7 (SS7) protocols. The applicable layers are the Signaling Connection Control Part (SCCP), the Message Transfer Part (MTP3-b) and Signaling ATM Adaptation Layer for Network to Network

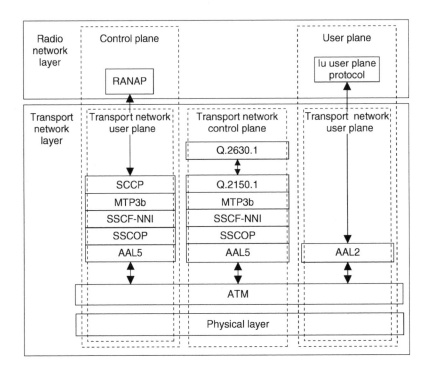

**Figure 5.6**   Iu CS protocol structure

Interfaces (SAAL-NNI). SAAL-NNI is further divided into Service Specific Co-ordination Function (SSCF), Service Specific Connection Oriented Protocol (SSCOP) and ATM Adaptation Layer 5 (AAL) layers. SSCF and SSCOP layers are specifically designed for signaling transport in ATM networks, and take care of such functions as signaling connection management. AAL5 is used to segment the data into ATM cells.

#### 5.4.1.2  Iu CS Transport Network Control Plane Protocol Stack

The Transport Network Contro Plane protocol stack consists of the Signaling Protocol for setting up AAL2 connections (Q.2630.1 and adaptation layer Q.2150.1), on top of BB SS7 protocols. The applicable BB SS7 protocols are those described above without the SCCP layer.

#### 5.4.1.3  Iu CS User Plane Protocol Stack

A dedicated AAL2 connection is reserved for each individual CS service. The Iu User Plane Protocol residing directly on top of AAL2 is described in more detail in Section 5.4.4.

### 5.4.2  Protocol Structure for Iu PS

The Iu PS protocol structure is depicted in Figure 5.7. Again, a common ATM transport is applied for both the User and the Control Planes. Also, the physical layer is as specified for Iu CS.

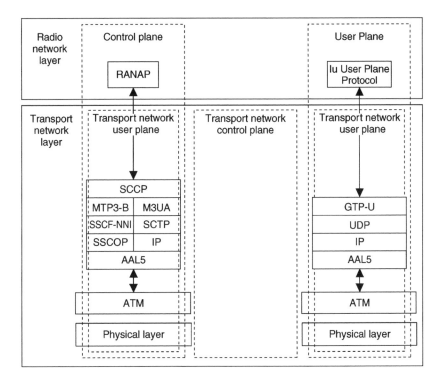

**Figure 5.7**  Iu PS protocol structure

### 5.4.2.1  Iu PS Control Plane Protocol Stack

The Control Plane protocol stack again consists of RANAP, and the same BB SS7-based signaling bearer as described in Section 5.4.1.1. Also as an alternative, an IP-based signaling bearer is specified. The SCCP layer is also used commonly for both. The IP-based signaling bearer consists of SS7 MTP3 – User Adaptation Layer (M3UA), Simple Control Transmission Protocol (SCTP), IP, and AAL5 which is common to both alternatives. The SCTP layer is specifically designed for signaling transport in the Internet. Specific adaptation layers are specified for different kinds of signaling protocols, such as M3UA for SS7-based signaling.

### 5.4.2.2  Iu PS Transport Network Control Plane Protocol Stack

The Transport Network Control Plane is not applied to Iu PS. The setting up of the GTP tunnel requires only an identifier for the tunnel, and the IP addresses for both directions, and these are already included in the RANAP RAB Assignment messages. The same information elements that are used in Iu CS for addressing and identifying the AAL2 signaling are used for the User Plane data in Iu CS.

### 5.4.2.3  Iu PS User Plane Protocol Stack

In the Iu PS User Plane, multiple packet data flows are multiplexed on one or several AAL5 Pre-defined Virtual Connections. The User Plane part of the GPRS Tunnelling Protocol (GTP-U) is the multiplexing layer that provides identities for individual packet data flow. Each flow uses UDP connectionless transport and IP addressing.

## 5.4.3  RANAP Protocol

RANAP is the signaling protocol in Iu that contains all the control information specified for the Radio Network Layer. The functionality of RANAP is implemented by various RANAP Elementary Procedures (EPs). Each RANAP function may require the execution of one or more EPs. Each EP consists of either just the request message (class 2 EP), the request and response message pair (class 1 EP), or one request message and one or more response messages (class 3 EP). The following RANAP functions are defined:

- *Relocation*. This function handles both SRNS relocation and hard handover, including intersystem case to/from GSM:
- *SRNS relocation*. The SRNS functionality is relocated from one RNS to another without changing the radio resources and without interrupting the user data flow. The prerequisite for SRNS relocation is that all Radio Links are already in the same DRNC that is the target for the relocation.
- *Inter-RNS hard handover*. This is used to relocate the serving RNS functionality from one RNS to another and to change the radio resources correspondingly by a hard handover in the Uu interface. The prerequisite for hard handover is that the UE is at the border of the source and target cells.
- *RAB management*. This function combines all RAB handling:
  - RAB set-up, including the possibility for queuing the setup;
  - modification of the characteristics of an existing RAB;
  - clearing an existing RAB, including the RAN-initiated case.
- *Iu release*. Releases all resources (Signaling link and U-plane) from a given instance of Iu related to the specified UE. Also includes the RAN-initiated case.
- *Reporting unsuccessfully transmitted data*. This function allows the CN to update its charging records with information from UTRAN if part of the data sent was not successfully sent to the UE.

- *Common ID management*. In this function the permanent identification of the UE is sent from the CN to UTRAN to allow paging coordination from possibly two different CN domains.
- *Paging*. This is used by CN to page an idle UE for a UE terminating service request, such as a voice call. A paging message is sent from the CN to UTRAN with the UE common identification (permanent ID) and the paging area. UTRAN will either use an existing signaling connection, if one exists, to send the page to the UE or broadcast the paging in the requested area.
- *Management of tracing*. The CN may, for O&M purposes, request UTRAN to start recording all activity related to a specific UE–UTRAN connection.
- *UE–CN signaling transfer*. This functionality provides transparent transfer of UE–CN signaling messages that are not interpreted by UTRAN in two cases.
- *Transfer of the first UE message from UTRAN to UE*: this may be, for example, a response to paging, a request of a UE-originated call, or just registration to a new area. It also initiates the signaling connection for the Iu.
- *Direct transfer*: used for carrying all consecutive signaling messages over the Iu signaling connection in both the uplink and downlink directions.
- *Security rode control*. This is used to set the ciphering or integrity checking on or off. When ciphering is on, the signaling and user data connections in the radio interface are encrypted with a secret key algorithm. When integrity checking is on, an integrity checksum, further secured with a secret key, is added to some or all of the radio interface signaling messages. This ensures that the communication partner has not changed, and the content of the information has not been altered.
- *Management of overload*. This is used to control the load over the Iu interface against overload due, for example, to processor overload at the CN or UTRAN. A simple mechanism is applied that allows stepwise reduction of the load and its stepwise resumption, triggered by a timer.
- *Reset*. This is used to reset the CN or the UTRAN side of the Iu interface in error situations. One end of the Iu may indicate to the other end that it is recovering from a restart, and the other end can remove all previously established connections.
- *Location reporting*. This functionality allows the CN to receive information on the location of a given UE. It includes two elementary procedures, one for controlling the location reporting in the RNC and the other to send the actual report to the CN.

## 5.4.4   Iu User Plane Protocol

The Iu User Plane protocol is in the Radio Network Layer of the Iu User Plane. It has been defined so that it would be, as much as possible, independent of the CN domain that it is used for. The purpose of the User Plane protocol is to carry user data related to RABs over the Iu interface. Each RAB has its own instance of the protocol. The protocol performs either a fully transparent operation, or framing for the user data segments and some basic control signaling to be used for initialization and online control. Based on these cases, the protocol has two modes:

1. *Transparent mode*. In this mode of operation the protocol does not perform any framing or control. It is applied for RABs that do not require such features but that assume fully transparent operation.
2. *Support mode for predefined (service data unit) SDU sizes*. In this mode the User Plane performs framing of the user data into segments of predefined size. The SDU sizes typically correspond to Adaptive Multirate Codec speech frames, or to the frame sizes derived from the data rate of a CS data call. Also, control procedures for initialization and rate control are defined, and a functionality is specified to indicate the quality of the frame based, for example, on a cyclic redundancy check from the radio interface.

### 5.4.5  Protocol Structure of Iu BC, and the Service Area Broadcast Protocol

The Iu BC [2] interface connects the RNC in UTRAN with the broadcast domain of the CN, namely with the Cell Broadcast Center. It is used to define the Cell Broadcast information that is transmitted to the mobile user via the Cell Broadcast Service (e.g. name of city/region visualized on the mobile phone display). Note that this should not be confused with the UTRAN or CN information broadcast on the broadcast common control channel. Iu BC is a control-plane-only interface. The protocol structure of Iu BC is shown in Figure 5.8.

#### 5.4.5.1  Service Area Broadcast Protocol

The Service Area Broadcast protocol (SABP) [23] provides the capability for the Cell Broadcast Center in the CN to define, modify and remove cell broadcast messages from the RNC. RNC uses them and the NBAP protocol and RRC signaling to transfer such messages to the mobile. The SABP has the following functions:

- *Message handling*. This function is responsible for the broadcast of new messages, amending existing broadcasted messages and stopping the broadcasting of specific messages.

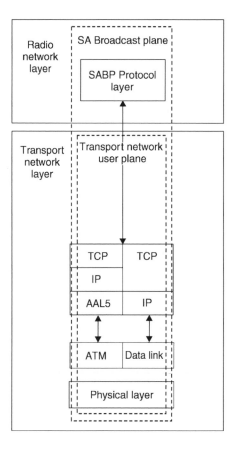

**Figure 5.8**  Iu BC protocol structure

- *Load handling*. This function is responsible for determining the loading of the broadcast channels at any particular point in time.
- *Reset*. This function permits the Cell Broadcast Center to end broadcasting in one or more Service Areas.

## 5.5  UTRAN Internal Interfaces

### 5.5.1  RNC–RNC Interface (Iur Interface) and the RNSAP Signaling

The protocol stack of the RNC-to-RNC interface (Iur interface) is shown in Figure 5.9. Although this interface was initially designed in order to support the inter-RNC soft handover (shown on the left-hand side of Figure 5.4), more features were added during the development of the standard and, currently, the Iur interface provides four distinct functions:

1. Support of basic inter-RNC mobility.
2. Support of dedicated channel traffic.
3. Support of common channel traffic.
4. Support of global resource management.

For this reason, the Iur signaling protocol itself (RNSAP) is divided into four different *modules* (to be intended as groups of procedures). In general, it is possible to implement only part of the four Iur modules between two RNCs, according to the operator's need.

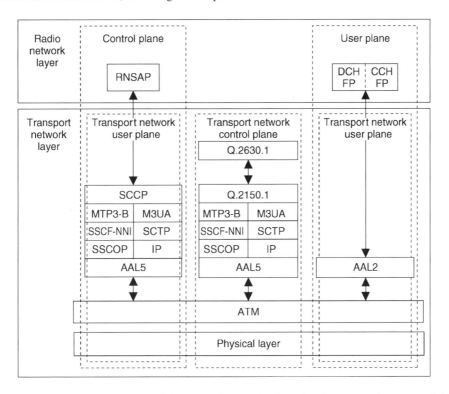

**Figure 5.9**  Release 99 protocol stack of the Iur interface. As for the Iu interface, two options are possible for the transport of the RNSAP signaling: the SS7 stack (SCCP and MTP3b) and the new SCTP/IP-based transport. Two User Plane protocols are defined (DCH: dedicated channel; CCH: common channel)

#### 5.5.1.1  Iur1: Support of the Basic Inter-RNC Mobility

This functionality requires the *basic* module of RNSAP signaling as described in [11]. This first brick for the construction of the Iur interfaces provides on its own the functionality needed for the mobility of the user between the two RNCs, but does not support the exchange of any user data traffic. If this module is not implemented, then the Iur interface as such does not exist, and the only way for a user connected to UTRAN via the RNS1 to utilize a cell in RNS2 is to disconnect itself temporarily from UTRAN (release the RRC connection).

The functions offered by the Iur basic module include:

- Support of SRNC relocation.
- Support of inter-RNC cell and UTRAN registration area update.
- Support of inter-RNC packet paging.
- Reporting of protocol errors.

Since this functionality does not involve user data traffic across Iur, the User Plane and the Transport Network Control Plane protocols are not needed.

#### 5.5.1.2  Iur2: Support of Dedicated Channel Traffic

This functionality requires the Dedicated Channel module of RNSAP signaling and allows the dedicated and shared channel traffic between two RNCs. Even if the initial need for this functionality is to support the inter-RNC soft handover state, it also allows the anchoring of the SRNC for all the time the user is utilizing dedicated channels (dedicated resources in the Node B), commonly for as long as the user has an active connection to the CS domain.

This functionality requires also the User Plane Frame Protocol for the dedicated and shared channel, plus the Transport Network Control Plane protocol (Q.2630.1) used for the set-up of the transport connections (AAL2 connections). Each dedicated channel is conveyed over one transport connection, except the coordinated DCH used to obtain unequal error protection in the air interface.

The Frame Protocol for dedicated channels, in short DCH FP [15], defines the structure of the data frames carrying the user data and the control frames used to exchange measurements and control information. For this reason, the Frame Protocol also specifies simple messages and procedures. The user data frames are normally routed transparently through the DRNC; thus, the Iur frame protocol is used also in Iub and referred to as Iur/Iub DCH FP. The user plane procedure for shared channels is described in the Frame Protocol for a common channel in an Iur interface, in short, Iur CCH FP [13].

The functions offered by the Iur DCH module are:

- Establishment, modification and release of the dedicated and shared channel in the DRNC due to handovers in the dedicated channel state.
- Set-up and release of dedicated transport connections across the Iur interface.
- Transfer of DCH Transport Blocks between SRNC and DRNC.
- Management of the radio links in the DRNS, via dedicated measurement report procedures, power setting procedures and compress mode control procedures.

#### 5.5.1.3  Iur3: Support of Common Channel Traffic

This functionality allows the handling of common channel (i.e. RACH, FACH and CPCH) data streams across the Iur interface. It requires the Common Transport Channel module of the RNSAP protocol and the Iur Common Transport Channel Frame Protocol (in short, CCH FP). The Q.2630.1 signaling protocol of the Transport Network Control Plane is also needed if signaled AAL2 connections are used.

If this functionality is not implemented, then every inter-RNC cell update always triggers an SRNC relocation, i.e. the serving RNC is always the RNC controlling the cell used for common or shared channel transport.

The identification of the benefits of this feature caused a long debate in the relevant standardization body. On the one hand, this feature allows the implementation of the total anchor RNC concept, avoiding the SRNC relocation procedure (via the CN); on the other hand, it requires the splitting of the Medium Access Control (MAC) layer functionality into two network elements, generating inefficiency in the utilization of the resources and complexity in the Iur interface. The debate could not reach an agreement; thus, the feature is supported by the standard but is not essential for the operation of the system.

The functions offered by the Iur common transport channel module are:

- Set-up and release of the transport connection across the Iur for common channel data streams.
- Splitting of the MAC layer between the SRNC (MAC-d) and the DRNC (MAC-c). The scheduling for downlink data transmission is performed in the DRNC.
- Flow control between the MAC-d and MAC-c.

### 5.5.1.4  Iur4: Support of Global Resource Management

This functionality provides signaling to support enhanced radio resource management and O&M features across the Iur interface. It is implemented via the global module of the RNSAP protocol, and does not require any User Plane protocol, since there is no transmission of user data across the Iur interface. The function is considered optional. This function has been introduced in subsequent releases for the support of common radio resource management between RNCs, advanced positioning methods and Iur optimization purposes.

The functions offered by the Iur global resource module are:

- Transfer of cell information and measurements between two RNCs.
- Transfer of positioning parameters between controllers.
- Transfer of Node B timing information between two RNCs.

## 5.5.2   RNC–Node B Interface and the NBAP Signaling

The protocol stack of the RNC–Node B interface (Iub interface) is shown, with the typical triple plane notation, in Figure 5.10. In order to understand the structure of the interface, it is necessary briefly to introduce the logical model of the Node B, depicted in Figure 5.11. This consists of a common control port (a common signaling link) and a set of traffic termination points each controlled by a dedicated control port (dedicated signaling link). One traffic termination point controls a number of mobiles having dedicated resources in the Node B, and the corresponding traffic is conveyed through dedicated data ports. Common data ports outside the traffic termination points are used to convey RACH, FACH and PCH traffic.

Note that there is no relation between the traffic termination point and the cells, i.e. one traffic termination point can control more than one cell, and one cell can be controlled by more than one traffic termination point.

The Iub interface signaling (NBAP) is divided into two essential components: the common NBAP (C-NBAP), which defines the signaling procedures across the common signaling link, and the dedicated NBAP (D-NBAP), used in the dedicated signaling link.

The User Plane Iub frame protocols define the structures of the frames and the basic in-band control procedures for every type of transport channel (i.e. for every type of data port of the model). The Q.2630.1 signaling is used for the dynamic management of the AAL2 connections used in the User Plane.

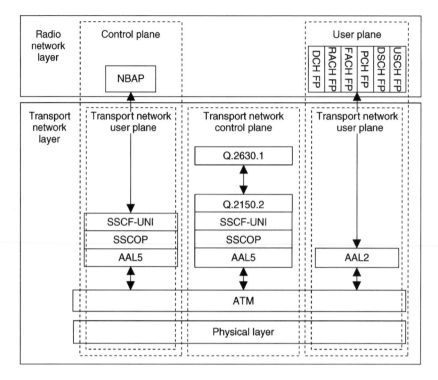

**Figure 5.10**  Release 99 protocol stack of the Iub interface. This is similar to the Iur interface protocol, the main difference being that in the Radio Network and Transport Network Control Planes the SS7 stack is replaced by the simpler SAAL-UNI as signaling bearer. Note also that the SCTP/IP option is not present here

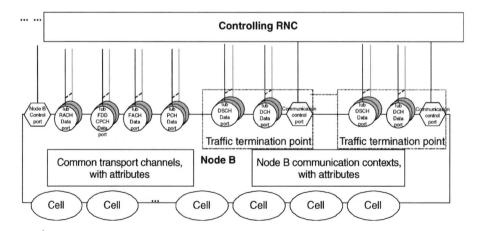

**Figure 5.11**  Logical model of the Node B for FDD

### 5.5.2.1   C-NBAP and the Logical O&M

The C-NBAP procedures are used for the signaling that is not related to one specific UE context already existing in the Node B. In particular, the C-NBAP defines all the procedures for the logical O&M of the Node B, such as configuration and fault management.

The main functions of the C-NBAP are:

- Set-up of the first radio link of one UE, and selection of the traffic termination point.
- Cell configuration.
- Handling of the RACH/FACH/CPCH and PCH channels.
- Initialization and reporting of Cell or Node B specific measurement.
- Location Measurement Unit (LMU) control.
- Fault management.

### 5.5.2.2   Dedicated NBAP

When the RNC requests the first radio link for one UE via the C-NBAP Radio Link Set-up procedure, the Node B assigns a traffic termination point for the handling of this UE context, and every subsequent signaling related to this mobile is exchanged with D-NBAP procedures across the dedicated control port of the given Traffic Termination Point.

The main functions of the D-NBAP are:

- Addition, release and reconfiguration of radio links for one UE context.
- Handling of dedicated and shared channels.
- Handling of softer combining.
- Initialization and reporting of radio-link-specific measurement.
- Radio-link fault management.

## 5.6   UTRAN Enhancements and Evolution

The Release 99 UTRAN architecture described in Chapter 4 defines the basic set of network elements and interface protocols for the support of the Release 99 WCDMA radio interface. Since then, enhancement of the architecture and related specification were needed in order to support new WCDMA radio interface features, but also as result of the necessity to provide a more efficient, scalable and robust 3GPP system architecture. The four most significant additions to the UTRAN architecture introduced in Release 5 are described in subsequent sections.

### 5.6.1   IP Transport in UTRAN

ATM is the transport technology used in the first release of the UTRAN. Even before the completion of the specification it was clear that 3GPP could not stay immune from the increasing popularity of IP technology, and a second option for the transport, 'IP transport', was introduced in the specification in Release 5. Accordingly, user plane FP frames can also be conveyed over UDP/IP protocols on Iur/Iub, and over RTP/UDP/IP protocols in an Iu CS interface, in addition to the initially defined option of AAL2/ATM. A second option for the Iub control plane, using SCTP directly below the application part, is also introduced. The protocols to be used to convey IP frames are, in general, left unspecified in order not to limit the use of layer 2 and physical layer interfaces available in the operator networks. Although the IP transport requires small changes in the specification (and almost none in the control plane application parts), the adoption of IP technology is a relevant step for both the operator and the vendors, changing the way the network itself is managed and, in some cases, the way the network elements are implemented.

## 5.6.2   Iu Flex

The Release 99 architecture presented in Figure 5.3 is characterized by having only one MSC and one SGSN connected to the RNC, i.e. only one Iu PS and Iu CS interface in the RNC. This limitation is overcome in the Release 5 specification with the introduction of the Iu flex (from the word 'flexible') concept, which allows one RNC to have more than one Iu PS and Iu CS interface instances with the core. The main benefits of this feature are to introduce the possibility of load sharing between the CN nodes and to increase the possibility of anchoring the MSC and SGSN in case of SRNS relocation. Iu flex has limited impact in the UTRAN specification, since the CN node to be used is negotiated between the UE and the CN.

## 5.6.3   Stand-Alone SMLC and Iupc Interface

Location-based services are expected to be a very important source of revenue for the mobile operators, and a number of different applications are expected to be available and largely used. Following the example of the GSM BSS, the UTRAN architecture also includes a stand-alone Serving Mobile Location Center (stand-alone SMLC, or, simply, SAS), which is a new network element for the handling of positioning measurements and the calculation of the mobile station position. The SAS is connected to the RNC via the Iupc interface and the Positioning Calculation Application Part (PCAP) is the L3 protocol used for the RNC-SAS signaling. Stand-alone SMLC and Iu PC interface are optional elements, since SMLC functionality can be integrated in the RNC as well; thus, it depends on the individual network implementation whether to use it or not. The first version of the Iu PC supported only Assisted GPS, but then for later versions support for other positioning methods was added. The latest methods being included in Release 7 specifications are Assisted Galileo and Uplink TDOA methods.

## 5.6.4   Interworking between GERAN and UTRAN, and the Iur-g Interface

The Iu interface has also been scheduled to be part of the GSM/EDGE Radio Access Network (GERAN) in GERAN Release 5. This allows reuse of the 3G CN also for the GSM/EDGE radio interface (and frequency band), but also allows more optimized interworking between the two radio technologies. As an effect of this, the RNSAP basic mobility module (described in Section 5.5.1.1) is enhanced to allow also the mobility to and from GERAN cells in the target and the source, and the RNSAP global module (see Section 5.5.1.4) is enhanced in order to allow the GERAN cells' measurements to be exchanged between controllers. The last feature allows a Common Radio Resource Management (CRRM) between UTRAN and GERAN radios. The term Iur-g interface is often used to refer to the above-mentioned set of Iur functionalities that are utilized also by the GERAN.

## 5.6.5   IP-Based RAN Architecture

The increasing role of IP technology in modern telecoms and IT networks has already been mentioned in the previous sections to motivate the introduction of the IP Transport option in UTRAN. We will see in the next section how the need to provide optimized support IP services leads to sensible changes in the architecture of the CN with the introduction of a new subsystem, the IP Multimedia Subsystem (IMS), to form what is now commonly referred to as the All IP CN. Is the introduction of IP Transport in UTRAN enough to provide the most suitable RAN architecture to be implemented with IP technology, integrated with the always more commonly used IP networks and platforms, and utilized by IP packet services? In 3GPP, work has been done to investigate new architecture alternatives aiming for a more distributed operation from a centralized network structure, with the motivation to achieve a flat architecture similar to the case of LTE with good scalability for handling increased data rates. The developments from Release 99 with features like HSDPA in Release 5 (Chapter 12) or HSUPA in Release 6 (Chapter 13) have taken the first steps in taking the scheduling and retransmissions (MAC layer) closer to the air interface to improve system performance.

The term All IP RAN is nowadays often used to refer to this IP-Optimized RAN architecture concept and implementation, but is currently not yet associated with any 3GPP standard feature. For this reason, this term is sometimes used to refer to a RAN implementation based on the current architecture but using IP Transport. The flat architecture having all radio-related protocols terminated in the Node B site (with RNC functionality co-located with Node B) is discussed further in Chapter 15.

## 5.7   UMTS CN Architecture and Evolution

While the UMTS radio interface, WCDMA, represented a bigger step in the radio access evolution from GSM networks, the UMTS CN did not experience major changes in the 3GPP Release 99 specification. The Release 99 structure was inherited from the GSM CN and, as stated also earlier, both UTRAN- and GERAN-based RANs connect to the same CN.

### 5.7.1   Release 99 CN Elements

The Release 99 CN has two domains, a CS domain and a PS domain, to cover the need for different traffic types. The division comes from the different requirements of the data, depending on whether it is real time (circuit switched) or non-real time (packet data). We now present the functional split in the CN side; however, it should be understood that several functionalities can be implemented in a single physical entity and all entities do not necessarily exist as separate physical units in real networks. Figure 5.12 illustrates the Release 99 CN structure with both CS and PS domains shown. Figure 5.12

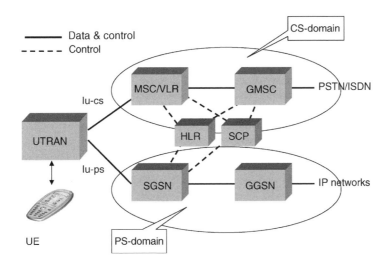

**Figure 5.12**   Release 99 UMTS CN structure

also contains registers as well as the Service Control Point (SCP) to indicate the link for providing a particular service to the end user.

The CS domain has the following elements as introduced in Section 5.1:

- MSC, including VLR
- GMSC.

The PS domain has the following elements as introduced in Section 5.1:

- SGSN, which covers similar functions to the MCS for the packet data, including VLR-type functionality.
- GGSN connects PS CN to other networks, for example, to the internet.

In addition to the two domains, the network needs various registers for proper operation:

- HLR with the functionality as covered in Section 5.1.
- Equipment Identity Register contains the information related to the terminal equipment and can be used, for example, to prevent a specific terminal from accessing the network.

## 5.7.2   Release 5 CN and IP Multimedia Subsystem

The Release 5 CN has many additions compared with Release 99 CNs. Release 4 already included the change in the CN CS domain when the MSC was divided into the MSC server and the Media Gateway (MGW). Also, the GMSC was divided into the GMSC server and the MGW. Release 5 contains the first phase of the IMS, which will enable a standardized approach for IP-based service provision via the PS domain as discussed in Chapter 2. The capabilities of the IMS will be further enhanced in Release 6. The Release 6 IMS will allow the provision of services similar to the CS domain services from the PS domain. The following summarizes the elements in the Release 5-based architecture, added on top of Release 99 and Release 4 architectures. The Release 5 architecture is presented in Figure 5.13, with the simplification that the registers, now part of Home Subscriber Server (HSS), are shown only as independent items without all the connections to the other elements shown.

From a protocols perspective, the key protocol between the terminal and the IMS is the Session Initiation Protocol SIP, which is the basis for IMS-related signaling, with the contents as described in Chapter 2.

The following elements have experienced changes in the CS domain for Release 4:

- The MSC or GMSC server takes care of the control functionality as MSC or GMSC respectively, but the user data goes via the MGW. One MSC/GMSC server can control multiple MGWs, which allows better scalability of the network when, for example, the data rates increase with new data services. In that case, only the number of MGWs needs to be increased.
- The MGW performs the actual switching for user data and network interworking processing, e.g. echo cancellation or speech decoding/encoding.

In the PS domain, the SGSN and GGSN are as in Release 99 with some enhancements, but for the IP-based service delivery the IMS now has the following key elements included:

- Media Resource Function (MRF), which, for example, controls media stream resources or can mix different media streams. The standard defines further the detailed functional split for the MRF.
- Call Session Control Function (CSCF), which acts as the first contact point to the terminal in the IMS (as a proxy). The CSCF covers several functionalities, from handling of the session states to being a contact point for all IMS connections intended for a single user and acting as a firewall towards other operator's networks.

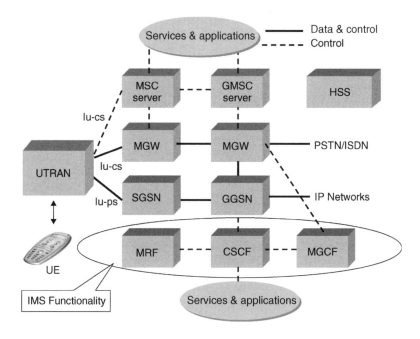

**Figure 5.13** Release 5 UMTS CN architecture

- MGW Control Function, to handle protocol conversions. This may also control a service coming via the CS domain and perform processing in an MGW, e.g. for echo cancellation.

An overview of the different elements and their interfaces can be found in [22] and further details of the CN protocols are given in [25].

## References

[1] 3GPP Technical Specification 25.401 UTRAN Overall Description.
[2] 3GPP Technical Specification 25.410 UTRAN Iu Interface: General Aspects and Principles.
[3] 3GPP Technical Specification 25.411 UTRAN Iu Interface: Layer 1.
[4] 3GPP Technical Specification 25.412 UTRAN Iu Interface: Signalling Transport.
[5] 3GPP Technical Specification 25.413 UTRAN Iu Interface: RANAP Signalling.
[6] 3GPP Technical Specification 25.414 UTRAN Iu Interface: Data Transport and Transport Signalling.
[7] 3GPP Technical Specification 25.415 UTRAN Iu Interface: CN-RAN User Plane Protocol.
[8] 3GPP Technical Specification 25.420 UTRAN Iur Interface: General Aspects and Principles.
[9] 3GPP Technical Specification 25.421 UTRAN Iur Interface: Layer 1.
[10] 3GPP Technical Specification 25.422 UTRAN Iur Interface: Signalling Transport.
[11] 3GPP Technical Specification 25.423 UTRAN Iur Interface: RNSAP Signalling.
[12] 3GPP Technical Specification 25.424 UTRAN Iur Interface: Data Transport and Transport Signalling for CCH Data Streams.
[13] 3GPP Technical Specification 25.425 UTRAN Iur Interface: User Plane Protocols for CCH Data Streams.
[14] 3GPP Technical Specification 25.426 UTRAN Iur and Iub Interface Data Transport and Transport Signalling for DCH Data Streams.
[15] 3GPP Technical Specification 25.427 UTRAN Iur and Iub Interface User Plane Protocols for DCH Data Streams.

[16] 3GPP Technical Specification 25.430 UTRAN Iub Interface: General Aspects and Principles.
[17] 3GPP Technical Specification 25.431 UTRAN Iub Interface: Layer 1.
[18] 3GPP Technical Specification 25.432 UTRAN Iub Interface: Signalling Transport.
[19] 3GPP Technical Specification 25.433 UTRAN Iub Interface: NBAP Signalling.
[20] 3GPP Technical Specification 25.434 UTRAN Iub Interface: Data Transport and Transport Signalling for CCH Data Streams.
[21] 3GPP Technical Specification 25.435 UTRAN Iub Interface: User Plane Protocols for CCH Data Streams.
[22] 3GPP Technical Specification 23.002 Network Architecture, Version 5.5.0, January 2002.
[23] 3GPP Technical Specification 25.419 UTRAN Iu Interface: Service Area Broadcast Protocol (SABP).
[24] 3GPP Technical Specification 25.450 UTRAN Iupc Interface: General Aspects and Principles.
[25] Kaaranen, H., Naghian, S., Laitinen, L., Ahtiainen, A. and Niemi, V., *UMTS Networks: Architecture, Mobility and Services*, Chichester: John Wiley & Sons, Ltd, 2001.

# 6

# Physical Layer

Antti Toskala

## 6.1   Introduction

In this chapter the Wideband Code Division Multiple Access (WCDMA – Universal Terrestrial Radio Access (UTRA) Frequency Division Duplex (FDD)) physical layer is described. The physical layer of the radio interface typically has been the main discussion topic when different cellular systems were compared. The physical layer structures naturally relate directly to the achievable performance issues, when observing a single link between a terminal station and a base station. For the overall system performance the protocols in the other layers, such as handover protocols, also have a great deal of impact. Naturally it is essential to have low Signal-to-Interference Ratio (SIR) requirements for sufficient link performance with various coding and diversity solutions in the physical layer, since the physical layer defines the fundamental capacity limits. The performance of the WCDMA physical layer is described in detail in Chapter 11.

The physical layer has a major impact on equipment complexity with respect to the required baseband processing power in the terminal station and base station equipment. As well as the diversity benefits on the performance side, the wideband nature of WCDMA also offers new challenges in its implementation. As third-generation systems are wideband from the service point of view as well, the physical layer cannot be designed around only a single service, such as speech; more flexibility is needed for future service introduction. The new requirements of third-generation systems for the air interface are summarized in Section 6.3. This chapter presents the WCDMA physical layer solutions to meet those requirements. This chapter uses the term 'terminal' for the user equipment. The UTRA FDD physical layer specifications are contained in [1–5].

This chapter has been structured as follows. First, the transport channels are described together with their mapping to different physical channels in Section 6.2. Spreading and modulation for uplink and downlink are presented in Section 6.3, and the physical channels for user data and control data are described in Sections 6.4 and 6.5. In Section 6.6, the key physical layer procedures, such as power control and handover measurements, are covered. The biggest change in Release 5 affecting the physical layer is the addition of the high-speed downlink packet access (HSDPA) feature. As there are significant differences in HSDPA when compared with Release 99-based operation (which is naturally retained as well),

*WCDMA for UMTS: HSPA Evolution and LTE, Fifth Edition*   Edited by Harri Holma and Antti Toskala
© 2010 John Wiley & Sons, Ltd

the HSDPA details are covered in a separate chapter to maintain clear separation between the first-phase WCDMA standard and the first evolution step of the radio interface development. For further details on HSDPA, refer to Chapter 12. From Release 6, the biggest impact for the physical layer was the high-speed uplink packet access (HSUPA), as covered in Chapter 13, and the beamforming on the network side has been made more complete by defining the related measurements in the network side. The introduction of Multimedia Broadcast Multicast Service (MBSM), as discussed in Chapter 14, has some similarities to the physical layer operation, though it uses the existing physical channels. For Releases 7, 8 and 9, further enhancements were included in the specification, and those are covered in Chapter 15, including higher-order modulation and discontinuous reception and transmission as part of the continuous packet connectivity topic as well as mapping the CS voice service on top of HSDPA and HSUPA channels.

## 6.2   Transport Channels and Their Mapping to the Physical Channels

In UTRA, the data generated at higher layers is carried over the air through transport channels, which are mapped in the physical layer to different physical channels. The physical layer is required to support variable bit-rate transport channels to offer bandwidth-on-demand services, and be able to multiplex several services to one connection. This section presents the mapping of the transport channels to the physical channels, and how those two requirements are taken into account in the mapping.

Each transport channel is accompanied by the Transport Format Indicator (TFI) at each time event at which data is expected to arrive for the specific transport channel from the higher layers. The physical layer combines the TFI information from different transport channels to the Transport Format Combination Indicator (TFCI). The TFCI is transmitted in the physical control channel to inform the receiver which transport channels are active for the current frame; the exception to this is the use of Blind Transport Format Detection (BTFD) that will be covered in connection with the Downlink Dedicated Channels (Downlink DCHs). The TFCI is decoded appropriately in the receiver and the resulting TFI is given to higher layers for each of the transport channels that can be active for the connection. In Figure 6.1, two transport channels are mapped to a single physical channel, and also the error indication is provided for

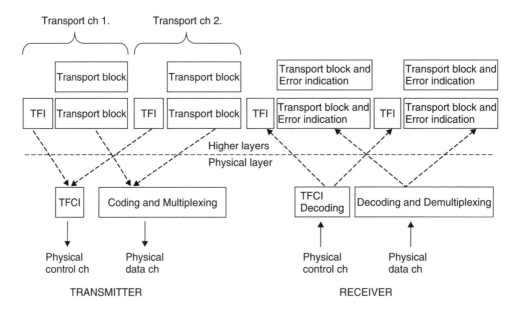

**Figure 6.1**   The interface between higher layers and the physical layer

each transport block. The transport channels may have a different number of blocks and at any moment not all the transport channels are necessarily active.

One physical control channel and one or more physical data channels form a single Coded Composite Transport Channel (CCTrCh). There can be more than one CCTrCh on a given connection, but only one physical layer control channel is transmitted in such a case.

The interface between higher layers and the physical layer is less relevant for terminal implementation, since basically everything takes place within the same equipment; thus, the interfacing here is rather a tool for specification work. For the network side, the division of functions between physical and higher layers is more important, since there the interface between physical and higher layers is represented by the Iub-interface between the base station and Radio Network Controller (RNC) as described in Chapter 5. In the 3GPP specification the interfacing between physical layer and higher layers is covered in [6].

Two types of transport channel exist: dedicated channels (DCHs) and common channels. The main difference between them is that a common channel is a resource divided between all or a group of users in a cell, whereas a DCH resource, identified by a certain code on a certain frequency, is reserved for a single user only. The transport channels are compared in Section 10.3 for the transmission of packet data.

## 6.2.1   Dedicated Transport Channel

The only dedicated transport channel is the DCH, which is the term used in the 25-series of the UTRA specification. The dedicated transport channel carries all the information intended for the given user coming from layers above the physical layer, including data for the actual service as well as higher layer control information. The content of the information carried on the DCH is not visible to the physical layer; thus, higher layer control information and user data are treated in the same way. Naturally, the physical layer parameters set by UTRAN may vary between control and data.

The familiar GSM channels, the traffic channel (TRCH) or associated control channel (ACCH), do not exist in the UTRA physical layer. The dedicated transport channel carries both the service data, such as speech frames, and higher layer control information, such as handover commands or measurement reports from the terminal. In WCDMA, a separate transport channel is not needed because of the support of variable bit rate and service multiplexing.

The dedicated transport channel is characterized by features such as fast power control, fast data rate change on a frame-by-frame basis, and the possibility of transmission to a certain part of the cell or sector with varying antenna weights with adaptive antenna systems. The DCH supports soft handover.

## 6.2.2   Common Transport Channels

There are six different common transport channel types defined for UTRA in Release 99, which are introduced in the following sections. There are a few differences from second-generation systems, e.g. transmission of packet data on the common channels and a downlink shared channel (DSCH) for transmitting packet data. Common channels do not have soft handover, but some of them can have fast power control. The new transport channel in Release 5, High-Speed DSCH (HS-DSCH) is covered in Chapter 12.

### 6.2.2.1   Broadcast Channel

The Broadcast Channel (BCH) is a transport channel that is used to transmit information specific to the UTRA network or for a given cell. The most typical data needed in every network is the available random access codes and access slots in the cell, or the types of transmit diversity methods used with other channels for that cell. As the terminal cannot register to the cell without the possibility of decoding the BCH, this channel is needed for transmission with relatively high power in order to reach all the users within the intended coverage area. From a practical point of view, the information rate on the BCH is limited by the ability of low-end terminals to decode the data rate of the BCH, resulting in a low and fixed data rate for the UTRA BCH.

### 6.2.2.2  Forward Access Channel

The Forward Access Channel (FACH) is a downlink transport channel that carries control information to terminals known to be located in the given cell. This is used, for example, after a random access message has been received by the base station. It is also possible to transmit packet data on the FACH. There can be more than one FACH in a cell. One of the FACHs must have such a low bit rate that it can be received by all the terminals in the cell area. With more than one FACH, the additional channels can have a higher data rate. The FACH does not use fast power control, and the messages transmitted need to include in-band identification information to ensure their correct receipt.

### 6.2.2.3  Paging Channel

The Paging Channel (PCH) is a downlink transport channel that carries data relevant to the paging procedure, i.e. when the network wants to initiate communication with the terminal. The simplest example is a speech call to the terminal: the network transmits the paging message to the terminal on the PCH of those cells belonging to the location area that the terminal is expected to be in. The identical paging message can be transmitted in a single cell or in up to a few hundred cells, depending on the system configuration. The terminals must be able to receive the paging information in the whole cell area. The design of the PCH also affects the terminal's power consumption in the standby mode. The less often the terminal has to tune the receiver in to listen for a possible paging message, the longer the terminal's battery will last in standby mode.

### 6.2.2.4  RACH

The RACH is an uplink transport channel intended to be used to carry control information from the terminal, such as requests to set up a connection. It can also be used to send small amounts of packet data from the terminal to the network. For proper system operation the RACH must be heard from the whole desired cell coverage area, which also means that practical data rates have to be rather low, at least for the initial system access and other control procedures. The coverage of the RACH compared with the DCH is presented in Section 13.2.

### 6.2.2.5  Uplink Common Packet Channel

The uplink common packet channel (CPCH) is an extension to the RACH channel that is intended to carry packet-based user data in the uplink direction. The reciprocal channel providing the data in the downlink direction is the FACH. In the physical layer, the main differences to the RACH are the use of fast power control, a physical layer-based collision-detection mechanism and a CPCH status monitoring procedure. The uplink CPCH transmission may last several frames, in contrast with one or two frames for the RACH message. As CPCH was not implemented in any of the networks, 3GPP decided to remove that from Release 5 onwards; thus, CPCH can only be found from Release 99 and Release 4 specifications.

### 6.2.2.6  DSCH

The DSCH is a transport channel intended to carry dedicated user data and/or control information; it can be shared by several users. In many respects it is similar to the FACH, though the shared channel supports the use of fast power control as well as variable bit rate on a frame-by-frame basis. The DSCH does not need to be heard in the whole cell area and can employ the different modes of transmit antenna diversity methods that are used with the associated downlink DCH. The DSCH is always associated with a downlink DCH. As DSCH was replaced in practice with HSDPA, 3GPP decided to take DSCH away from Release 5 specifications onwards.

### 6.2.2.7  Required Transport Channels

The common transport channels needed for basic network operation are RACH, FACH and PCH, while the use of DSCH and CPCH is optional and can be decided by the network.

## 6.2.3  Mapping of Transport Channels onto the Physical Channels

The different transport channels are mapped to different physical channels, though some of the transport channels are carried by identical (or even the same) physical channel. The transport channel to physical channel mapping is illustrated in Figure 6.2.

In addition to the transport channels introduced earlier, there exist physical channels to carry only information relevant to physical layer procedures. The Synchronization Channel (SCH), the Common Pilot Channel (CPICH) and the Acquisition Indication Channel (AICH) are not directly visible to higher layers and are mandatory from the system function point of view, to be transmitted from every base station. The CPCH Status Indication Channel (CSICH) and the Collision Detection/Channel Assignment Indication Channel (CD/CA-ICH) are needed if CPCH is used.

A DCH is mapped onto two physical channels. The Dedicated Physical Data Channel (DPDCH) carries higher layer information, including user data, while the Dedicated Physical Control Channel (DPCCH) carries the necessary physical layer control information. These two dedicated physical

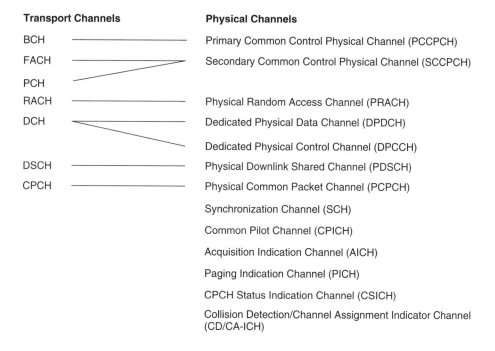

**Figure 6.2**   Transport-channel to physical-channel mapping

channels are needed to support efficiently the variable bit rate in the physical layer. The bit rate of the DPCCH is constant, while the bit rate of DPDCH can change from frame to frame.

### 6.2.4   Frame Structure of Transport Channels

UTRA channels use a 10 ms radio frame structure. The frame structure also employs a longer period, called the system frame period. The System Frame Number (SFN) is a 12-bit number, and is used by procedures that span more than a single frame. Physical layer procedures, such as the paging procedure or random access procedure, are examples of procedures that need a longer period than 10 ms for correct definition.

## 6.3   Spreading and Modulation

### 6.3.1   Scrambling

The concept of spreading the information in a CDMA system is introduced in Chapter 3. In addition to spreading, part of the process in the transmitter is the scrambling operation. This is needed to separate terminals or base stations from each other. Scrambling is used on top of spreading, so it does not change the signal bandwidth but only makes the signals from different sources separable from each other. With scrambling, it would not matter if the actual spreading were performed with identical codes for several transmitters. Figure 6.3 shows the relationship of the chip rate in the channel to spreading and scrambling in UTRA. As the chip rate is already achieved in spreading by the channelization codes, the symbol rate is not affected by the scrambling. The concept of channelization codes is covered in the following section.

### 6.3.2   Channelization Codes

Transmissions from a single source are separated by channelization codes, i.e. downlink connections within one sector and the dedicated physical channel in the uplink from one terminal. The spreading/channelization codes of UTRA are based on the Orthogonal Variable Spreading Factor (OVSF) technique, which was originally proposed in [7].

The use of OVSF codes allows the spreading factor to be changed and orthogonality between different spreading codes of different lengths to be maintained. The codes are picked from the code tree, which is illustrated in Figure 6.4. When a connection uses a variable spreading factor, the proper use of the code tree also allows despreading according to the smallest spreading factor. This requires only that channelization codes are used from the branch indicated by the code used for the smallest spreading factor.

There are certain restrictions as to which of the channelization codes can be used for a transmission from a single source. Another physical channel may use a certain code in the tree if no other physical

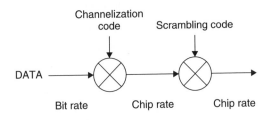

**Figure 6.3**   Relation between spreading and scrambling

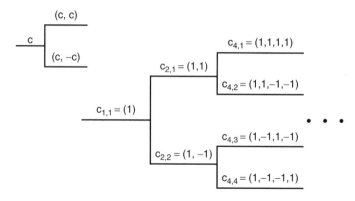

**Figure 6.4**  Beginning of the channelization code tree

channel to be transmitted using the same code tree is using a code that is on an underlying branch, i.e. using a higher spreading factor code generated from the intended spreading code to be used. Neither can a smaller spreading factor code on the path to the root of the tree be used. The downlink orthogonal codes within each base station are managed by the RNC in the network.

The functionality and characteristics of the scrambling and channelization codes are summarized in Table 6.1. Their usage will be described in more detail in Section 6.3.3.

The definition for the same code tree means that for transmission from a single source, from either a terminal or a base station, one code tree is used with one scrambling code on top of the tree. This means that different terminals and different base stations may operate their code trees totally independently of each other; there is no need to coordinate the code tree resource usage between different base stations or terminals.

**Table 6.1**  Functionality of the channelization and scrambling codes

|  | Channelization code | Scrambling code |
|---|---|---|
| Usage | Uplink: Separation of physical data (DPDCH) and control channels (DPCCH) from same terminal | Uplink: Separation of terminal |
|  | Downlink: Separation of downlink connections to different users within one cell | Downlink: Separation of sectors (cells) |
| Length | 4–256 chips (1.0–66.7 µs) | Uplink: (1) 10 ms = 38 400 chips or (2) 66.7 µs = 256 chips. Option (2) can be used with advanced base station receivers |
|  | Downlink also 512 chips | Downlink: 10 ms = 38 400 chips |
| Number of codes | Number of codes under one scrambling code = spreading factor | Uplink: several million |
|  |  | Downlink: 512 |
| Code family | OVSF | Long 10 ms code: Gold code Short code: Extended S(2) code family |
| Spreading | Yes, increases transmission bandwidth | No, does not affect transmission bandwidth |

## 6.3.3   Uplink Spreading and Modulation

### 6.3.3.1   Uplink Modulation

In the uplink direction there are basically two additional terminal-oriented criteria that need to be taken into account in the definition of the modulation and spreading methods. The uplink modulation should be designed so that the terminal amplifier efficiency is maximized and/or the audible interference from the terminal transmission is minimized.

Discontinuous uplink transmission can cause audible interference to audio equipment that is very close to the terminal, such as hearing aids. This is a completely separate issue from the interference in the air interface. The audible interference is only a nuisance for the user and does not affect network performance, such as its capacity. With GSM operation we are familiar with the occasional audible interference with audio equipment that is not properly protected. The interference from GSM has a frequency of 217 Hz, which is determined by the GSM frame frequency. This interference falls into the band that can be heard by the human ear. With a CDMA system, the same issues arise when discontinuous uplink transmission is used, e.g. with a speech service. During the silent periods no information bits need to be transmitted, only the information for link maintenance purposes, such as power control with a 1.5 kHz command rate. With such a rate the transmission of the pilot and the power control symbols with time multiplexing in the uplink direction would cause audible interference in the middle of the telephony voice frequency band. Therefore, in a WCDMA uplink the two dedicated physical channels are not time multiplexed, rather, in-phase (I)–quadrature (Q)/code multiplexing is used.

The continuous transmission achieved with an I–Q/code multiplexed control channel is shown in Figure 6.5. Now, as the pilot and the power control signaling are maintained on a separate continuous channel, no pulsed transmission occurs. The only pulse occurs when the data channel DPDCH is switched on and off, but such switching happens quite seldom. The average interference to other users and the cellular capacity remain the same as in the time-multiplexed solution. In addition, the link level performance is the same in both schemes if the energy allocated to the pilot and the power control signaling is the same.

For the best possible power amplifier efficiency, the terminal transmission should have as low peak-to-average ratio (PAR) as possible to allow the terminal to operate with a minimal amplifier back-off requirement, mapping directly to the amplifier power conversion efficiency, which in turn is directly proportional to the terminal talk time. With the I–Q/code multiplexing, called also dual-channel Quadrature Phase-Shift Keying (QPSK) modulation, the power levels of the DPDCH and DPCCH are typically different, especially as data rates increase and would lead in extreme cases to Binary Phase-Shift Keying (BPSK)-type transmission when transmitting the branches independently. This has been avoided by using a complex-valued scrambling operation after the spreading with channelization codes.

The signal constellation of the I–Q/code multiplexing before complex scrambling is shown in Figure 6.6. The same constellation is obtained after descrambling in the receiver for the data detection.

The transmission of two parallel channels, DPDCH and DPCCH, leads to multicode transmission, which increases the peak-to-average power ratio (crest factor). In Figure 6.6, the PAR changes when $G$ (the relative strengths of the DPDCH and DPCCH) is changed. By using the spreading modulation solution shown in Figure 6.7, the transmitter power amplifier efficiency remains the same as for normal balanced QPSK transmission in general. The complex scrambling codes are formed in such a way that

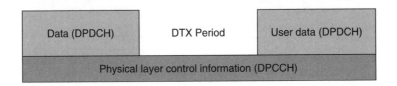

**Figure 6.5**   Parallel transmission of DPDCH and DPCCH when data is present/absent (DTX)

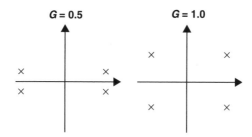

**Figure 6.6** Constellation of I–Q/code multiplexing before complex scrambling. $G$ denotes the relative gain factor between DPCCH and DPDCH branches

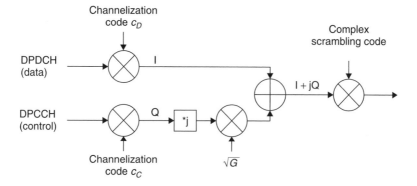

**Figure 6.7** I–Q/code multiplexing with complex scrambling

the rotations between consecutive chips within one symbol period are limited to $\pm 90°$. The full $180°$ rotation can happen only between consecutive symbols. This method further reduces the PAR of the transmitted signal from the normal QPSK transmission.

The efficiency of the power amplifier remains constant irrespective of the power difference $G$ between DPDCH and DPCCH. This can be explained with Figure 6.8, which shows the signal constellation for the I–Q/code multiplexed control channel with complex spreading. In the middle constellation with $G = 0.5$ the possible constellation points are only circles or only crosses during one symbol period. Their constellation is the same as for rotated QPSK. Thus, the signal envelope variations with complex

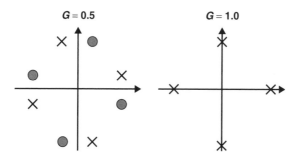

**Figure 6.8** Signal constellation for I–Q/code multiplexed control channel with complex scrambling. $G$ denotes the power difference between DPCCH and DPDCH

spreading are very similar to QPSK transmission for all values of $G$. The I–Q/code multiplexing solution with complex scrambling results in power amplifier output back-off requirements that remain constant as a function of the power difference between DPDCH and DPCCH.

The power difference between DPDCH and DPCCH has been quantified in UTRA physical layer specifications to 4-bit words, i.e. 16 different values. At a given point in time the gain value for either DPDCH or DPCCH is set to unity and then for the other channel a value between zero and one is applied to reflect the desired power difference between the channels. Limiting the number of possible values to 4-bit representation is necessary to make the terminal transmitter implementation simple. The power differences can have 15 different values between $-23.5\,dB$ and $0.0\,dB$ and one bit combination for no DPDCH when there is no data to be transmitted.

UTRA will face challenges in amplifier efficiency when compared with GSM. The GSM modulation is Gaussian Minimum Shift Keying (GMSK), which has a constant envelope and is thus optimized for amplifier PAR. As a narrowband system, the GSM signal can be spread relatively more widely in the frequency domain. This allows the use of a less linear amplifier with better power conversion efficiency. Narrowband amplifiers are also easier to linearize if necessary. In practice, the efficiency of a WCDMA power amplifier is slightly lower than that of the GSM power amplifier. On the other hand, WCDMA uses fast power control in the uplink, which reduces the average required transmission power.

Instead of applying combined I–Q and code multiplexing with complex scrambling, it would be possible to use pure code multiplexing. With code multiplexing, multicode transmission occurs with parallel control and data channels. This approach increases the transmitted signal envelope variations and sets higher requirements for power amplifier linearity. Especially for low bit rates, as for speech, the control channel can have an amplitude more than 50% of the data channel, which causes more envelope variations than the combined I–Q/code multiplexing solution.

### 6.3.3.2  Uplink Spreading

For the uplink DPCCH spreading code, there is an additional restriction. The same code cannot be used by any another code channel even on a different I or Q branch. The reason for this restriction is that physical channels transmitted with the same channelization codes on I and Q branches with the dual-channel QPSK principle cannot be separated before the DPCCH has been detected and channel phase estimates are available. This causes the restriction that, with multicode transmission for DPDCH, the number of possible parallel spreading codes to be allocated to DPDCH is six and not eight, when considering the spreading factor of 4 (which would be used in the case of DPDCH multicode transmission).

In the uplink direction the spreading factor on the DPDCH may vary on a frame-by-frame basis. The spreading codes are always taken from the code tree described earlier. When the channelization code used for spreading is always taken from the same branch of the code tree, the despreading operation can take advantage of the code tree structure and avoid chip-level buffering. The terminal provides data rate information, or more precisely the TFCI, on the DPCCH, to allow data detection with a variable spreading factor on the DPDCH.

### 6.3.3.3  Uplink Scrambling Codes

The transmissions from different sources are separated by the scrambling codes. In the uplink direction there are two alternatives: short and long scrambling codes. The long codes with 25-degree generator polynomials are truncated to the 10 ms frame length, resulting in 38400 chips with 3.84 Mcps. The short scrambling code length is 256 chips. The long scrambling codes are used if the base station uses a Rake receiver. The Rake receiver is described in Section 3.4. If advanced multiuser detectors or interference cancellation receivers are used in the base station, then short scrambling codes can be used to make the implementation of the advanced receiver structures easier. The base station multiuser detection algorithms are introduced in Section 11.5.2. Both of the two scrambling code families contain millions of scrambling codes; thus, in the uplink direction the code planning is not needed.

The short scrambling codes have been chosen from the extended S(2) code family. The long codes are Gold codes. The complex-valued scrambling sequence is formed in the case of short codes by combining two codes and in the case of long codes from a single sequence, where the other sequence is the delayed version of the first one.

The complex-valued scrambling code can be formed from two real-valued codes $c_1$ and $c_2$ with the decimation principle as

$$c_{\text{scrambling}} = c_1[w_0 + jc_2(2k)w_1], \quad k = 0, 1, 2 \tag{6.1}$$

with sequences $w_0$ and $w_1$ given as chip rate sequences:

$$w_0 = \{11\}, \quad w_1 = \{1 - 1\} \tag{6.2}$$

The decimation factor with the second code is 2. This way of creating the scrambling codes will reduce the zero crossings in the constellation and will further reduce the amplitude variations in the modulation process.

#### 6.3.3.4 Spreading and Modulation on Uplink Common Channels

The RACH contains preambles that are sent using the same scrambling code sequence as with the uplink transmission, the difference being that only 4096 chips from the beginning of the code period are needed and the modulation state transitions are limited in a different way. The spreading and scrambling process on the RACH is BPSK-valued; thus, only one sequence is used to spread and scramble both the I and Q branches. This has been chosen to reduce the complexity of the required matched filter in the base station receivers for RACH reception.

The RACH message part spreading and modulation, including scrambling, is identical to that for the DCH. The codes available for RACH scrambling use are transmitted on the BCH of each cell.

For the peak-to-average reduction, an additional rotation function is used on the RACH preamble, given as:

$$b(k) = a(k)e^{j(\pi 4 + \pi 2k)}, \quad k = 0, 1, 2, \ldots, 4095 \tag{6.3}$$

where $a(k)$ is the binary preamble and $b(k)$ is the resulting complex-valued preamble with limited 90° phase transition between chips. The autocorrelation properties are not affected by this operation.

The RACH preambles have a modulation pattern on top of them, called signature sequences. These have been defined by taking the higher Doppler frequencies as well as frequency errors into account. The sequences have been generated from 16 symbols, which have additionally been interleaved over the preamble duration to avoid large inter-sequence cross-correlations in case of large frequency errors that could otherwise severely degrade the cross-correlation properties between the signature sequences. The 16 signature sequences have been specified for RACH use, but there can be multiple scrambling codes each using the same set of signatures.

The CPCH spreading and modulation are identical to those of the RACH in order to maximize the commonality for both terminal and base station implementation when supporting CPCH. RACH and CPCH processes will be described in more detail in connection with the physical layer procedures.

### 6.3.4 Downlink Spreading and Modulation

#### 6.3.4.1 Downlink Modulation

In the downlink direction, normal QPSK modulation has been chosen with time-multiplexed control and data streams. The time-multiplexed solution is not used in the uplink because it would generate audible interference during discontinuous transmission (DTX). The audible interference generated with DTX is not a relevant issue in the downlink since the common channels have continuous transmission in any case. Also, as several parallel code transmissions exist in the downlink, similar optimization for PAR as with single code (pair) transmission is not relevant. Also, reserving a channelization code just for DPCCH purposes results in slightly worse code resource utilization when sending several transmissions from a single source.

Since the I and Q branches have equal power, the scrambling operation does not provide a similar difference to the envelope variations as in the uplink. The DTX is implemented by gating the transmission on and off.

### 6.3.4.2  Downlink Spreading

The spreading in the downlink is based on the channelization codes, as in the uplink. The code tree under a single scrambling code is shared by several users; typically, only one scrambling code, and thus only one code tree, is used per sector in the base station. The common channels and DCHs share the same code tree resource. There is one exception for the physical channels: the SCH, which is not under a downlink scrambling code. The SCH spreading codes are covered in a later section.

In the downlink, the DCH spreading factor does not vary on a frame-by-frame basis; the data rate variation is taken care of either with a rate-matching operation or with DTX, where the transmission is off during part of the slot.

In the case of multicode transmission for a single user, the parallel code channels have different channelization codes and are under the same scrambling code as normally are all the code channels transmitted from the base station. The spreading factor is the same for all the codes with multicode transmission. Each CCTrCh may have a different spreading factor even if received by the same terminal. As in the downlink normal QPSK modulation is used, the number of spreading codes available (under the same scrambling code) is equal to the spreading factor. If we consider the smallest spreading factor of 4, then at most four of those codes would be available, but due to the common channel requirements for code space, then at most three codes could be allocated for a particular terminal. The number of bits then would be roughly equal to the six codes possible for the uplink, as each QPSK symbol carries 2 bits.

The downlink codes are typically considered to be in the code tree. When a particular spreading factor is allocated, e.g. spreading factor 8, then the codes as part of the same branch (sub-tree) can no longer be used. Thus, from the code tree, booking a spreading factor 8 will occupy 1/8th of the total code tree resource (Figure 6.9). Note here the two fundamental differences between uplink and downlink. In the downlink direction the code tree is shared by the users of the cell or sector; in the uplink, as each user has

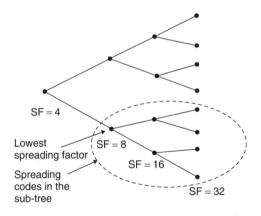

**Figure 6.9**  Code tree example

their own scrambling code, they all have independent code trees. Further, in the uplink there are separate code trees on I and Q branches, whereas in a downlink with QPSK modulation there is only one code tree (but each symbol was then carrying 2 bits).

### 6.3.4.3  Downlink Scrambling

The downlink scrambling uses long codes, the same Gold codes as in the uplink. The complex-valued scrambling code is formed from a single code by simply having a delay between the I and Q branches. The code period is truncated to 10 ms; no short codes are used in the downlink direction. The downlink set of the (primary) scrambling codes is limited to 512 codes, otherwise the cell search procedure described in Section 6.6.5 would become too excessive. The scrambling codes must be allocated to the sectors in the network planning. Because the number of scrambling codes is so high, the scrambling code planning is a trivial task and can be done automatically by the network planning tool. The 512 primary scrambling codes are expected to be enough from the cell planning perspective, especially as the secondary scrambling codes can be used in the case of beam steering as used on DCHs. This allows the capacity to evolve with adaptive antenna techniques without consuming extra primary scrambling codes and causing problems for downlink code planning.

The actual code period is very long with the 18-degree code generator, but only the first 38 400 chips are used. Limiting the code period was necessary from the system perspective: the terminals would have difficulty in finding the correct code phase with a code period spanning several frames and 512 different codes to choose from.

The secondary downlink scrambling codes can be applied with the exception of those common channels that need to be heard in the whole cell and/or prior to the initial registration. Only one scrambling code should be used per cell or sector to maintain the orthogonality between different downlink code channels. With adaptive antennas the beams provide additional spatial isolation and the orthogonality between different code channels is less important. However, in all cases the best strategy is still to keep as many users as possible under a single scrambling code to minimize downlink interference. If a secondary scrambling code needs to be introduced in the cell, then only those users not fitting under the primary scrambling code should use the secondary code. The biggest loss in orthogonality occurs when the users are shared evenly between two different scrambling codes.

### 6.3.4.4  SCH Spreading and Modulation

The downlink SCH is a special type of physical channel that is not visible above the physical layer. It contains two channels, primary and secondary SCHs. These channels are utilized by the terminal to find the cells, and are not under the cell-specific primary scrambling code. The terminal must be able to synchronize to the cell before knowing the downlink scrambling code.

The primary SCH contains a code word with 256 chips, with an identical code word in every cell. The primary SCH code word is sent without modulation on top. The code word is constructed from shorter 16-chip sequences in order to optimize the required hardware at the terminal. When detecting this sequence there is normally no prior timing information available and, typically, a matched filter is needed for detection. Therefore, for terminal complexity and power consumption reasons, it was important to optimize this synchronization sequence for low-complexity matched filter implementation.

The secondary SCH code words are similar sequences, but they vary from one base station to another, with a total of 16 sequences in use. These 16 sequences are used to generate a total of 64 different code words that identify to which of the 64 code groups a base station belongs. Like the primary SCH, the secondary SCH is not under the base-station-specific scrambling code, but the code sequences are sent without scrambling on top. The SCH code words contain modulation to indicate the use of open-loop transmit diversity on the BCH. The SCH itself can use time-switched transmit antenna diversity (TSTD) and is the only channel in UTRA FDD that uses TSTD.

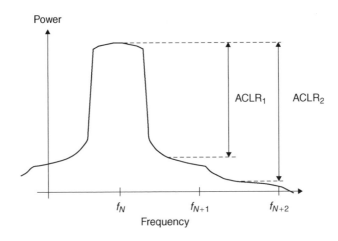

**Figure 6.10**  ACLR for the first and second adjacent carriers

## 6.3.5  Transmitter Characteristics

The pulse shaping method applied to the transmitted symbols is root-raised cosine filtering with a roll-off factor of 0.22. The same roll-off is valid for both the terminals and the base stations. The key radio-frequency (RF) parameter impacts to terminal design are introduced in Chapter 20 and have a significant impact not only on the implementation, but also on the system performance.

The nominal carrier spacing in WCDMA is 5 MHz, but the carrier frequency in WCDMA can be adjusted with a 200 kHz raster. The central frequency of each WCDMA carrier is indicated with an accuracy of 200 kHz. The target of this adjustment is to provide more flexibility for channel spacing within the operator's band. This is also an important possibility when considering sharing the spectrum with GSM, for example, where one does not necessary need full 5 MHz reservation for WCDMA.

The Adjacent Channel Leakage Ratio (ACLR) determines how much of the transmitted power is allowed to leak into the first or second neighboring carrier, as shown in Figure 6.10.

In the Base Transceiver Station (BTS) transmitter the ACLR values have been set to 45 dB and 50 dB for $ACLR_1$ and $ACLR_2$ respectively. The impact of the ACLR on system performance is studied in Section 8.5.

The terminal frequency accuracy requirement has been defined to be $\pm 0.1$ ppm when compared with the received carrier frequency. On the base station side, the requirement is tighter: $\pm 0.05$ ppm. The baseband timing is tied to the same timing reference as RF. The base station value needs to be tighter than the terminal value, since the base station carrier frequency is the reference for the terminal accuracy. The terminal needs also to be able to search the total frequency uncertainty area caused by the base station frequency error tolerance on top of the terminal tolerances and the error caused by terminal movement. With the 200 kHz carrier raster, the looser base station frequency accuracy would start to cause problems. In 3GPP, the RF parameters for terminals are specified in [8] and for base stations in [9].

## 6.4  User Data Transmission

For user data transmission in second-generation systems, such as the first versions of GSM, typically only one service has been active at a time: either voice or low-rate data. From the beginning, the technology base has required that the physical layer implementation be defined to the last detail without real flexibility. For example, puncturing patterns in GSM have been defined bit by bit, whereas such a definition for all possible service combinations and data rates is simply not possible for UTRA. Instead, algorithms for generating such patterns are defined. Signal processing technology has also evolved greatly; thus, there is no longer a need to have items like puncturing on hardware as in the early days of GSM hardware development. For the circuit-switched traffic (e.g. speech and video), a transmission-DCH needs to be used, whereas for packet data there are additional choices available: RACH and CPCH for the uplink and FACH and DSCH respectively for the downlink.

### 6.4.1 Uplink Dedicated Channel

As described earlier, the uplink direction uses I–Q/code multiplexing for user data and physical layer control information. The physical layer control information is carried by the DPCCH with a fixed spreading factor of 256. The higher layer information, including user data, is carried on one or more DPDCHs, with a possible spreading factor ranging from 256 down to 4. The uplink transmission may consist of one or more DPDCHs with a variable spreading factor, and a single DPCCH with a fixed spreading factor.

The DPDCH data rate may vary on a frame-by-frame basis. Typically, with a variable rate service, the DPDCH data rate is informed on the DPCCH. The DPCCH is transmitted continuously and rate information is sent with TFCI, the DPCCH information on the data rate on the current DPDCH frame. If the TFCI is not decoded correctly, then the whole data frame is lost. Because the TFCI indicates the transport format of the same frame, the loss of the TFCI does not affect any other frames. The reliability of the TFCI is higher than the reliability of the user data detection on the DPDCH. Therefore, the loss of the TFCI is a rare event. Figure 6.11 illustrates the uplink DCH structure in more detail.

The uplink DPCCH uses a slot structure with 15 slots over the 10 ms radio frame. This results in a slot duration of 2560 chips or about 666 µs. This is actually rather close to the GSM burst duration of 577 µs. Each slot has four fields to be used for pilot bits, TFCI, Transmission Power Control (TPC) bits and Feedback Information (FBI) bits. The pilot bits are used for the channel estimation in the receiver, and the TPC bits carry the power control commands for the downlink power control. The FBI bits are used when closed-loop transmission diversity is used in the downlink. The use of FBI bits is covered in Section 6.6.6. A total of six slot structures exist for uplink DPCCH. The different options are 0, 1 or 2 bits for FBI bits and these same alternatives with and without TFCI bits. The TPC and pilot bits are always present and their number varies in such a way that the DPCCH slot is always fully used.

It is beneficial to transmit with a single DPDCH for as long as possible, for reasons of terminal amplifier efficiency, because multicode transmission increases the PAR of the transmission, which reduces the efficiency of the terminal power amplifier. The maximum user data rate on a single code is derived

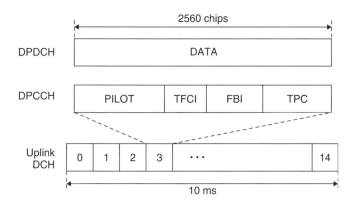

**Figure 6.11**  Uplink DCH structure

**Table 6.2** Uplink DPDCH data rates

| DPDCH spreading factor | DPDCH channel bit rate (kbps) | Max. user data rate with 1/2-rate coding (approx.) |
|---|---|---|
| 256 | 15 | 7.5 kbps |
| 128 | 30 | 15 kbps |
| 64 | 60 | 30 kbps |
| 32 | 120 | 60 kbps |
| 16 | 240 | 120 kbps |
| 8 | 480 | 240 kbps |
| 4 | 960 | 480 kbps |
| 4, with 6 parallel codes | 5740 | 2.8 Mbps |

from the maximum channel bit rate, which is 960 kbps with spreading factor 4. With channel coding, the practical maximum user data rate for the single code case is in the order of 400–500 kbps.

When higher data rates are needed, parallel code channels are used. This allows up to six parallel codes to be used (as explained in Section 6.3.3.2), raising the channel bit rate for data transmission up to 5740 kbps, which can accommodate 2 Mbps user data or an even higher data rate if the coding rate is 1/2. Therefore, it is possible to offer a user data rate of 2 Mbps even after retransmission. The achievable data rates with different spreading factors are presented in Table 6.2. The rates given assume 1/2-rate coding and do not include bits taken for coder tail bits or the Cyclic Redundancy Check (CRC). The relative overhead due to tail bits and CRC bits has significance only with low data rates.

The uplink receiver in the base station needs to perform typically the following tasks when receiving the transmission from a terminal:

- The receiver starts receiving the frame and despreading the DPCCH and buffering the DPDCH according to the maximum bit rate, corresponding to the smallest spreading factor.
- For every slot:
  - obtain the channel estimates from the pilot bits on the DPCCH;
  - estimate the SIR from the pilot bits for each slot;
  - send the TPC command in the downlink direction to the terminal to control its uplink transmission power;
  - decode the TPC bit in each slot and adjust the downlink power of that connection accordingly.
- For every second or fourth slot:
  - decode the FBI bits, if present, over two or four slots and adjust the diversity antenna phases, or phases and amplitudes, depending on the transmission diversity mode.
- For every 10 ms frame:
  - decode the TFCI information from the DPCCH frame to obtain the bit rate and channel decoding parameters for DPDCH.
- For Transmission Time Intervals (TTIs, interleaving period) of 10, 20, 40 or 80 ms:
  - decode the DPDCH data.

The same functions are valid for the downlink as well, with the following exceptions:

- In the downlink the DCH spreading factor is constant, as well as with the common channels. The only exception is the DSCH, which also has a varying spreading factor.
- The FBI bits are not in use in the downlink direction.
- There is a CPICH available in addition to the pilot bits on DPCCH. The common pilot can be used to aid the channel estimation.

- In the downlink, transmission may occur from two antennas in the case of transmission diversity. The receiver does the channel estimation from the pilot patterns sent from two antennas and, consequently, accommodates the despread data sent from two different antennas. The overall impact on the complexity is small, however.

## 6.4.2  Uplink Multiplexing

In the uplink direction the services are multiplexed dynamically so that the data stream is continuous with the exception of zero rate. The symbols on the DPDCH are sent with equal power level for all services. This means, in practice, that the service coding and channel multiplexing needs in some cases to adjust the relative symbol rates for different services in order to balance the power level requirements for the channel symbols. The rate-matching function in the multiplexing chain in Figure 6.12 can be used for such quality balancing operations between services on a single DPDCH. For the uplink DPDCH, fixed positions for different services do not exist, but the frame is filled according to the outcome of the rate matching and interleaving operation(s). The uplink multiplexing is done in 11 steps, as illustrated in Figure 6.12.

After receiving a transport block from higher layers, the first operation is CRC attachment. The CRC is used for error checking of the transport blocks at the receiving end. The CRC length that can be inserted has four different values: 8, 12, 16 and 24 bits. The more bits the CRC contains, the lower is the probability of an undetected error in the transport block in the receiver. The physical layer provides the transport block to higher layers, together with the error indication from the CRC check.

After the CRC attachment, the transport blocks are either concatenated or segmented to different coding blocks. This depends on whether the transport block fits the available code block size as defined for the channel coding method. The benefit of the concatenation is better performance in terms of lower

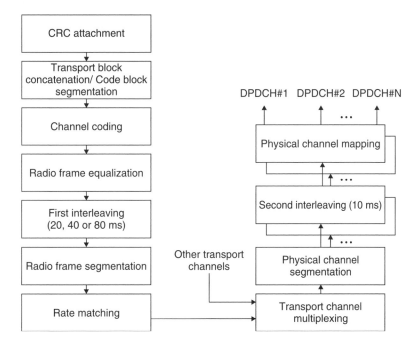

**Figure 6.12**   Uplink multiplexing and channel coding chain

overhead due to encoder tail bits and in some cases due to better channel coding performance because of the larger block size. On the other hand, code block segmentation allows the avoidance of excessively large code blocks that could also be a complexity issue. If the transport block with CRC attached does not fit into the maximum available code block, then it will be divided into several code blocks.

The channel encoding is performed on the coding blocks after the concatenation or segmentation operation. Originally it was considered to have the possibility to send data without any channel coding, as is done with AMR class C bits in GSM, but that was removed at a later stage as there was no real need identified.

The function of radio frame equalization is to ensure that data can be divided into equal-sized blocks when transmitted over more than a single 10 ms radio frame. This is done by padding the necessary number of bits until the data can be in equal-sized blocks per frame.

The first interleaving or inter-frame interleaving is used when the delay budget allows more than 10 ms of interleaving. The interlayer length of the first interleaving has been defined to be 20, 40 and 80 ms. The interleaving period is directly related to the TTI, which indicates how often data arrives from higher layers to the physical layer. The start positions of the TTIs for different transport channels multiplexed together for a single connection are time aligned. The TTIs have a common starting point, i.e. a 40 ms TTI goes in twice, even for an 80 ms TTI on the same connection. This is necessary to limit the possible transport format combinations from the signaling perspective. The timing relation with different TTIs is illustrated in Figure 6.13. If the first interleaving is used, then the frame segmentation will distribute the data coming from the first interleaving over two, four or eight consecutive frames in line with the interleaving length.

Rate matching is used to match the number of bits to be transmitted to the number available on a single frame. This is achieved either by puncturing or by repetition. In the uplink direction, repetition is preferred, and basically the only reason why puncturing is used is when facing the limitations of the terminal transmitter or base station receiver. Another reason for puncturing is to avoid multicode transmission. The rate-matching operation in Figure 6.12 needs to take into account the number of bits coming from the other transport channels that are active in that frame. The uplink rate matching is a dynamic operation that may vary on a frame-by-frame basis. When the data rate of the service with lowest TTI varies as in Figure 6.13, the dynamic rate matching adjusts the rate-matching parameters for other transport channels as well, so that all the symbols in the radio frame are used. For example, if with two transport channels the other momentarily has zero rate, then rate matching increases the symbol rate for the other service sufficiently so that all uplink channel symbols are used, assuming that the spreading factor would stay the same.

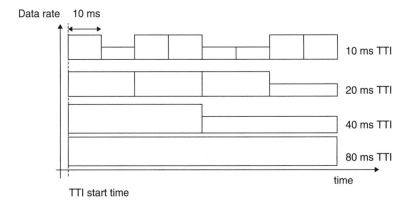

**Figure 6.13**    TTI start time relation with different TTIs on a single connection

The higher layers provide a semi-static parameter, the rate-matching attribute, to control the relative rate matching between different transport channels. This is used to calculate the rate-matching value when multiplexing several transport channels for the same frame. When this rule is applied as specified, with the aid of the rate-matching attribute and TFCI the receiver can calculate backwards the rate-matching parameters used and perform the inverse operation. By adjusting the rate-matching attribute, the quality of different services can be fine-tuned to reach an equal or near-equal symbol power level requirement.

The different transport channels are multiplexed together by the transport channel multiplexing operation. This is a simple serial multiplexing on a frame-by-frame basis. Each transport channel provides data in 10 ms blocks for this multiplexing. When more than one physical channel (spreading code) is used, physical channel segmentation is used. This operation simply divides the data evenly on the available spreading codes, as currently no cases have been specified where the spreading factors would be different in multicode transmissions. The use of serial multiplexing also means that with multicode transmission the lower rates can be implemented by sending fewer codes than with the full rate.

The second interleaving performs 10 ms radio-frame interleaving, sometimes called intra-frame interleaving. This is a block interleaver with intercolumn permutations applied to the 30 columns of the interleaver. It is worth noting that the second interleaving is applied separately for each physical channel, in case more than a single code channel is used. From the output of the second interleaver the bits are mapped on the physical channels. The number of bits given for a physical channel at this stage is exactly the number that the spreading factor of that frame can transmit. Alternatively, the number of bits to transmit is zero and the physical channel is not transmitted at all.

## 6.4.3 User Data Transmission with the Random Access Channel

In addition to the uplink DCH, user data can be sent on the RACH, mapped on the Physical RACH (PRACH). This is intended for low data-rate operation with packet data where continuous connection is not maintained. In the RACH message it will be possible to transmit with a limited set of data rates based on prior negotiations with the UTRA network. The RACH operation does not include power control; thus, the validity of the power level obtained with the PRACH power ramping procedure will be valid only for a short period, over one or two frames at most, depending on the environment.

The PRACH has, as a specific feature, preambles that are sent prior to data transmission. These use a spreading factor of 256 and contain a signature sequence of 16 symbols, resulting in a total length of 4096 chips for the preamble. Once the preamble has been detected and acknowledged with the Acquisition Indicator Channel (AICH), the 10 ms (or 20 ms) message part is transmitted. The spreading factor for the message part may vary from 256 to 32, depending on the transmission needs, but is subject to prior agreement with the UTRA network. Additionally, the 20 ms message length has been defined for range improvement reasons; this is studied in detail in Section 11.2.2. The AICH structure is covered in Section 6.5.6, whereas the RACH procedure is covered in detail in Section 6.6.4.

## 6.4.4 Uplink Common Packet Channel

As well as the previously covered user data transmission methods, an extension for RACH was defined in Release 99. The main differences in the uplink from RACH data transmission are the reservation of the channel for several frames and the use of fast power control, which is not needed with RACH when sending only one or two frames. The uplink CPCH has as a pair the DPCCH in the downlink direction, providing fast power control information.

The higher layer downlink signaling to a terminal using uplink CPCH is provided by the FACH. In case the RNC wants to send a signaling message for a terminal as a response to CPCH activity, e.g. an ARQ message, the CPCH connection might have already been terminated by the base station. CPCH was removed from Release 5 onwards from 3GPP specifications due to lack of interest in the market place; thus, it is not covered in further detail.

## 6.4.5 Downlink Dedicated Channel

The Downlink DCH is transmitted on the Downlink Dedicated Physical Channel (Downlink DPCH). The Downlink DPCH applies time multiplexing for physical control information and user data transmission, as shown in Figure 6.14. As in the uplink, the terms DPDCH and DPCCH are used in the 3GPP specification for the Downlink DCHs.

The spreading factor for the highest transmission rate determines the channelization code to be reserved from the given code tree. The variable data rate transmission may be implemented in two ways:

- If TFCI is not present, the positions for the DPDCH bits in the frame are fixed. As the spreading factor is also always fixed in the Downlink DPCH, the lower rates are implemented with DTX by gating the transmission on/off. Since this is done on the slot interval, the resulting gating rate is 1500 Hz. As in the uplink, there are 15 slots per 10 ms radio frame; this determines the gating rate. The data rate, if there is more than one alternative, is determined with BTFD, which is based on the use of a guiding transport channel or channels that have different CRC positions for different Transport Format Combinations (TFCs). For a terminal it is mandatory to have BTFD capability with relatively low rates only, such as with AMR speech service. With higher data rates, the benefits from avoiding the TFCI overhead are also insignificant and the complexity of BTFD rates starts to increase.
- With TFCI available it is also possible to use flexible positions, and it is up to the network to select which mode of operation is used. With flexible positions it is possible to keep continuous transmission and implement the DTX with repetition of the bits. In such a case the frame is always filled as in the uplink direction.
- The downlink multiplexing chain in Figure 6.15 (Section 6.4.6) is also impacted by the DTX, the DTX indication having been inserted before the first interleaving.

In the downlink the spreading factors range from 4 to 512, with some restrictions on the use of spreading factor 512 in the case of soft handover. The restrictions are due to the timing adjustment step of 256 chips in soft handover operation; but in any case, the use of a spreading factor of 512 for soft handover is not expected to occur very often. Typically, such a spreading factor is used to provide information on power control, etc., when providing services with minimal downlink activity, as with file uploading and so on.

Modulation causes some differences between the uplink and downlink data rates. While the uplink DPDCH consists of BPSK symbols, the downlink DPDCH consists of QPSK symbols each carrying 2 bits. As the BPSK symbols carry only 1 bit per symbol, use of same spreading ratio in uplink and

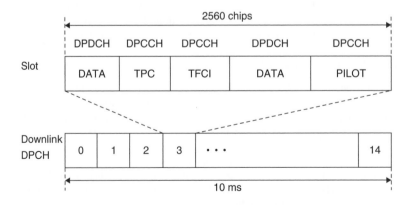

**Figure 6.14**   Downlink DPCH control/data multiplexing

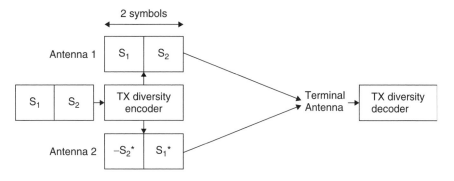

**Figure 6.15**   Open-loop transmit diversity encoding

**Table 6.3**   Downlink DCH symbol and bit rates

| Spreading factor | Channel symbol rate (kbps) | Channel bit rate (kbps) | DPDCH channel bit rate range (kbps) | Max. user data rate with 1/2-rate coding (approx.) |
|---|---|---|---|---|
| 512 | 7.5 | 15 | 3–6 | 1–3 kbps |
| 256 | 15 | 30 | 12–24 | 6–12 kbps |
| 128 | 30 | 60 | 42–51 | 20–24 kbps |
| 64 | 60 | 120 | 90 | 45 kbps |
| 32 | 120 | 240 | 210 | 105 kbps |
| 16 | 240 | 480 | 432 | 215 kbps |
| 8 | 480 | 960 | 912 | 456 kbps |
| 4 | 960 | 1920 | 1872 | 936 kbps |
| 4, with 3 parallel codes | 2880 | 5760 | 5616 | 2.8 Mbps |

downlink DPDCH gives a double data rate in the downlink direction, especially at higher data rates where time-multiplexed DPCCH overhead is very small. These downlink data rates are given in Table 6.3 with raw bit rates calculated from the QPSK-valued symbols in the downlink reserved for data use.

The Downlink DPCH can use either open-loop or closed-loop transmit diversity to improve performance. The use of such enhancements is not required from the network side, but is mandatory in terminals. It was made mandatory as it was felt that this kind of feature has a strong relation to such issues as network planning and system capacity, so it was made a baseline implementation capability. The open-loop transmit diversity coding principle is shown in Figure 6.16, where the information is coded to be sent from two antennas. The method is also denoted in the 3GPP specification as space–time block-coding-based transmit diversity. Another possibility is to use feedback mode transmit diversity, where the signal is sent from two antennas based on the FBI from the terminal. The feedback mode uses phase offsets between the antennas. The feedback mode of transmit diversity is covered in Section 6.6.6.

## 6.4.6   Downlink Multiplexing

The multiplexing chain in the downlink is mainly similar to that in the uplink, but there are also some functions that are done differently.

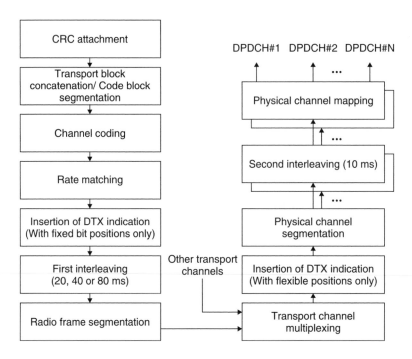

**Figure 6.16**  Downlink multiplexing and channel coding chain

As in the uplink, the interleaving is implemented in two parts, covering both intra-frame and inter-frame interleaving. Also, the rate matching allows one to balance the required channel symbol energy for different service qualities. The services can be mapped to more than one code as well, which is necessary if the single code capability in either the terminal or base station is exceeded.

There are differences in the order in which rate matching and segmentation functions are performed, as shown in Figure 6.16. Whether fixed or flexible bit positions are used determines the DTX indication insertion point. The DTX indication bits are not transmitted over the air; they are just inserted to inform the transmitter at which bit positions the transmission should be turned off. They were not needed in the uplink where the rate matching was done in a more dynamic way, always filling the frame when there was something to transmit on the DPDCH.

The use of fixed positions means that, for a given transport channel, the same symbols are always used. If the transmission rate is below the maximum, then DTX indication bits are used for those symbols. The different transport channels do not have a dynamic impact on the rate-matching values applied for another channel, and all transport channels can use the maximum rate simultaneously as well. The use of fixed positions is partly related to the possible use of blind rate detection. When a transport channel always has the same position regardless of the data rate, the channel decoding can be done with a single decoding 'run' and the only thing that needs to be tested is which position of the output block is matched with the CRC check results. This naturally requires that different rates have different numbers of symbols.

With flexible positions the situation is different, since now the channel bits unused by one service may be utilized by another service. This is useful when it is possible to have such a transport channel combination that they do not all need to be able to reach the full data rate simultaneously, but can alternate with the need for full rate transmission. This allows the necessary spreading code occupancy in the downlink to be reduced. The concept of flexible versus fixed positions in the downlink is illustrated in

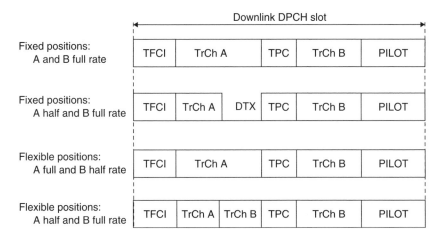

**Figure 6.17**    Flexible and fixed transport channel slot positions in the downlink

Figure 6.17. The use of blind rate detection is also possible in principle with flexible positions, but is not required by the specifications. If the data rate is not too high and the number of possible data rates is not very high, then the terminal can run channel decoding for all the combinations and check which of the cases comes out with the correct CRC result.

## 6.4.7    Downlink Shared Channel

Transmitting data with high peak rate and low activity cycle in the downlink quickly causes the channelization codes under a single scrambling code to start to run out. To avoid this problem, in Release 99 3GPP introduced the DSCH to be operated together with a Downlink DCH. The DCH provided an indication to the terminal when it has to decode the DSCH and which spreading code from the DSCH it has to despread. Thus, code-tree capacity was shared on a frame-by-frame basis. In Release 99, time DSCH was expected to be the approach for supporting 2 Mbps, but it became unnecessary with Release 5 HSDPA introduction, which allowed one to have the same benefits, such as dynamic code sharing, but enabled further improved performance. Thus, DSCH was removed from the 3GPP specification from Release 5 onwards.

## 6.4.8    Forward Access Channel for User Data Transmission

The FACH can be used for transmission of user (packet) data. The channel is typically multiplexed with the PCH to the same physical channel, but can exist as a standalone channel as well. The main difference with the dedicated and shared channels is that FACH does not allow the use of fast power control and applies either slow power control or no power control at all. Slow power control is possible if a lot of data is transmitted between the base station and the terminal and the latter provides feedback on the quality of the received packets. This type of power control cannot combat the effect of the fading channel, but combats more the longer-term changes in the propagation environment. For less frequent transmission, FACH needs to use more or less the full power level. The power control for FACH is also typically very slow, since the FACH data transmission is controlled by RNC, which means rather a large delay for any FBI from the base station.

Whether the FACH contains pilot symbols or not depends on whether it applies beamforming techniques. Normally, FACH does not contain pilot symbols and the receiver uses the CPICH as phase reference.

As FACH needs to be received by all terminals, the primary FACH cannot use high data rates. If higher data rates were desired of FACH, then this would require a separate physical channel where only the capabilities in terms of maximum data rates of those terminals allocated to that channel need to be taken into account. The necessary configuration would become rather complicated when terminals with different capabilities are included. The FACH has a fixed spreading factor, and reserving FACH for very high data rates is not optimized from the code resource point of view, especially if not all the terminals can decode the high data rate FACH.

Messages on FACH normally need in-band signaling to tell for which user the data was intended. In order to read such information, the terminal must decode FACH messages first. Running such decoding continuously is not desirable due to power consumption, especially with higher FACH rates.

## 6.4.9   Channel Coding for User Data

In UTRA, two channel-coding methods have been defined. Half-rate and 1/3-rate convolutional coding are intended to be used with relatively low data rates, equivalent to the data rates provided by second-generation cellular networks today, though an upper limit has not been specified. For higher data rates, 1/3-rate turbo coding can be applied and typically brings performance benefits when large enough block sizes are achieved. It has been estimated that roughly 300 bits should be available per TTI in order to give turbo coding some gain over convolutional coding. This also depends on the required quality level and operational environment.

The convolutional coding is based on constraint length 9 coding with the use of tail bits. The selected turbo encoding/decoding method is 8-state parallel concatenated convolutional code. The main motivation for turbo coding for higher bit rates has been performance, while for low rates the main reason not to use it has been both low rate or low block length performance and the desire to allow the use of simple blind rate detection with low-rate services such as speech. Blind rate detection with turbo coding typically requires detection of all transmission rates, while with convolutional coding trial methods can allow only a single Viterbi pass for determining which transmission rate was used. This is performed together with the help of CRC and applying a proper interleaving technique.

Turbo coding has specific interleaving that has been designed with a large variety of data rates in mind. The maximum turbo coding block size has been limited to 5114 information bits, since after that block size only the memory requirements increase and there is no significant effect on the performance side observed. For the higher amount of data per interleaving period, several blocks are used, with a block size as equal as possible at or below 5114 bits. The actual block size for data is a little smaller, since the tail bits and CRC bits are to be accommodated in the block size.

The minimum block size for turbo coding was initially defined to be 320 bits, which corresponds to 32 kbps with 10 ms interleaving or down to 4 kbps with 80 ms interleaving. The possible range of block sizes was, however, extended down to 40 bits, since with variable rate connection it is not desirable to change the codec on the fly when coming down from the maximum rate. Nor may a transport channel change the channel coding method on a frame-by-frame basis. Data rates below 40 bits can be transmitted with turbo coding as well, but in such a case padding with dummy bits is used to fill the 40 bits minimum size interleaver.

With speech service, AMR coding uses an unequal error protection scheme. This means that the three different classes of bits have different protection. Class A bits (those that contribute the most to voice quality) have the strongest protection, whereas class C bits are sent with weaker channel coding. This gives around 1 dB gain in $E_b/N_0$ compared with the equal error protection scheme. The coding methods usable by different channels are summarized in Table 6.4. Although the FACH has two options given, the cell access use of FACH is based on convolutional coding, as not all terminals support turbo coding.

**Table 6.4**   Channel coding options with different channels

| | |
|---|---|
| DCH | Turbo coding or convolutional coding |
| FACH | Turbo coding or convolutional coding |
| Other common channels | $\frac{1}{2}$-rate convolutional coding |

### 6.4.10   Coding for TFCI Information

The TFCI may carry from 1 to 10 bits of transport format information. The coding in the normal mode is second-order Reed–Muller code punctured from 32 bits to 30 bits, carrying up to 10 bits of information. The TFCI coding is illustrated in Figure 6.18.

## 6.5   Signaling

For signaling purposes, a lot of information needs to be transmitted between the network and the terminals. The following sections describe the methods used for transmitting signaling messages generated above the physical layer, as well as the required physical layer control channels needed for system operation but not necessarily visible for higher layer functionality.

### 6.5.1   Common Pilot Channel (CPICH)

The CPICH is an unmodulated code channel, which is scrambled with the cell-specific primary scrambling code. The function of the CPICH is to aid the channel estimation at the terminal for the DCH and to provide the channel estimation reference for the common channels when they are not associated with the DCHs or not involved in the adaptive antenna techniques.

   UTRA has two types of CPICH: primary and secondary. The difference is that the Primary CPICH is always under the primary scrambling code with a fixed channelization code allocation and there is only one such channel for a cell or sector. The Secondary CPICH may have any channelization code of length 256 and may be under a secondary scrambling code as well. The typical area of Secondary CPICH usage would be operations with narrow antenna beams intended for service provision at specific 'hot spots' or places with high traffic density.

   An important area for the Primary CPICH is the measurements for the handover and cell selection/reselection. The use of CPICH reception level at the terminal for handover measurements has the consequence that, by adjusting the CPICH power level, the cell load can be balanced between different cells. Reducing the CPICH power causes part of the terminals to hand over to other cells, while increasing it invites more terminals to hand over to the cell as well as to make their initial access to the network in that cell.

   The CPICH does not carry any higher layer information; neither is there any transport channel mapped to it. The CPICH uses the spreading factor of 256. It may be sent from two antennas in case

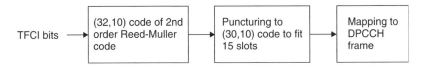

**Figure 6.18**   TFCI information coding

transmission diversity methods are used in the base station. In this case, the transmissions from the two antennas are separated by a simple modulation pattern on the CPICH transmitted from the diversity antenna, called diversity CPICH. The diversity pilot is used with both open-loop and closed-loop transmission diversity schemes.

## 6.5.2   Synchronization Channel (SCH)

The SCH is needed for the cell search. It consists of two channels, i.e. the Primary and Secondary SCHs.

The Primary SCH uses a 256-chip spreading sequence identical in every cell. The system-wide sequence has been optimized for matched filter implementations, as described in connection with SCH spreading and modulation in Section 6.3.4.4.

The Secondary SCH uses sequences with different code word combination possibilities representing different code groups. Once the terminal has identified the Secondary SCH, it has obtained frame and slot synchronization as well as information on the group the cell belongs to. There are 64 different code groups in use, pointed out by the 256 chip sequences sent on the Secondary SCHs. Such a full cell search process with a need to search for all groups is needed naturally only at the initial search upon terminal power-on or when entering a coverage area, otherwise a terminal has more information available on the neighboring cells and not all the steps are always necessary.

As with the CPICH, no transport channel is mapped on the SCH, as the code words are transmitted for cell search purposes only. The SCH is time multiplexed with the Primary Common Control Physical Channel (Primary CCPCH). For the SCH there are always 256 chips out of 2560 chips from each slot. The Primary and Secondary SCHs are sent in parallel, as illustrated in Figure 6.19. Further details on the cell search procedure are covered in Section 6.6.

## 6.5.3   Primary Common Control Physical Channel (Primary CCPCH)

The Primary CCPCH is the physical channel carrying the BCH. It needs to be demodulated by all the terminals in the system. As a result, the parameters with respect to, for example, the channel coding and spreading code contain no flexibility, as they need to be known by all terminals made since the publication of the Release 99 specifications. The contents of the signaling messages have room for flexibility as long as the new message structures are such that they do not cause unwanted or unpredictable behavior in the terminals deployed in the network.

The Primary CCPCH contains no Layer 1 control information, as it is fixed rate and does not carry power control information for any of the terminals. The pilot symbols are not used, since the Primary

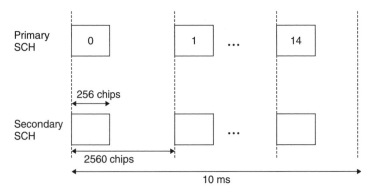

**Figure 6.19**   Primary and Secondary SCH principles

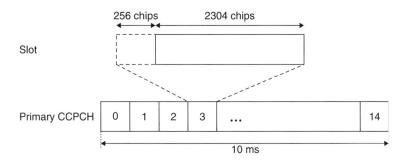

**Figure 6.20**   Primary CCPCH frame structure

CCPCH needs to be available over the whole cell area and does not use specific antenna techniques; rather, it is sent with the same antenna radiation pattern as the CPICH. This allows the CPICH to be used for channel estimation with coherent detection in connection with the Primary CCPCH.

The channel bit rate is 30 kbps with a spreading ratio of the permanently allocated channelization code of 256. The total bit rate is reduced further as the Primary CCPCH alternates with the SCH, reducing the bit rate without coding available for system information to 27 kbps. This is illustrated in Figure 6.20, where the 256-chip idle period on the Primary CCPCH is shown.

The channel coding with the Primary CCPCH is 1/2-rate convolutional coding with 20 ms interleaving over two consecutive frames. It is important to keep the data rate with the Primary CCPCH low, as, in practice, it will be transmitted with very high power from the base station to reach all terminals, having a direct impact on system capacity. If Primary CCPCH decoding fails, then the terminals cannot access the system if they are unable to obtain the critical system parameters, such as random access codes or code channels used for other common channels.

As a performance improvement method, the Primary CCPCH may apply open-loop transmission diversity. In such a case, the use of open transmission diversity on the Primary CCPCH is indicated in the modulation of the Secondary SCH. This allows the terminals to have the information before attempting to decode the BCH with the initial cell search.

### 6.5.4   Secondary Common Control Physical Channel (Secondary CCPCH)

The Secondary Common Control Physical Channel (Secondary CCPCH) carries two different common transport channels: the FACH and the PCH. The two channels can share a single Secondary CCPCH or can use different physical channels. This means that, in the minimum configuration, each cell has at least one Secondary CCPCH. In case of a single Secondary CCPCH, fewer degrees of freedom exist in terms of data rates, and so on, since again all the terminals in the network need to be able to detect the FACH and PCH. Since there can be more than one FACH or PCH, however, for the additional Secondary CCPCHs the data rates can vary more, as long as the terminals not capable of demodulating higher data rates are using another, lower data rate Secondary CCPCH.

The spreading factor used in a Secondary CCPCH is fixed and determined according to the maximum data rate. The data rate may vary with DTX or rate-matching parameters, but the channelization code is always reserved according to the maximum data rate. The maximum data rate usable is naturally dependent on the terminal capabilities. As with the Primary CCPCH, the channel coding method is $\frac{1}{2}$-rate convolutional coding when carrying the channels used for cell access, FACH or PCH. When used to carry PCH, the interleaving period is always 10 ms. For data transmission with FACH, turbo coding or 1/3-rate convolutional coding may also be applied.

The Secondary CCPCH does not contain power control information, and for other layer 1 control information the following combinations can be used:

- Neither pilot symbols nor rate information (TFCI). Used with PCH and FACH when no adaptive antennas are in use and a channel needs to be detected by all terminals.
- No pilot symbols, but rate information with TFCI. Used typically with FACH when it is desired to use FACH for data transmission with variable transport format and data rate. In such a case, variable transmission rates are implemented by DTX or repetition.
- Pilot symbol with or without rate information (TFCI). Typical for the case when an uplink channel is used to derive information for adaptive antenna processing purposes and user-specific antenna radiation patterns or beams are used.

The FACH and PCH can be multiplexed to a single Secondary CCPCH, as the paging indicators (PIs) used together with the PCH are multiplexed to a different physical channel, called the PI Channel (PICH). The motivation for multiplexing the channels together is base station power budget. Since both of the channels need to be transmitted at full power for all the terminals to receive, avoiding the need to send them simultaneously obviously reduces base station power level variations. In order to enable this multiplexing, it has been necessary to terminate both FACH and PCH at RNC.

As a performance improvement method, open-loop transmission diversity can be used with a Secondary CCPCH as well. The performance improvement of such a method is higher for common channels in general, as neither Primary nor Secondary CCPCH can use fast power control. Also, since they are often sent with full power to reach the cell edge, reducing the required transmission power level improves downlink system capacity.

## 6.5.5 Random Access Channel (RACH) for Signaling Transmission

The RACH is typically used for signaling purposes, to register the terminal after power-on to the network or to perform location update after moving from one location area to another or to initiate a call. The structure of the physical RACH for signaling purposes is the same as when using the RACH for user data transmission, as described in connection with the user data transmission. With signaling use, the major difference is that the data rate needs to be kept relatively low, otherwise the range achievable with RACH signaling starts to limit the system coverage. This is more critical the lower the data rates used as a basis for network coverage planning. RACH range issues are studied in detail in Chapter 11. The detailed RACH procedure will be covered in Section 6.6.4.

The RACH that can be used for initial access has a relatively low payload size, since it needs to be usable by all terminals. The ability to support a 16 kbps data rate on RACH is a mandatory requirement for all terminals, regardless of what kind of services they provide.

## 6.5.6 Acquisition Indicator Channel (AICH)

In connection with the RACH, the AICH is used to indicate from the base station the reception of the RACH signature sequence. The AICH uses an identical signature sequence as the RACH on one of the downlink channelization codes of the base station to which the RACH belongs. Once the base station has detected the preamble with the random access attempt, then the same signature sequence that has been used on the preamble will be echoed back on AICH. As the structure of AICH is the same as with the RACH preamble, it also uses a spreading factor of 256 and 16 symbols as the signature sequence. There can be up to 16 signatures, acknowledged on the AICH at the same time. Both signature sets can be used with AICH. The procedure with AICH and RACH is described in Section 6.6.4.

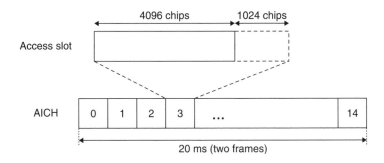

**Figure 6.21**  AICH access slot structure

For the detection of AICH, the terminal needs to obtain the phase reference from the CPICH. The AICH also needs to be heard by all terminals and needs to be sent typically at high power level without power control.

The AICH is not visible to higher layers, but is controlled directly by the physical layer in the base station, as operation via an RNC would make the response time too slow for a RACH preamble. There are only a few timeslots to detect the RACH preamble and to transmit the response to the terminal on AICH. The AICH access slot structure is shown in Figure 6.21.

## 6.5.7  Paging Indicator Channel (PICH)

The PCH is operated together with the PICH to provide terminals with efficient sleep mode operation. The PIs use a channelization code of length 256. The PIs occur once per slot on the corresponding physical channel, the PICH. Each PICH frame carries 288 bits to be used by the PI bit, and 12 bits are left idle. Depending on the PI repetition ratio, there can be 18, 36, 72 or 144 PIs per PICH frame. How often a terminal needs to listen to the PICH is parameterized, and the exact moment depends on running the SFN.

For detection of the PICH the terminal needs to obtain the phase reference from the CPICH; and, as with the AICH, the PICH needs to be heard by all terminals in the cell and, thus, needs to be sent at high power level without power control. The PICH frame structure with different PI repetition factors is illustrated in Figure 6.22.

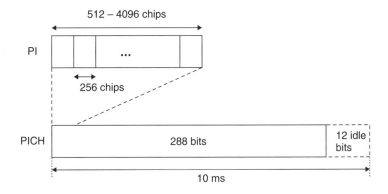

**Figure 6.22**  PICH structure with different PI repetition rates

## 6.6    Physical Layer Procedures

In the physical layer of a CDMA system there are many procedures essential for system operation. Examples include the fast power control and random access procedures. Other important physical layer procedures are paging, handover measurements and operation with transmit diversity. These procedures have been naturally shaped by the CDMA-specific properties of the UTRA FDD physical layer.

### 6.6.1    Fast Closed-Loop Power Control Procedure

The fast closed-loop power control procedure is denoted in the UTRA specifications as inner loop power control. It is known to be essential in a CDMA-based system due to the uplink near–far problem illustrated in Chapter 3. The fast power control operation operates on a basis of one command per slot, resulting in a 1500 Hz command rate. The basic step size is 1 dB. Additionally, multiples of that step size can be used and smaller step sizes can be emulated. The emulated step size means that the 1 dB step is used, for example, only every second slot, thus emulating the 0.5 dB step size. 'True' step sizes below 1 dB are difficult to implement with reasonable complexity, as the achievable accuracy over the large dynamic range is difficult to ensure. The specifications define the relative accuracy for a 1 dB power control step to be ±0.5 dB. The other 'true' step size specified is 2 dB.

Fast power control operation has two special cases: operation with soft handover and with compressed mode in connection with handover measurements. Soft handover needs special concern, as there are several base stations sending commands to a single terminal; with compressed mode operation, breaks in the command stream are periodically provided to the terminal.

In soft handover the main issue for terminals is how to react to multiple power control commands from several sources. This has been solved by specifying the operation such that the terminal combines the commands but also takes the reliability of each individual command decision into account in deciding whether to increase or decrease the power.

In the compressed mode case, the fast power control uses a larger step size for a short period after a compressed frame. This allows the power level to converge more quickly to the correct value after a break in the control stream. The need for this method depends heavily on the environment and it is not relevant for the lower terminal or very short transmission gap lengths (TGLs).

The SIR target for closed-loop power control is set by the outer loop power control. The latter power control is introduced in Section 3.5 and described in detail in Section 9.2.2.

On the terminal side it is specified rather strictly what is expected to be done inside a terminal in terms of (fast) power control operation. On the network side there is much greater freedom to decide how a base station should behave upon reception of a power control command, as well as the basis on which the base station should tell a terminal to increase or decrease the power.

### 6.6.2    Open-Loop Power Control

In UTRA FDD there is also open-loop power control, which is applied only prior to initiating the transmission on the RACH (or CPCH). Open-loop power control is not very accurate, since it is difficult to measure large power dynamics accurately in the terminal equipment. The mapping of the actual received absolute power to the absolute power to be transmitted shows large deviations, due to variation in the component properties and to the impact of environmental conditions, mainly temperature. Also, the transmission and reception occur at different frequencies, but the internal accuracy inside the terminal is the main source of uncertainty. The requirement for open-loop power control accuracy is specified to be within ±9 dB in normal conditions.

Open-loop power control was used in earlier CDMA systems, such as IS-95, being active in parallel with closed-loop power control. The motivation for such usage was to allow corner effects or other sudden environmental changes to be covered. As the UTRA fast power control has almost double the command rate, it was concluded that a 15 dB adjustment range does not need open-loop power control to be operated simultaneously. Additionally, the fast power control step size can be increased from 1 dB to 2 dB, which would allow a 30 dB correction range during a 10 ms frame.

The use of open-loop power control while in active mode also has some impact on link quality. The huge inaccuracy of open-loop power control can cause it to make adjustments to the transmitted power level even when they are not needed. As such behavior depends on terminal unit tolerances and on various environmental variables, running open-loop power control makes it more difficult from the network side to predict how a terminal will behave in different conditions.

### 6.6.3 Paging Procedure

The PCH operation is organized as follows. A terminal, once registered to a network, has been allocated a paging group. For the paging group there are PIs which appear periodically on the PICH when there are paging messages for any of the terminals belonging to that paging group.

Once a PI has been detected, the terminal decodes the next PCH frame transmitted on the Secondary CCPCH to see whether there was a paging message intended for it. The terminal may also need to decode the PCH in case the PI reception indicates low reliability of the decision. The paging interval is illustrated in Figure 6.23.

The less often the PIs appear, the less often the terminal needs to wake up from the sleep mode and the longer the battery life. The trade-off is obviously the response time to the network-originated call. An infinite PI interval does not lead to infinite battery duration, as there are also other tasks that the terminal needs to perform during idle mode.

### 6.6.4 RACH Procedure

The Random Access procedure in a CDMA system has to cope with the near–far problem, as when initiating the transmission there is no exact knowledge of the required transmission power. The open-loop power control creates large uncertainty in terms of absolute power values from the received power measurement to the transmitter power level setting value, as stated in connection with the open-loop description. In UTRA, the RACH procedure has the following phases:

- The terminal decodes the BCH to find out the available RACH sub-channels and their scrambling codes and signatures.
- The terminal randomly selects one of the RACH sub-channels from the group its access class allows it to use. Furthermore, the signature is also selected randomly from among the available signatures.
- The downlink power level is measured and the initial RACH power level is set with the proper margin due to the open-loop inaccuracy.
- A 1 ms RACH preamble is sent with the selected signature.
- The terminal decodes AICH to see whether the base station has detected the preamble.
- If no AICH is detected, the terminal increases the preamble transmission power by a step given by the base station, as multiples of 1 dB. The preamble is retransmitted in the next available access slot.
- When an AICH transmission is detected from the base station, the terminal transmits the 10 ms or 20 ms message part of the RACH transmission.

The RACH procedure is illustrated in Figure 6.24, where the terminal transmits the preamble until acknowledgement is received on AICH, and then the message part follows.

In the case of data transmission on RACH, the spreading factor and, thus, the data rate may vary; this is indicated by the TFCI on the DPCCH on PRACH. Spreading factors from 256 to 32 have been defined to be possible; thus, a single frame on RACH may contain up to 1200 channel symbols, which,

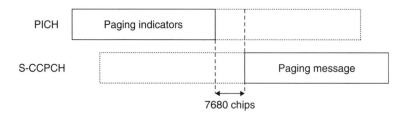

**Figure 6.23**   PICH relation to PCH

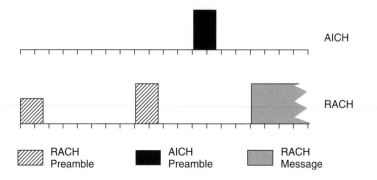

**Figure 6.24**   PRACH ramping and message transmission

depending on the channel coding, maps to around 600 or 400 bits. For the maximum number of bits the achievable range is naturally less than what can be achieved with the lowest rates, especially as RACH messages do not use methods such as macro-diversity as in the DCH.

## 6.6.5   Cell Search Procedure

The cell search procedure or synchronization procedure in an asynchronous CDMA system differs greatly from the procedure in a synchronous system like IS-95. Since the cells in an asynchronous UTRA CDMA system use different scrambling codes and not just different code phase shifts, terminals with today's technology cannot search for 512 codes of 10 ms duration without any prior knowledge. There would be too many comparisons to make and users would experience too long an interval from power-on to the service availability indication in the terminal.

The cell search procedure using the SCH has basically three steps, though from the standards point of view there will be no requirements as to which steps to perform and when. Rather, the standard will set requirements for performance in terms of maximum search duration in reference test conditions. The basic steps for the initial cell search are typically as follows:

1. The terminal searches the 256-chip primary synchronization code, being identical for all cells. As the primary synchronization code is the same in every slot, the peak detected corresponds to the slot boundary.
2. Based on the peaks detected for the primary synchronization code, the terminal seeks the largest peak from the Secondary SCH code word. There are 64 possibilities for the secondary synchronization code word. The terminal needs to check all 15 positions, as the frame boundary is not available before Secondary SCH code word detection.

3. Once the Secondary SCH code word has been detected, the frame timing is known. The terminal then seeks the primary scrambling codes that belong to that particular code group. Each group consists of eight primary scrambling codes. These need to be tested for a single position only, as the starting point is known already.

When setting the network parameters, the properties of the synchronization scheme need to be taken into account for optimum performance. For the initial cell search there is no practical impact, but the target cell search in connection with handover can be optimized. Basically, since there are rather a large number of code groups, in a practical planning situation one can, in most cases, implement the neighboring cell list so that all the cells in the list for one cell belong to a different code group. Thus, the terminal can search for the target cell and skip step 3 totally, just confirming detection without needing to compare the different primary scrambling codes for that step. Alternatively, as has been shown in practical networks resulting to similar performance, one can aim to have all neighboring cells under one code group as well. Initially, having more groups was expected not to be the optimal case, but as experience from the field has shown, using one group only is the preferred alternative.

Further ways of improving cell search performance include the possibility of providing information on the relative timing between cells. This kind of information, which is being measured by the terminals for soft handover purposes in any case, can be used to improve especially the step 2 performance. The more accurate the relative timing information, the fewer slot positions need to be tested for the Secondary SCH code word, and the better is the probability of correct detection.

## 6.6.6   *Transmit Diversity Procedure*

As was mentioned in connection with the downlink channels, UTRA uses two types of transmit diversity transmission for user data performance improvement, as studied in Chapter 11. These are classified as open-loop and closed-loop methods. In this section, the feedback procedure for closed-loop transmit diversity is described. The open-loop method was covered in connection with the Downlink DCH description (Section 6.4.5).

In the case of closed-loop transmit diversity the base station uses two antennas to transmit the user information. The use of these two antennas is based on the feedback from the terminal, transmitted in the FBI bits in the uplink DPCCH. The closed-loop transmit diversity itself has two modes of operation.

In mode 1, the terminal feedback commands control the phase adjustments that are expected to maximize the power received by the terminal. The base station thus maintains the phase with antenna 1 and then adjusts the phase of antenna 2 based on the sliding averaging over two consecutive feedback commands. Thus, with this method, four different phase settings are applied to antenna 2.

In mode 2, the amplitude is adjusted in addition to the phase adjustment. The same signaling rate is used, but now the command is spread over 4 bits in four uplink DPCCH slots, with a single bit for amplitude and 3 bits for phase adjustment. This gives a total of eight different phases and two different amplitude combinations, i.e. 16 combinations for signal transmission from the base station. The amplitude values have been defined to be 0.2 and 0.8, while the phase values are naturally distributed evenly for the antenna phase offsets, from $-135°$ to $+180°$ phase offset. In this mode the last three slots of the frame contain only phase information, while amplitude information is taken from the previous four slots. This allows the command period to go even with 15 slots as with mode 1, where the average at the frame boundary is slightly modified by averaging the commands from slot 13 and slot 0 to avoid discontinuities in the adjustment process.

The closed-loop method may be applied only on the DCHs or with an HS-DSCH together with a DCH. The open-loop method may be used on both the common and dedicated channels. As part of the feature clean-up, 3GPP decided to remove mode 2 from Release 5 onwards, as that had not been implemented in the networks deployed so far.

## 6.6.7    Handover Measurements Procedure

Within the UTRA FDD the possible handovers are as follows:

- Intra-mode handover, which can be soft handover, softer handover or hard handover. Hard handover may take place as intra- or inter-frequency handover.
- Inter-mode handover as handover to the UTRA TDD mode, this, however, has not been implemented anywhere in real networks. Since the operators running TDD (including the biggest network using TD-SCDMA in China) do not normally have also WCDMA network, there has not been need for dual mode equipment running both FDD and TDD but the devices are then running either WCDMA with GSM or TDD (TD-SCDMA) with GSM.
- Inter-system handover, which in Release 99 means only GSM handover. The GSM handover may take place to a GSM system operating at 850 MHz, 900 MHz, 1800 MHz and 1900 MHz.
- In Release 8, together with the introduction of Long Term Evolution (LTE), the necessary elements for WCDMA specifications were added to the specifications to enable handover to (and from) LTE. 3GPP soecifications use also the term Evolved UTRA (E-UTRA) or E-UTRAN when covering the whole radio access network.

The main relevance of the handover to the physical layer is what to measure for handover criteria and how to obtain the measurements.

### 6.6.7.1    Intra-Mode Handover

The UTRA FDD intra-mode handover relies on the $E_c/N_0$ measurement performed from the CPICH. The quantities defined that can be measured by the terminal from the CPICH are as follows:

- Received Signal Code Power (RSCP), which is the received power on one code after despreading, defined on the pilot symbols.
- Received Signal Strength Indicator (RSSI), which is the wideband received power within the channel bandwidth.
- $E_c/N_0$, representing the RSCP divided by the total received power in the channel bandwidth, i.e. RSCP/RSSI.

There are also other items that can be used as a basis for handover decisions in UTRAN, as the actual handover algorithm decisions are left as an implementation issue. One such parameter mentioned in the standardization discussions was the DCH SIR, giving information on the cell orthogonality and being measured in any case for power control purposes.

Additional essential information for soft handover purposes is the relative timing information between the cells. As in an asynchronous network, there is a need to adjust the transmission timing in soft handover to allow coherent combining in the Rake receiver, otherwise the transmissions from the different base stations would be difficult to combine; in particular, the power control operation in soft handover would suffer additional delay. The timing measurement in connection with the soft handover operation is illustrated in Figure 6.25. The new base station adjusts the downlink timing in steps of 256 chips based on the information it receives from the RNC. When a new cell is added to active set, the UE is also provided with the information what is the downlink timing actually used as there could be ambiguity otherwise to which direction the network would be placed the timing (as the timing step is 256 chips in setting the timing for a radio link to be added).

When the cells are within the 10 ms window, the relative timing can be found from the primary scrambling code phase, since the code period used is 10 ms. If the timing uncertainty is larger, then

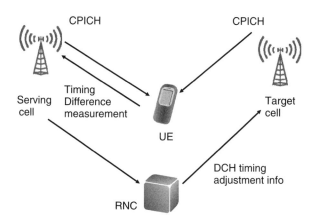

**Figure 6.25**   Timing measurement for soft handover

the terminal needs to decode the SFN from the Primary CCPCH. This always takes time and may suffer from errors, which also requires a CRC check to be made on the SFN. The 10 ms window has no relevance when the timing information is provided in the neighboring cell list. In such a case only the phase difference of the scrambling codes needs to be considered, unless the base stations are synchronized to chip level.

For the hard handover between frequencies, such accurate timing information on chip level is not needed. Obtaining the other measurements is slightly more challenging, as the terminal must make the measurements on a different frequency. This is typically done with the aid of compressed mode, which is described later in this chapter.

#### 6.6.7.2   Inter-Mode Handover

On request from UTRAN, the dual-mode FDD–TDD terminals operating in FDD measure the power level from the TDD cells available in the area. The TDD CCPCH bursts sent twice during the 10 ms TDD frame can be used for measurement, since they are always guaranteed to exist in the downlink. The TDD cells in the same coverage area are synchronized; thus, finding one slot with the reference mid-amble means that other TDD cells have roughly the same timing for their burst with reference power. UTRA TDD mode, with a focus on TD-SCDMA, is covered in detail in Chapter 18.

#### 6.6.7.3   Inter-System Handover

For UTRA–GSM handover, similar requirements as for GSM–GSM handover are basically valid. Normally the terminal receives the GSM SCH during compressed frames in UTRA FDD to allow measurements from other frequencies. The measurement quantity is GSM carrier RSSI. GSM 1800 set special requirements for compressed mode and required that compressed mode was specified for the uplink also. This was also needed for TDD measurements.

For the UTRA to E-UTRA (thus WCDMA to LTE), the compressed mode measurement is also the method used as LTE is operated either on a different frequency band or even if in the same band, then on a different part of the spectrum. The measurement quentities are E-UTRA Reference Signal Receiver Quality (RSRQ) or E-UTRA Reference Signal Received Power (RSRP), covered in Release 8 versions of [5].

## 6.6.8  Compressed Mode Measurement Procedure

The compressed mode, often referred to as the slotted mode, is needed when making measurements from another frequency in a CDMA system without a full dual receiver terminal. The compressed mode means that transmission and reception are halted for a short time, of the order of a few milliseconds, in order to perform measurements on the other frequencies. The aim is not to lose data but to compress the data transmission in the time domain. Frame compression can be achieved through three different methods:

- Lowering the data rate from higher layers, as higher layers have knowledge of the compressed mode schedule for the terminal.
- Increasing the data rate by changing the spreading factor. For example, using spreading factor 64 instead of spreading factor 128 doubles the number of available symbols and makes it very straight-forward to achieve the desired compression ratio for the frame.
- Reducing the symbol rate by puncturing the multiplexing chain at the physical layer. In practice, this is limited to the rather short TGLs, since puncturing has some practical limits. The benefit is obviously in keeping the existing spreading factor and not causing new requirements for channelization code usage. This approach was never implemented in the networks; thus, 3GPP decided to remove compressed mode by puncturing from the specifications from Release 5 onwards.

The compressed frames are provided normally in the downlink and in some cases in the uplink as well. If they appear in the uplink, then they need to be simultaneous with the downlink frames, as illustrated in Figure 6.26.

The specified TGLs are 3, 4, 5, 7, 10 and 14 slots. TGLs of 3, 4 and 7 can be obtained with both single- and double-frame methods. For TGLs of 10 or 14, only the double-frame method can be used. An example of the double-frame method is illustrated in Figure 6.27, where the idle slots are divided between two frames. This allows minimization of the impact during a single frame and keeping, for example, the required increment in the transmission power lower than with the single-frame method.

The case when uplink compressed frames are always needed with UTRA is the GSM 1800 measurements, where the close proximity of the GSM 1800 downlink frequency band to the core UTRA FDD uplink frequency band at 1920 MHz and upwards is too close to allow simultaneous transmission and reception.

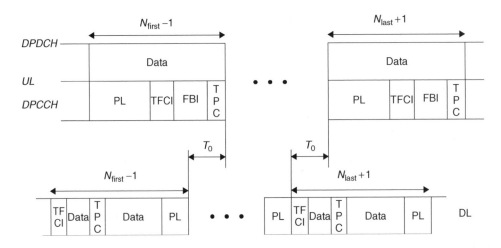

**Figure 6.26**  Compressed frames in the uplink and downlink

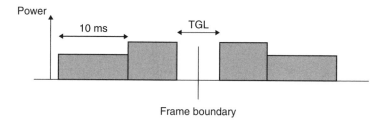

**Figure 6.27**    Compressed mode with the double-frame method

Use of the compressed mode in the uplink with GSM 900 measurements or UTRA inter-frequency handover depends on terminal capability. To maintain the continuous uplink, the terminal needs to have a means of generating the additional frequency in parallel while maintaining the existing frequency. In practice, this means additional oscillators for frequency generation as well as some other duplicated components, which add to terminal power consumption. Also the use of receiver diversity in the devices means that in order to maintain the performance with receiver diversity, one would either need to use a compressed or a third receiver. Dropping the diversity branch only when reaching the cell edge to do measurements on other carriers/system could cause additional call drops, as then there would be additional reduction in the downlink link budget when starting the measurements.

The use of compressed mode has an inevitable impact on link performance, as studied in [10] for the uplink compressed mode and in [11] for the downlink. Link performance does not deteriorate very much if the terminal is not at the cell edge, since there is room to compensate the momentary performance loss with fast power control. The impact is largest at the cell edge; the difference in uplink performance between compressed mode and non-compressed mode is very slight until headroom is less than 4 dB. At 0 dB headroom, the difference from normal transmission is between 2 and 4 dB, depending on the transmission gap duration with compressed frames. The 0 dB headroom corresponds to terminal operation at full power at the cell edge with no possibility of (soft) handover and with no room to run fast power control any more. The use of soft handover (or handover in general) will improve the situation, since low headroom values are less likely to occur, as with typical planning there is some overlap in the cell coverage area and the 0 dB headroom case should occur only when leaving the coverage area. The compressed mode performance is analysed in Section 9.3.2.

The actual time available for sampling on another frequency is reduced from the above values, due to the time taken by the hardware to switch the frequency; thus, very short values of 1 or 2 slots have been excluded, since there is no really practical time available for measurements. The smallest value used in the specifications is 3, which itself allows only a very short measurement time window and should be considered for use only in specific cases.

## 6.6.9   Other Measurements

In the base station, other measurements are needed to give RNC sufficient information on uplink status and base station transmission power resource usage. The following have been specified for the base station, to be supported by signaling between base station and RNC:

- RSSI, to give information on the uplink load.
- Uplink SIR on the DPCCH.
- Total transmission power on a single carrier at a base station transmitter, giving information on the available power resources at the base station.

- The transmission code on a single code for one terminal. This is used, for example, in balancing power between radio links in soft handover.
- Block Error Rate (BLER) and Bit Error Rate estimates for different physical channels.

The BLER measurement is to be supported by the terminals as well. The main function of terminal BLER measurement is to provide feedback for the outer loop power control operation in setting the SIR target for fast power control operation.

Support of position location functionality needs measurements from the physical layer. For that purpose, a second type of timing measurement has been specified that gives the timing difference between the primary scrambling codes of different cells with 1/4-chip resolution for improved position location accuracy. The achievable position accuracy in theory can thus be estimated from the fact that a single chip corresponds to a distance of roughly 70 m. In a cellular environment there are obviously further factors contributing to the achievable accuracy. To alleviate the impact of the near–far problem for a terminal that is very close to a base station, the specifications also contain a method of introducing idle periods in base station transmission. This enables timing measurements from base stations that would otherwise be too weak due to close proximity of the serving base station.

## 6.6.10   Operation with Adaptive Antennas

UTRA has been designed to allow the use of adaptive antennas, also known as beamforming, both in the uplink and downlink directions. Basically, there are two types of beamforming that one may use. Either a beam may use the secondary CPICH (S-CPICH) or then any may use only the dedicated pilot symbols. However, the use of dedicated pilots as a sole phase reference was removed from Release 5 specifications onwards, as it had not been implemented in any of the networks. From the physical layer point of view, the use of adaptive antennas is fully covered with Release 99, but the exact performance requirements for the terminals in different operation scenarios are covered in Release 5.

What kind of beamforming may be applied with different channels depends on whether the channel contains dedicated pilot symbols or not. For the S-CCPCH, beamforming could be used in theory, but in reality it is not practical as the channel is intended to be received by several terminals and thus the slot formats with pilots are not supported by the terminals. Also, the typically short-duration communication would be difficult, as it takes some time to adapt to the receiver at the terminal for a different delay profile than the P-CPICH transmission. The summary of the use of beamforing on different downlink channel types is given in Table 6.5.

If it is desired to use beamforming together with any of the transmit diversity modes, then S-CPICH needs to be transmitted, including the diversity pilot, in the same antenna beam.

**Table 6.5**   Application of beamforming concepts on downlink physical channel types

| Physical channel type | Beamforming with S-CPICH | Beamforming without S-CPICH |
| --- | --- | --- |
| P-CCPCH | No | No |
| SCH | No | No |
| S-CCPCH | No | No |
| DPCH | Yes | Yes, but removed in Release 5 |
| PICH | No | No |
| PDSCH, HS-PDSCH & HS-SCCH (with associated DPCH) | Yes | No |
| AICH | No | No |
| CSICH | No | No |

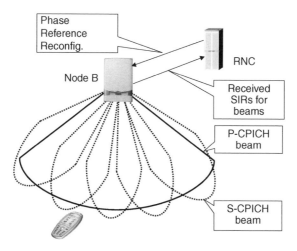

**Figure 6.28**   Release 6 beamforming enhancement method

In the UTRAN side the full support for beamforming parameterization was completed for Release 5. This includes phase reference change signaling from RNC to Node B. To improve radio resource management with beamforming further, Release 6 contains added functionality for this. The uplink SIR measurement has been extended to be possible for all the 'beams' in the uplink direction. The practical example of having four beams with S-PCICH per sector allows the Node B to report for each terminal from which beam the best signal is received and, thus, RNC can make decisions to reconfigure the phase reference (i.e. change the fixed beam) of the terminal. The enhancement is illustrated in Figure 6.28. The physical layer modifications are covered in the Release 6 version of [5].

## 6.6.11   Site Selection Diversity Transmission

The Site Selection Diversity Transmission (SSDT) is a specific mode of soft handover which was included in Release 99 specifications but was completed from the UTRAN point of view only in the Release 5 specification. The main principle of the SSDT feature is that, based on the feedback signaling from the terminal, the Node Bs in the active set may transmit only the DPCCH (control) part of the transmission and use DTX for the data part.

The terminal will send in the uplink the ID of the strongest Node B, based on the measurements on the CPICH, with the intervals between 2 and 10 ms, depending on the length of the ID codeword selected and the number of feedback bits allocated for SSDT use in the uplink direction. On the network side, all Node Bs received the data from RNC but only the one receiving the correct code word with sufficient quality sends the data onwards on the DPDCH of the downlink DCH. The $Q$th parameter determines the minimum quality (SIR) level that Node B must receive in order to consider the non-primary commands valid. The SSDT principle is illustrated in Figure 6.29, which shows the example case of two Node Bs in the active set.

As SSDT had not been implemented in any of the networks, 3GPP approved its removal in June 2005 from Release 5 onwards. The practical use of SSDT was also challenging, as the operation was fully UE-based; thus, using the power unused for other purposes was difficult, as the transmission was not under BTS control. With HSDPA, the trend was also to move away from macro-diversity for user data, which made SSDT less relevant as well.

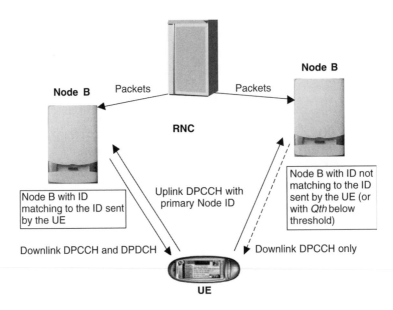

**Figure 6.29**  SSDT principle

## 6.7   Terminal Radio Access Capabilities

As explained in Chapter 2, the class mark approach of GSM is not applied in the same way with UMTS. Instead, a terminal upon connection establishment informs a network of a large set of capability parameters and not only one or more class mark values. The reason for this approach has been the large variety of capabilities and data rates with UMTS terminals, which would have resulted in a very high number of different class marks. For practical guidance, reference classes were specified anyway.

The reference classes in [12] have a few common values as well, which are not covered here. For example, the support for spreading factor 512 is not expected to be covered by any of the classes by default. For the channel coding methods, turbo coding is supported with classes above 32 kbps class and with higher classes the higher data rates above 64 kbps are supported with turbo coding only as can be seen in Tables 6.6 and 6.7. For the convolutional coding, all the classes have the value of 640 bits at an arbitrary time instant for both encoding and decoding. This is needed in any case for decoding of BCHs. All the classes, except the 32 kbps uplink, support at least eight parallel transport channels.

The value given for the number of bits received at an arbitrary time instant needs to be converted to the maximum data rate supported by considering at the same time the interleaving length (or TTI length with 3GPP terminology). For example, the value 6400 bits for the 384 kbps class can be converted to the maximum data rate with a particular TTI as follows. The data rate of the application is 256 kbps; thus, the number of bits per 10 ms is 2560 bits. With 10 or 20 ms TTI lengths the number of bits per interleaving period stays below 6400 bits, but with 40 ms TTI the 6400 limit would be exceeded and the terminal would not have enough memory to operate with such a configuration. Similarly, a 384 kbps data rate with a terminal of a same class could be maintained with a 10 ms TTI, but a 20 ms TTI would exceed the limit.

The value ranges given in [12] range from beyond what the classes contain; for example, it is possible for a terminal to indicate values allowing 2 Mbps with 80 ms TTI. The minimum values have been determined by the necessary capabilities needed to access the system, e.g. to listen the BCH or to access the RACH.

**Table 6.6** Terminal radio access capability parameter combinations for downlink decoding

| Reference combination | 32 kbps class | 64 kbps class | 128 kbps class | 384 kbps class | 768 kbps class | 2048 kbps class |
|---|---|---|---|---|---|---|
| *Transport channel parameters* | | | | | | |
| Maximum sum of number of bits of all transport blocks being received at an arbitrary time instant | 640 | 3840 | 3840 | 6400 | 10 240 | 20 480 |
| Maximum sum of number of bits of all turbo coded transport blocks being received at an arbitrary time instant | Not supported | 3840 | 3840 | 6400 | 10 240 | 20 480 |
| Maximum number of simultaneous CCTrCHs, higher value with PDSCH support | 1 | 2/1 | 2/1 | 2/1 | 2 | 2 |
| Maximum total number of transport blocks received within TTIs that end at the same time | 8 | 8 | 16 | 32 | 64 | 96 |
| Maximum number of TFCs in the TFC Set (TFCS) | 32 | 48 | 96 | 128 | 256 | 1024 |
| Maximum number of Transport Formats | 32 | 64 | 64 | 64 | 128 | 256 |
| *Physical channel parameters* | | | | | | |
| Maximum number of DPCH/PDSCH codes simultaneously received, higher value with DSCH support | 1 | 2/1 | 2/1 | 3 | 3 | 3 |
| Maximum number of physical channel bits received in any 10 ms interval (DPCH, PDSCH, S-CCPCH), higher value with DSCH support. | 1200 | 3600/2400 | 7200/4800 | 19 200 | 28 800 | 57 600 |
| Support of Physical DSCH | No | Yes/No | Yes/No | Yes/No | Yes | Yes |

The key physical channel parameter is the maximum number of physical channel bits received/transmitted per 10 ms interval. This determines which spreading factors are supported. For example, a value of 1200 bits for the 32 kbps class indicates that in the downlink the spreading factors supported are 256, 128 and 64, whereas in the uplink the smallest value supported would be 64. The difference arises from the use of QPSK modulation in the downlink and BPSK modulation in the uplink, as explained in Section 6.3.3.1.

There are also parameters that are not dependent on a particular reference combination. Such parameters indicate, for example, support for a particular terminal position location method. In the RF side, the class-independent parameters allow one to indicate, for example, supported frequency bands or the terminal power class.

The parameters in Table 6.6 and 6.7 cover the UTRA FDD, whereas for UTRA TDD there are a few additional TDD-specific parameters in the complete tables [12], such as the number of slots to be received, etc.

**Table 6.7** Terminal radio access capability parameter combinations for uplink encoding

| Reference combination | 32 kbps class | 64 kbps class | 128 kbps class | 384 kbps class | 768 kbps class |
|---|---|---|---|---|---|
| *Transport channel parameters* | | | | | |
| Maximum sum of number of bits of all transport blocks being transmitted at an arbitrary time instant | 640 | 3840 | 3840 | 6400 | 10 240 |
| Maximum sum of number of bits of all turbo coded transport blocks being transmitted at an arbitrary time instant | Not Supported | 3840 | 3840 | 6400 | 10 240 |
| Maximum total number of transport blocks transmitted within TTIs that start at the same time | 4 | 8 | 8 | 16 | 32 |
| Maximum number of TFCs in the TFCS | 16 | 32 | 48 | 64 | 128 |
| Maximum number of Transport Formats | 32 | 32 | 32 | 32 | 64 |
| *Physical channel parameters* | | | | | |
| Maximum number of DPDCH bits transmitted per 10 ms | 1200 | 2400 | 4800 | 9600 | 19 200 |

The first terminals on the market were typically providing data rates of 384 kbps in the downlink and 64 kbps in the uplink, such as the Nokia 7600. The next generation of devices, such as the Nokia 6630, provided 384 kbps in both uplink and downlink directions [13], when in WCDMA mode, it is also (like most of the UMTS terminals on the market), a dual-mode GSM/WCDMA terminal and this 384 kbps uplink and downlink capability has been available in many other networks as well. The terminal offering is obviously evolving: on the downlink side especially, the data rates have exceeded the original 2 Mbps limit with HSDPA, as described in Chapter 12; for example, the Nokia N95 (released in 2007) provided 3.6 Mbps in the downlink direction and 384 kbps in the uplink. The latest devices have reached 10 Mbps, as shown in the measurements in Chapter 2. Also the uplink data rates have now increased along with the introduction of HSUPA, have reached 2 Mbps and beyond, as covered in Chapter 13.

## 6.8 Conclusion

This chapter covered the WCDMA physical layer mainly based on the first WCDMA specification release, i.e. Release 99. At the time of writing, the physical layer as described has been implemented in more than 450 million devices and, thus, is a well-proven technology, and has been shown to enable such an implementation which is also competitive with second-generation systems from a power consumption point of view, as covered in Chapter 18. While Releases 5 and 6 with HSDPA and HSUPA will take the WCDMA a great step forward, the Release 5 and 6 devices shall still support the basic physical layer operation from Release 99 with a few relaxations as feature removal activity in 3GPP removed several unused features from the specifications. Also, while packet data are moving heavily towards HSDPA and HSUPA, the basic speech service has been relying on the DCH of Release 99. Now with the Release 8, this aspect is also changing with the introduction of the CS voice over HSPA as covered in Chapter 15. With this development it seems that only CS video may eventually remain as the service using

Release 99 channels, though most of the basic Release 99 physical procedures will still remain valid even if the channels carrying the actual user data change.

# References

[1] 3GPP Technical Specification 25.211, Physical Channels and Mapping of Transport Channels onto Physical Channels (FDD).
[2] 3GPP Technical Specification 25.212, Multiplexing and Channel Coding (FDD).
[3] 3GPP Technical Specification 25.213, Spreading and Modulation (FDD).
[4] 3GPP Technical Specification 25.214, Physical Layer Procedures (FDD).
[5] 3GPP Technical Specification 25.215, Physical Layer – Measurements (FDD).
[6] 3GPP Technical Specification 25.302, Services Provided by the Physical Layer.
[7] Adachi, F., Sawahashi, M. and Okawa, K., 'Tree-Structured Generation of Orthogonal Spreading Codes with Different Lengths for Forward Link of DS-CDMA Mobile', *Electronics Letters*, Vol. 33, No. 1, 1997, pp. 27–28.
[8] 3GPP Technical Specification 25.101, UE Radio Transmission and Reception (FDD).
[9] 3GPP Technical Specification 25.104, UTRA (BS) FDD; Radio Transmission and Reception.
[10] Toskala, A., Lehtinen, O. and Kinnunen, P., 'UTRA GSM Handover from Physical Layer Perspective', *Proc. ACTS Summit 1999*, Sorrento, Italy, June 1999.
[11] Gustafsson, M., Jamal, K. and Dahlman, E., 'Compressed Mode Techniques for Inter-Frequency Measurements in a Wide-Band DS-CDMA System', *Proc. IEEE Int. Conf. on Personal Indoor and Mobile Radio Communications*, PIMRC'97, Helsinki, Finland, 1–4 September 1997, Vol. 1, pp. 231–235.
[12] 3GPP Technical Specification 25.306, UE Radio Access Capabilities.
[13] www.nokia.com.

# 7

# Radio Interface Protocols

Jukka Vialén and Antti Toskala

## 7.1   Introduction

The radio interface protocols are needed to set up, reconfigure and release the Radio Bearer (RB) services (including the Universal Terrestrial Radio Access (UTRA) Frequency Division Duplex (FDD)/Time Division Duplex (TDD) service), which were discussed in Chapter 2.

The protocol layers above the physical layer are called the data link layer (layer 2) and the network layer (layer 3). In the UTRA FDD radio interface, layer 2 is split into sublayers. In the control plane, layer 2 contains two sublayers: the Medium Access Control (MAC) protocol and the Radio Link Control (RLC) protocol. In the user plane, in addition to MAC and RLC, two additional service-dependent protocols exist: the Packet Data Convergence Protocol (PDCP) and the Broadcast/Multicast Control (BMC) protocol. Layer 3 consists of one protocol, called the Radio Resource Control (RRC), which belongs to the control plane. The other network layer protocols, such as Call Control, Mobility Management, Short Message Service (SMS), and so on, are transparent to the Universal Mobile Telecommunication Services (UMTS) Terrestrial Radio Access Network (UTRAN) and are not described in this book.

In this chapter, the general radio interface protocol architecture is first described before going into deeper details of each protocol. For each protocol, the logical architecture and main functions are described. In the MAC section (Section 7.3) the logical channels (services offered by MAC) and mapping between logical channels and transport channels are also explained. For MAC and RLC, an example layer model is defined to describe what happens to a data packet passing through these protocols. In the RRC section (Section 7.8), the RRC service states are described together with the main (RRC) functions and signaling procedures. Releases 4 and 5 have not resulted in major modifications to the Layer 2/3 protocols principles, the Release 5 High-Speed Downlink Packet Access (HSDPA) features have resulted in a new MAC entity in Node B, as presented in Chapter 12. As a new feature for Release 6, the Multimedia Broadcast Multicast Service (MBMS, covered in detail in Chapter 14) principles were introduced, as well as the High-Speed Uplink Packet Access (HSUPA), along with necessary changes, especially in the MAC layer, as presented in Chapter 13. This chapter is then concluded by an introduction to the early User Equipment (UE) handling principles and enhancements for the call set-up time reduction.

*WCDMA for UMTS: HSPA Evolution and LTE, Fifth Edition*   Edited by Harri Holma and Antti Toskala
© 2010 John Wiley & Sons, Ltd

## 7.2 Protocol Architecture

The overall radio interface protocol architecture [1] is shown in Figure 7.1. Figure 7.1 contains only those protocols that are visible in UTRAN.

The physical layer offers services to the MAC layer via transport channels [2] that were characterized by how and with what characteristics data are transferred (transport channels were discussed in Chapter 6).

The MAC layer, in turn, offers services to the RLC layer by means of logical channels. The logical channels are characterized by what type of data is transmitted. Logical channels are described in detail in Section 7.3.

The RLC layer offers services to higher layers via service access points (SAPs), which describe how the RLC layer handles the data packets and if, for example, the automatic repeat request (ARQ) function is used. On the control plane, the RLC services are used by the RRC layer for signaling transport. On the user plane, the RLC services are used either by the service-specific protocol layers PDCP or BMC or by other higher-layer u-plane functions (e.g. speech codec). The RLC services are called Signaling RBs (SRBs) in the control plane and RBs in the user plane for services not utilizing the PDCP or BMC protocol. The RLC protocol can operate in three modes: transparent, unacknowledged and acknowledged. These are discussed further in Section 7.4.

The PDCP exists only for the Packet-Switched (PS) domain services. Its main function is header compression. Services offered by PDCP are called RBs.

The BMC protocol is used to convey over the radio interface messages originating from the Cell Broadcast (CB) Centre (CBC). In Release 99 of the 3GPP specifications, the only specified broadcasting service is the SMS CB Service (CBS), which is derived from the Global System for Mobile Communications (GSM). The service offered by the BMC protocol is also called a RB.

The RRC layer offers services to higher layers (to the Non-Access Stratum (NAS)) via SAPs, which are used by the higher layer protocols in the UE side and by the Iu RAN Application Part protocol in the

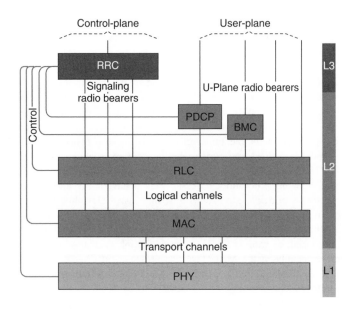

**Figure 7.1**   UTRA FDD radio interface protocol architecture

UTRAN side. All higher layer signaling (mobility management, call control, session management, and so on) is encapsulated into RRC messages for transmission over the radio interface.

The control interfaces between the RRC and all the lower layer protocols are used by the RRC layer to configure characteristics of the lower layer protocol entities, including parameters for the physical, transport and logical channels. The same control interfaces are used by the RRC layer, e.g. to command the lower layers to perform certain types of measurement and by the lower layers to report measurement results and errors to the RRC.

## 7.3   The Medium Access Control Protocol

In the MAC layer [3] the logical channels are mapped to the transport channels. The MAC layer is also responsible for selecting an appropriate transport format (TF) for each transport channel depending on the instantaneous source rate(s) of the logical channels. The TF is selected with respect to the TF combination set (TFCS), which is defined by the admission control for each connection.

### 7.3.1   MAC Layer Architecture

The MAC layer logical architecture is shown in Figure 7.2. The MAC layer consists of three logical entities:

- *MAC-b* handles the broadcast channel (BCH). There is one MAC-b entity in each UE and one MAC-b in the UTRAN (located in Node B) for each cell.
- *MAC-c/sh* handles the common channels and shared channels: paging channel (PCH), forward link access channel (FACH), random access channel (RACH), uplink Common Packet Channel (CPCH) and Downlink Shared Channel (DSCH). There is one MAC-c/sh entity in each UE that is using shared channel(s) and one MAC-c/sh in the UTRAN (located in the controlling RNC) for each cell. Note that the Broadcast Control Channel (BCCH) logical channel can be mapped to either the BCH or FACH transport channel. Since the MAC header format for the BCCH depends on the transport channel used, two BCCH instances are shown in Figure 7.2. For the Paging Control Channel (PCCH), there is no MAC header; thus, the only function of the MAC layer is to forward the data received from the PCCH to the PCH at the time instant determined by the RRC.

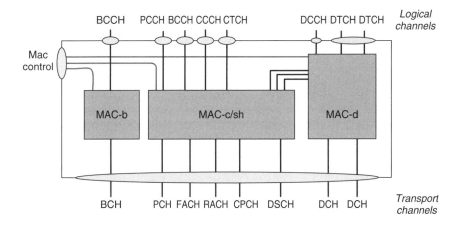

**Figure 7.2**   MAC layer architecture

- *MAC-d* is responsible for handling dedicated channels (DCHs) allocated to a UE in connected mode. There is one MAC-d entity in the UE and one MAC-d entity in the UTRAN (in the serving RNC) for each UE.

## 7.3.2 MAC Functions

The functions of the MAC layer include:

- *Mapping* between logical channels and transport channels.
- *Selection of appropriate TF* (from the TFCS) for each Transport Channel, depending on the instantaneous source rate.
- *Priority handling between data flows of one UE*. This is achieved by selecting 'high bit rate' and 'low bit rate' TFs for different data flows.
- *Priority handling between UEs by means of dynamic scheduling*. A dynamic scheduling function may be applied for common and shared downlink transport channels FACH and DSCH.
- *Identification of UEs on common transport channels*. When a common transport channel (RACH, FACH or CPCH) carries data from dedicated-type logical channels (Dedicated Control Channel (DCCH), Dedicated Traffic Channel (DTCH)), the identification of the UE (Cell Radio Network Temporary Identity (C-RNTI) or UTRAN Radio Network Temporary Identity (U-RNTI)) is included in the MAC header.
- *Multiplexing/demultiplexing of higher layer protocol data units* (PDUs) into/from transport blocks delivered to/from the physical layer on common transport channels. MAC handles service multiplexing for common transport channels (RACH/FACH/CPCH). This is necessary, since it cannot be done in the physical layer.
- *Multiplexing/demultiplexing of higher layer PDUs into/from transport block sets delivered to/from the physical layer on dedicated transport channels*. MAC also allows service multiplexing for dedicated transport channels. While the physical layer multiplexing makes it possible to multiplex any type of service, including services with different quality of service (QoS) parameters, MAC multiplexing is possible only for services with the same QoS parameters. Physical layer multiplexing is described in Chapter 6.
- *Traffic volume monitoring*. MAC receives RLC PDUs together with status information on the amount of data in the RLC transmission buffer. MAC compares the amount of data corresponding to a transport channel with the thresholds set by RRC. If the amount of data is too high or too low, MAC sends a measurement report on traffic volume status to RRC. The RRC can also request MAC to send these measurements periodically. The RRC use these reports for triggering reconfiguration of RBs and/or Transport Channels.
- *Dynamic Transport Channel type switching*. Execution of the switching between common and dedicated transport channels is based on a switching decision derived by RRC.
- *Ciphering*. If an RB is using transparent RLC mode, ciphering is performed in the MAC sub-layer (MAC-d entity). Ciphering is an XOR operation (as in GSM and a General Packet Radio System (GPRS)) where data are XORed with a ciphering mask produced by a ciphering algorithm. In MAC ciphering, the time-varying input parameter (COUNT-C) for the ciphering algorithm is incremented at each transmission time interval (TTI), i.e. once every 10, 20, 40 or 80 ms, depending on the transport channel configuration. Each RB is ciphered separately. The ciphering details are described in 3GPP specification TS 33.102 [4].
- *Access Service Class (ASC) selection for RACH transmission*. The PRACH resources (i.e. access slots and preamble signatures for FDD) may be divided between different ASCs in order to provide different priorities of RACH usage. The maximum number of ASCs is eight. MAC indicates the ASC associated with a PDU to the physical layer.

## 7.3.3   Logical Channels

The data transfer services of the MAC layer are provided on logical channels. A set of logical channel types is defined for the different kinds of data transfer services offered by MAC. A general classification of logical channels is into two groups: control channels and traffic channels. Control channels are used to transfer control-plane information, and traffic channels for user-plane information.

The control channels are:

- *BCCH*: a downlink channel for broadcasting system control information.
- *PCCH*: a downlink channel that transfers paging information.
- *DCCH*: a point-to-point bidirectional channel that transmits dedicated control information between a UE and the RNC. This channel is established during the RRC connection establishment procedure.
- *Common Control Channel (CCCH)*: a bidirectional channel for transmitting control information between the network and UEs. This logical channel is always mapped onto RACH/FACH transport channels. A long UTRAN UE identity is required (U-RNTI, which includes the SRNC address), so that the uplink messages can be routed to the correct serving RNC even if the RNC receiving the message is not the RNC serving this UE.

The traffic channels are:

- *DTCH*: a DTCH is a point-to-point channel, dedicated to one UE, for the transfer of user information. A DTCH can exist in both uplink and downlink.
- *Common Traffic Channel (CTCH)*: a point-to-multipoint downlink channel for transfer of dedicated user information for all or a group of specified UEs.

## 7.3.4   Mapping between Logical Channels and Transport Channels

The mapping between logical channels and transport channels is shown in Figure 7.3. The following connections between logical channels and transport channels exist:

- PCCH is connected to PCH.
- BCCH is connected to BCH and may also be connected to FACH.
- DCCH and DTCH can be connected to either RACH and FACH, to CPCH and FACH, to RACH and DSCH, to DCH and DSCH, or to a DCH and DCH.
- CCCH is connected to RACH and FACH.
- CTCH is connected to FACH.

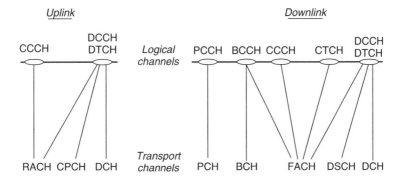

**Figure 7.3**   Mapping between logical channels and transport channels, uplink and downlink directions

## 7.3.5   Example Data Flow Through the MAC Layer

To illustrate the operation of the MAC layer, Figure 7.4 shows the MAC functions when data are processed through the layer. To keep Figure 7.4 readable, the viewpoint is selected to be a network-side transmitting entity, and uplink transport channels RACH and CPCH are omitted. The right-hand side of Figure 7.4 describes the building of a MAC PDU when a packet received from a DCCH or DTCH logical channel is processed by the MAC functions, which are shown in the left-hand side of Figure 7.4. In this example, the MAC PDU is forwarded to the FACH transport channel.

A data packet arriving from the DCCH/DTCH logical channel first triggers the transport channel type selection in the MAC layer. In this example, the FACH transport channel is selected. In the next phase, the multiplexing unit adds a C/T field indicating the logical channel instance where the data originate. For common transport channels, such as FACH, this field is always needed. For dedicated transport channels (DCH), it is needed only if several logical channel instances are configured to use the same transport channel. The C/T field is 4 bits, allowing up to 15 simultaneous logical channels per transport channel (the value '1111' for the C/T field is reserved for future use). The priority tag (not part of the MAC PDU) for FACH and DSCH is set in MAC-d and used by MAC-c/sh when scheduling data onto transport channels. Priority for FACH can be set per UE; for DSCH it can be set per PDU. A flow control function in the Iur interface (Chapter 5) is needed to limit buffering between MAC-d and MAC-c/sh (which can be located in different RNCs). After receiving the data from MAC-d, the MAC-c/sh entity first adds the UE identification type (2 bits), the actual UE identification (C-RNTI 16 bits, or U-RNTI 32 bits), and the Target Channel Type Field (TCTF, 2 bits in this example), which is needed to separate the logical

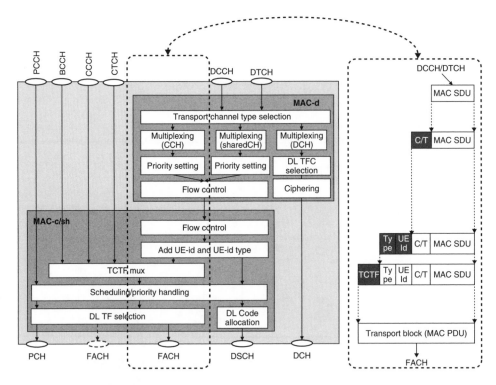

**Figure 7.4**   UTRAN side MAC entity (left) and building of MAC PDU when data received from DTCH or DCCH is mapped to FACH (right)

channel type using the transport channel (for FACH, the possible logical channel types could be BCCH, CCCH, CTCH or DCCH/DTCH). Now the MAC PDU is ready and the task for the scheduling/priority handling function is to decide the exact timing when the PDU is passed to layer 1 via the FACH transport channel (with an indication of the TF to be used).

## 7.4 The Radio Link Control Protocol

The RLC protocol [5] provides segmentation and retransmission services for both user and control data. Each RLC instance is configured by RRC to operate in one of three modes: transparent mode (Tr), unacknowledged mode (UM) or acknowledged mode (AM). The service the RLC layer provides in the control plane is called the SRB. In the user plane, the service provided by the RLC layer is called an RB only if the PDCP and BMC protocol are not used by that service, otherwise the RB service is provided by the PDCP or BMC.

### 7.4.1 RLC Layer Architecture

The RLC layer architecture is shown in Figure 7.5. All three RLC entity types and their connection to RLC-SAPs and to logical channels (MAC-SAPs) are shown. Note that the Tr and UM RLC entities are defined to be unidirectional, whereas the AM-mode entities are described as bidirectional.

For all RLC modes, the cyclic redundancy check (CRC) error detection is performed on physical layer and the result of the CRC check is delivered to RLC together with the actual data.

In Tr, no protocol overhead is added to higher layer data. Erroneous PDUs can be discarded or marked as erroneous. Transmission can be of the streaming type, in which higher layer data are not segmented, though in special cases transmission with limited segmentation/reassembly capability can be accomplished. If segmentation/reassembly is used, it has to be negotiated in the RB setup procedure. The UMTS QoS classes, including the streaming class, were introduced in Chapter 2.

In UM, no retransmission protocol is in use and data delivery is not guaranteed. Received erroneous data are either marked or discarded, depending on the configuration. On the sender side, a timer-based discard without explicit signaling function is applied; thus, RLC service data units (SDUs) which are not transmitted within a specified time are simply removed from the transmission buffer. The PDU structure includes sequence numbers (SNs) so that the integrity of higher layer PDUs can be observed. Segmentation and concatenation are provided by means of header fields added to the data. An RLC entity in UM is defined as unidirectional, because no association between uplink and downlink is needed.

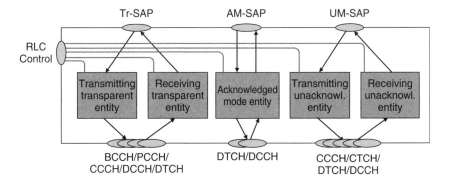

**Figure 7.5** RLC layer architecture

The UM is used, for example, for certain RRC signaling procedures, where the acknowledgement and retransmissions are part of the RRC procedure. Examples of user services that could utilize UM RLC are the CB service (see Section 7.6) and Voice-over-IP.

In the AM, an ARQ mechanism is used for error correction. The quality versus delay performance of the RLC can be controlled by RRC through the configuration of the number of retransmissions provided by RLC. If RLC is unable to deliver the data correctly (maximum number of retransmissions reached or the transmission time exceeded), the upper layer is notified and the RLC SDU is discarded. Also, the peer entity is informed of an SDU discard operation by sending a Move Receiving Window command (in a STATUS message), so that the receiver also removes all AM data (AMD) PDUs belonging to the discarded RLC SDU. An AM RLC entity is bidirectional and capable of 'piggybacking' an indication of the status of the link in the opposite direction into user data. RLC can be configured for both in-sequence and out-of-sequence delivery. With in-sequence delivery the order of higher layer PDUs is maintained, whereas out-of-sequence delivery forwards higher layer PDUs as soon as they are completely received. In addition to data PDU delivery, *status* and *reset* control procedures can be signaled between peer RLC entities. The control procedures can even use a separate logical channel; thus, one AM RLC entity can use either one or two logical channels. The AM is the normal RLC mode for packet-type services, such as internet browsing and email downloading, for example.

## 7.4.2  RLC Functions

The functions of the RLC layer are:

- *Segmentation and reassembly*. This function performs segmentation/reassembly of variable-length higher layer PDUs into/from smaller RLC Payload Units (PUs). One RLC PDU carries one PU. The RLC PDU size is set according to the smallest possible bit rate for the service using the RLC entity. Thus, for variable-rate services, several RLC PDUs need to be transmitted during one transmission time interval when any bit rate higher than the lowest one is used.
- *Concatenation*. If the contents of an RLC SDU do not fill an integral number of RLC PUs, then the first segment of the next RLC SDU may be put into the RLC PU in concatenation with the last segment of the previous RLC SDU.
- *Padding*. When concatenation is not applicable and the remaining data to be transmitted do not fill an entire RLC PDU of given size, the remainder of the data field is filled with padding bits.
- *Transfer of user data*. RLC supports acknowledged, unacknowledged and transparent data transfer. Transfer of user data is controlled by QoS setting.
- *Error correction*. This function provides error correction by retransmission in the acknowledged data transfer mode.
- *In-sequence delivery of higher layer PDUs*. This function preserves the order of higher layer PDUs that were submitted for transfer by RLC using the acknowledged data transfer service. If this function is not used, then out-of-sequence delivery is provided.
- *Duplicate detection*. This function detects duplicated received RLC PDUs and ensures that the resultant higher layer PDU is delivered only once to the upper layer.
- *Flow control*. This function allows an RLC receiver to control the rate at which the peer RLC transmitting entity may send information.
- *SN check (unacknowledged data transfer mode)*. This function guarantees the integrity of reassembled PDUs and provides a means of detecting corrupted RLC SDUs through checking the SN in RLC PDUs when they are reassembled into an RLC SDU. A corrupted RLC SDU is discarded.
- *Protocol error detection and recovery*. This function detects and recovers from errors in the operation of the RLC protocol.
- *Ciphering* is performed in the RLC layer for acknowledged and unacknowledged RLC modes. The same ciphering algorithm is used as for MAC layer ciphering, the only difference being the

time-varying input parameter (COUNT-C) for the algorithm, which for RLC is incremented together with the RLC PDU numbers. For retransmission, the same ciphering COUNT-C is used as for the original transmission (resulting in the same ciphering mask); this would not be so if ciphering were on the MAC layer. An identical ciphering mask for retransmissions is essential from Release 5 onwards, when the HSDPA feature with physical layer retransmission combining (as described in Chapter 11) is used. The ciphering details are described in 3GPP specification TS 33.102 [4].

- *Suspend/resume function for data transfer.* Suspension is needed during the security mode control procedure so that the same ciphering keys are always used by the peer entities. Suspensions and resumptions are local operations commanded by RRC via the control interface.

## 7.4.3   *Example Data Flow Through the RLC Layer*

This section takes a closer look at how data packets pass through the RLC layer. Figure 7.6 shows a simplified block diagram of an AM-RLC entity. Figure 7.6 shows only how an AMD PDU can be constructed. It does not show how separate control PDUs (*status, reset*) between RLC entities are built.

Data packets (RLC SDUs) received from higher layers via AM-SAP are segmented and/or concatenated to PUs of fixed length. The PU length is a semi-static value that is decided in the RB set-up and can only be changed through the (RRC) RB reconfiguration procedure. For concatenation or padding purposes, bits carrying information on the length and extension are inserted into the beginning of the last

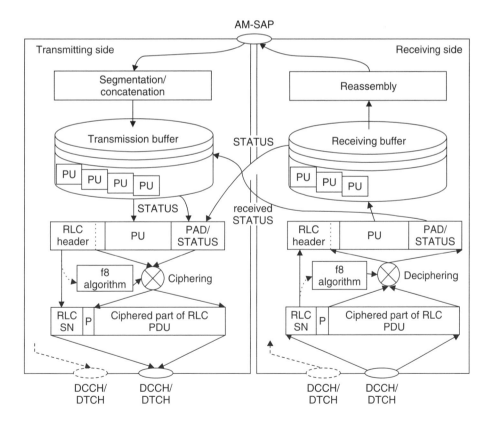

**Figure 7.6**   A simplified block diagram of an RLC AM entity

PU where data from an SDU is included. If several SDUs fit into one PU, then they are concatenated and the appropriate length indicators are inserted into the beginning of the PU. The PUs are then placed in the transmission buffer, which, in this example, also takes care of retransmission management.

An RLC AMD PDU is constructed by taking one PU from the transmission buffer, adding a header for it and, if the data in the PU do not fill the whole RLC AMD PDU, a PADding field or piggybacked STATUS message is appended. The piggybacked STATUS message can originate either from the receiving side (if the peer entity has requested a status report) or from the transmitting side to indicate an RLC SDU discard. The header contains the RLC PDU SN (12 bits for AM-RLC), poll bit P (which is used to request STATUS from the peer entity) and optionally a length indicator (7 or 15 bits), which is used if concatenation of SDUs, padding or a piggybacked STATUS PDU takes place in the RLC PDU.

Next, the AM RLC PDU is ciphered, excluding the first two octets which comprise the PDU SN and the poll bit (P). The PDU SN is one input parameter to the ciphering algorithm (forming the least significant bits of a COUNT-C parameter), and it must be readable by the peer entity to be able to perform deciphering. The details of the ciphering process are described in 3GPP specification TS 33.102 [4].

After this the PDU is ready to be forwarded to the MAC layer via a logical channel. In Figure 7.6, extra logical channels are shown by dashed lines, indicating that one RLC entity can be configured to send control PDUs and data PDUs using different logical channels. Note, however, that Figure 7.6 does not describe how the separate control PDUs are constructed.

The receiving side of the AM entity receives RLC AMD PDUs through one of the logical channels from the MAC sublayer. Errors are checked with the (physical layer) CRC, which is calculated over the whole RLC PDU. The actual CRC check is performed in the physical layer and the RLC entity receives result of this CRC check together with the data. After deciphering, the whole header and possible piggybacked status information can be extracted from the RLC PDU. If the received PDU was a control message or if status information was piggybacked to an AMD PDU, the control information (STATUS message) is passed to the transmitting side, which will check its retransmission buffer against the received status information. The PDU number from the RLC header is needed for deciphering and also when storing the deciphered PU into the receiving buffer. Once all PUs belonging to a complete SDU are in the receiving buffer, the SDU is reassembled. After this (not shown in Figure 7.6), the checks for in-sequence delivery and duplicate detection are performed before the RLC SDU is delivered to the higher layer.

As described in Chapter 12, the same RLC is used with HSDPA as well. The packets not successfully transmitted from the MAC-hs, when the discard timer expires in Node B, will be retransmitted again to the Node B from RLC layer when AM is used. Also, in connection with the various HSDPA mobility cases, there can be packets that are not transmitted from the Node B and the Node B MAC-hs buffer will be flushed, and the RLC layer is used to recover the lost data.

## 7.5   The Packet Data Convergence Protocol

The PDCP [6] exists only in the user plane and only for services from the PS domain. The PDCP contains compression methods, which are needed to get better spectral efficiency for services requiring Internet Protocol (IP) packets to be transmitted over the radio. For 3GPP Release 99 standards, a header compression method is defined, for which several header compression algorithms can be used. As an example of why header compression is valuable, the size of the combined Real-Time Protocol (RTP)/User Datagram Protocol (UDP)/IP headers is at least 40 bytes for IPv4 and at least 60 bytes for IPv6, while the payload, e.g. for IP voice service, can be about 20 bytes or less.

### 7.5.1   PDCP Layer Architecture

An example of the PDCP layer architecture is shown in Figure 7.7. Multiplexing of RBs in the PDCP layer is not part of 3GPP Release 99, but is one possible feature for future releases. The multiplexing possibility is illustrated in Figure 7.7 with the two PDCP SAPs (one with dashed lines) provided by one

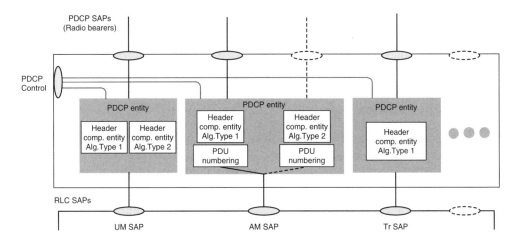

**Figure 7.7**   The PDCP layer architecture

PDCP entity using AM RLC. Every PDCP entity uses zero, one or several header compression algorithm types with a set of configurable parameters. Several PDCP entities may use the same algorithm types. The algorithm types and their parameters are negotiated during the RRC RB establishment or reconfiguration procedures and indicated to the PDCP through the PDCP Control SAP.

## 7.5.2   PDCP Functions

The main PDCP functions are:

- Compression of redundant protocol control information (e.g. Transport Control Protocol (TCP)/IP and RTP/UDP/IP headers) at the transmitting entity, and decompression at the receiving entity. The header compression method is specific to the particular network layer, transport layer or upper layer protocol combinations, e.g. TCP/IP and RTP/UDP/IP. The only compression method that is mentioned in the PDCP Release 99 specification is RFC2507 [7].
- Transfer of user data. This means that the PDCP receives a PDCP SDU from the non-access stratum and forwards it to the appropriate RLC entity and vice versa.
- Support for lossless Serving Radio Network Subsystem (SRNS) relocation. In practice, this means that those PDCP entities that are configured to support lossless SRNS relocation have PDU SNs, which, together with unconfirmed PDCP packets, are forwarded to the new SRNC during relocation. This is only applicable when PDCP is using AM RLC with in-sequence delivery.

## 7.6   The Broadcast/Multicast Control Protocol

The other service-specific layer 2 protocol, i.e. the BMC protocol [8], exists also only in the user plane. This protocol is designed to adapt broadcast and multicast services, originating from the Broadcast domain, on the radio interface. In Release 99 of the standard, the only service utilizing this protocol is the SMS CB service. This service is directly taken from GSM. It utilizes UM RLC using the CTCH logical channel which is mapped into the FACH transport channel. Each SMS CB message is targeted to a geographical area, and RNC maps this area into cells.

### 7.6.1  BMC Layer Architecture

The BMC protocol, shown in Figure 7.8, does not have any special logical architecture.

### 7.6.2  BMC Functions

The main functions of the BMC protocol are:

- *Storage of CB messages*. The BMC in RNC stores the CB messages received over the CBC–RNC interface for scheduled transmission.
- *Traffic volume monitoring and radio resource request for CBS*. On the UTRAN side, the BMC calculates the required transmission rate for the CBS based on the messages received over the CBC–RNC interface, and requests appropriate CTCH/FACH resources from RRC.
- *Scheduling of BMC messages*. The BMC receives scheduling information together with each CB message over the CBC–RNC interface. Based on this scheduling information, on the UTRAN side the BMC generates schedule messages and schedules BMC message sequences accordingly. On the UE side, the BMC evaluates the schedule messages and indicates scheduling parameters to RRC, which are used by RRC to configure the lower layers for CBS discontinuous reception.
- *Transmission of BMC messages to UE*. This function transmits the BMC messages (Scheduling and CB messages) according to the schedule.
- *Delivery of CB messages to the upper layer*. This UE function delivers the received non-corrupted CB messages to the upper layer.

When sending SMS CB messages to a cell for the first time, appropriate capacity has to be allocated in the cell. The CTCH has to be configured and the transport channel used has to be indicated to all UEs via (RRC) system information broadcast on the BCH. The capacity allocated for SMS CB is cell-specific and may vary over time to allow efficient use of the radio resources.

## 7.7  Multimedia Broadcast Multicast Service

In Release 6 the MBMS is being added to the standard, as covered in detail in Chapter 14. The principle is, similar to the CBS, to enable transmission of content to multiple users in a point to multipoint manner. The difference to CBS is that MBMS also enables UTRAN to control and monitor the users receiving the data and thus also enables charging for the content being delivered via MBMS. Typically, CBS has been used for low-rate information, such as sending cell location name, etc., but with MBMS the most quoted data rate has been 64 kbps, which enables more sophisticated content to be distributed.

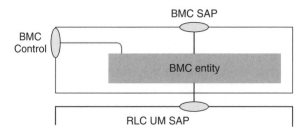

**Figure 7.8**   The BMC layer architecture

# 7.8 The Radio Resource Control Protocol

The major part of the control signaling between UE and UTRAN is RRC [9, 10] messages. RRC messages carry all parameters required to set up, modify and release layer 2 and layer 1 protocol entities. RRC messages also carry in their payload all higher layer signaling (mobility management (MM), connection management (CM), session management (SM), etc.). The mobility of user equipment in the connected mode is controlled by RRC signaling (measurements, handovers, cell updates, etc.).

## 7.8.1 RRC Layer Logical Architecture

The RRC layer logical architecture is shown in Figure 7.9. The RRC layer can be described through four *functional entities*:

- The Dedicated Control Function Entity (DCFE) handles all functions and signaling specific to one UE. In the SRNC, there is one DCFE entity for each UE having an RRC connection with this RNC. DCFE uses mostly AM RLC (AM-SAP), but some messages are sent using UM SAP (e.g. RRC Connection Release) or transparent SAP (e.g. Cell Update). DCFE can utilize services from all SRBs, which are described in Section 7.8.3.4.
- The Paging and Notification control Function Entity (PNFE) handles paging of idle-mode UE(s). There is at least one PNFE in the RNC for each cell controlled by this RNC. The PNFE uses the PCCH logical channel normally via transparent SAP of RLC. However, the specification mentions that PNFE could also utilize UM-SAP. In this example architecture, the PNFE in RNC, when receiving a paging message from an Iu interface, needs to check with the DCFE whether or not this UE already has an RRC connection (signaling connection with another CN domain); if it does, then the paging message is sent (by the DCFE) using the existing RRC connection.
- The broadcast control function entity (BCFE) handles the system information broadcasting. There is at least one BCFE for each cell in the RNC. The BCFE uses either BCCH or FACH logical channels, normally via transparent SAP. The specification mentions that BFCE could also utilize UM-SAP.
- The final entity is normally drawn outside of the RRC protocol, but still belonging to access stratum and 'logically' to the RRC layer, since the information required by this entity is part of RRC messages.

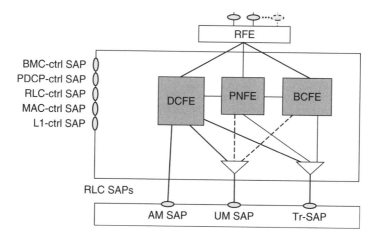

**Figure 7.9** The RRC layer architecture

The entity is called the Routing Function Entity (RFE) and its task is the routing of higher layer (NAS) messages to different MM/CM entities (UE side) or different CN domains (UTRAN side). Every higher layer message is piggybacked onto the RRC Direct Transfer messages (three types of Direct Transfer message are specified: Initial Direct Transfer (uplink), Uplink Direct Transfer and Downlink Direct Transfer).

## 7.8.2   RRC Service States

The two basic operational modes of a UE are idle mode and connected mode. The connected mode can be further divided into service states, which define what kind of physical channels a UE is using. Figure 7.10 shows the main RRC service states in the connected mode. It also shows the transitions between idle mode and connected mode and the possible transitions within the connected mode.

In the idle mode [11], after the UE is switched on, it selects (either automatically or manually) a public land mobile network (PLMN) to contact. The UE looks for a suitable cell of the chosen PLMN, chooses that cell to provide available services, and tunes to its control channel. This choice is known as 'camping on a cell'. The cell search procedure described in Chapter 6 is part of this camping process. After camping on a cell in idle mode, the UE is able to receive system information and CB messages. The UE stays in idle mode until it transmits a request to establish an RRC connection (Section 7.8.3.4). In idle mode the UE is identified by NAS identities such as international mobile subscriber identity (IMSI), temporary mobile subscriber identity (TMSI) and packet TMSI (P-TMSI). In addition, the UTRAN has no information of its own about the individual idle-mode UEs and can only address, for example, all UEs in a cell or all UEs monitoring a paging occasion.

In the Cell_DCH state, a dedicated physical channel is allocated to the UE and the UE is known by its serving RNC on a cell or active set level. The UE performs measurements and sends measurement reports according to measurement control information received from RNC. The DSCH can also be used in this state, and UEs with certain capabilities are also able to monitor the FACH for system information messages.

In the Cell_FACH state, no dedicated physical channel is allocated for the UE, but RACHs and FACHs are used instead, for transmitting both signaling messages and small amounts of user-plane data. In this state the UE is also capable of listening to the BCH to acquire system information. The CPCH can also be used when instructed by UTRAN. In this state the UE performs cell reselections, and after a reselection always sends a Cell Update message to the RNC, so that the RNC knows the UE location on a cell level. For identification, a C-RNTI in the MAC PDU header separates UEs from each other in a cell. When the UE performs cell reselection, it uses the U-RNTI when sending the Cell Update message, so that UTRAN can route the Cell Update message to the current serving RNC of the UE, even if the first RNC receiving the message is not the current SRNC. The U-RNTI is part of the RRC message, not in the MAC header. If the new cell belongs to another radio access system, such as GPRS, then the UE enters idle mode and accesses the other system according to that system's access procedure.

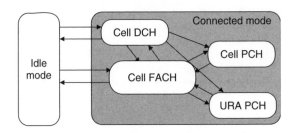

**Figure 7.10**   UE modes and RRC states in connected mode

In the Cell_PCH state the UE is still known on a cell level in SRNC, but it can be reached only via the PCH. In this state the UE battery consumption is less than in the Cell_FACH state, since the monitoring of the PCH includes a discontinuous reception (DRX) functionality. The UE also listens to system information on BCH. A UE supporting the CBS is also capable of receiving BMC messages in this state. If the UE performs a cell reselection, then it moves autonomously to the Cell_FACH state to execute the Cell Update procedure, after which it re-enters the Cell_PCH state if no other activity is triggered during the Cell Update procedure. If a new cell is selected from another radio access system, then the UTRAN state is changed to idle mode and access to the other system is performed according to that system's specifications.

The URA_PCH state is very similar to the Cell_PCH, except that the UE does not execute Cell Update after each cell reselection, but instead reads UTRAN Registration Area (URA) identities from the BCH, and only if the URA changes (after cell reselection) does UE inform its location to the SRNC. This is achieved with the URA Update procedure, which is similar to the Cell Update procedure (the UE enters the Cell_FACH state to execute the procedure and then reverts to the URA_PCH state). One cell can belong to one or many URAs, and only if the UE cannot find its latest URA identification from the list of URAs in a cell does it need to execute the URA Update procedure. This 'overlapping URA' feature is needed to avoid ping-pong effects in a possible network configuration, where geographically succeeding base stations are controlled by different RNCs.

The UE leaves the connected mode and returns to idle mode when the RRC connection is released or at RRC connection failure.

### 7.8.2.1  Enhanced State Model for Multimode Terminals

Figure 7.11 presents an overview of the possible state transitions of a multimode terminal, in this example, a UTRA FDD–GSM/GPRS terminal. With these terminal types it is possible to perform

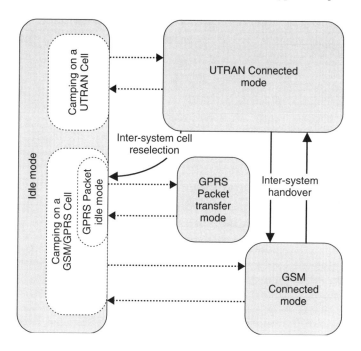

**Figure 7.11**   UE RRC states for a dual-mode UTRA FDD–GSM/GPRS terminal

inter-system handover between UTRA FDD and GSM, and inter-system cell reselection from UTRA FDD to GPRS. The actual signaling procedures that relate to the black arrows in Figure 7.11 are described in Section 7.8.3.

### 7.8.2.2  Example State Transition Cases with Packet Data

Understanding what is involved in the signaling for the RRC state changes is essential when analysing the system performance in the case of packet data operation. When sending or receiving reasonable amounts of data, UE will stay in the Cell_DCH state, but once the data runs out and timers have elapsed, the UE will be removed from the Cell_DCH state. Moving back to the Cell_DCH state always requires signaling between UE and SRNC, as well as for the network to set up the necessary links to Node B. The use of the Cell_DCH or Cell_FACH state is always a trade-off between terminal power consumption, service delay, signaling load and network resource utilization. The timing impacts from state changes are analysed in Chapter 10.

The first case is based on the UE-initiated state change, where an application has created data to be transmitted to the network and the amount is such that going to the Cell_FACH state and sending the data on RACH is not sufficient but a DCH is needed to be set up. The signaling flow is illustrated in Figure 7.12. To change to the Cell_FACH state there is no need to send signaling to the network. Once in the Cell_FACH state the UE initiates signaling on the RACH and, after the network has received the measurement report on RACH and the radio link has been set up between Node B and RNC, the reconfiguration message is sent on FACH to inform the DCH parameters to be used.

The network-initiated RRC state change obviously occurs when there is too much downlink data to be transmitted, so that using FACH is not enough. The network first transmits the paging message in the cell where the terminal is located (as the terminal location is known at the cell level in the Cell_PCH state). Upon reception of the paging message the terminal moves to the Cell_FACH state and initiates signaling on the RACH as illustrated in Figure 7.13. Now there is no need for any measurement report, as transition is initiated by the network. The response from the terminal in the example case of Figure 7.13 is reconfiguration of the complete message, assuming that the DCH parameters have been altered in connection with the state transition.

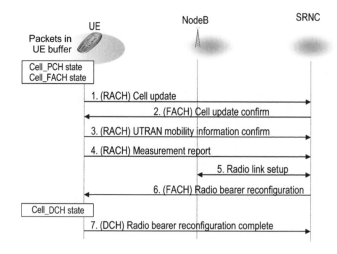

**Figure 7.12**   UE-initiated RRC state transition

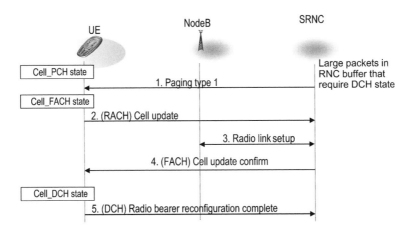

**Figure 7.13**  Network-initiated state transition

## 7.8.3  *RRC Functions and Signaling Procedures*

Since the RRC layer handles the main part of the control signaling between the UEs and UTRAN, it has a long list of functions to perform. Most of these functions are part of the RRM algorithms, which are discussed in Chapters 9 and 10; but since the information is carried in RRC layer messages, the specifications list the functions as part of the RRC protocol. The main RRC functions are:

- broadcast of system information, related to access stratum and NAS;
- paging;
- initial cell selection and reselection in idle mode;
- establishment, maintenance and release of an RRC connection between the UE and UTRAN;
- control of RBs, transport channels and physical channels;
- control of security functions (ciphering and integrity protection);
- integrity protection of signaling messages;
- UE measurement reporting and control of the reporting;
- RRC connection mobility functions;
- support of SRNS relocation;
- support for downlink outer loop power control in the UE;
- open-loop power control;
- CBS-related functions;
- support for UE Positioning functions.

In the following sections, these functions and related signaling procedures are described in more detail.

### 7.8.3.1  Broadcast of System Information

The broadcast system information originates from the CN, from RNC and from Node Bs. The System Information messages are sent on a BCCH logical channel, which can be mapped to the BCH or FACH transport channel. A System Information message carries system information blocks (SIBs), which group together system information elements of the same nature. Dynamic (i.e. frequently changing) parameters are grouped into different SIBs from the more static parameters. One System Information message can carry either several SIBs or only part of one SIB, depending on the size of the SIBs to be transmitted.

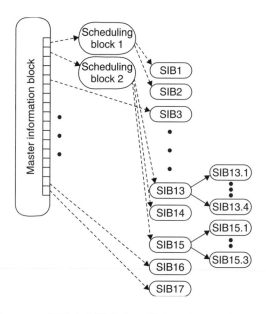

**Figure 7.14**   The overall structure of SIBs in 3GPP Release 99. Dotted arrows show an example where scheduling information for each SIB could be included

One System Information message will always fit into the size of a BCH or FACH transport block. If padding is required, it is inserted by the RRC layer.

The SIBs are organized as a tree (Figure 7.14). A master information block (MIB) gives references and scheduling information to a number of SIBs in a cell. It may also include reference and scheduling information to one or two scheduling blocks, which give references and scheduling information for all additional SIBs. The MIB is sent regularly on the BCH and its scheduling is static. In addition to scheduling information of other SIBs and scheduling blocks, the MIB contains only the parameters 'Supported PLMN Types' and, depending on which PLMN types are supported, either 'PLMN identity' (GSM MAP) or 'ANSI-41 Core Network Information'. The SIBs contain all the other actual system information.

The scheduling information (included in the MIB or in scheduling blocks) for SIBs containing frequently changing parameters contains SIB-specific timers (value in frames), which can be used by the UE to trigger re-reading of each block.

For the other SIBs (with more 'static' parameters) the MIB, or the 'parent' SIB, contains, as part of the scheduling information, a 'value tag' that the UE compares with the latest read 'value tag' of this SIB. Only if the value tag has changed after the last reading of the SIB in question does the UE re-read it. Thus, by monitoring the MIB and the scheduling blocks, the UE can notice if any of the SIBs (of the more 'static' nature) have changed. UTRAN can also inform the change in system information by Paging messages sent on the PCH transport channel (see Section 7.8.3.2) or by a System Information Change Indication message on the FACH transport channel. With these two messages, all UEs needing information about a change in the system information (all UEs in the Cell_FACH, Cell_PCH and URA_PCH states) can be reached.

The number of SIBs in 3GPP Release 99 is one MIB, two scheduling blocks and 17 SIBs. Only SIB number 10, containing information needed only in Cell_DCH state, is sent using FACH transport channel, all other SIBs (including MIB and scheduling blocks) are sent on BCH. Scheduling information for each SIB can be included only in one place, either in MIB or in one of the scheduling blocks.

### 7.8.3.2   Paging

The RRC layer can broadcast paging information on the PCCH from the network to selected UEs in a cell. The paging procedure can be used for three reasons:

- *At CN-originated call or session se-tup*. In this case the request to start paging comes from the CN via the Iu interface.
- *To change the UE state from Cell_PCH or URA_PCH to Cell_FACH*. This can be initiated, for example, by downlink packet data activity.
- *To indicate a change in the system information*. In this case RNC sends a paging message with no paging records, but with information describing a new 'value tag' for the MIB. This type of paging is targeted to all UEs in a cell.

### 7.8.3.3   Initial Cell Selection and Reselection in Idle Mode

The most suitable cell is selected, based on idle-mode measurements and cell selection criteria. The cell search procedure described in Chapter 6 is part of the cell selection process.

### 7.8.3.4   Establishment, Maintenance and Release of RRC Connection

The establishment of an RRC connection and SRBs between UE and UTRAN (RNC) is initiated by a request from higher layers (NAS) on the UE side. In a network-originated case the establishment is preceded by an RRC Paging message. The request from NAS is actually a request to set up a Signaling Connection between UE and CN (a Signaling Connection consists of an RRC Connection and an Iu Connection). Only if the UE is in idle mode (thus no RRC connection exists), does the UE initiate an RRC Connection Establishment procedure. There can always be only zero or one RRC connection between one UE and UTRAN. If more than one signaling connection between UE and CN nodes exists, then they all 'share' the same RRC connection.

The 'maintenance' of RRC connection refers to the RRC Connection Re-establishment functionality, which can be used to re-establish a connection after radio link failure. Timers are used to control the allowed time for a UE to return to 'in-service-area' and to execute the re-establishment. The re-establishment functionality is included in the Cell Update procedure (Section 7.8.3.9).

The RRC connection establishment procedure is shown in Figure 7.15. There is no need for a contention resolution step such as in GSM [12], since the UE identifier used in the connection request and set-up messages is a unique UE identity (for GSM-based CN P-TMSI + RAI, TMSI + LAI or IMSI). In the RRC connection establishment procedure, this initial UE identifier is used only for the purpose of uniqueness and can be discarded by UTRAN after the procedure ends. Thus, when these UE identities are needed later for the higher layer (NAS) signaling, they must be re-sent (in the higher layer messages). The RRC Connection Set-up message may include a dedicated physical channel assignment for the UE (move to Cell_DCH state), or it can command the UE to use common channels (move to Cell_FACH state). In the latter case, a radio network temporary identity (U-RNTI and possibly C-RNTI) to be used as a UE identity on common transport channels is allocated to the UE. The channel names in Figure 7.15 indicate either the logical channel or logical/transport channel used for each message.

The RRC connection establishment procedure creates three (optionally four) SRBs designated by the RB identities #1 . . . #4 (RB identity #0 is reserved for signaling using CCCH). The SRBs can later be created, reconfigured or even deleted with the normal RB control procedures. The SRBs are used for RRC signaling according to the following rules:

- RB#1 is used for all messages sent on the DCCH and RLC-UM.
- RB#2 is used for all messages sent on the DCCH and RLC-AM, except for the Direct Transfer messages.

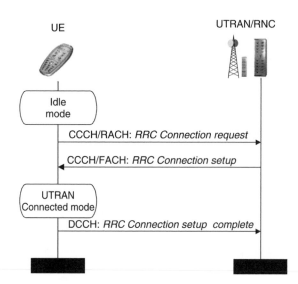

**Figure 7.15**   RRC connection establishment procedure

- RB#3 is used for the Direct Transfer messages (using DCCH and RLC-AM), which carries higher layer signaling. The reason for reserving a dedicated SRB for the Direct Transfer is to enable prioritization of UE-UTRAN signaling over the UE-CN signaling by using the RLC services (no need for extra RRC functionality).
- RB#4 is optional and, if it exists, is also used for the Direct Transfer messages (using DCCH and RLC-AM). With two SRBs carrying higher layer signaling, UTRAN can handle prioritization on signaling, RB#4 being used for 'low priority' and RB#3 for 'high priority' NAS signaling. The priority level is indicated to RRC through the actual NAS message to be carried over the radio. An example of low-priority signaling could be the SMS.
- For RRC messages utilizing Tr RLC and CCCH logical channel (e.g. Cell Update, URA Update), RB identity #0 is used. A special function required in the RRC layer for these messages is padding, because RLC in Tr neither imposes size requirements nor performs padding, but the message size must still be equal to a Transport Block size.

### 7.8.3.5   Control of RBs, Transport Channels and Physical Channels

On request from higher layers, RRC performs the establishment, reconfiguration and release of RBs. At establishment and reconfiguration, UTRAN (RNC) performs admission control and selects parameters describing the RB processing in layer 2 and layer 1. The SRBs are normally set up during the RRC Connection Establishment procedure (Section 7.8.3.4), but can also be controlled through the normal RB procedures.

The transport channel and physical channel parameters are included in the RB procedures, but can also be configured separately with transport channel and physical channel dedicated procedures. These are needed, for example, if temporary congestion occurs in the network or when switching the UE between Cell_DCH and Cell_FACH states.

### 7.8.3.6   Control of Security Functions

The RRC Security Mode Control procedure is used to start ciphering and integrity protection between the UE and UTRAN and to trigger the change of the ciphering and integrity keys during the connection.

The ciphering key is CN-domain specific; thus, in a typical network configuration (see Chapter 5), two ciphering keys can be used simultaneously for one UE: one for the PS domain services and one for the Circuit-Switched (CS) domain services. For the signaling (which uses common RB(s) for both CN domains) the newer of these two keys is used. Ciphering is executed on the RLC layer for services using unacknowledged or acknowledged RLC and on the MAC layer for services using transparent RLC.

Integrity protection (see Section 7.8.3.7) is used only for signaling. In a typical network configuration, two integrity keys would be available, but since only one RRC Connection can exist per UE, all signaling is protected with one and the same integrity key (IK), which is always the newer of the keys $IK_{CS}$ and $IK_{PS}$.

### 7.8.3.7 Integrity Protection of Signaling

The RRC layer inserts a 32-bit integrity checksum, called a Message Authentication Code MAC-I, into most RRC PDUs. The integrity checksum is used by the receiving RRC entity to verify the origin and integrity of the messages. The receiving entity also calculates MAC-I and compares it with the one received with the signaling message. Messages received with a wrong or missing message authentication code are discarded. Since all higher layer (NAS) signaling is carried in RRC Direct Transfer messages, all higher layer messages are automatically also integrity protected. The only exception to this is the initial higher layer message carried in Initial Direct Transfer message.

The checksum is calculated using a UMTS integrity algorithm (UIA) that uses a secret 128-bit IK as one input parameter. The key is generated together with the ciphering key (CK) during the authentication procedure [13]. Figure 7.16 illustrates the calculation of MAC-I using the integrity algorithm f9 [14]. In addition to the IK, other parameters used as input to the algorithm are COUNT-I, which is incremented by one for each integrity-protected message, a random number FRESH generated by RNC, DIRECTION bit (uplink/downlink), and the actual signaling message. Also, the SRB identity should affect the calculation of MAC-I. Since the algorithm f9 was ready when this requirement was identified, no new input parameter could be added to the f9 algorithm. The SRB identity is inserted into the MESSAGE before it is given to the integrity algorithm.

Only a few RRC messages cannot be integrity-protected; examples are the messages exchanged during the RRC Connection Establishment procedure, since the algorithms and parameters are not yet negotiated when these messages are sent.

### 7.8.3.8 UE Measurement Reporting and Control

The measurements performed by the UE are controlled by the RNC using RRC protocol messages, in terms of what to measure, when to measure and how to report, including both UTRA radio interface and other systems. RRC signaling is also used to report the measurements from the UE to the UTRAN (RNC).

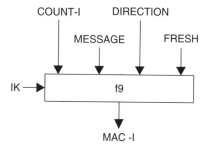

**Figure 7.16**   Calculation of message authentication code MAC-I

## Measurement Control

The measurement control (and reporting) procedure is designed to be very flexible. Serving RNC may start, stop or modify a number of parallel measurements in the UE and each of these measurements (including how they are reported) can be controlled independently of each other. The measurement control information is included in SIB Type 12 and SIB Type'11. For UEs in the Cell_DCH state, a dedicated Measurement Control message can also be used. This is illustrated in Figure 7.17.

The measurement control information includes:

- *Measurement identity number*. A reference number that is used by the UTRAN at modification or release of the measurement and by the UE in the measurement report.
- *Measurement command*. May be either setup, modify or release.
- *Measurement type*. One of the seven types from a predefined list, where each type describes what the UE measures. The seven types of measurement are defined as:
  - Intra-frequency measurements: measurements on downlink physical channels at the same frequency as the active set.
  - Inter-frequency measurements: measurements on downlink physical channels at frequencies that differ from the frequency of the active set.
  - Inter-system measurements: measurements on downlink physical channels belonging to a radio access system other than UTRAN, e.g. GSM.
  - Traffic volume measurements: measurements on uplink traffic volume, e.g. RLC buffer payload for each RB.
  - Quality measurements: measurements of quality parameters, e.g. downlink transport channel block error rate.
  - Internal measurements: measurements of UE transmission power and UE received signal level.
  - Measurements for Location Services (LCS) [15]. The basic measurement provided by the UE for the network-based OTDOA-IPDL positioning method is Observed Time Difference of system frame numbers (SFNs) between measured cells.
- *Measurement objects*. The objects the UE will measure, and corresponding object information. In handover measurements this is the cell information needed by the UE to make measurements on certain intra-frequency, inter-frequency or inter-system cells. In traffic volume measurements, this parameter contains transport channel identification.
- *Measurement quantity*. The quantity the UE measures.
- *Measurement reporting quantities*. The quantities the UE includes in the report.
- *Measurement reporting criteria*. The criteria that trigger the measurement report, such as periodic or event-triggered reporting.
- *Reporting mode*. This specifies whether the UE transmits the measurement report using acknowledged or unacknowledged data transfer of RLC.

## Measurement Reporting

The measurement reporting procedure (shown in Figure 7.18) is initiated from the UE side when the reporting criteria are met. The UE sends a Measurement Report message, including the measurement identity number and the measurement results.

The Measurement Report message is used in the Cell_DCH and Cell_FACH states. In the Cell_FACH state, it is used only for a traffic volume measurement report. Traffic volume measurements may be triggered also in Cell_PCH and URA_PCH states, but the UE first has to change to Cell_FACH state before being able to send a measurement report. In order to receive measurement information needed for the immediate establishment of macrodiversity when establishing a dedicated physical channel, the UTRAN may also request the UE to append radio link-related (intra-frequency) measurement reports to the following messages when they are sent on the RACH:

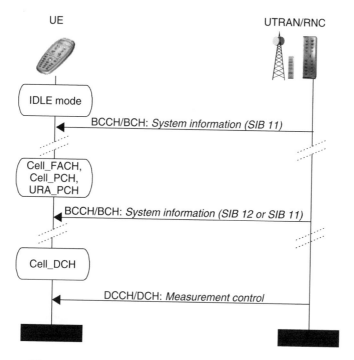

**Figure 7.17** Measurement control procedures in different UE states

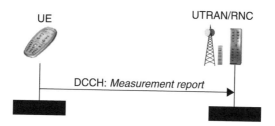

**Figure 7.18** Measurement reporting procedure

- *RRC Connection Request* message sent to establish an RRC connection.
- *Initial Direct Transfer* and *Uplink Direct Transfer* messages.
- *Cell Update* message.
- *Measurement Report* message sent to report uplink traffic volume in Cell_FACH state.

### 7.8.3.9 RRC Connection Mobility Functions

RRC 'connection mobility' means keeping track of a UE's location (on a cell or active set level) while the UE is in UTRAN Connected mode. For this, a number of RRC procedures are defined. When DCHs are allocated to a UE, a normal way to perform mobility control is to use an Active Set Update and Hard Handover procedures. When the UE is using only common channels (RACH/FACH/PCH) while in the

UTRAN Connected mode, specific procedures are used to keep track of UE location either on cell or on URA level.

The UE mobility-related RRC procedures include:

- *Active Set Update* to update the UE's active set while in the Cell_DCH state.
- *Hard Handover* to make inter-frequency or intra-frequency hard handovers while in the Cell_DCH state.
- *Inter-system handover* between UTRAN and another radio access system (e.g. GSM).
- *Inter-system cell reselection* between UTRAN and another radio access system (e.g. GPRS).
- *Inter-system cell change order* between UTRAN and another radio access system (e.g. GPRS).
- *Cell Update* to report the UE location into RNC while in the Cell_FACH or Cell_PCH state.
- *URA Update* to report the UE location into RNC while in the URA_PCH state.

These procedures are described in the following sections.

### Active Set Update

The purpose of the active set update procedure is to update the active set of the connection between the UE and UTRAN while the UE is in the Cell_DCH state. The procedure (shown in Figure 7.19) can have one of the following three functions: radio link addition, radio link removal, or combined radio link addition and removal. The maximum number of simultaneous radio links is eight, and it is even possible to remove all of them with one Active Set Update command. The soft handover algorithm and its performance are discussed in Section 9.3.1.

### Hard Handover

The Hard Handover procedure can be used to change the radio-frequency band of the connection between the UE and UTRAN or to change the cell on the same frequency when no network support of macrodiversity exists. It can also be used to change the mode between FDD and TDD. This procedure is used only in the Cell_DCH state. No dedicated signaling messages have been defined for the Hard Handover, but the functionality can be performed as part of the following RRC procedures: Physical channel reconfiguration, RB establishment, RB reconfiguration, RB release and Transport channel reconfiguration.

### Inter-System Handover from UTRAN

The inter-system handover from the UTRAN procedure is shown in Figure 7.20. This procedure is used for handover from UTRAN to another radio access system when the UE has at least one RAB in use for a CS domain service. For Release 99 UE, only support of handover of one RAB is expected, although the specification also allows handover of multiple RABs and even RABs from CS and PS domains

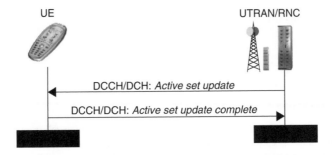

**Figure 7.19** Active set update procedure

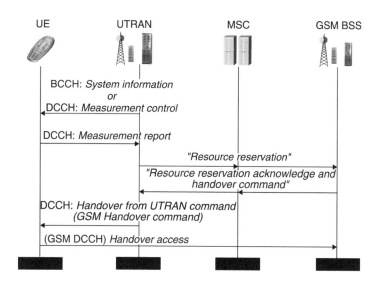

**Figure 7.20**    Inter-system handover procedure from UTRAN to GSM

simultaneously. In this example the target system is GSM, but the specifications also support handover to PCS 1900 and cdma2000 radio access systems. This procedure may be used in the Cell_DCH and Cell_FACH states. The UE receives the GSM neighbor cell parameters [12] either on System Information or in a Measurement Control message. These parameters are required to be able to measure candidate GSM cells. Based on the measurement report from UE, including GSM measurements, RNC makes a handover decision. After resources have been reserved from the GSM base station subsystem (BSS), the RNC sends a Handover From UTRAN Command message that carries a piggybacked GSM Handover Command. At this point the GSM round-robin protocol in UE takes control and sends a GSM Handover Access message to the GSM base station controller (BSC). After successful completion of the handover procedure, GSM BSS initiates resource release from UTRAN which will release the radio connection and remove all context information for the UE concerned.

*Inter-System Handover to UTRAN*
The inter-system handover to the UTRAN procedure is shown in Figure 7.21. This procedure is used for handover from a non-UTRAN system to UTRAN. In this example, the other system is again GSM. The dual-mode UE receives the UTRAN neighbour cell parameters on GSM System Information messages. The parameters required to be able to measure UTRA FDD cells include Downlink Centre frequency or UTRA Absolute Radio Frequency Channel Number (UARFCN), Downlink Bandwidth (in 3GPP Release 99, only 5 MHz are allowed, but other bandwidths may appear in future), Downlink Scrambling Code or scrambling code group for the Primary Common Pilot Channel (CPICH), and reference time difference for UTRA cell (timing between the current GSM cell and the UMTS cell that is to be measured).

After receiving a measurement report from the GSM mobile station, including UTRA measurements, and after making a handover decision, the GSM BSC initiates resource reservation from UTRAN RNC. In the next phase, GSM BSC sends a GSM Inter-System Handover Command [12] including a piggy-backed UMTS Handover To UTRAN Command message, which contains all the information required to set up the connection to a UTRA cell. The GSM handover message (Inter-System Handover Command) must fit into one 23-octet data link layer PDU. Since the amount of information that could be included in the Handover To UTRAN Command is large, a preconfiguration mechanism is included in the standards. The preconfiguration means that only a reference number to a predefined set of UTRA parameters

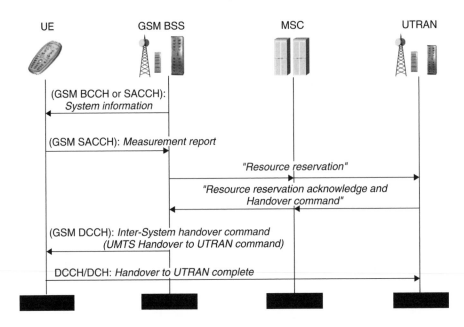

**Figure 7.21**   Inter-system handover procedure from GSM to UTRAN

(RB, Transport Channel and Physical Channel parameters) is included in the message. Naturally, the preconfiguration has to be transmitted to the UE beforehand. This can be done by GSM signaling or if the UE has been earlier in UMTS mode and has been able to read the preconfiguration information from SIB type 16. The UE completes the procedure with a Handover to UTRAN Complete message to RNC. After successful completion of the handover procedure, RNC initiates resource release from GSM BSS.

### Inter-System Cell Reselection from UTRAN
The inter-system cell reselection procedure from UTRAN is used to transfer a connection between the UE and UTRAN to another radio access system, such as GSM/GPRS. This procedure may be initiated in states Cell_FACH, Cell_PCH or URA_PCH. It is controlled mainly by the UE, but to some extent also by UTRAN. When UE has initiated an establishment of a connection to the other radio access system, it releases all UTRAN-specific resources.

### Inter-System Cell Reselection to UTRAN
The inter-system cell reselection procedure to UTRAN is used to transfer a connection between the UE and another radio access system, such as GSM/GPRS, to UTRAN. This procedure is controlled mainly by the UE, but to some extent also by the other radio access system. The UE initiates an RRC connection establishment procedure to UTRAN with cause value 'Inter-system cell reselection', and releases all resources specific to the other radio access system.

### Inter-System Cell Change Order from UTRAN
The inter-system cell change order procedure (illustrated in Figure 7.22) can be used by the UTRAN to order UE to another radio access system. This procedure is used for UEs having at least one RAB for PS domain services. This procedure may be used in Cell_DCH and Cell_FACH states. As in the case of inter-system handover from UTRAN, Release 99 UE is expected to be able to perform inter-system cell change with only one PS domain RAB, but the specification does not restrict this.

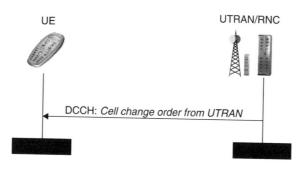

**Figure 7.22** Inter-system cell change order from UTRAN

The procedure is initiated by UTRAN with the Cell Change Order from UTRAN message. The message contains at least the required information of the target cell. After successful establishment of connection between UE and the other radio access system (e.g. GSM/GPRS), the other radio access system initiates release of the used UTRAN radio resources and UE context information.

### Inter-System Cell Change Order to UTRAN

This procedure is used by the other radio access system (e.g. GSM/GPRS) to command UE to move to the UTRAN cell. The 'cell change order' message in the other radio access system includes the identity of the target UTRAN cell. On the UTRAN side, the UE initiates an RRC connection establishment procedure with 'establishment cause' set to 'Inter-RAT cell change order'.

### Cell Update

The Cell Update procedure can be triggered for several reasons, including cell reselection, expiry of periodic cell update timer, initiation of uplink data transmission, UTRAN-originated paging and radio link failure in Cell_DCH state.

The Cell Update Confirm may include UTRAN mobility information elements (new U-RNTI and C-RNTI) for the UE. In this case, it responds with a UTRAN Mobility Information Confirm message so that the RNC knows that the new identities are taken into use.

The Cell Update Confirm may also include an RB release, RB reconfiguration, transport channel reconfiguration or physical channel reconfiguration. In these cases, the UE responds with a suitable 'complete' message; see Figure 7.23.

### URA Update

The URA Update procedure is used in the URA_PCH state. It can be triggered either after cell reselection, if the new cell does not broadcast the URA identifier that the UE is following, or by expiry of a periodic URA Update timer. Since no uplink activity is possible in the URA_PCH state, the UE has to switch temporarily to the Cell_FACH state to execute the signal processing procedure, as shown in Figure 7.24.

UTRAN registration areas may be hierarchical to avoid excessive signaling. This means that several URA identifiers may be broadcast in one cell and that different UEs in one cell may reside in different URAs. A UE in the URA_PCH state always has one and only one valid URA. If a cell broadcasts several URAs, then the RNC assigns one URA to a UE in the URA Update Confirm message.

The URA Update Confirm may assign a new URA Identity that the UE has to follow. It may also assign new RNTIs for the UE. In these cases, UE responds with a UTRAN Mobility Information Confirm message so that the RNC knows that the new identities are taken into use.

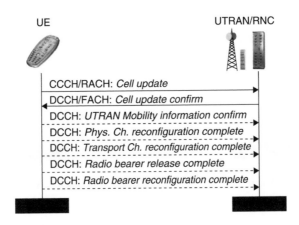

**Figure 7.23**    Cell update procedure

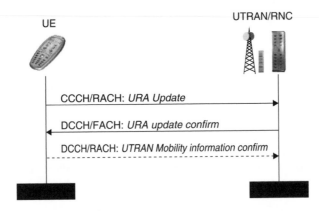

**Figure 7.24**    URA update procedure

### 7.8.3.10   Support of SRNS Relocation

In the SRNS relocation procedure (see Chapter 5), the SRNC RRC layer builds a special RRC message – RRC Information to Target RNC. The issue that makes this message a 'special' one is it is not targeted for UE but for the new SRNC. Thus, this message is not sent over the air, but carried from the old SRNC to the new one via the CN. The initialization information contains, for example, RRC state information and all the required protocol parameters (RRC, RLC, MAC, PDCP, PHY) that are needed to set up the UE context in the new SRNC. In addition, the expected PDCP SNs (which are normally maintained locally in UE and UTRAN) need to be sent between UE and UTRAN, in any RRC messages that are sent during the SRNS relocation.

### 7.8.3.11   Support for Downlink Outer Loop Power Control

All RRC messages that can be used to add or reconfigure downlink transport channels (e.g. RB Setup/Reconfiguration/Release, Transport Channel Reconfiguration) include a parameter 'Quality Target' (BLER quality value) that is used to configure the quality requirement (initial downlink SIR target) for each downlink transport channel separately.

The outer loop power control algorithm and its performance are discussed in Section 9.2.

### 7.8.3.12 Open-Loop Power Control

Prior to PRACH transmission (see Chapter 6), the UE calculates the power for the first preamble as:

$$\text{Preamble\_Initial\_Power} = \text{Primary CPICH DL TX power} - \text{CPICH\_RSCP}$$
$$+ \text{UL interference} + \text{constant value} \qquad (7.1)$$

The value for the CPICH_RSCP is measured by the UE, all other parameters being received on System Information.

As long as the physical layer is configured for PRACH transmission, the UE continuously recalculates the Preamble_Initial_Power when any of the broadcast parameters used in the above formula changes. The new Preamble_Initial_Power is then resubmitted to the physical layer.

When establishing the first DPCCH the UE shall start the UL inner loop power control at a power level according to

$$\text{DPCCH\_Initial\_Power} = \text{DPCCH\_Power\_offset} - \text{CPICH\_RSCP} \qquad (7.2)$$

The value for the DPCCH_Power_offset is received from UTRAN on various signaling messages. The value for the CPICH_RSCP shall be measured by the UE. The open-loop power control is not used once the inner loop power control is running.

### 7.8.3.13 CBS-Related Functions

The CBS-related functions of the RRC layer are as follows:

- Initial configuration of the BMC layer.
- Allocation of radio resources for CBS, in practice allocating the schedule for mapping the CTCH logical channel into the FACH transport channel and further into the S-CCPCH physical channel.
- Configuration of layer 1 and layer 2 for CBS discontinuous reception in the UE.

### 7.8.3.14 UE Positioning-Related Functions

Although the full set of Release 99 UTRAN specifications supports only the Cell_ID-based positioning method, the RRC protocol is already also capable of supporting both UE-based and UE-assisted OTDOA and GPS methods [15]. This includes the capability to transfer positioning-related UE measurements to UTRAN and delivery of assistance data for OTDOA and/or GPS from UTRAN to UE, which can be done either with System Information or with a dedicated message, called Assistance Data Delivery.

For Release 7, work is being done to add the necessary support for the Galileo satellite positioning system as well as for the uplink Time Difference Of Arrival (TDOA) methods. The work on Galileo will allow provide for assistance data for the support of assisted Galileo satellite operation. The necessary signaling details can be found from June 2007 and later versions of the 25.331. Further details on Galileo can be found in [16]. The 3GPP specifications are using the term Global Navigation Satellite System (GNSS) or A-GNSS for the case with assistance data for the Galileo operation. The uplink TDOA is based on the use of Location Management Units at the base station sites, which try to measure the signal from a particular UE in at least three different locations and use signal processing to extract off-line the UE signal timing (difference) at each observation point to enable calculation of the UE location.

## 7.9 Early UE Handling Principles

The topic that was discussed energetically during 2002 and the first part of 2003 in 3GPP standardization was how to handle the potential problems with terminals that are launched on the networks before full testing coverage exists. An example of the problem is a feature that is mandatory for the terminals but which is not implemented on any of the networks or on available test equipment before a terminal is put onto the market. Thus, in such a case, there can potentially be problems detected in the field. In order to ensure smooth system evolution, therefore, it was necessary to have a method to cope with such terminals without being forced to deactivate a particular feature where only a single UE is having problems.

The basic method chosen was the approach where a bit map is created in the CN side from the NAS signaling which contains the IMEISV information of a particular UE. The 3GPP standards define the meaning of the bit map generated by either the SGSN or the MSC. Based on the bit map, RNC can, for example, decide not to activate a particular feature for a given UE while allowing that to be used by other UEs. The method is illustrated in Figure 7.25.

If there are problems related to the first phase of the call set-up, there are few bits received in the initial access signaling to cope with that. However, those bits have not been needed in the field so far. In general, the first approach familiar from the GSM experiences is to see whether general parameterization can avoid the problem before using the bit-map solution. In 3GPP, preparation has been made to create the technical reports which can describe either the actions recommended by the bits in the bit map or the recommended specific parameterization. References [17] and [18] have not been proposed yet (March 2010) for any specific cases. There have been a few instances of UEs with strange behavior reported to 3GPP, the latest issue being the fast dormancy (pre-Release 8) where the faulty implementation causes heavy signaling load for the network, as discussed in Chapter 15. In the recent releases, Releases 7, 8 and 9, the issue has been addressed by making more or less all new additions to the release optional for the device. In this case a device that wants to implement, for example, CS over HSPA to save power and improve network capacity is not forced to wait for other Release 7 features (unless those are directly linked in implementing the CS over HSPA) to be ready on the network side for the inter-operability testing.

## 7.10 Improvements for Call Set-up Time Reduction

During the course of the Release 6 work, improvements were made to enable faster call set-up times. Obviously, the set-up times can also be improved by design choices in the actual implementations. The

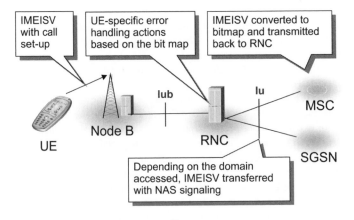

**Figure 7.25**  Early UE handling principle

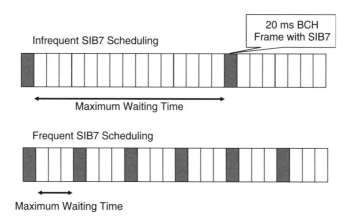

**Figure 7.26**   Impact for SIB7 scheduling on waiting time

following types of method were identified in 3GPP Release 6 to address the issue:

- Avoiding reading SIBs before RACH procedure (SIBs 11 and 12).
- Making the necessary system information available more frequently by scheduling the access proce-dure critical information on BCH more often, as illustrated in Figure 7.26 for the case of SIB7, which is carrying information necessary for RACH operation.
- Combining of different signaling steps to allow parallel progress of different parts of the call set-up process.
- Avoiding the use of activation time for reconfigurations (setting up a channel), but rather using acti-vation time 'now'.
- Making the signaling faster with higher data rate and especially with a shorter TTI channel, in case of HSDPA and HSUPA mapping the signaling on them.
- Using default configurations.

Further details of the methods can be found in [19], but note that all methods described were not included in the specifications. Some of the methods are applicable for the Release 99 scenario, while some of them require support of Releases 5 or 6 protocols and features like HSDPA and HSUPA. Additionally, eliminating the need for the Iub interface signaling by having RNC functionality in the Node B, as covered in Chapter 15, will obviously improve the situation as well for the PS call set-up delay, as the flat architecture is intended for PS traffic handling.

# References

[1]  3G TS 25.301, Radio Interface Protocol Architecture.
[2]  3G TS 25.302, Services Provided by the Physical Layer.
[3]  3G TS 25.321, MAC Protocol Specification.
[4]  3G TS 24.008, Mobile Radio Interface Layer 3 Specification, Core Network Proto cols – Stage 3.
[5]  3G TS 25.322, RLC Protocol Specification.
[6]  3G TS 25.323, PDCP Protocol Specification.
[7]  IETF RFC 2507 IP Header Compression.
[8]  3G TS 25.324, Broadcast/Multicast Control Protocol (BMC) Specification.
[9]  3G TS 25.303, UE Functions and Interlayer Procedures in Connected Mode.
[10]  3G TS 25.331, RRC Protocol Specification.

[11]  3G TS 25.304, UE Procedures in Idle Mode.
[12]  GSM 04.18 Digital Cellular Telecommunications System (Phase 2+); Mobile Radio Interface Layer 3 Spec-
      ification, Radio Resource Control Protocol.
[13]  3G TS 33.102 3G, Security; Security Architecture.
[14]  3G TS 33.105 3G, Security; Cryptographic Algorithm Requirements.
[15]  3G TS 25.305, Stage 2 Functional Specification of Location Services in UTRAN.
[16]  Galileo OS Signal in Space ICD (OS SIS ICD), Draft 0, Galileo Joint Undertaking, 23May 2006.
[17]  3GPP TR 25.994, 'Measures Employed by the UMTS Radio Access Network (UTRAN) to Overcome Early
      User Equipment (UE) Implementation Faults'.
[18]  3GPP TR 25.995, 'Measures Employed by the UMTS Radio Access Network (UTRAN) to Cater for Legacy
      User Equipment (UE) which Conforms to Superseded Versions of the RAN Interface Specification'.
[19]  3GPP TR 25.815, Signalling Enhancements for Circuit-Switched (CS) and Packet-Switched (PS) Connec-
      tions; Analyses and recommendations, (Release 7), September 2006.

# 8

# Radio Network Planning

Harri Holma, Zhi-Chun Honkasalo, Seppo Hämäläinen, Jaana Laiho, Kari Sipilä and Achim Wacker

## 8.1   Introduction

This chapter presents Wideband Code Division Multiple Access (WCDMA) radio network planning, including dimensioning, detailed capacity and coverage planning, and network optimization. The WCDMA radio network planning process is shown in Figure 8.1. In the dimensioning phase an approximate number of base station sites, base stations and their configurations and other network elements is estimated, based on the operator's requirements and the radio propagation in the area. The dimensioning must fulfil the operator's requirements for coverage, capacity and quality of service. Capacity and coverage are closely related in WCDMA networks, and therefore both must be considered simultaneously in the dimensioning of such networks. The dimensioning of WCDMA networks is introduced in Section 8.2.

In Section 8.3, detailed capacity and coverage planning are presented, together with a WCDMA planning tool. In detailed planning, real propagation maps and operator's traffic estimates in each area are needed. The base station locations and network parameters are selected by the planning tool and/or the planner. Capacity and coverage can be analysed for each cell after the detailed planning. One case study of the detailed planning is presented in Section 8.3 with capacity and coverage analysis. When the network is in operation, its performance can be observed by measurements, and the results of those measurements can be used to visualize and optimize network performance. The planning and the optimization process can also be automated with intelligent tools and network elements. The optimization is introduced in Section 8.3.

As most WCDMA-based networks will be deployed on top of the Global System for Mobile Communications (GSM) network, the GSM co-planning issues need to be considered. Co-planning is discussed in Section 8.4.

The adjacent channel interference must be considered in designing any wideband systems where large guard bands are not possible. In Section 8.5, the effect of interference between operators is analysed and network planning solutions are presented.

Section 8.6 presents the WCDMA frequency variants and their differences. These frequency variants are needed in the first place to deploy WCDMA in the USA.

*WCDMA for UMTS: HSPA Evolution and LTE, Fifth Edition*   Edited by Harri Holma and Antti Toskala
© 2010 John Wiley & Sons, Ltd

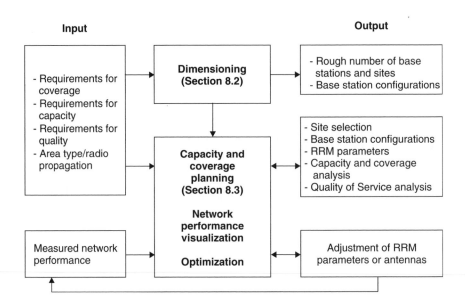

**Figure 8.1**  WCDMA radio network planning process

## 8.2  Dimensioning

WCDMA radio network dimensioning is a process through which possible configurations and the amount of network equipment are estimated, based on the operator's requirements related to the following:

Coverage:

- coverage regions;
- area type information;
- propagation conditions.

Capacity:

- spectrum available;
- subscriber growth forecast;
- traffic density information.

Quality of Service:

- area location probability (coverage probability);
- blocking probability;
- end user throughput.

Dimensioning activities include radio link budget and coverage analysis, capacity estimation and, finally, estimations of the number of sites and base station hardware, radio network controllers (RNCs), equipment at different interfaces, and core network elements (i.e. Circuit-Switched (CS) Domain and Packet-Switched (PS) Domain Core Networks).

## 8.2.1  Radio Link Budgets

The link budget of the WCDMA uplink is presented in this section. There are some WCDMA-specific parameters in the link budget that are not used in a Time Division Multiple Access (TDMA)-based radio access system such as GSM. The most important ones are as follows:

- *Interference margin.* The interference margin is needed in the link budget because the loading of the cell, the load factor, affects the coverage (see Section 8.2.2). The more loading is allowed in the system, the larger is the interference margin needed in the uplink, and the smaller is the coverage area. For coverage-limited cases, a smaller interference margin is suggested, while in capacity-limited cases, a larger interference margin should be used. In the coverage-limited cases, the cell size is limited by the maximum allowed path loss in the link budget, and the maximum air interface capacity of the base station site is not used. Typical values for the interference margin in the coverage-limited cases are 1.0–3.0 dB, corresponding to 20–50% loading.
- *Fast fading margin (=power control headroom).* Some headroom is needed in the mobile station transmission power to maintain adequate closed loop fast power control. This applies especially to slow-moving pedestrian mobiles where fast power control is able to effectively compensate the fast fading. The power control headroom was studied in [1]. The performance of fast power control is discussed in Section 9.2.1. Typical values for the fast fading margin are 2.0–5.0 dB for slow-moving mobiles.
- *Soft handover gain.* Handovers – soft or hard – give a gain against slow fading (=log-normal fading) by reducing the required log-normal fading margin. This is because the slow fading is partly uncorrelated between the base stations, and by making a handover the mobile can select a better base station. Soft handover gives an additional macro diversity gain against fast fading by reducing the required $E_b/N_0$ relative to a single radio link, due to the effect of macro diversity combining. The total soft handover gain is assumed to be between 2.0 and 3.0 dB in the examples below, including the gain against slow and fast fading. The handovers are discussed in Section 9.3 and the macro diversity gain for the coverage in Section 12.2.

Other parameters in the link budget are discussed in Chapter 7 in [2]. Below, three examples of link budgets are given for typical Universal Mobile Telecommunication Services (UMTS) services: 12.2 kbps voice service using Adaptive Multirate (AMR) speech codec, 144 kbps real-time data and 384 kbps non-real-time data, in an urban macro-cellular environment at the planned uplink noise rise of 3 dB. An interference margin of 3 dB is reserved for the uplink noise rise. The assumptions that have been used in the link budgets for the receivers and transmitters are shown in Tables 8.1 and 8.2.

The link budget in Table 8.3 is calculated for 12.2 kbps speech for in-car users, including 8.0 dB in-car loss. No fast fading margin is reserved in this case, since at 120 km/h the fast power control is unable to compensate for the fading. The required $E_b/N_0$ is assumed to be 5.0 dB. The $E_b/N_0$ requirement depends on the bit rate, service, multipath profile, mobile speed, receiver algorithms and base station antenna structure. For low mobile speeds, the $E_b/N_0$ requirement is low but, on the other

**Table 8.1**  Assumptions for the mobile station

|                              | Speech terminal | Data terminal |
| ---------------------------- | --------------- | ------------- |
| Maximum transmission power   | 21 dBm          | 24 dBm        |
| Antenna gain                 | 0 dBi           | 2 dBi         |
| Body loss                    | 3 dB            | 0 dB          |

**Table 8.2**   Assumptions for the base station

| | |
|---|---|
| Noise figure | 5.0 dB |
| Antenna gain | 18 dBi (3-sector base station) |
| $E_b/N_0$ requirement | Speech: 5.0 dB |
| | 144 kbps real-time data: 1.5 dB |
| | 384 kbps non-real-time data: 1.0 dB |
| Cable loss | 2.0 dB |

**Table 8.3**   Reference link budget of AMR 12.2 kbps voice service (120 km/h, in-car users, Vehicular A type channel, with soft handover)

| *Transmitter (mobile)* | | |
|---|---|---|
| Max. mobile transmission power [W] | 0.125 | |
| As above in dBm | 21.0 | a |
| Mobile antenna gain [dBi] | 0.0 | b |
| Body loss [dB] | 3.0 | c |
| Equivalent Isotropic Radiated Power (EIRP) [dBm] | 18.0 | d = a + b − c |
| *Receiver (base station)* | | |
| Thermal noise density [dBm/Hz] | −174.0 | e |
| Base station receiver noise figure [dB] | 5.0 | f |
| Receiver noise density [dBm/Hz] | −169.0 | g = e + f |
| Receiver noise power [dBm] | −103.2 | h = g + 10 log (3 840 000) |
| Interference margin [dB] | 3.0 | I |
| Total effective noise + interference [dBm] | −100.2 | j = h + i |
| Processing gain [dB] | 25.0 | k = 10 log (3840/12.2) |
| Required $E_b/N_0$ [dB] | 5.0 | l |
| Receiver sensitivity [dBm] | −120.2 | m = l − k + j |
| Base station antenna gain [dBi] | 18.0 | n |
| Cable loss in the base station [dB] | 2.0 | o |
| Fast fading margin [dB] | 0.0 | p |
| Max. path loss [dB] | **154.2** | q = d − m + n − o − p |
| Log-normal fading margin [dB] | 7.3 | r |
| Soft handover gain [dB], multicell | 3.0 | s |
| In-car loss [dB] | 8.0 | t |
| Allowed propagation loss for cell range [dB] | **141.9** | u = q − r + s − t |

hand, a fast fading margin is required. Typically, the low mobile speeds are the limiting factor in the coverage dimensioning because of the required fast fading margin. Table 8.4 shows the link budget for a 144 kbps real-time data service when an indoor location probability of 80% is provided by the outdoor base stations. The main differences between Tables 8.3 and 8.4 are the different processing gain, a higher mobile transmission power and a lower $E_b/N_0$ requirement. Additionally, a headroom of 4.0 dB is reserved for the fast power control to be able to compensate for the fading at 3 km/h. An average building penetration loss of 15 dB is assumed here.

The value on row $q$ gives the maximum path loss between the mobile and the base station antennas. The additional margins on rows $r$ and $t$ are needed to guarantee indoor coverage in the presence of shadowing. The shadowing is caused by buildings, hills, etc. and is modelled as log-normal fading. The value on row $u$ is used in the calculation of the cell range.

**Table 8.4** Reference link budget of AMR 144 kbps real time data service (3 km/h, indoor user covered by outdoor base station, Vehicular A type channel, with soft handover)

*Transmitter (mobile)*

| | | |
|---|---|---|
| Max. mobile transmission power [W] | 0.25 | |
| As above in dBm | 24.0 | a |
| Mobile antenna gain [dBi] | 2.0 | b |
| Body loss [dB] | 0.0 | c |
| Equivalent Isotropic Radiated Power (EIRP) [dBm] | 26.0 | d = a + b − c |
| *Receiver (base station)* | | |
| Thermal noise density [dBm/Hz] | −174.0 | e |
| Base station receiver noise figure [dB] | 5.0 | f |
| Receiver noise density [dBm/Hz] | −169.0 | g = e + f |
| Receiver noise power [dBm] | −103.2 | h = g + 10 log (3 840 000) |
| Interference margin [dB] | 3.0 | i |
| Total effective noise + interference [dBm] | −100.2 | j = h + i |
| Processing gain [dB] | 14.3 | k = 10 log (3840/144) |
| Required $E_b/N_0$ [dB] | 1.5 | l |
| Receiver sensitivity [dBm] | −113.0 | m = l − k + j |
| Base station antenna gain [dBi] | 18.0 | n |
| Cable loss in the base station [dB] | 2.0 | o |
| Fast fading margin [dB] | 4.0 | p |
| Max. path loss [dB] | **151.0** | q = d − m + n − o − p |
| Log-normal fading margin [dB] | 4.2 | r |
| Soft handover gain [dB], multicell | 2.0 | s |
| Indoor loss [dB] | 15.0 | t |
| Allowed propagation loss for cell range [dB] | **133.8** | u = q − r + s − t |

Table 8.5 presents a link budget for a 384 kbps non-real-time data service for outdoors. The processing gain is lower than in the previous tables because of the higher bit rate. Also, the $E_b/N_0$ requirement is lower than that of the lower bit rates. The effect of the bit rate on the coverage is discussed in Section 11.2. This link budget is calculated assuming no soft handover.

From the link budgets above, the cell range $R$ can be readily calculated for a known propagation model, for example the Okumura–Hata model or the Walfish–Ikegami model. For more on propagation models, see e.g. [3]. The propagation model describes the average signal propagation in that environment, and it converts the maximum allowed propagation loss in dB on the row $u$ to the maximum cell range in kilometres. As an example we can take the Okumura–Hata propagation model for an urban macro cell with base station antenna height of 30 m, mobile antenna height of 1.5 m and carrier frequency of 1950 MHz [4]:

$$L = 137.4 + 35.2 \log_{10}(R) \qquad (8.1)$$

where $L$ is the path loss in dB and $R$ is the range in km. For suburban areas we assume an additional area correction factor of 8 dB and obtain the path loss as:

$$L = 129.4 + 35.2 \log_{10}(R) \qquad (8.2)$$

According to Equation (8.2), the cell range of 12.2 kbps speech service with 141.9 dB path loss in Table 8.3 in a suburban area would be 2.3 km. The range of 144 kbps indoors would be 1.4 km. Once the cell range $R$ is determined, the site area, which is also a function of the base station sectorization configuration, can then be derived. For a cell of hexagonal shape covered by an omnidirectional antenna, the coverage area can be approximated as $2.6R^2$.

**Table 8.5** Reference link budget of non-real-time 384 kbps data service (3 km/h, outdoor user, Vehicular A type channel, no soft handover)

| *Transmitter (mobile)* | | |
|---|---|---|
| Max. mobile transmission power [W] | 0.25 | |
| As above in dBm | 24.0 | a |
| Mobile antenna gain [dBi] | 2.0 | b |
| Body loss [dB] | 0.0 | c |
| Equivalent Isotropic Radiated Power (EIRP) [dBm] | 26.0 | d = a + b − c |
| *Receiver (base station)* | | |
| Thermal noise density [dBm/Hz] | −174.0 | e |
| Base station receiver noise figure [dB] | 5.0 | f |
| Receiver noise density [dBm/Hz] | −169.0 | g = e + f |
| Receiver noise power [dBm] | −103.2 | h = g + 10 log (3 840 000) |
| Interference margin [dB] | 3.0 | i |
| Total effective noise + interference [dBm] | −100.2 | j = h + i |
| Processing gain [dB] | 10.0 | k = 10 log (3840/384) |
| Required $E_b/N_0$ [dB] | 1.0 | l |
| Receiver sensitivity [dBm] | −109.2 | m = l − k + j |
| Base station antenna gain [dBi] | 18.0 | n |
| Cable loss in the base station [dB] | 2.0 | o |
| Fast fading margin [dB] | 4.0 | p |
| Max. path loss [dB] | **147.2** | q = d − m + n − o − p |
| Log-normal fading margin [dB] | 7.3 | r |
| Soft handover gain [dB], multicell | 0.0 | s |
| Indoor loss [dB] | 0.0 | t |
| Allowed propagation loss for cell range [dB] | **139.9** | u = q − r + s − t |

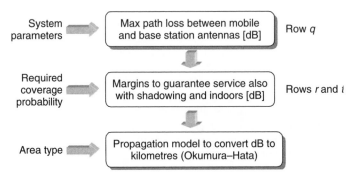

**Figure 8.2** Cell range calculation

The process of evaluating the cell range is summarized in Figure 8.2.

## 8.2.2   *Load Factors*

The second phase of dimensioning is estimating the amount of supported traffic per base station site. When the frequency reuse of a WCDMA system is 1, the system is typically interference-limited and the amount of interference and delivered cell capacity must thus be estimated.

### 8.2.2.1 Uplink Load Factor

The theoretical spectral efficiency of a WCDMA cell can be calculated from the load equation whose derivation is shown below. We first define the $E_b/N_0$, energy per user bit divided by the noise spectral density:

$$(E_b/N_0)_j = \text{Processing gain of user } j \cdot \frac{\text{Signal of user } j}{\text{Total received power (excl.own signal)}} \tag{8.3}$$

This can be written:

$$(E_b/N_0)_j = \frac{W}{\upsilon_j R_j} \cdot \frac{P_j}{I_{\text{total}} - P_j} \tag{8.4}$$

where $W$ is the chip rate, $P_j$ is the received signal power from user $j$, $\upsilon_j$ is the activity factor of user $j$, $R_j$ is the bit rate of user $j$, and $I_{\text{total}}$ is the total received wideband power including thermal noise power in the base station. Solving for $P_j$ gives

$$P_j = \frac{1}{1 + \frac{W}{(E_b/N_0)_j \cdot R_j \cdot \upsilon_j}} I_{\text{total}} \tag{8.5}$$

We define $P_j = L_j \cdot I_{\text{total}}$ and obtain the load factor $L_j$ of one connection

$$L_j = \frac{1}{1 + \frac{W}{(E_b/N_0)_j \cdot R_j \cdot \upsilon_j}} \tag{8.6}$$

The total received interference, excluding the thermal noise $P_N$, can be written as the sum of the received powers from all $N$ users in the same cell

$$I_{\text{total}} - P_N = \sum_{j=1}^{N} P_j = \sum_{j=1}^{N} L_j \cdot I_{\text{total}} \tag{8.7}$$

The noise rise is defined as the ratio of the total received wideband power to the noise power

$$\text{Noise rise} = \frac{I_{\text{total}}}{P_N} \tag{8.8}$$

and using Equation (8.7) we can obtain

$$\text{Noise rise} = \frac{I_{\text{total}}}{P_N} = \frac{1}{1 - \sum_{j=1}^{N} L_j} = \frac{1}{1 - \eta_{\text{UL}}} \tag{8.9}$$

where we have defined the load factor $\eta_{\text{UL}}$ as

$$\eta_{\text{UL}} = \sum_{j=1}^{N} L_j \tag{8.10}$$

When $\eta_{\text{UL}}$ becomes close to 1, the corresponding noise rise approaches infinity and the system has reached its pole capacity.

Additionally, in the load factor the interference from the other cells must be taken into account by the ratio of other cell to own cell interference, $i$:

$$i = \frac{\text{other cell interference}}{\text{own cell interference}} \tag{8.11}$$

The uplink load factor can then be written as

$$\eta_{UL} = (1+i) \cdot \sum_{j=1}^{N} L_j = (1+i) \cdot \sum_{j=1}^{N} \frac{1}{1 + \frac{W}{(E_b/N_0)_j \cdot R_j \cdot v_j}} \tag{8.12}$$

The load equation predicts the amount of noise rise over thermal noise due to interference. The noise rise is equal to $E_b/N_0$. The interference margin on row $i$ in the link budget must be equal to the maximum planned noise rise.

The required $E_b/N_0$ can be derived from link level simulations, from measurements and from the 3GPP performance requirements. It includes the effect of the closed loop power control and soft handover. The effect of soft handover is measured as the macro diversity combining gain relative to the single link $E_b/N_0$ result. The other cell to own (serving) cell interference ratio $i$ is a function of cell environment or cell isolation (e.g. macro/micro, urban/suburban) and antenna pattern (e.g. omni, 3-sector or 6-sector [5]). The parameters are further explained in Table 8.6.

The load equation is commonly used to make a semi-analytical prediction of the average capacity of a WCDMA cell, without going into system-level capacity simulations. This load equation can be used for the purpose of predicting cell capacity and planning noise rise in the dimensioning process.

For a classical all-voice-service network, where all $N$ users in the cell have a low bit rate of $R$, we can note that

$$\frac{W}{E_b/N_0 \cdot R \cdot v} \gg 1 \tag{8.13}$$

and the above uplink load equation can be approximated and simplified to

$$\eta_{UL} = \frac{E_b/N_0}{W/R} \cdot N \cdot v \cdot (1+i). \tag{8.14}$$

An example uplink noise rise is shown in Figure 8.3 for data service, assuming an $E_b/N_0$ requirement of 1.5 dB and $i = 0.65$. The noise rise of 3.0 dB corresponds to a 50% load factor, and the noise rise of 6.0 dB to a 75% load factor. Instead of showing the number of users $N$, we show the total data throughput per cell of all simultaneous users. In this example, a throughput of 860 kbps can be supported with 3.0 dB noise rise, and 1300 kbps with 6.0 dB noise rise.

**Table 8.6** Parameters used in uplink load factor calculation

|  | Definitions | Recommended values |
|---|---|---|
| $N$ | Number of users per cell | |
| $v_j$ | Activity factor of user $j$ at physical layer | 0.67 for speech, assumed 50% voice activity and DPCCH overhead during DTX 1.0 for data |
| $E_b/N_0$ | Signal energy per bit divided by noise spectral density that is required to meet a predefined Block error rate, *BLER*. Noise includes both thermal noise and interference | Dependent on service, bit rate, multipath fading channel, receive antenna diversity, mobile speed, etc. See Section 12.5. |
| $W$ | WCDMA chip rate | 3.84 Mcps |
| $R_j$ | Bit rate of user $j$ | Dependent on service |
| $i$ | Other cell to own cell interference ratio seen by the base station receiver | Macro cell with omnidirectional antennas: 55%. Macro cell with 3 sectors: 65%. |

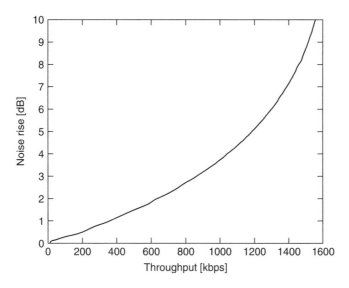

**Figure 8.3**  Uplink noise rise as a function of uplink data throughput

#### 8.2.2.2   Downlink Load Factor

The downlink load factor, $\eta_{DL}$, can be defined based on a similar principle as for the uplink, although the parameters are slightly different [6]:

$$\eta_{DL} = \sum_{j=1}^{N} \upsilon_j \cdot \frac{(E_b/N_0)_j}{W/R_j} \cdot [(1 - \alpha_j) + i_j] \tag{8.15}$$

where $-10\log_{10}(1 - \eta_{DL})$ is equal to the noise rise over thermal noise due to multiple access interference. The parameters are further explained in Table 8.7. Compared to the uplink load equation, the most important new parameter is $\alpha_j$, which represents the orthogonality factor in the downlink. WCDMA employs orthogonal codes in the downlink to separate users, and without any multipath propagation the orthogonality remains when the base station signal is received by the mobile. However, if there is sufficient delay spread in the radio channel, the mobile will see part of the base station signal as multiple access interference. The orthogonality of 1 corresponds to perfectly orthogonal users. Typically, the orthogonality is between 0.4 and 0.9 in multipath channels.

In the downlink, the ratio of other cell to own cell interference, $i$, depends on the user location and is therefore different for each user $j$. The load factor can be approximated by its average value across the cell, that is

$$\overline{\eta_{DL}} = \sum_{j=1}^{N} \upsilon_j \cdot \frac{(E_b/N_0)_j}{W/R_j} \cdot [(1 - \overline{\alpha}) + \overline{i}] \tag{8.16}$$

In downlink interference modelling, the effect of soft handover transmission can be modelled in two different ways:

1. Increase the number of connections by soft handover overhead, and reduce the $E_b/N_0$ requirement per link with soft handover gain.
2. Keep the number of connections fixed, i.e. equal to the number of users, and use the combined $E_b/N_0$ requirement.

**Table 8.7** Parameters used in downlink load factor calculation

|  | Definitions | Recommended values for dimensioning |
|---|---|---|
| $N$ | Number of users per cell | |
| $\upsilon_j$ | Activity factor of user $j$ at physical layer | 0.58 for speech, assumed 50% voice activity and DPCCH overhead during DTX 1.0 for data |
| $E_b/N_0$ | Signal energy per bit divided by noise spectral density, required to meet a predefined Block error rate, *BLER*. Noise includes both thermal noise and interference | Dependent on service, bit rate, multipath fading channel, transmit antenna diversity, mobile speed, etc. See Section 12.5. |
| $W$ | WCDMA chip rate | 3.84 Mcps |
| $R_j$ | Bit rate of user $j$ | Dependent on service |
| $\alpha_j$ | Orthogonality of channel of user $j$ | Dependent on the multipath propagation 1: fully orthogonal 1-path channel 0: no orthogonality |
| $i_j.$ | Ratio of other cell to own cell base station power, received by user $j$ | Each user sees a different $i_j$, depending on its location in the cell and log-normal shadowing |
| $\overline{\alpha}$ | Average orthogonality factor in the cell | ITU Vehicular A channel: '50% ITU Pedestrian A channel: ~90% |
| $\overline{i}$ | Average ratio of other cell to own cell base station power received by user. Own cell interference is here wideband | Macro cell with omnidirectional antennas: 55%. Macro cell with 3 sectors: 65%. |

*Note*: The own cell is defined as the best serving cell. If a user is in soft handover, all the other base stations in the active set are counted as part of the 'other cell'.

If the soft handover gain per link is assumed to be 3 dB, the combined $E_b/N_0$ is the same both with and without soft handover. In that case we do not need to include the effect of soft handover in the air interface dimensioning. This simplified approach is used in the examples later in this chapter. Figure 8.4 illustrates the soft handover modelling in dimensioning with two cells.

The downlink load factor $\eta_{DL}$ exhibits very similar behavior to the uplink load factor $\eta_{UL}$, in the sense that when approaching unity, the system reaches its pole capacity and the noise rise over thermal goes to infinity.

For downlink dimensioning, it is important to estimate the total amount of base station transmission power required. This should be based on the *average* transmission power for the user, not the *maximum*

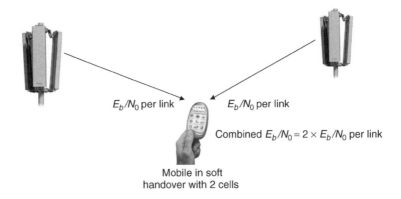

**Figure 8.4** Soft handover modelling with two cells

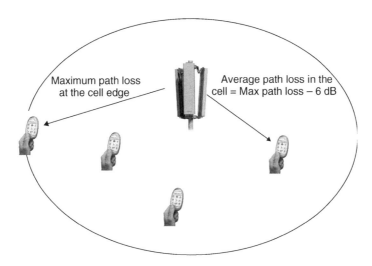

**Figure 8.5** Maximum and average path loss in macro cells

transmission power for the cell edge shown by the link budget. The reason is that the wideband technology gives trunking gain in the power amplifier dimensioning: while some users at the cell edge require high power, other users close to the base station need much less power at the same time. The difference between the maximum and the average path loss is typically 6 dB in macro cells and it is illustrated in Figure 8.5. This effect can be considered as power trunking gain of the wideband technology allowing use of a smaller base station power amplifier than in narrowband technologies.

The minimum required transmission power for each user is determined by the average attenuation between base station transmitter and mobile receiver, that is $\overline{L}$, and the mobile receiver sensitivity, in the absence of multiple access interference (intra- or inter-cell). Then the effect of noise rise due to interference is added to this minimum power and the total represents the transmission power required for a user at an 'average' location in the cell. Mathematically, the total base station transmission power can be expressed by the following equation:

$$BS\_TxP = \frac{N_{\mathrm{rf}} \cdot W \cdot \overline{L} \cdot \sum_{j=1}^{N} \upsilon_j \frac{(E_b/N_0)_j}{W/R_j}}{1 - \overline{\eta}_{\mathrm{DL}}} \tag{8.17}$$

where $N_{\mathrm{rf}}$ is the noise spectral density of the mobile receiver front-end. The value of $N_{\mathrm{rf}}$ can be obtained from

$$N_{\mathrm{rf}} = \mathrm{k} \cdot T + NF = -174.0\,\mathrm{dBm} + NF \text{ (assuming } T = 290\,K) \tag{8.18}$$

where k is the Boltzmann constant of $1.381 \times 10^{-23}$ J/K, $T$ is temperature in Kelvin and $NF$ is the mobile station receiver noise figure with typical values of 5–9 dB.

### *Downlink Common Channels*
Part of the downlink power has to be allocated for the common channels that are transmitted independently of the traffic channels. The common channels were introduced in Section 6.5. The amount of power of the common channels affects synchronization time, channel estimation accuracy, and the reception quality of the broadcast channel. On the other hand, the common channels eat up the capacity of the cell that could otherwise be allocated for the traffic channels. Typical power allocations for the common channels are shown in Table 8.8.

**Table 8.8**  Typical powers for the downlink common channels [7]

| Downlink common channel | Relative to CPICH | Activity | Average power allocation with 20 W maximum power |
|---|---|---|---|
| Common pilot channel CPICH | 0 dB | 100% | 2.0 W |
| Primary synchronisation channel SCH | −3 dB | 10% | 0.1 W |
| Secondary synchronization channel SCH | −3 dB | 10% | 0.1 W |
| Primary common control physical channel P-CCPCH | −5 dB | 90% | 0.6 W |
| Paging indicator channel PICH | −8 dB | 100%[1] | 0.3 W |
| Acquisition indicator channel AICH | −8 dB | 100%[1] | 0.3 W |
| Secondary common control physical channel S-CCPCH | 0 dB[2] | 10%[3] | 0.2 W |
| Total common channel powers | | | 3.6 W |

[1]Worst case
[2]Depends on the FACH bit rate, 32 kbps assumed here
[3]Depends on the amount of PCH and FACH traffic

### 8.2.2.3  Example Load Factor Calculation

Example downlink load factor calculations are demonstrated in this section. The assumptions are shown in Table 8.9.

The results are obtained as follows:

1. Assume the required aggregate cell throughput in kbps.
2. Calculate load factor $\overline{\eta_{DL}}$ from Equation (8.16). Throughput is equal to the number of users $N \times$ bit rate $R \times (1 - \text{BLER})$.
3. Calculate average path loss from Equation (8.17).
4. Calculate maximum path loss by adding 6 dB.

**Table 8.9**  Assumptions in example calculation

| Parameter | Data | Voice |
|---|---|---|
| Activity factor $\upsilon_j$ | 1.0 | Downlink: 0.58 Uplink: 0.67 |
| $E_b/N_0$ | 5.0 dB | 7.0 dB |
| Block error rate BLER | 10% | 1% |
| Bit rate of user $R_j$ | 64 kbps | 12.2 kbps |
| Mobile antenna gain | 2 dBi | 0 dBi |
| WCDMA chip rate $W$ | 3.84 Mcps | |
| Orthogonality $\overline{\alpha}$ | 0.5 | |
| Other cell to own cell interference ratio $i$ | 0.65 | |
| Base station output power | 20 W | |
| Common channel power allocation | 15% of base station max power $= 3$ W | |
| Base station cable loss in downlink | 3 dB | |
| Base station cable loss in uplink | 0 dB, cable loss compensated with mast head amplifier | |
| Average mobile noise figure $N_{rf}$ | 7 dB | |
| Maximum vs. average path loss | 6 dB | |

**Table 8.10**  Maximum path loss calculations for data

| Throughput $N \cdot R \cdot (1 - \text{BLER})$ | Load factor $\overline{\eta_{DL}}$ | Average path loss $\overline{L}$ | Max path loss |
|---|---|---|---|
| 100 kbps | 12% | 170.7 dB | 176.7 dB |
| 200 kbps | 25% | 167.1 dB | 173.1 dB |
| 300 kbps | 37% | 164.5 dB | 170.5 dB |
| 400 kbps | 50% | 162.3 dB | 168.3 dB |
| 500 kbps | 62% | 160.1 dB | 166.6 dB |
| 600 kbps | 74% | 157.7 dB | 164.7 dB |
| 700 kbps | 87% | 154.1 dB | 160.1 dB |
| 800 kbps | 99% | 142.1 dB | 148.1 dB |
| 808 kbps | 100% = pole capacity | – | – |

The results are shown for data in Table 8.10. It is assumed that the soft handover gain is 3 dB per link, i.e. the total transmission power is the same with and without soft handover. In this case we need not consider the soft handover in the air interface dimensioning.

The results in Table 8.10 are plotted in Figure 8.6 together with the corresponding uplink calculations. The uplink is calculated for 64 kbps data and the link budget is shown in Table 8.19. In both uplink and downlink the air interface load affects the coverage but the effect is not exactly the same. In the downlink, the coverage depends more on the load than in the uplink, according to Figure 8.6. The reason is that in the downlink the power of 20 W is shared between the downlink users: the more users, the less power per user. Therefore, even with low load in the downlink, the coverage decreases as a function of the number of users.

We note that with the above assumptions the coverage is clearly limited by the uplink for loads below 760 kbps, while the capacity is downlink limited. Therefore, in Chapter 11 the coverage discussion concentrates on the uplink, while the capacity discussion concentrates on downlink.

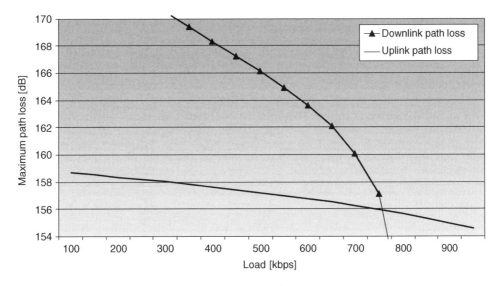

**Figure 8.6**  Example coverage vs. capacity relationship in downlink and uplink in macro cells

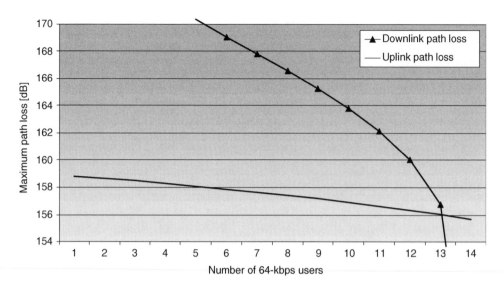

**Figure 8.7**   The same as Figure 8.6 for 64-kbps users with BLER of 10%

Figure 8.7 presents the same curves as Figure 8.6 but with number of simultaneous 64-kbps users. Figure 8.8 presents the load curves for voice users.

We need to further remember that in third generation networks the data traffic can be asymmetric between uplink and downlink, and the load can be different in uplink and in downlink.

The WCDMA load equations assume that all users are allocated the same bit rate which corresponds to real-time service with a guaranteed bit rate. If we allocate the same power, instead of the same bit rate, to all users, the cell throughput would be increased by 30–40%.

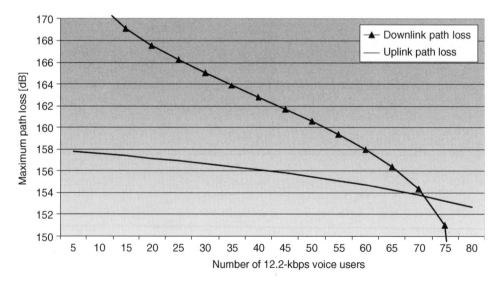

**Figure 8.8**   Example coverage vs. capacity relationship in macro cells for voice users

### Power Splitting Between Frequencies

In Figure 8.6 a base station maximum power of 20 W is assumed. What would happen to the downlink performance if the maximum power was lowered to 10 W? The difference in downlink coverage and capacity between 10 W and 20 W base station output powers is shown in Figure 8.9. If we lower the downlink power by 3.0 dB, the maximum allowed path loss is 3 dB lower. The effect on the capacity is smaller than the effect on the coverage because of the load curve. If we now keep the downlink path loss fixed at 156 dB, which is the maximum uplink path loss with 3 dB interference margin, the downlink capacity is decreased by only 5% (0.2 dB) from 760 kbps to 720 kbps. Increasing downlink transmission power is an inefficient approach to increasing the interference limited downlink capacity, since the available power does not affect the pole capacity.

Assume we had 20 W downlink transmission power available. Splitting the downlink power between two frequencies would increase downlink capacity from 760 kbps to $2 \times 720$ kbps $= 1440$ kbps, i.e. by 90%. The splitting of the downlink power between two carriers is an efficient approach to increasing the downlink capacity without any extra investment in power amplifiers. The power splitting approach requires that the operator's frequency allocation allows the use of two carriers in the base station.

The advantages of the wideband technology in WCDMA can be seen in the example above. It is possible to make a trade-off between the downlink capacity and coverage: if there are fewer users, more power can be allocated for one user allowing a higher path loss. The wideband power amplifier also allows the addition of a second carrier without adding a power amplifier. On the other hand, WCDMA requires very linear base station power amplifiers, which is challenging for the implementation.

### Power Splitting Between Sectors

In the initial deployment phase the sectorized uplink is needed to improve the coverage, but the sectorization may bring more capacity than is required by the initial traffic density. In WCDMA it is possible to use sectorized uplink reception while having only one common wideband power amplifier for all the sectors. This solution is illustrated in Figure 8.10.

This solution is low-cost compared to the real sectorization, where one wideband power amplifier is needed for each sector. The performance of this low-cost solution can be estimated from Figure 8.9.

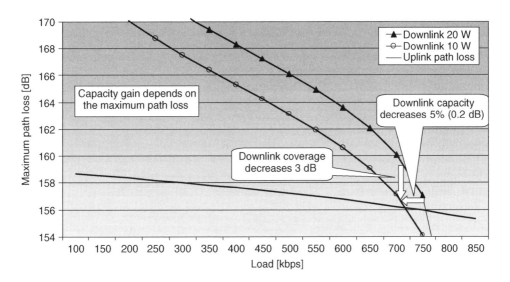

**Figure 8.9**  Effect of base station output power on downlink capacity and coverage

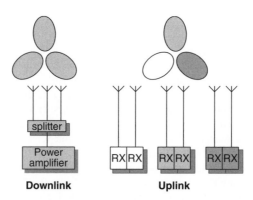

**Figure 8.10**   3-sector uplink with receive diversity, single power amplifier in downlink

Let's take an example where the uplink uses three sectors while only one 20 W power amplifier is used in downlink: the available power per sector is $20/3 = 6.7$ W. That power gives 680 kbps capacity in Figure 8.9 with 156 dB maximum path loss. The curve for 6.7 W can be obtained by moving the 20 W curve down by $10\log_{10}(6.7/20) = -4.7$ dB. Since the same power amplifier is shared between all sectors, the capacity of 680 kbps is the total capacity of the site. This low-cost solution provides $680/(3 \times 760) = 30\%$ of the capacity of the real sectorized solution, and from RNC point of view it is equal to a single sector solution. The performance of the solution is summarized in Table 8.11. RNC is not involved in softer handovers between the sectors in this solution, but it is the base station baseband that decides which uplink sectors are used in receiving the user signal.

From the downlink coverage point of view, the downlink power splitting has less effect, see the downlink coverage discussion in Section 11.2.2. The downlink coverage is only slightly reduced as the power is reduced from 20 W to 6.7 W. We can conclude that power splitting is a feasible option to start the network operation: the uplink coverage is equal to the real sectorized solution and the downlink coverage is only slightly reduced.

## 8.2.3   Capacity Upgrade Paths

When the amount of traffic increases, the downlink capacity can be upgraded in a number of different ways. The most typical upgrade options are:

- more power amplifiers if initially the power amplifier is split between sectors;
- two or more carriers if the operator's frequency allocation permits;
- transmit diversity with a 2nd power amplifier per sector.

The availability of these capacity upgrade solutions depends on the base station manufacturer. All these capacity upgrade options may not be available in all base station types. Chapters 6 and 7 in [3] provide an extensive overview of coverage and capacity enhancement methods.

**Table 8.11**   3-sector uplink, omni downlink

| | |
|---|---|
| Uplink coverage | Equal to normal 3-sector configuration |
| Downlink capacity | 30% of the 3-sector downlink |
| Number of logical sectors (RNC) | One |

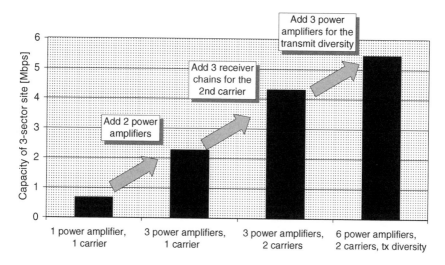

**Figure 8.11**   An example capacity upgrade path for 3-sector macro site

An example capacity upgrade path is shown in Figure 8.11. The initial solution with one power amplifier and one carrier gives 680 kbps capacity while the two-carrier three-sector transmit diversity solution gives $2 \times 3 \times 760 \times 1.2 = 5.5$ Mbps assuming 20% capacity increase from the transmit diversity. The transmit diversity procedure is described in Section 6.6.7 and its performance is discussed in Section 11.3.2.

These capacity upgrade solutions do not require any changes to the antenna configurations, only upgrades within the base station cabinet are needed on the site. The uplink coverage is not affected by these upgrades.

The capacity can be improved also by increasing the number of antenna sectors, for example, starting with omni-directional antennas and upgrading to 3-sector and finally to 6-sector antennas. The drawback of increasing the number of sectors is that the antennas must be replaced. The sectorization is analysed in more detail in [5]. The increased number of sectors also brings improved coverage through a higher antenna gain.

## 8.2.4   Capacity per km²

Providing high capacity will be challenging in urban areas where the offered amount of traffic per km² can be very high. In this section we evaluate the maximal capacity that can be provided per km² using macro and micro sites. The results are shown in Table 8.12 and illustrated in Figure 8.12.

**Table 8.12**   Capacities per km² with macro and micro layers in an urban area

|  | Macro cell layer | Micro cell layer |
|---|---|---|
| Capacity per site per carrier with one power amplifier | 630 kbps | – |
| Maximum capacity per site per carrier | 3 Mbps with three sectors | 2 Mbps |
| Capacity per site with three UMTS frequencies | 9 Mbps | 6 Mbps |
| Initial sparse site density | 0.5 sites/km² | – |
| Maximum dense site density | 5 sites/km² | 30 sites/km² |
| Maximum capacity | 45 Mbps/km² | 180 Mbps/km² |

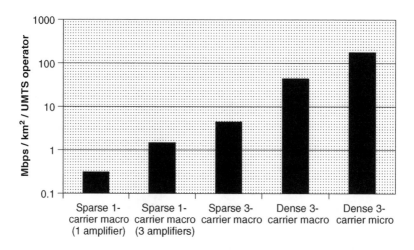

**Figure 8.12** Capacity per km² for one UMTS operator with macro and micro layers

Let us assume that the maximum capacity per sector per carrier is 1 Mbps in macro cells and 2 Mbps in micro cells with transmit diversity. If the UMTS operation is started using 1-carrier macro cells with one power amplifier per site with 2 km² site area, the available capacity will be 630 kbps/carrier/2 km² = 315 kbps/km²/carrier. If more capacity is needed, the operator can install more power amplifiers and deploy several frequencies per site, his frequency license permitting. With three frequencies, the capacity will be 3 carriers × 3.0 Mbps/carrier/2 km² = 4.5 Mbps/km².

When more capacity is needed, the operator can add more macro sites. If we assume a maximum macro site density of 5 sites per km², the capacity of the macro cell layer is 3 Mbps/carrier × 3 carriers/site × 5 site/km² = 45 Mbps/km². If clearly more capacity is needed, the micro cell layer needs to be deployed.

For the micro cell layer we assume a maximum site density of 30 sites per km². Having an even higher site density is challenging because the other-to-own cell interference tends to increase and the capacity per site decreases. Also, the site acquisition may be difficult if more sites are needed.

Using three carriers, the micro cell layer would offer 2 Mbps/carrier × 3 carriers/site × 30 site/km² = 180 Mbps/km². The total capacity of the UMTS band of 12 carriers could be up to 700 Mbps/km² if all operators deploy dense micro cell networks. Using indoor pico cells with frequency division duplex or with time division duplex can further enhance the capacity. The new frequency bands shown in Chapter 1 will allow further capacity enhancements in the future. Chapter 2 discusses the achievable capacity per subscriber in terms of MB/sub/month.

## 8.2.5 Soft Capacity

### 8.2.5.1 Erlang Capacity

In the dimensioning in Section 8.2 the number of channels per cell was calculated. Based on those figures, we can calculate the maximum traffic density that can be supported with a given blocking probability. The traffic density can be measured in Erlang and is defined ([8], p. 270) as:

$$\text{Traffic density[Erlang]} = \frac{\text{Call arrival rate[calls/s]}}{\text{Call departure rate[calls/s]}} \qquad (8.19)$$

Equally loaded cells          Less interference in the neighbouring cells
→ higher capacity in the middle cell

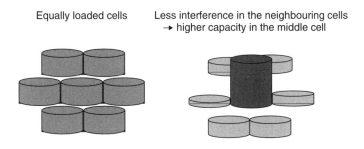

**Figure 8.13**   Interference sharing between cells in WCDMA

If the capacity is hard blocked, i.e. limited by the amount of hardware, the Erlang capacity can be obtained from the Erlang B model [8]. If the maximum capacity is limited by the amount of interference in the air interface, it is by definition a soft capacity, since there is no single fixed value for the maximum capacity. For a soft capacity limited system, the Erlang capacity cannot be calculated from the Erlang B formula, since it would give too pessimistic results. The total channel pool is larger than just the average number of channels per cell, since the adjacent cells share part of the same interference, and therefore more traffic can be served with the same blocking probability. The soft capacity can be explained as follows. The less interference is coming from the neighboring cells, the more channels are available in the middle cell, as shown in Figure 8.13. With a low number of channels per cell, i.e. for high bit rate real time data users, the average loading must be quite low to guarantee low blocking probability. Since the average loading is low, there is typically extra capacity available in the neighboring cells. This capacity can be borrowed from the adjacent cells, therefore the interference sharing gives soft capacity. Soft capacity is important for high bit rate real time data users, e.g. for video connections. It can also be obtained in GSM if the air interface capacity is limited by the amount of interference instead of the number of time slots; this assumes low frequency reuse factors in GSM with fractional loading.

In the soft capacity calculations below it is assumed that the number of subscribers is the same in all cells but the connections start and end independently. In addition, the call arrival interval follows a Poisson distribution. This approach can be used in dimensioning when calculating Erlang capacities. There is an additional soft capacity in WCDMA if the number of users in the neighboring cells is smaller.

The difference between hard blocking and soft blocking is shown with a few examples in the uplink below. WCDMA soft capacity is defined in this section as the increase of Erlang capacity with soft blocking compared to that with hard blocking with the same maximum number of channels per cell on average with both soft and hard blocking:

$$Soft\ capacity = \frac{\text{Erlang capacity with soft blocking}}{\text{Erlang capacity with hard blocking}} - 1 \qquad (8.20)$$

Uplink soft capacity can be approximated based on the total interference at the base station. This total interference includes both own cell and other cell interference. Therefore, the total channel pool can be obtained by multiplying the number of channels per cell in the equally loaded case by $1 + i$, which gives the single isolated cell capacity, since

$$i + 1 = \frac{\text{other cell interference}}{\text{own cell interference}} + 1$$

$$= \frac{\text{other cell interference} + \text{own cell interference}}{\text{own cell interference}} = \frac{\text{isolated cell capacity}}{\text{multi cell capacity}} \qquad (8.21)$$

The basic Erlang B formula is then applied to this larger channel pool (= interference pool). The Erlang capacity obtained is then shared equally between the cells. The procedure for estimating the soft capacity is summarized below:

1. Calculate the number of channels per cell, $N$, in the equally loaded case, based on the uplink load factor, Equation (8.12).
2. Multiply that number of channels by $1 + i$ to obtain the total channel pool in the soft blocking case.
3. Calculate the maximum offered traffic from the Erlang B formula.
4. Divide the Erlang capacity by $1 + i$.

### 8.2.5.2  Uplink Soft Capacity Examples

A few numerical examples of soft capacity calculations are given, with the assumptions shown in Table 8.13.

The capacities obtained, in terms of both channels based on Equation (8.12) and Erlang per cell, are shown in Table 8.14. The trunking efficiency shown in Table 8.14 is defined as the hard blocked capacity divided by the number of channels. The lower the trunking efficiency, the lower is the average loading, the more capacity can be borrowed from the neighboring cells, and the more soft capacity is available.

We note that there is more soft capacity for higher bit rates than for lower bit rates. This relationship is shown in Figure 8.14.

**Table 8.13**  Assumptions in soft capacity calculations

| | |
|---|---|
| Bit rates | Speech: 7.95 and 12.2 kbps |
| | Real time data: 16–144 kbps |
| Voice activity | Speech 67% |
| | Data 100% |
| $E_b/N_0$ | Speech: 4 dB |
| | Data 16–32 kbps: 3 dB |
| | Data 64 kbps: 2 dB |
| | Data 144 kbps: 1.5 dB |
| $i$ | 0.55 |
| Noise rise | 3 dB (=50% load factor) |
| Blocking probability | 2% |

**Table 8.14**  Soft capacity calculations in the uplink

| Bit rate (kbps) | Channels per cell | Hard blocked capacity | Trunking efficiency | Soft blocked capacity | Soft capacity |
|---|---|---|---|---|---|
| 7.95 | 92.7 | 80.9 Erl | 87% | 84.2 Erl | 4% |
| 12.2 | 60.5 | 50.1 Erl | 83% | 52.8 Erl | 5% |
| 16 | 39.0 | 30.1 Erl | 77% | 32.3 Erl | 7% |
| 32 | 19.7 | 12.9 Erl | 65% | 14.4 Erl | 12% |
| 64 | 12.5 | 7.0 Erl | 56% | 8.2 Erl | 17% |
| 144 | 6.4 | 2.5 Erl | 39% | 3.2 Erl | 28% |

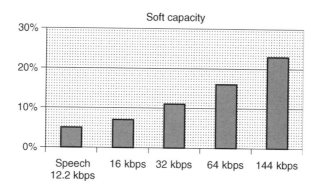

**Figure 8.14**  Soft capacity as a function of bit rate for real-time connections

It should be noted that the amount of soft capacity depends also on the propagation environment and on the network planning which affect the value of $i$. The soft capacity can be obtained only if the radio resource management algorithms can utilize a higher capacity in one cell if the adjacent cells have lower loading. This can be achieved if the radio resource management algorithms are based on the wideband interference, not on the throughput or the number of connections. Interference-based admission control is presented in Section 9.5.

Similar soft capacity is also available in the WCDMA downlink as well as in GSM if interference-based radio resource management algorithms are applied.

## 8.2.6  Network Sharing

The cost of the network deployment can be reduced by network sharing. An example of a network sharing approach is illustrated in Figure 8.15 where both operators have their own core networks and share a common radio access network, RAN. This solution offers cost savings in site acquisition, civil works, transmission, RAN equipment costs and operation expenses. Both operators can still keep their full independence in core network, services and have dedicated radio carrier frequencies. When the amount of traffic increases in the future, the operators can exit the shared RAN and continue with separate RANs.

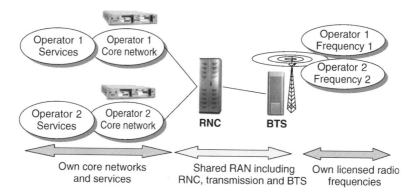

**Figure 8.15**  Sharing of a WCDMA radio access network

In the case of shared RAN, the RNC is connected with multiple Iu-CS and Iu-PS interfaces to each operator's core network. The subscribers of different operators are separated to their own licensed frequencies. This feature is supported in 3GPP Release '99.

The base station configuration of a shared 3-sector site is typically $2 + 2 + 2$, both operators using $1 + 1 + 1$ layers. The radio coverage will be the same for both operators and they share the radio capacity and the hardware resources including base station, transmission and RNC resources. Consequently, a combined radio network dimensioning and planning is needed, based on the total capacity figures from both operators.

The shared RAN allows some flexibility in setting the parameters for the radio resource management algorithms. For example, the maximum allowed load threshold for the packet scheduling can be different, and handover parameters and neighbor lists can be optimized separately.

Other levels of network sharing are also possible. One geographical area could be covered by only one operator that allows roaming for the other operators' users. It is also possible to share the sites and the related costs while having independent RANs.

## 8.3 Capacity and Coverage Planning and Optimization

### 8.3.1 Iterative Capacity and Coverage Prediction

In this section, detailed capacity and coverage planning are presented. In the detailed planning phase real propagation data from the planned area is needed, together with the estimated user density and user traffic. Also, information about the existing base station sites is needed in order to utilize the existing site investments. The output of the detailed capacity and coverage planning are the base station locations, configurations and parameters. A comprehensive treatment of the topic can be found in Chapter 3 in [3].

Since, in WCDMA, all users are sharing the same interference resources in the air interface, they cannot be analysed independently. Each user is influencing the others and causing their transmission powers to change. These changes themselves again cause changes, and so on. Therefore, the whole prediction process has to be done iteratively until the transmission powers stabilize. This iterative process is illustrated in Figure 8.16. Also, the mobile speeds, multipath channel profiles, and bit rates and type of services used play a more important role than in second generation TDMA/Frequency Division Multiple Access (FDMA) systems. Furthermore, in WCDMA fast power control in both uplink and downlink, soft/softer handover and orthogonal downlink channels are included, which also impact on system performance. The main difference between WCDMA and TDMA/FDMA coverage prediction is that the interference estimation is already crucial in the coverage prediction phase in WCDMA. In the current GSM coverage planning processes the base station sensitivity is typically assumed to be constant and the coverage threshold is the same for each base station. In the case of WCDMA, the base station sensitivity depends on the number of users and used bit rates in all cells, thus it is cell- and service-specific. Note also that in third generation networks, the downlink can be loaded higher than the uplink or vice versa.

### 8.3.2 Planning Tool

In second generation systems, detailed planning concentrated strongly on coverage planning. In third generation systems, a more detailed interference planning and capacity analysis than simple coverage optimization is needed. The tool should aid the planner to optimize the base station configurations, the antenna selections and antenna directions and even the site locations, in order to meet the quality of service and the capacity and service requirements at minimum cost. To achieve the optimum result the tool must have knowledge of the radio resource algorithms. Uplink and downlink coverage probability is determined for a specific service by testing the service availability in each location of the plan. A

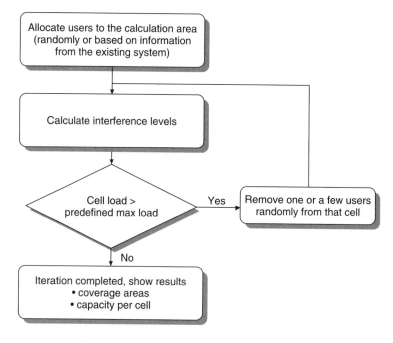

**Figure 8.16**  Iteration capacity and coverage calculations

detailed description of the planning tool can be found in [9]. An example of a commercial WCDMA planning tool is shown in Figure 8.17.

The actual detailed planning phase does not differ very much from second generation planning. The sites and sectors are placed in the tool. The main difference is the importance of the traffic layer. The proposed detailed analysis methods (see the following sections) use discrete mobile stations in the WCDMA analysis. The mobile station density in different cells should be based on actual traffic information. The hotspots should be identified as an input for accurate analysis. One source of information concerning user density would be the data from the operator's second generation network or later from the third generation.

### 8.3.2.1  Uplink and Downlink Iterations

The target in the uplink iteration is to allocate the simulated mobile stations' transmission powers so that the interference levels and the base station sensitivity values converge. The base station sensitivity level is corrected by the estimated uplink interference level (noise rise) and therefore is cell-specific. The impact of the uplink loading on the sensitivity is taken into account with a term $-10\log_{10}(1 - \eta_{\mathrm{UL}})$, where $\eta_{\mathrm{UL}}$ is given by Equation (8.12). In the uplink iteration the transmission powers of the mobile stations are estimated based on the sensitivity level of the best server, the service, the speed and the link losses. Transmission powers are then compared to the maximum allowed transmission power of the mobile stations, and mobile stations exceeding this limit are put to outage. The interference can then be re-estimated and new loading values and sensitivities for each base station assigned. If the uplink load factor is higher than the set limit, the mobile stations are randomly moved from the highly loaded cell to another carrier (if the spectrum allows) or to outage.

The aim of the downlink iteration is to allocate correct base station transmission powers to each mobile station until the received signal at the mobile station meets the required $E_b/N_0$ target.

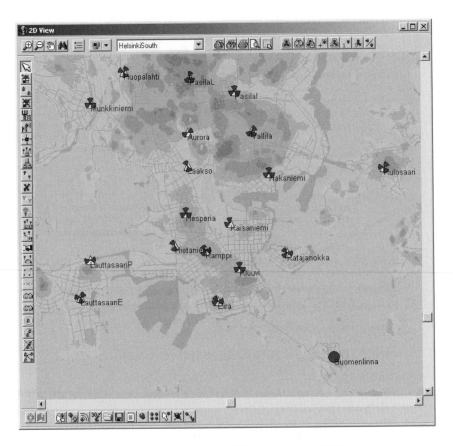

**Figure 8.17**   Commercial WCDMA planning tool [10]

### 8.3.2.2   Modelling of Link Level Performance

In radio network dimensioning and planning it is necessary to make simplifying assumptions concerning the multipath propagation channel, transmitter and receiver. A traditional model is to use the average received $E_b/N_0$ ensuring the required quality of service as the basic number, which includes the effect of the power delay profile. In systems using fast power control the average received $E_b/N_0$ is not enough to characterize the influence of the radio channel on network performance. Also, the transmission power distribution must be taken into account when modelling link level performance in network level calculations. An appropriate approach is presented in [1] for the WCDMA uplink. It has been demonstrated that, due to the fast power control in the multipath fading environment, in addition to the average received $E_b/N_0$ requirement, an average transmission power rise is needed in interference calculations. The power rise is presented in detail in Section 9.2.1.2. Furthermore, a power control headroom must be included in the link budget estimation to allow power control to follow the fast fading at the cell edge.

Multiple links are taken into account in the simulator when estimating the soft handover gains in the average received and transmitted power and also in the required power control headroom. During the simulations the transmission powers are corrected by the voice activity factor, soft handover gain and average power rise for each mobile station.

## 8.3.3   Case Study

In this case study an area in Espoo, Finland, was planned, comprising roughly $12 \times 12$ km², as shown in Figure 8.18. The network planning tool described in Section 8.3.2 was utilized in this case study.

The operator's coverage probability requirement for the 8 kbps, 64 kbps and 384 kbps services was set, respectively, to 95%, 80% and 50%, or better. The planning phase started with radio link budget estimation and site location selections. In the next planning step the dominance areas for each cell were optimized. In this context the dominance is related only to the propagation conditions. Antenna tilting, bearing and site locations can be tuned to achieve clear dominance areas for the cells. Dominance area optimization is crucial for interference and soft handover area and soft handover probability control. The improved soft/softer handover and interference performance is automatically seen in the improved network capacity. The plan consists of 19 three-sectored macro sites, and the average site area is 7.6 km². In the city area, the uplink loading limitation was set to 75%, corresponding to a 6 dB noise rise. If the loading was exceeded, the necessary number of mobile stations was randomly set to outage (or moved to another carrier) from the highly loaded cells. Table 8.15 shows the user distribution in the simulations and the other simulation parameters are listed in Table 8.16.

In all three simulation cases the cell throughput in kbps and the coverage probability for each service were of interest. Furthermore, the soft handover probability and loading results were collected.

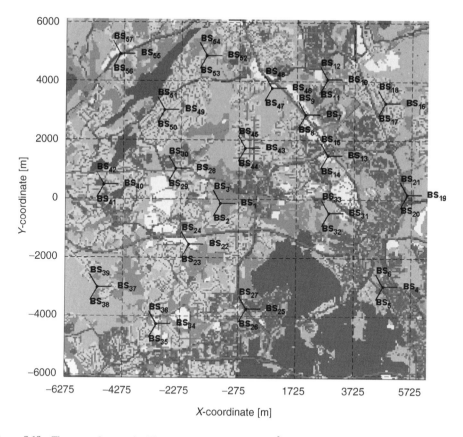

**Figure 8.18**   The network scenario. The area measures $12 \times 12$ km² and is covered with 19 sites, each with three sectors

**Table 8.15** The user distribution

| Service in kbps | Users per service |
|---|---|
| 8 | 1735 |
| 64 | 250 |
| 384 | 15 |

**Table 8.16** Parameters used in the simulator

| | |
|---|---|
| Uplink loading limit | 75% |
| Base station maximum transmission power | 20 W (43 dBm) |
| Mobile station maximum transmission power | 300 mW (= 25 dBm) |
| Mobile station power control dynamic range | 70 dB |
| Slow (log-normal) fading correlation between base stations | 50% |
| Standard deviation for the slow fading | 6 dB |
| Multipath channel profile | ITU Vehicular A |
| Mobile station speeds | 3 km/h and 50 km/h |
| Mobile/base station noise figures | 7 dB/5 dB |
| Soft handover addition window | −6 dB |
| Pilot channel power | 30 dBm |
| Combined power for other common channels | 30 dBm |
| Downlink orthogonality | 0.5 |
| Activity factor speech/data | 50%/100% |
| Base station antennas | 65°/17 dBi |
| Mobile antennas speech/data | Omni/1.5 dBi |

Tables 8.17 and 8.18 show the simulation results for cell throughput and coverage probabilities. The maximum uplink loading was set to 75% according to Table 8.16. Note that in Table 8.17 in some cells the loading is lower than 75%, and, correspondingly, the throughput is also lower than the achievable maximum value. The reason is that there was not enough offered traffic in the area to fully load the cells. The loading in cell 5 was 75%. Cell 5 is located in the lower right corner in Figure 8.18, and there is no other cell close to cell 5. Therefore, that cell can collect more traffic than the other cells. For example, cells 2 and 3 are in the middle of the area and there is not enough traffic to fully load the cells.

Table 8.18 shows that mobile station speed has an impact on both throughput and coverage probability. When mobile stations are moving at 50 km/h, fewer can be served, the throughput is lower and the resulting loading is higher than when mobile stations are moving at 3 km/h. If the throughput values are normalized to correspond to the same loading value, the difference between the 3 km/h and 50 km/h cases is more than 20%. The better capacity with the slower-moving mobile stations can be explained by the better $E_b/N_0$ performance. The fast power control is able to follow the fading signal and the required $E_b/N_0$ target is reduced. The lower target value reduces the overall interference level and more users can be served in the network.

Comparing coverage probability, the faster-moving mobile stations experience better quality than the slow-moving ones, because for the latter headroom is needed in the mobile transmission power to be able to maintain the fast power control – see Section 8.2.1. The impact of the speed can be seen, especially if the bit rates used are high, because for low bit rates the coverage is better due to a larger processing gain. The coverage is tested in this planning tool by using a test mobile after the uplink iterations have converged. It is assumed that this test mobile does not affect the loading in the network.

**Table 8.17**  The cell throughput, loading and soft handover (SHO) overhead. UL = uplink, DL = downlink

| Cell ID | Throughput UL (kbps) | Throughput DL (kbps) | UL loading | SHO overhead |
|---|---|---|---|---|
| Basic loading: mobile speed 3 km/h, served users: 1805 | | | | |
| cell 1 | 728.00 | 720.00 | 0.50 | 0.34 |
| cell 2 | 208.70 | 216.00 | 0.26 | 0.50 |
| cell 3 | 231.20 | 192.00 | 0.24 | 0.35 |
| cell 4 | 721.60 | 760.00 | 0.43 | 0.17 |
| cell 5 | 1508.80 | 1132.52 | 0.75 | 0.22 |
| cell 6 | 762.67 | 800.00 | 0.53 | 0.30 |
| MEAN (all cells) | 519.20 | 508.85 | 0.37 | 0.39 |

| Cell ID | Throughput UL (kbps) | Throughput DL (kbps) | UL loading | SHO overhead |
|---|---|---|---|---|
| Basic loading: mobile speed 50 km/h, served users: 1777 | | | | |
| cell 1 | 672.00 | 710.67 | 0.58 | 0.29 |
| cell 2 | 208.70 | 216.00 | 0.33 | 0.50 |
| cell 3 | 226.67 | 192.00 | 0.29 | 0.35 |
| cell 4 | 721.60 | 760.00 | 0.50 | 0.12 |
| cell 5 | 1101.60 | 629.14 | 0.74 | 0.29 |
| cell 6 | 772.68 | 800.00 | 0.60 | 0.27 |
| MEAN | 531.04 | 506.62 | 0.45 | 0.39 |

| Cell ID | Throughput UL (kbps) | Throughput DL (kbps) | UL loading | SHO overhead |
|---|---|---|---|---|
| Basic loading: mobile speed 50 km/h and 3 km/h, served users: 1802 | | | | |
| cell 1 | 728.00 | 720.00 | 0.51 | 0.34 |
| cell 2 | 208.70 | 216.00 | 0.29 | 0.50 |
| cell 3 | 240.00 | 200.00 | 0.25 | 0.33 |
| cell 4 | 730.55 | 760.00 | 0.44 | 0.20 |
| cell 5 | 1162.52 | 780.92 | 0.67 | 0.33 |
| cell 6 | 772.68 | 800.00 | 0.55 | 0.32 |
| MEAN | 525.04 | 513.63 | 0.40 | 0.39 |

This example case demonstrates the impact of the user profile, i.e. the service used and the mobile station speed, on network performance. It is shown that the lower mobile station speed provides better capacity: the number of mobile stations served and the cell throughput are higher in the 3 km/h case than in the 50 km/h case. Comparing coverage probability, the impact of the mobile station speed is different. The higher speed reduces the required fast fading margin and thus the coverage probability is improved when the mobile station speed is increased.

## 8.3.4  Network Optimization

Network optimization is a process to improve the overall network quality as experienced by the mobile subscribers and to ensure that network resources are used efficiently. Optimization includes:

1. Performance measurements.
2. Analysis of the measurement results.
3. Updates in the network configuration and parameters.

The optimization process is shown in Figure 8.19.

**Table 8.18**  The coverage probability results

| Basic loading: mobile speed 3 km/h | Test mobile speed: | |
|---|---|---|
| | 3 km/h | 50 km/h |
| 8 kbps | 96.6% | 97.7% |
| 64 kbps | 84.6% | 88.9% |
| 384 kbps | 66.9% | 71.4% |
| Basic loading: mobile speed 50 km/h | Test mobile speed: | |
| | 3 km/h | 50 km/h |
| 8 kbps | 95.5% | 97.1% |
| 64 kbps | 82.4% | 87.2% |
| 384 kbps | 63.0% | 67.2% |
| Basic loading: mobile 3 and 50 km/h | Test mobile speed: | |
| | 3 km/h | 50 km/h |
| 8 kbps | 96.0% | 97.5% |
| 64 kbps | 83.9% | 88.3% |
| 384 kbps | 65.7% | 70.2% |

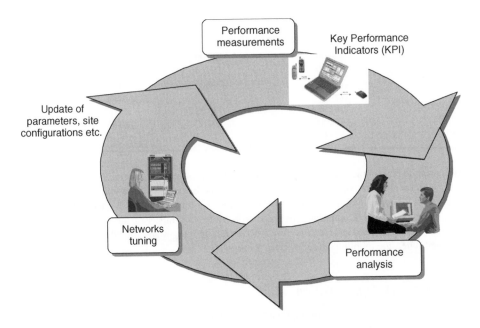

**Figure 8.19**  Network optimization process

A clear picture of the current network performance is needed for the performance optimization. Typical measurement tools are shown in Figure 8.20. The measurements can be obtained from the test mobile and from the radio network elements. The WCDMA mobile can provide relevant measurement data, e.g. uplink transmission power, soft handover rate and probabilities, CPICH $E_b/N_0$ and downlink

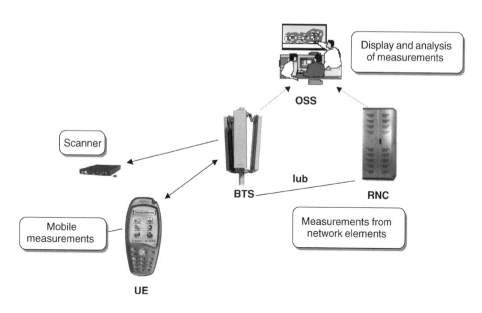

**Figure 8.20** Network performance measurements

BLER. Also, scanners can be used to provide some of the downlink measurements, like CPICH measurements for the neighbor list optimization.

The radio network can typically provide connection level and cell level measurements. Examples of the connection measurements include uplink BLER and downlink transmission power. The connection level measurements both from the mobile and from the network are important to get the network running and provide the required quality for the end users. The cell level measurements become more important in the capacity optimization phase. The cell level measurements may include total received power and total transmitted power, the same parameters that are used by the radio resource management algorithms.

The measurement tools can provide lots of results. In order to speed up the measurement analysis it is beneficial to define those measurement results that are considered the most important ones, Key Performance Indicators, KPIs. Examples of KPIs are total base station transmission power, soft handover overhead, drop call rate and packet data delay. The comparison of KPIs and desired target values indicates the problem areas in the network where the network tuning can be focused.

The network tuning can include updates of RRM parameters, e.g. handover parameters, common channel powers or packet data parameters. The tuning can also include changes of antenna directions. It may be possible to adjust the antenna tilts remotely without any site visits. An example case is illustrated in Figure 8.21. If there is too much overlapping of the adjacent cells, the other cell interference is high and the system capacity is low. The effect of other cell interference is represented with the parameter other cell to own cell interference ratio, $i$, in the load equations of Section 8.2, see Equation (8.16). The importance of the other cell interference is illustrated in Figure 8.22: if the other cell interference can be decreased by 50%, the capacity can be increased by 57%. The large overlapping can be seen from the high number of users in soft handover between these cells.

With advanced Operations Support System (OSS) the network performance monitoring and optimization can be automated. OSS can point out the performance problems, propose corrective actions and even make some tuning actions automatically.

The network performance can be best observed when the network load is high. With low load some of the problems may not be visible. Therefore, we need to consider artificial load generation to

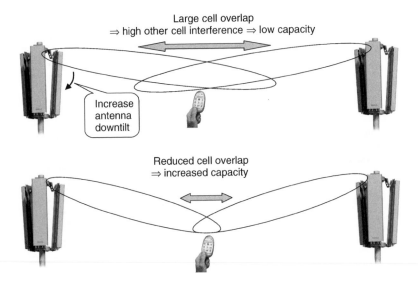

**Figure 8.21**  Network tuning with antenna tilts

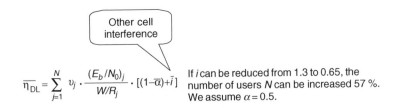

**Figure 8.22**  Importance of other cell interference for WCDMA downlink capacity

emulate high loading in the network. A high uplink load can be generated by increasing the $E_b/N_0$ target of the outer loop power control. In the normal operation the outer loop power control provides the required quality with minimum $E_b/N_0$. If we manually increase the $E_b/N_0$ target, e.g. 10 dB higher than the normal operation point, that uplink connection will cause 10 times more interference and converts 32 kbps connection into 320 kbps high bit rate connection from the interference point of view. The effect of higher $E_b/N_0$ can be seen in the uplink load equation of Equation (8.12). The same approach can be applied in the downlink as well in Equation (8.16). Another load generation approach in downlink is to transmit dummy data in downlink with a few code channels, even if there are no mobiles receiving that data. That approach is called Orthogonal Channel Noise Source (OCNS).

For more information on the radio network optimization process, please refer to [3], Chapter 8, and for advanced monitoring and network tuning see [3], Chapter 10.

## 8.4  GSM Co-planning

Utilization of existing base station sites is important in speeding up WCDMA deployment and in sharing sites and transmission costs with the existing second generation system. The feasibility of sharing sites depends on the relative coverage of the existing network compared to WCDMA. In this section we compare the relative uplink coverage of existing GSM900 and GSM1800 full rate speech services and WCDMA speech and 64 kbps and 144 kbps data services. Table 8.19 shows the assumptions made and the results of the comparison of coverage. The maximum path loss of the WCDMA 144 kbps here is 3 dB greater than in Table 8.4. The difference comes because of a smaller interference margin, a lower base station receiver noise figure, and no cable loss. Note also that the soft handover gain is included in the fast fading margin in Table 8.19 and the mobile station power class is here assumed to be 21 dBm.

Table 8.19 shows that the maximum path loss of the 144 kbps data service is the same as for speech service of GSM1800. Therefore, a 144 kbps WCDMA data service can be provided when using GSM1800 sites, with the same coverage probability as GSM1800 speech. If GSM900 sites are used for WCDMA and 64 kbps full coverage is needed, a 3 dB coverage improvement is needed in WCDMA. Section 11.2.1 analyses the uplink coverage of WCDMA and presents a number of solutions for improving WCDMA coverage to match GSM site density. The comparison in Table 8.19 assumes that GSM900 sites are planned as coverage-limited. In densely populated areas, however, GSM900 cells are typically smaller to provide enough capacity, and WCDMA co-siting is feasible.

The downlink coverage of WCDMA is discussed in Section 11.2.2 and is shown to be better than the uplink coverage. Therefore, it is possible to provide full downlink coverage for bit rates 144 to 384 kbps using GSM1800 sites.

Any comparison of the coverage of WCDMA and GSM depends on the exact receiver sensitivity values and on system parameters such as handover parameters and frequency hopping. The aim of this exercise is to compare the coverage of the GSM base station systems that have been deployed up to the present with WCDMA coverage in the initial deployment phase during 2002. The sensitivity of the latest GSM base stations is better than the one assumed in Table 8.19.

**Table 8.19**  Typical maximum path losses with existing GSM and with WCDMA

| | GSM900/ speech | GSM1800/ speech | WCDMA/ speech | WCDMA/ 64 kbps | WCDMA/ 144 kbps |
|---|---|---|---|---|---|
| Mobile transmission power | 33 dBm | 30 dBm | 21 dBm | 21 dBm | 21 dBm |
| Receiver sensitivity[1] | −110 dBm | −110 dBm | −125 dBm | −120 dBm | −117 dBm |
| Interference margin[2] | 1.0 dB | 0.0 dB | 2.0 dB | 2.0 dB | 2.0 dB |
| Fast fading margin[3] | 2.0 dB | 2.0 dB | 2.0 dB | 2.0 dB | 2.0 dB |
| Base station antenna gain[4] | 16.0 dBi | 18.0 dBi | 18.0 dBi | 18.0 dBi | 18.0 dBi |
| Body loss[5] | 3.0 dB | 3.0 dB | 3.0 dB | – | – |
| Mobile antenna gain[6] | 0.0 dBi | 0.0 dBi | 0.0 dBi | 2.0 dBi | 2.0 dBi |
| Relative gain from lower frequency compared to UMTS frequency[7] | 7.0 dB | 1.0 dB | – | – | – |
| Maximum path loss | 160.0 dB | 154.0 dB | 157.0 dB | 157.0 dB | 154.0 dB |

[1]WCDMA sensitivity assumes 4.0 dB base station noise figure and $E_b/N_0$ of 4.0 dB for 12.2 kbps speech, 2.0 dB for 64 kbps and 1.5 dB for 144 kbps data. For the $E_b/N_0$ values see Section 12.5. GSM sensitivity is assumed to be −110 dBm with receive antenna diversity.

[2]The WCDMA interference margin corresponds to 37% loading of the pole capacity: see Figure 8.3. An interference margin of 1.0 dB is reserved for GSM900 because the small amount of spectrum in 900 MHz does not allow large reuse factors.

[3]The fast fading margin for WCDMA includes the macro diversity gain against fast fading.

[4]The antenna gain assumes three-sector configuration in both GSM and WCDMA.

[5]The body loss accounts for the loss when the terminal is close to the user's head.

[6]A 2.0 dBi antenna gain is assumed for the data terminal.

[7]The attenuation in 900 MHz is assumed to be 7.0 dB lower than in UMTS band and in GSM1800 band 1.0 dB lower than in UMTS band.

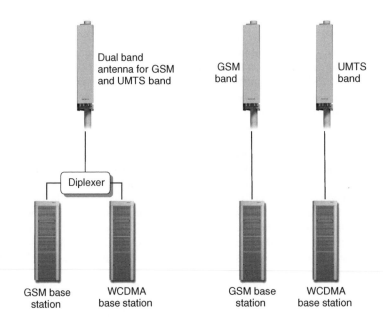

**Figure 8.23**  Co-siting of GSM and WCDMA

Since the coverage of WCDMA typically is satisfactory when reusing GSM sites, GSM site reuse is the preferred solution in practice. Let us consider next the practical co-siting of the system. Co-sited WCDMA and GSM systems can share the antenna when a dual band or wideband antenna is used. The antenna needs to cover both the GSM band and UMTS band. GSM and WCDMA signals are combined with a diplexer to the common antenna feeder. The shared antenna solution is attractive from the site solution point of view but it limits the flexibility in optimizing the antenna directions of GSM and WCDMA independently. Another co-siting solution is to use separate antennas for the two networks. That solution gives full flexibility in optimizing the networks separately. These two solutions are shown in Figure 8.23. The co-siting of GSM and WCDMA is taken into account in 3GPP performance requirements and the interference between the systems can be avoided.

## 8.5   Inter-Operator Interference

### 8.5.1   Introduction

In this section, the effect of adjacent channel interference between two operators on adjacent frequencies is studied. Adjacent channel interference needs to be considered, because it will affect all wideband systems where large guard bands are not possible, and WCDMA is no exception. If the adjacent frequencies are isolated in the frequency domain by large guard bands, spectrum is wasted due to the large system bandwidth. Tight spectrum mask requirements for a transmitter and high selectivity requirements for a receiver, in the mobile station and in the base station, would guarantee low adjacent channel interference. However, these requirements have a huge impact, especially on the implementation of a small WCDMA mobile station.

Adjacent Channel Interference power Ratio (ACIR) is defined as the ratio of the transmission power to the power measured after a receiver filter in the adjacent channel(s). Both the transmitted and the

**Table 8.20**   Requirements for adjacent channel performance [11]

| Frequency separation | Required attenuation |
| --- | --- |
| Adjacent carrier (5 MHz separation) | 33 dB both uplink and downlink |
| Second adjacent carrier (10 MHz separation) | 43 dB in uplink, 40 dB in downlink (estimated from in-band blocking) |

received power are measured with a filter that has a Root-Raised Cosine filter response with roll-off of 0.22 and a bandwidth equal to the chip rate [11]. The adjacent channel interference is caused by transmitter non-idealities and imperfect receiver filtering. In both uplink and downlink, the adjacent channel performance is limited by the performance of the mobile. In the uplink the main source of adjacent channel interference is the non-linear power amplifier in the mobile station, which introduces adjacent channel leakage power. In the downlink the limiting factor for adjacent channel interference is the receiver selectivity of the WCDMA terminal. The requirements for adjacent channel performance are shown in Table 8.20.

Such an interference scenario, where the adjacent channel interference could affect network performance, is illustrated in Figure 8.24. Operator 1's mobile is connected to a far-away base station and is at the same time located close to Operator 2's base station on the adjacent frequency. The mobile will receive interference from Operator 2's base station which may – in the worst case – block the reception of its own weak signal.

In the following sections the effect of the adjacent channel interference in this interference scenario is analysed by worst-case calculations and by system simulations. It will be shown that the worst-case calculations give very bad results but also that the worst-case scenario is extremely unlikely to happen in real networks. Therefore, simulations are also used to study this interference scenario. Finally, conclusions are drawn regarding adjacent channel interference and implications for network planning are discussed.

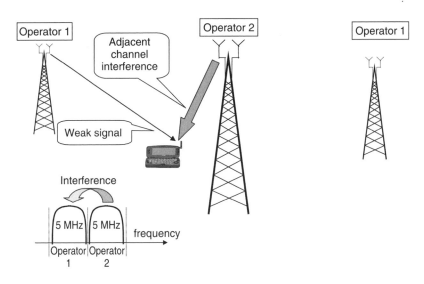

**Figure 8.24**   Adjacent channel interference in downlink

## 8.5.2   Uplink Versus Downlink Effects

While the mobile in Figure 8.24 receives interference, it will also cause interference in uplink to Operator 2's base station. In this section we analyse the differences between uplink and downlink in the worst-case scenario. The worst-case adjacent channel interference occurs when a mobile in uplink and a base station in downlink are transmitting on full power, and the mobile is located very close to a base station that is receiving on the adjacent carrier. A minimum coupling loss of 70 dB is assumed here. The minimum coupling loss is defined as the minimum path loss between mobile and base station antenna connectors. The level of the adjacent channel interference is calculated in Table 8.21 and it is compared to the receiver thermal noise level of Table 8.22, both in uplink and in downlink. The worst-case increase in the receiver interference level is calculated in Table 8.23.

The maximum desensitization in downlink is 41 dB and in uplink 22 dB, which indicates that the downlink direction will be affected before the mobile is able to cause high interference levels in uplink. This is mainly because of higher base station power compared to the mobile power. It is also preferable to cause interference to one connection in downlink than to allow that mobile to interfere with all uplink connections of one cell. In the following sections we concentrate on the downlink analysis.

## 8.5.3   Local Downlink Interference

The adjacent channel interference in downlink may cause dead zones around interfering base stations. In this section we evaluate the sizes of these dead zones as a function of the coverage of the own

**Table 8.21**   Worst-case adjacent channel interference level

|  | Downlink | Uplink |
|---|---|---|
| Interferer power | 43 dBm (base station) | 21 dBm (mobile) |
| Minimum coupling loss between mobile and | 70 dB | 70 dB interfering base station in Figure 8.24 |
| Adjacent channel attenuation | 33 dB | 33 dB |
| Adjacent channel interference | 43 dBm − 70 dB − 33 dB<br>= −60 dBm | 21 dBm − 70 dB − 33 dB<br>= −82 dBm |

**Table 8.22**   Receiver thermal noise level

|  | Downlink | Uplink |
|---|---|---|
| Thermal noise level kTB | −108 dBm | −108 dBm |
| Receiver noise figure | 7 dB | 4 dB |
| Receiver noise level | −108 dBm + 7 dB = −101 dBm | −108 dBm + 4 dB = −104 dBm |

**Table 8.23**   Worst-case desensitization

| Downlink | Uplink |
|---|---|
| −60 dBm − (−101 dBm)<br>= 41 dB | −82 dBm − (−104 dBm)<br>= 22 dB |

**Table 8.24**   Assumptions for dead zone calculation for 12.2 kbps voice

| Parameter | Value |
|---|---|
| Transmission power of Operator 2's base station | 33–43 dBm |
| Pilot power from Operator 1's base station | 33 dBm |
| Maximum allocated power per voice connection from | 33 dBm Operator 1's base station |
| Required $E_b/N_0$ for voice connection | 7 dB |
| Required $E_c/N_0$ for voice connection | 7 dB $-$ 10 log 10(3.84 e 6/12.2 e 3) $= -$ 18 dB |
| Path loss calculation to the interfering Operator 2's base | 37 dB $+$ 20 log 10 ($d$) station with distance $d$ [metres] in line-of-sight |

signal. The coverage is defined as the received pilot power level. The assumptions in the calculations are shown in Table 8.24.

The dead zones are evaluated as follows.

1. Assume received pilot power level from Operator 1's base station.
2. Calculate maximum received signal power level for the voice connection. In this case it is equal to the pilot power level since the maximum transmission power for voice is assumed to be equal to the pilot power of 33 dBm.
3. Calculate maximum tolerated interference level $I_0$ on the same carrier based on the required $E_c/I_0$.
4. Calculate maximum tolerated interference level on the adjacent carrier based on the adjacent channel attenuation.
5. Calculate minimum required path loss to the interfering base station.
6. Calculate minimum required distance to the interfering base station.

An example calculation is shown below assuming pilot power coverage of $-90$ dBm:

1. Assume pilot power level of $-90$ dBm.
2. Maximum received power for voice connection $-90$ dBm.
3. Maximum tolerated interference level $I_0 = -90$ dBm $+ 18$ dB $= -72$ dBm.
4. Maximum tolerated interference level on the adjacent carrier $-72$ dBm $+ 33$ dB $= -39$ dBm.
5. Minimum required path loss 43 dBm $- (-39$ dBm$) = 82$ dB when Operator 2's base station transmits with 43 dBm. The required path loss is reduced to 33 dBm $- (-39$ dBm$) = 72$ dB when operator 2's base station transmits only common channels with 33 dBm.
6. Minimum required distance $d = 10^\wedge((82 - 37)/20) = 178$ m or $d = 10^\wedge((72 - 37)/20) = 56$ m.

The results of the calculations are plotted in Figure 8.25. The results show that the dead zones can occur only if the following conditions take place at the same time: own network coverage is weak, the mobile is located close to the interfering base station that is operating on the adjacent frequency with maximum power, and User Equipment performance is just meeting 3GPP selectivity requirements.

## 8.5.4   Average Downlink Interference

Since the probability of the adjacent channel interference is low, we need to resort to system simulations to evaluate the effect on the average performance. More transmission power is needed because of adjacent channel interference which leads to a lower capacity. The simulations show the reduction in average capacity when the same outage probability is maintained, with and without adjacent channel interference. The simulation results and assumptions are presented in [12]. The worst-case scenario

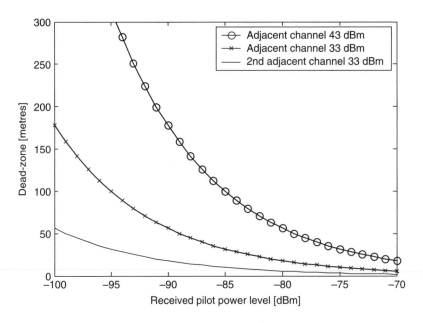

**Figure 8.25**   Dead zone sites as a function of own network coverage

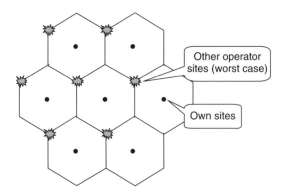

**Figure 8.26**   Worst case simulation scenario

is shown in Figure 8.26 where the site distance is 1 km and the interfering sites are just between our own sites. The best case is when the operators' sites are co-located.

The simulation results are shown in Table 8.25. The worst-case capacity loss is 2.0–3.5%. These capacity loss figures can be reduced with the solutions shown in the following section.

**Table 8.25**   Capacity loss because of adjacent channel interference

|               | Worst-case | Intermediate case | Co-siting |
| ------------- | ---------- | ----------------- | --------- |
| Capacity loss | 3.5%       | 2.5%              | No loss   |

## 8.5.5    Path Loss Measurements

The adjacent channel interference is basically about power competition between operators. The interference problems hit the connection if the interfering signal is strong at the same time as the own signal is weak. We can calculate the maximum tolerable power difference between own signal and the interfering signal in Figure 8.24. When the maximum power difference is known, we can go and measure the power differences between two operators' networks and find the locations where the interference could cause problems. We show an example for WCDMA voice service in downlink with the following assumptions:

- The required $E_c/I_0$ for WCDMA voice = $E_c/I_0$ − processing gain = −18 dB from Table 8.24.
- The maximum transmission power per WCDMA connection is assumed to be 33 dBm.
- WCDMA mobile selectivity is 33 dB.
- The base stations' transmit power is 43 dBm.

The maximum allowed signal power difference between two operators can be estimated as follows:

$$= -E_c/I_0 + \text{mobile selectivity} - \text{downlink power allocation} = 18\,\text{dB} + 33\,\text{dB} - 10\,\text{dB}$$

$$\text{(the power for a connection is 10 dB below the base station max power)} = 41\,\text{dB}$$

When the frequency separation is 10 MHz, the allowed signal power difference increases to 51 dB. Relative signal power measurements from today's network show that the probability of a larger power difference than 41 dB is typically <1–2% and larger than 51 dB is practically non-existent. This is the probability that counter-measures are needed against interference. The measurement results are in line with the simulation results.

## 8.5.6    Solutions to Avoid Adjacent Channel Interference

This section presents a few network planning and radio resource management solutions that make sure that adjacent channel interference does not affect WCDMA network performance.

If the operators using adjacent frequency bands co-locate their base stations, either in the same sites or using the same masts, adjacent channel interference problems can be avoided, since the received power levels from both operators' transmissions are then very similar. Since there are no large power differences, the adjacent channel attenuation of 33 dB is enough to prevent any adjacent channel interference problems.

The nominal WCDMA carrier spacing is 5.0 MHz but can be adjusted with a 200 kHz raster according to the requirements of the adjacent channel interference. By using a larger carrier spacing, the adjacent channel interference can be reduced. If the operator has two carriers in the same base station, the carrier spacing between them could be as small as 4.0 MHz, because the adjacent channel interference problems are completely avoided if the two carriers use the same base station antennas. In that case, a larger carrier spacing can be reserved between operators, as shown in Figure 8.27.

In addition to the network planning solutions, the radio resource management can also be effectively utilized to avoid the problems from inter-operator interference. The calculations in the sections above suggest the following radio resource management solutions to avoid adjacent channel interference in addition to the network planning solutions:

- Make inter-frequency handover to another frequency to provide higher selectivity and more protection against interference.
- Allocate more power per connection in downlink to overcome the effect of the interference.

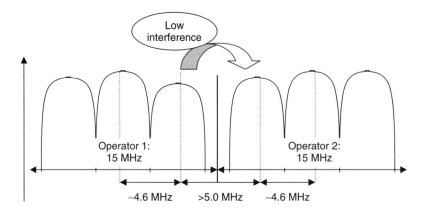

**Figure 8.27** Selection of carrier spacings within operator's bands and between operators

- Reduce the downlink instantaneous packet data bit rate to provide more processing gain to tolerate more interference.
- Reduce the downlink AMR voice bit rate to provide more processing gain.

## 8.6 WCDMA Frequency Variants

3GPP has produced WCDMA specifications for a number of cellular bands that have room for the WCDMA carrier. The frequency variants and their typical usage areas globally are listed in Figure 8.28. The frequency variants are 3GPP release independent, which means that, even if the frequency variant is added with 3GPP Release 7 schedule, the products for that band can use an earlier 3GPP release as the design basis.

The WCDMA deployment has started in Europe and Asia in the mainstream 2.1 GHz band I with a total 2 × 60, MHz allocation. The WCDMA terminals typically include WCDMA2100 together with a number of GSM bands. WCDMA deployment at 900 MHz band VIII starts during 2007 and requires, in practice, dual-band 900 + 2100 terminals.

| Operating band | 3GPP name | Total spectrum | Uplink [MHz] | Downlink [MHz] | |
|---|---|---|---|---|---|
| Band I | 2100 | 2×60 MHz | 1920–1980 | 2110–2170 | Mainstream WCDMA band |
| Band II | 1900 | 2×60 MHz | 1850–1910 | 1930–1990 | PCS band in Americas |
| Band III | 1800 | 2×75 MHz | 1710–1785 | 1805–1880 | Europe, Asia and Brazil |
| Band IV | 1700/2100 | 2×45 MHz | 1710–1755 | 2110–2155 | New 3G band in USA and Americas |
| Band V | 850 | 2×25 MHz | 824–849 | 869–894 | USA, Americas and Asia |
| Band VI | 800 | 2×10 MHz | 830–840 | 875–885 | Japan |
| Band VII | 2600 | 2×70 MHz | 2500–2570 | 2620–2690 | New 3G band |
| Band VIII | 900 | 2×35 MHz | 880–915 | 925–960 | Europe and Asia |
| Band IX | 1700 | 2×35 MHz | 1750–1785 | 1845–1880 | Japan |
| Band X | 1700/2100 | 2×60 MHz | 1710–1770 | 2110–2170 | Extended Band IV |

**Figure 8.28** WCDMA frequency bands in 3GPP

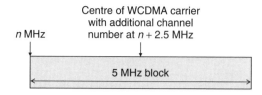

**Figure 8.29**   Additional channel numbers locate WCDMA carrier in the middle of 5 MHz block

The WCDMA networks in the USA started at 1.9 GHz band II and expanded to 850 MHz band V. The WCDMA terminals in the USA need, in practice, to have dual-band WCDMA 1900 + 850 MHz supported. There are also multiband terminals available supporting the 850, 1900 and 2100 bands. The new 3G band in the USA at 1.7/2.1 GHz is band IV and called Advanced Wireless Services (AWS). The WCDMA deployments started at the AWS band. Altogether, WCDMA is currently deployed in three different frequency bands in the USA.

The different frequency variants use exactly the same 3GPP WCDMA/HSPA specifications except for the differences in the RF parameters and requirements. The differences between the frequency variants are summarized below.

1. Additional channel numbers with 100 kHz offset are included for locating the WCDMA carrier exactly in the middle of a 5 MHz block for the bands (II, IV, V and VI); see Figure 8.29. The normal channel numbers are multiples of 200 kHz.
2. Narrowband blocking requirements for those bands (II, III, IV, V and VIII) where GSM may be deployed in the same band. The carrier separation between the WCDMA signal and narrowband GSM interference is 2.7 MHz in bands II, IV and V. That separation is the minimum possible when WCDMA is located in the middle of a 5 MHz block and the other operator's first GSM carrier is 0.2 MHz from the edge of the block, so in total $5.0/2 + 0.2 = 2.7$ MHz. For bands III and VIII the channel raster is 200 kHz without 100 kHz offset and, thus, the narrowband blocking distance is 2.8 MHz.
3. Relevant spurious emission requirements added.
4. Relaxed terminal sensitivity requirements for those bands (II, III, V and VIII) where the separation between uplink and downlink is only 20 MHz or less, and for band VII. Those requirements allow one to achieve a high enough duplex attenuation between transmission and reception in a small terminal. The relaxation is 2–3 dB compared with the other bands.

## 8.7   UMTS Refarming to GSM Band

UMTS has been deployed in the existing GSM bands at 850 MHz and 1900 MHz in North and South America and in some Asian countries. UMTS at 900 MHz is used in Europe and Asia. As the mobile broadband adoption accelerates globally, the refarming of existing bands brings more spectrum for UMTS. The low frequency at 850 and 900 MHz has also the benefit of providing better coverage which can be used to bring UMTS to new areas and to improve indoor coverage in the existing UMTS coverage areas. UMTS900 can deliver benefits with lower cost with less base station sites and with improved HSPA throughput.

Typical scenarios for UMTS refarming are depicted in Figure 8.30. Most typically the deployment of UMTS900 starts in the areas where the business need meets the practical possibilities, for example, in areas where low cost mobile broadband is needed and the amount of GSM traffic makes it possible to assign enough of the spectrum to UMTS.

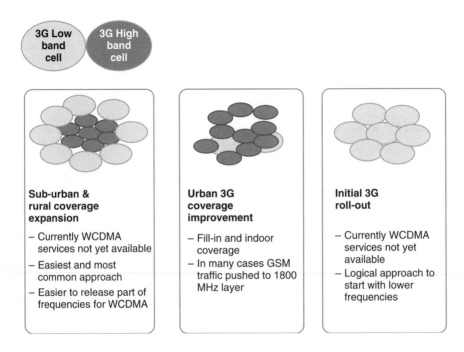

**Figure 8.30**  Use cases of low band refarming

The frequencies outside Europe are generally technology neutral which allows UMTS refarming without specific changes in the regulation. The UMTS900 deployment in Europe was limited by the GSM Directive that allowed only GSM technology at 900 MHz. That old directive was removed and the 900 MHz spectrum usage was liberated in July 2009.

All new UMTS terminals have dual band or triple band support which makes refarming an attractive solution since the terminal support will be there for UMTS900 by default. The triple band terminal variants are as follows:

- European model supporting 900 (Band VIII) + 1900 (Band II) + 2100 (Band I);
- Americas model 850 (Band V) + 1900 (Band II) + 2100 (Band I).

The UMTS terminals naturally also support GSM/EDGE in four bands.

This section considers the different deployment and performance aspects of UMTS refarming to GSM bands.

## 8.7.1  Coverage of UMTS900

UMTS900 improves coverage compared to UMTS2100 because of the lower frequency. The typical coverage area of a 3-sector macro cell in a suburban area is illustrated in Figure 8.31. The base station antenna height is assumed to be 30 meters and indoor penetration loss 10 dB. The maximum path loss is assumed to be 160 dB. The Okumura-Hata propagation model with a correction factor of $-5$ dB is applied. The mobile antenna gain is assumed to be 3 dB lower at 900 MHz compared to 2100 MHz, which makes the effective path loss at 900 MHz to be 157 dB. The coverage area of each base station at 900 MHz is three times larger than at 2100 MHz which make it feasible to provide coverage to larger areas.

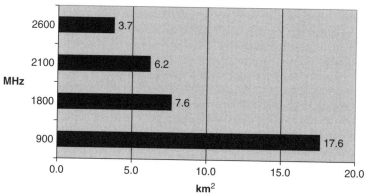

**Figure 8.31** Coverage area of 3-sector cell site in suburban area with 30 m mast height and 10 dB indoor penetration loss

**Table 8.26** Coverage benefit of UMTS900 compared to UMTS2100

|  | Urban | Rural | Rural fixed wireless |
|---|---|---|---|
| Propagation loss[1] | +11 dB | +11 dB | +11 dB |
| Base station antenna gain | −3 dB[2] | 0 dB[3] | 0 dB[3] |
| Mobile antenna gain | −3 dB | −3 dB | 0 dB[4] |
| Total gain of 900 MHz compared to 2100 MHz | +5 dB | +8 dB | +11 dB |

[1]According to Okumura-Hata
[2]Shared 1.3 m antenna for 900/1800/2100 giving 15 dBi gain at 900 MHz
[3]2.5 m antenna giving 18 dBi gain at 900 MHz
[4]External antenna assumed

The exact coverage benefit of the lower frequency depends on the site configuration and on the terminal antenna solution. Table 8.26 illustrates three different cases: urban, rural and fixed wireless rural deployments. The Okumura-Hata model shows 11 dB lower path loss for 900 MHz compared to 2100 MHz. Part of that benefit is lost because the antenna gains are lower at 900 MHz if the antenna size cannot be increased. If we use the shared base station antenna of 1.3 meter height for UMTS900 and UMTS2100 in urban areas, the antenna gain is 3 dB lower for 900 MHz. In the rural case, it is possible to use larger antennas for 900 MHz providing the same gain as 2100 MHz antennas. We also assume that the mobile antenna gain at 900 MHz is 3 dB lower than at 2100 MHz. The lower mobile antenna gain at 900 MHz can be compensated by using larger antenna which is feasible in fixed wireless installations. The base station cable losses are lower at 900 MHz, which improves the relative coverage for 900 MHz. The impact of the cable loss is negligible if remote RF modules are used close to the antennas. Therefore, we do not consider the impact of cable loss in this calculation. When these differences are combined together, 900 MHz gives 5 dB better coverage in urban areas, 8 dB in rural areas and up to 11 dB in rural areas with fixed directional antennas. We can conclude that 900 MHz offers largest coverage benefit in rural areas especially for wireless broadband applications. Figure 8.31 assumed total 8 dB coverage benefit for 900 MHz.

Currently there are already many UMTS900 deployments globally and the theoretical performance estimates have been proven in practice.

## 8.8   Interference between GSM and UMTS

3GPP narrowband blocking requirements are defined for the case where GSM and UMTS are deployed uncoordinated by different operators. The UMTS and GSM carrier spacing in 3GPP test cases is 2.8 MHz, which can be achieved when UMTS is deployed in a 5.0 MHz block and there is one GSM guard carrier between operators. In practice, it is possible to use smaller carrier spacing between GSM and UMTS. This section analyses the minimum required spectrum for UMTS refarming.

In case of co-sited GSM and UMTS base stations, the limiting factor for minimizing the GSM-UMTS carrier spacing is GSM mobile interference to UMTS base station reception. The reason is the limited power control dynamics in GSM mobile. The minimum transmit power is 5 dBm. The typical minimum coupling loss between mobile and base station is 70–80 dB, which makes the uplink received power level −65−−75 dBm. The interference scenario is illustrated in Figure 8.32. There is no interference from UMTS mobile to GSM base station because of fast and accuracy power control in UMTS mobiles.

In order to validate the interference assumptions, uplink power level measurements are collected from live GSM network. The number of samples is more than 25 billion. The distribution of the results is presented in Figure 8.33. The measurements show that there are practically no samples with

70–80 dB

Co-sited GSM900
+ UMTS900

GSM only terminal close
to BTS with transmit
power of 5 dBm

**Figure 8.32**   Interference scenario between co-sited GSM and UMTS on the same band

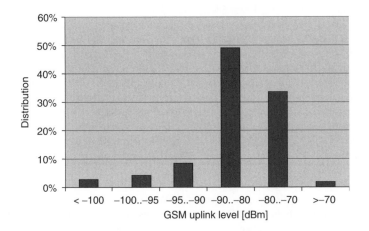

**Figure 8.33**   Uplink received power levels from GSM mobiles

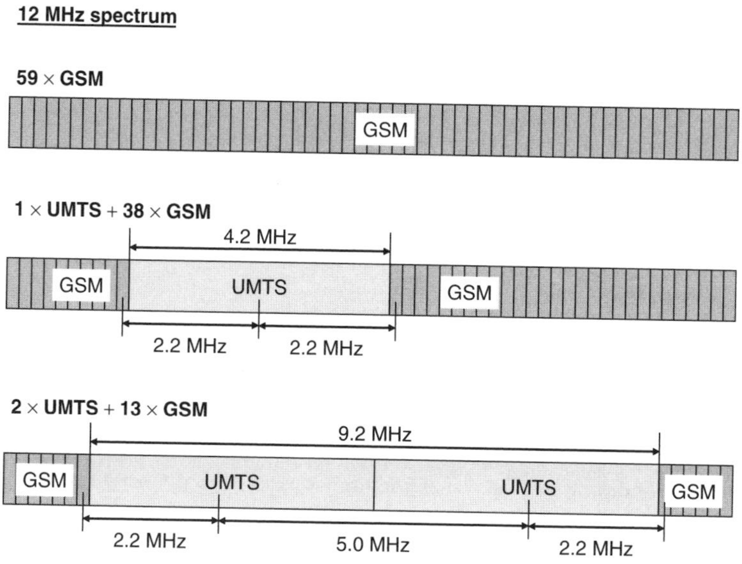

**12 MHz spectrum**

**59 × GSM**

**1 × UMTS + 38 × GSM**

**2 × UMTS + 13 × GSM**

**Figure 8.34**  UMTS refarming evolution with co-sited GSM and UMTS

levels above −70 dBm. The UMTS base station blocking measurements show that −70 dBm GSM interference can be tolerated with 2.2 MHz carrier separation when optimized base station filtering is applied. [13]. Therefore, we can conclude that 2.2 MHz carrier spacing can be used for UMTS refarming with co-sited GSM and UMTS.

If the systems are not co-sited, the power level differences are larger between GSM and UMTS. The GSM mobile can use a maximum transmit power of 33 dBm close to UMTS base station leading to received uplink power level up to −40 dBm. The required carrier spacing would then be 2.4 MHz from the uplink interference point of view with optimized UMTS base station filtering.

The downlink interference also needs to be considered when the systems are not co-sited. 3GPP has studied the interference between uncoordinated GSM and UMTS deployments at 900 MHz. The simulations show that UMTS does not cause interference to the GSM system due to the fast power control in UMTS and due to the narrowband GSM carrier. The interference caused by GSM base station to UMTS mobile has approximately 1% capacity loss in UMTS downlink with 2.8 MHz spacing. The downlink interference cases can be avoided by using quality-based inter-frequency or inter-system handovers.

The GSM-UMTS carrier spacing of 2.2 MHz corresponds to the total spectrum requirement of 4.2 MHz for UMTS. The potential spectrum evolution scenario is shown in Figure 8.34 with a total 12 MHz allocation with 59 GSM carriers plus 1 guard carrier. When one UMTS carrier is deployed with 4.2 MHz, there are 38 GSM carriers left. With a second UMTS carrier, there are still 13 GSM carriers left for GSM voice coverage. The total 900 MHz spectrum is 35 MHz. When there are three operators, the average allocation per operator is slightly below 12 MHz.

Two UMTS carriers enable use of a Dual cell HSPA (DC-HSPA) which improves the single user data rates up to 84 Mbps in downlink and 23 Mbps in uplink in Release 9.

## 8.9  Remaining GSM Voice Capacity

The remaining GSM voice capacity is estimated in this section. The key features for improving the GSM spectral efficiency are Adaptive Multirate (AMR) voice codec and Orthogonal Subchannels (OSC). AMR link adaptation allows the use of smaller frequency reuse factors, since higher

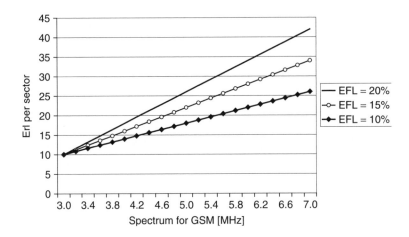

**Figure 8.35**   GSM capacity as a function of Effective Frequency Load (EFL) and available spectrum

interference levels can be tolerated. OSC allows squeezing up to four voice calls into single time slot. The broadcast channel (BCCH) requires higher frequency reuse than the hopping carriers. We assume in this analysis that BCCH reuse is 15 and it can carry 10 Erl of voice traffic. The reuse factor of hopping carriers is indicated as the effective frequency load (EFL) which is essentially the inverse of the reuse factor. EFL depends on the quality of the network planning and on the available features. Three different scenarios are considered here: EFL of 10%, 15% and 20%. The results are shown in Figure 8.35. The absolute minimum spectrum requirement for simultaneous GSM and UMTS deployment is $3.0 + 4.2 = 7.2$ MHz. If we need to carry on average 20 Erl of GSM voice traffic an EFL of 15% can be achieved, the required spectrum for GSM is 4.6 MHz, which makes the total 900 MHz spectrum requirement for UMTS and GSM $4.6 + 4.2 = 8.8$ MHz.

## 8.10   Shared Site Solutions with GSM and UMTS

Sharing sites between GSM and UMTS is preferred to save site acquisition and rental costs. This section illustrates the different site solution options for using the same sites for GSM900 and UMTS900 in Figure 8.36. A number of different options are needed since one solution is not enough to cover the different needs. Option 1 is to use separate antennas and antenna feeders for GSM and UMTS. That solution keeps those two systems completely independent and allows different optimization in terms of antenna tilting, but increasing the number of antennas is naturally a challenge for the site solution. Option 2 is to change the existing cross-polarized antenna into dual cross-polarized antenna having two antennas in the same radome and total four feeder ports. Options 1 and 2 also allow using RF module close to the antenna for UMTS while the GSM base station can still remain on the ground.

If the space at the site is very limited, then it may not be possible to add antennas or to have larger antennas. The same antenna can be used for GSM and for UMTS by using multiradio combiners as shown in Option 3. The losses in the combiners can be minimized in the uplink by using mast head amplifiers before the combiners. The downlink losses can be avoided if there is just one GSM carrier transmitted via one of the antenna branches and the UMTS carrier is transmitted via the other antenna branches. In case of multiple GSM carriers or in case of UMTS Multiple Input Multiple Output (MIMO) there will be some losses in the RF combining.

Option 4 uses the same multiradio base station both for GSM and for UMTS. There is no combining needed since GSM and UMTS signals are transmitted from the same RF module. Option 4 is attractive

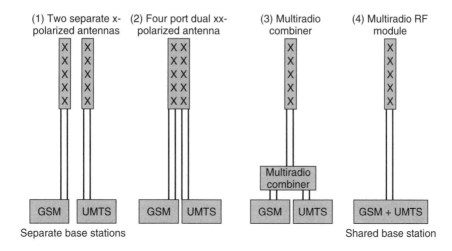

**Figure 8.36**  Options for combining GSM and UMTS in the same band

if the GSM base station equipment is old and needs to be modernized at the same time as UMTS900 rollout takes place.

## 8.11   Interworking of UMTS900 and UMTS2100

When UMTS900 is deployed in the same area as UMTS2100 (or UMTS850 and UMTS1900, or UMTS1500 and UMTS2100), the system parameters and algorithms need to be designed to make efficient use of the existing network resources on two bands. The system design must consider the facts that UMTS900 can provide better coverage, but on the other hand, UMTS2100 has more spectrum available and more capacity. UMTS900 must be utilized when the location does not have UMTS2100 coverage, but the system must be able to push mobiles to UMTS2100 network in order to avoid UMTS900 congestion. Typical spectrum availability is illustrated in Figure 8.37 where UMTS900 has just one carrier while UMTS2100 has three carriers.

The mobile can change the band in different phases on the connection:

1. *Idle mode*. The idle mode parameters are transmitted in System Information Blocks (SIB) on the broadcast channel. The mobile selects the band based on its measurements and based on the system parameters. The preferred layer for idle mode can be either UMTS2100 or UMTS900. It may be logical to use UMTS2100 as the preferred layer in urban areas where UMTS2100 has relatively good coverage and where the capacity requirements are high. That approach allows getting most traffic to the UMTS2100 layer. UMTS900 could be used as the preferred layer in rural areas where

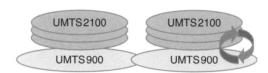

**Figure 8.37**  Interworking algorithms required between UMTS900 and UMTS2100

only UMTS900 provides full coverage and UMTS2100 is used only in some sites for capacity boost. There is no need for location or routing area update when changing between UMTS900 and UMTS2100 when the different bands are connected to the same RNC.

2. *Call set-up phase*. RNC can command the mobile to change its band in the Radio Bearer set-up phase. The decision can be based on mobile capability, source system Received Signal Code Power (RSCP) or based on service and load information. RNC knows the loading of both bands when both bands are connected to the same RNC. RNC would typically not have any information about the target band RSCP since that would require compressed mode measurements. Therefore, some safety margin is required especially when changing from a low band with better coverage to a high band with smaller coverage. That approach also requires that UMTS900 and UMTS2100 antennas are co-sited and use the same direction and tilting.

3. *RRC state transition*. The band can be changed also in the state transition from FACH to DCH. The algorithms can be similar to the call set-up phase.

4. *DCH state*. Compressed mode measurements of the target band can be used for handovers in DCH state. The handover decision in RNC can also take into account the load and service information.

# References

[1] Sipilä, K., Laiho-Steffens, J., Jäsberg, M. and Wacker, A., 'Modelling the Impact of the Fast Power Control on the WCDMA Uplink', *Proceedings of VTC'99*, Houston, Texas, May 1999, pp. 1266–1270.

[2] Ojanperä, T. and Prasad, R., *Wideband CDMA for Third Generation Mobile Communications*, New York: Artech House, 1998.

[3] Laiho, J., Wacker, A. and Novosad, T., *Radio Network Planning and Optimisation for UMTS*, New York: John Wiley & Sons, 2001.

[4] Saunders, S., *Antennas and Propagation for Wireless Communication Systems*, New York: John Wiley & Sons, 1999.

[5] Wacker, A., Laiho-Steffens, J., Sipilä, K. and Heiska, K., 'The Impact of the Base Station Sectorisation on WCDMA Radio Network Performance', *Proceedings of VTC'99*, Amsterdam, The Netherlands, September 1999, pp. 2611–2615.

[6] Sipilä, K., Honkasalo, Z., Laiho-Steffens, J. and Wacker, A., 'Estimation of Capacity and Required Transmission Power of WCDMA Downlink Based on a Downlink Pole Equation', *Proceedings of VTC2000*, Spring 2000.

[7] Wang, Y.-P. and Ottosson, T., 'Cell Search in W-CDMA', *IEEE J. Select. Areas Commun.*, Vol. 18, No. 8, 2000, pp. 1470–1482.

[8] Lee, J. and Miller, L., *CDMA Systems Engineering Handbook*, New York: Artech House, 1998.

[9] Wacker, A., Laiho-Steffens, J., Sipilä, K. and Jäsberg, M., 'Static Simulator for Studying WCDMA Radio Network Planning Issues', *Proceedings of VTC'99*, Houston, Texas, May 1999, pp. 2436–2440.

[10] Nokia NetAct™ Planner, http://www.nokia.com/networks/services/netact/netact_planner/.

[11] 3GPP Technical Specification 25.101, UE Radio Transmission and Reception (FDD).

[12] 3GPP Technical Report 25.942, RF System Scenarios.

[13] Holma, H., Ahonpaa, T. and Prieur, E. 'UMTS900 Co-Existence with GSM900', Vehicular Technology Conference, 2007.

# 9

# Radio Resource Management

Harri Holma, Klaus Pedersen, Jussi Reunanen, Janne Laakso and Oscar Salonaho

## 9.1  Introduction

Radio Resource Management (RRM) algorithms are responsible for efficient utilization of the air interface resources. RRM is needed to guarantee Quality of Service (QoS), to maintain the planned coverage area, and to offer high capacity. The family of RRM algorithms can be divided into handover control, power control, admission control, load control, and packet scheduling functionalities. Power control is needed to keep the interference levels at minimum in the air interface and to provide the required quality of service. WCDMA power control is described in Section 9.2. Handovers are needed in cellular systems to handle the mobility of the UEs across cell boundaries. Handovers are presented in Section 9.3. In third generation networks other RRM algorithms – such as admission control, load control and packet scheduling – are required to guarantee the quality of service and to maximize the system throughput with a mix of different bit rates, services and quality requirements. Admission control is presented in Section 9.5 and load control in Section 9.6. WCDMA packet scheduling is described in Chapter 10.

The RRM algorithms can be based on the amount of hardware in the network or on the interference levels in the air interface. Hard blocking is defined as the case where the hardware limits the capacity before the air interface gets overloaded. Soft blocking is defined as the case where the air interface load is estimated to be above the planned limit. The difference between hard blocking and soft blocking is analysed in Section 8.2.5. It is shown that soft blocking-based RRM gives higher capacity than hard blocking-based RRM. If soft blocking based RRM is applied, the air interface load needs to be measured. The measurement of the air interface load is presented in Section 9.4. In IS-95 networks RRM is typically based on the available channel elements (hard blocking), but that approach is not applicable in the third generation WCDMA air interface, where various bit rates have to be supported simultaneously.

Typical locations of the RRM algorithms in a WCDMA network are shown in Figure 9.1.

*WCDMA for UMTS: HSPA Evolution and LTE, Fifth Edition*   Edited by Harri Holma and Antti Toskala
© 2010 John Wiley & Sons, Ltd

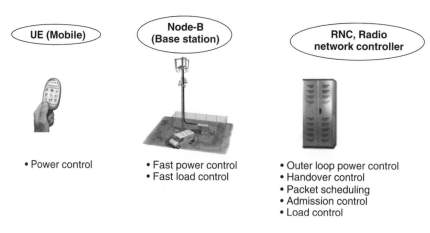

**Figure 9.1**  Typical locations of RRM algorithms in a WCDMA network

## 9.2  Power Control

Power control was introduced briefly in Section 3.5. In this chapter a few important aspects of WCDMA power control are covered. Some of these issues are not present in existing second generation systems, such as GSM and IS-95, but are new in third generation systems and therefore require special attention. In Section 9.2.1 fast power control is presented and in Section 9.2.2 outer loop power control is analysed. Outer loop power control sets the target for fast power control so that the required quality is provided.

In the following sections the need for fast power control and outer loop power control is shown using simulation results. Two special aspects of fast power control are presented in detail in Section 9.2.1: the relationship between fast power control and diversity, and fast power control in soft handover.

### 9.2.1  Fast Power Control

In WCDMA, fast power control with 1.5 kHz frequency is supported in both uplink and downlink. In GSM, only slow (frequency approximately 2 Hz) power control is employed. In IS-95, fast power control with 800 Hz frequency is supported only in the uplink.

#### 9.2.1.1  Gain of Fast Power Control

In this section, examples of the benefits of fast power control are presented. The simulated service is 8 kbps with BLER = 1% and 10 ms interleaving. Simulations are made with and without fast power control with a step size of 1 dB. Slow power control assumes that the average power is kept at the desired level and that the slow power control would be able to ideally compensate for the effect of path loss and shadowing, whereas fast power control can compensate also for fast fading. Two-branch receive diversity is assumed in the Node B. ITU Vehicular A is a five-tap channel with WCDMA resolution, and ITU Pedestrian A is a two-path channel where the second tap is very weak. The required $E_b/N_0$ with and without fast power control are shown in Table 9.1 and the required average transmission powers in Table 9.2.

Fast power control gives clear gain, which can be seen from Tables 9.1 and 9.2. The gain from the fast power control is larger:

- for low UE speeds than for high UE speeds;

**Table 9.1** Required $E_b/N_0$ values with and without fast power control

|  | Slow power control (dB) | Fast 1.5 kHz power control (dB) | Gain from fast power control (dB) |
|---|---|---|---|
| ITU Pedestrian A 3 km/h | 11.3 | 5.5 | 5.8 |
| ITU Vehicular A 3 km/h | 8.5 | 6.7 | 1.8 |
| ITU Vehicular A 50 km/h | 6.8 | 7.3 | −0.5 |

**Table 9.2** Required relative transmission powers with and without fast power control

|  | Slow power control (dB) | Fast 1.5 kHz power control (dB) | Gain from fast power control (dB) |
|---|---|---|---|
| ITU Pedestrian A 3 km/h | 11.3 | 7.7 | 3.6 |
| ITU Vehicular A 3 km/h | 8.5 | 7.5 | 1.0 |
| ITU Vehicular A 50 km/h | 6.8 | 7.6 | −0.8 |

- in required $E_b/N_0$ than in transmission powers;
- for those cases where only a little multipath diversity is available, as in the ITU Pedestrian A channel. The relationship between fast power control and diversity is discussed in Section 9.2.1.2.

In Tables 9.1 and 9.2 the negative gains at 50 km/h indicate that an ideal slow power control would give better performance than the realistic fast power control. The negative gains are due to inaccuracies in the SIR estimation, power control signaling errors, and the delay in the power control loop.

The gain from fast power control in Table 9.1 can be used to estimate the required fast fading margin in the link budget in Section 8.2.1. The fast fading margin is needed in the UE transmission power to maintain adequate closed loop fast power control. The maximum cell range is obtained when the UE is transmitting with full constant power, i.e. without the gain of fast power control. Typical values for the fast fading margin for low mobile speeds are 2–5 dB.

#### 9.2.1.2 Power Control and Diversity

In this section the importance of diversity is analysed together with fast power control. At low UE speed the fast power control can compensate for the fading of the channel and keep the received power level fairly constant. The main sources of errors in the received powers arise from inaccurate SIR estimation, signaling errors and delays in the power control loop. The compensation of the fading causes peaks in the transmission power. The received power and the transmitted power are shown as a function of time in Figures 9.2 and 9.3 with a UE speed of 3 km/h. These simulation results include realistic SIR estimation and power control signaling. A power control step size of 1.0 dB is used. In Figure 9.2 very little diversity is assumed, while in Figure 9.3 more diversity is assumed in the simulation. Variations in the transmitted power are higher in Figure 9.2 than in Figure 9.3. This is due to the difference in the amount of diversity. The diversity can be obtained with, for example, multipath diversity, receive antenna diversity, transmit antenna diversity or macro diversity.

With less diversity there are more variations in the transmitted power, but also the average transmitted power is higher. Here we define power rise to be the ratio of the average transmission power in a fading channel to that in a non-fading channel when the received power level is the same in both fading and non-fading channels with fast power control. The power rise is depicted in Figure 9.4.

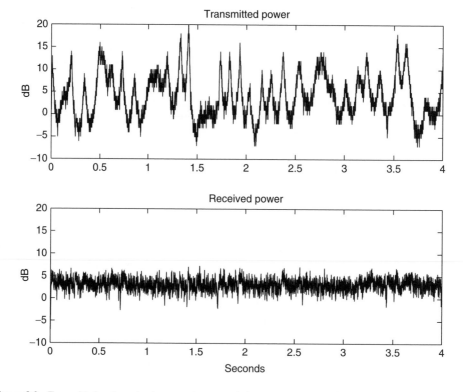

**Figure 9.2** Transmitted and received powers in two-path (average tap powers 0 dB, −10 dB) Rayleigh fading channel at 3 km/h

**Table 9.3** Simulated power rises. Multipath channel ITU Pedestrian A, antenna diversity assumed

| UE speed (km/h) | Average power rise (dB) |
| --- | --- |
| 3 | 2.1 |
| 10 | 2.0 |
| 20 | 1.6 |
| 50 | 0.8 |
| 140 | 0.2 |

The link level results for uplink power rise are presented in Table 9.3. The simulations are performed at different UE speeds in a two-path ITU Pedestrian A channel with average multipath component powers of 0.0 dB and −12.5 dB. In the simulations the received and transmitted powers are collected slot by slot. With ideal power control the power rise would be 2.3 dB. At low UE speeds the simulated power rise values are close to the theoretical value of 2.3 dB, indicating that fast power control works efficiently in compensating the fading. At high UE speeds (>100 km/h) there is only very little power rise since the fast power control cannot compensate for the fading.

More information about the uplink power control modelling can be found in [1].

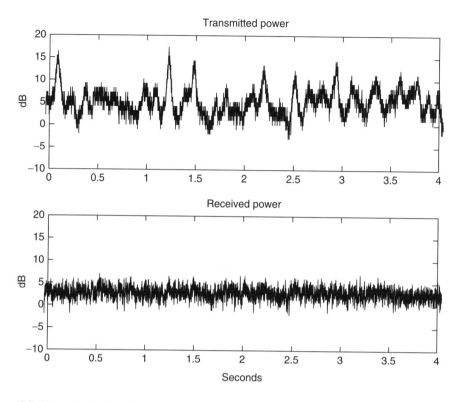

**Figure 9.3** Transmitted and received powers in three-path (equal tap powers) Rayleigh fading channel at 3 km/h

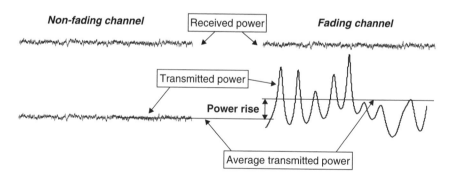

**Figure 9.4** Power rise in fading channel with fast power control

Why is the power rise important for WCDMA system performance? In the downlink, the air interface capacity is directly determined by the required transmission power, since that determines the transmitted interference. Thus, to maximize the downlink capacity the transmission power needed by one link should be minimized. In the downlink, the received power level in the UE does not affect the capacity.

In the uplink, the transmission powers determine the amount of interference to the adjacent cells, and the received powers determine the amount of interference to other UEs in the same cell. If, for

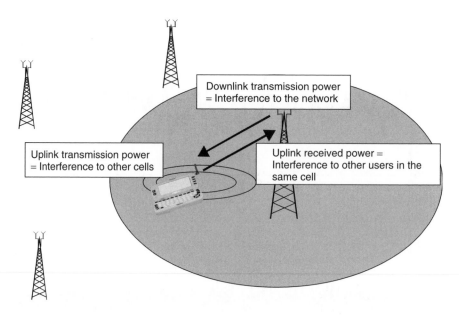

**Figure 9.5**   Effect of received and transmission powers on interference levels

example, there were only one WCDMA cell in one area, the uplink capacity of this cell would be maximized by minimizing the required received powers, and the power rise would not affect the uplink capacity. We are, however, interested in cellular networks where the design of the uplink diversity schemes has to take into account both the transmitted and received powers. The effect of received and transmission powers on network interference levels is illustrated in Figure 9.5.

### 9.2.1.3   Power Control in Soft Handover

Fast power control in soft handover has two major issues that are different from the single-link case: power drifting in the Node B powers in the downlink, and reliable detection of the uplink power control commands in the UE. These aspects are illustrated in Figure 9.6 and in Figure 9.7 and described in more detail in this section. A solution for improving the power control signaling quality is also presented in this section.

#### *Downlink Power Drifting*

The UE sends a single command to control the downlink transmission powers; this is received by all Node Bs in the active set. The Node Bs detect the command independently, since the power control commands cannot be combined in RNC because it would cause too much delay and signaling in the network. Due to signaling errors in the air interface, the Node Bs may detect this power control command in a different way. It is possible that one of the Node Bs lowers its transmission power to that UE while the other Node B increases its transmission power. This behavior leads to a situation where the downlink powers start drifting apart; this is referred to here as power drifting.

Power drifting is not desirable, since it mostly degrades the downlink soft handover performance. It can be controlled via RNC. The simplest method is to set relatively strict limits for the downlink power control dynamics. These power limits apply to the UE's specific transmission powers. Naturally, the smaller the allowed power control dynamics, the smaller the maximum power drifting. On the other hand, large power control dynamics typically improve power control performance, as shown in Table 9.2.

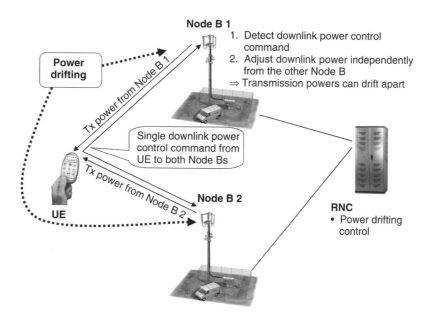

**Figure 9.6** Downlink power drifting in soft handover

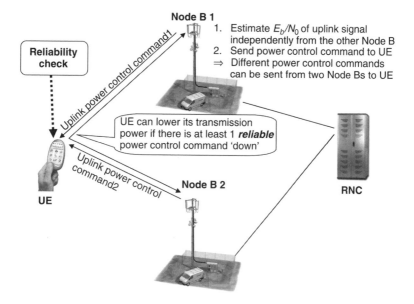

**Figure 9.7** Reliability check of the uplink power control commands in UE in soft handover

Another way to reduce power drifting is as follows. RNC can receive information from the Node Bs concerning the transmission power levels of the soft handover connections. These levels are averaged over a number of power control commands, e.g. over 500 ms or equivalently over 750 power control commands. Based on those measurements, RNC can send a reference value for the downlink transmission powers to the Node Bs. The soft handover Node Bs use that reference value in their downlink power control for that connection to reduce the power drifting. The idea is that a small correction is periodically performed towards the reference power. The size of this correction is proportional to the difference between the actual transmitted power and the reference power. This method will reduce the amount of power drifting. Power drifting can happen only if there is fast power control in the downlink. In IS-95, only slow power control is used in the downlink, and no method of controlling downlink power drifting is needed.

### Reliability of Uplink Power Control Commands

All the Node Bs in the active set send an independent power control command to the UEs to control the uplink transmission power. It is enough if one of the Node Bs in the active set receives the uplink signal correctly. Therefore, the UE can lower its transmission power if one of the Node Bs sends a power-down command. Maximal ratio combining can be applied to the data bits in soft handover in the UE, because the same data is sent from all soft handover Node Bs, but not to the power control bits because they contain different information from each of the Node Bs. Therefore, the reliability of the power control bits is not as good as for the data bits, and a threshold in the UE is used to check the reliability of the power control commands. Very unreliable power control commands should be discarded because they are corrupted by interference. The 3GPP specifications include this requirement for the UE in [2].

### Improved Power Control Signaling Quality

The power control signaling quality can be improved by setting a higher power for the dedicated physical control channel (DPCCH) than for the dedicated physical data channel (DPDCH) in the downlink if the UE is in soft handover. This power offset between DPCCH and DPDCH can be different for different DPCCH fields: power control bits, pilot bits and TFCI. The power offset is illustrated in Figure 9.8.

The reduction of the UE transmission power is typically up to 0.5 dB with power offsets. This reduction is obtained because of the improved quality of the power control signaling. [3]

## 9.2.2   Outer Loop Power Control

The outer loop power control is needed to keep the quality of communication at the required level by setting the target for the fast power control. The outer loop aims to provide the required quality: no worse, no better. Too high quality would waste capacity. The outer loop is needed in both uplink and downlink because there is fast power control in both uplink and downlink. In the following sections a few aspects of this control loop are described; these apply to both uplink and downlink. The uplink outer loop is located in RNC and the downlink outer loop in the UE. In IS-95, outer loop power control is used only in the uplink because there is no fast power control in the downlink.

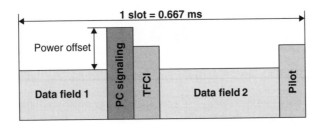

**Figure 9.8**   Power offset for improving downlink signaling quality

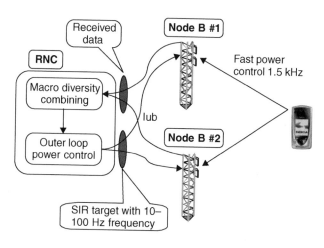

**Figure 9.9**   Uplink outer loop power control in RNC

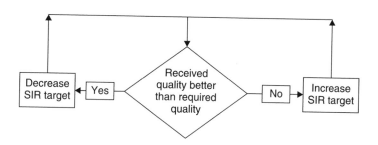

**Figure 9.10**   General outer loop power control algorithm

An overview of uplink outer loop power control is shown in Figure 9.9. The uplink quality is observed after macro diversity combining in RNC and the SIR target is sent to the Node Bs. The frequency of the fast power control is 1.5 kHz and frequency of the outer loop power control typically 10–100 Hz. A general outer loop power control algorithm is presented in Figure 9.10.

### 9.2.2.1   Gain of Outer Loop Power Control

In this section we analyse how much the SIR target needs to be adjusted when the UE speed or the multipath propagation environment changes. The terms SIR target and $E_b/N_0$ target are used interchangeably in this chapter. Simulation results with AMR speech service and BLER = 1% are shown in Table 9.4 with outer loop power control. Three different multipath profiles are used: non-fading channel corresponding to strong line-of-sight component, fading ITU Pedestrian A channel, and fading three-path channel with equal average powers of the multipath components. No antenna diversity is assumed here.

The lowest average $E_b/N_0$ target is needed in the non-fading channel and the highest target in the ITU Pedestrian A channel with high UE speed. This result indicates that the higher the variation in the received power, the higher the $E_b/N_0$ target needs to be to provide the same quality. If we were to select a fixed $E_b/N_0$ target of 5.3 dB according to the static channel, the frame error rate of the

**Table 9.4**  Average $E_b/N_0$ targets in different environments

| Multipath | UE speed (km/h) | Average $E_b/N_0$ target (dB) |
|---|---|---|
| Non-fading | – | 5.3 |
| ITU Pedestrian A | 3 | 5.9 |
| ITU Pedestrian A | 20 | 6.8 |
| ITU Pedestrian A | 50 | 6.8 |
| ITU Pedestrian A | 120 | 7.1 |
| 3-path equal powers | 3 | 6.0 |
| 3-path equal powers | 20 | 6.4 |
| 3-path equal powers | 50 | 6.4 |
| 3-path equal powers | 120 | 6.9 |

connection would be too high in fading channels and speech quality would be degraded. If we were to select a fixed $E_b/N_0$ target of 7.1 dB, the quality would be good enough but unnecessary high powers would be used in most situations. We can conclude that there is clearly a need to adjust the target of the fast closed loop power control by outer loop power control.

### 9.2.2.2  Estimation of Received Quality

A few different approaches to measuring the received quality are introduced in this section. A simple and reliable approach is to use the result of the error detection – cyclic redundancy check (CRC) – to detect whether there is an error or not. The advantages of using the CRC check are that it is a very reliable detector of frame errors, and it is simple. The CRC-based approach is well suited for those services where errors are allowed to occur fairly frequently, at least once every few seconds, such as the non-real time packet data service, where the block error rate (BLER) can be up to 10–20% before retransmissions, and the speech service, where typically BLER = 1% provides the required quality. With Adaptive Multirate (AMR) speech codec the interleaving depth is 20 ms and BLER = 1% corresponds to one error on average every 2 seconds.

The received quality can also be estimated based on soft frame reliability information. Such information could be, for example:

- Estimated bit error rate (BER) before channel decoder, called raw BER or physical channel BER.
- Soft information from Viterbi decoder with convolutional codes.
- Soft information from Turbo decoder, for example BER or BLER after an intermediate decoding iteration.
- Received $E_b/N_0$.

Soft information is needed for high quality services, see Section 9.2.2.4. Raw BER is used as soft information over the Iub interface. The estimation of the quality is illustrated in Figure 9.11.

### 9.2.2.3  Outer Loop Power Control Algorithm

One possible outer loop power control algorithm is presented in [4]. The algorithm is based on the result of a CRC check of the data and can be characterized in pseudocode as shown in Figure 9.12.

If the BLER of the connection is a monotonically decreasing function of the $E_b/N_0$ target, this algorithm will result in a BLER equal to the BLER target if the call is long enough. The step size parameter determines the convergence speed of the algorithm to the desired target and also defines

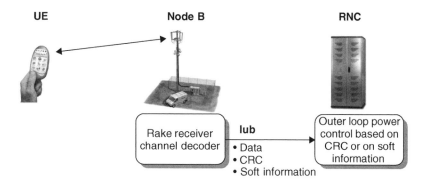

**Figure 9.11**   Estimation of quality in outer loop in RNC

**IF** CRC check OK
   *Step_down = BLER_target \* Step_size*;
   *Eb/N0_target (n + 1) = Eb/N0_target (n) − Step_down*;
**ELSE**
   *Step_up = Step_size − BLER_target \* Step_size*;
   *Eb/N0_target (n + 1) = Eb/N0_target (n) + Step_up*;
**END**

where

   *Eb/N0_target (n)* is the $E_b/N_0$ target in frame *n*,
   *BLER_target* is the BLER_target for the call and
   *Step_size* is a parameter, typically 0.3–0.5 dB

**Figure 9.12**   Pseudocode of one outer loop power control algorithm

the overhead caused by the algorithm. The principle is that the higher the step size, the faster the convergence and the higher the overhead. Figure 9.13 gives an example of the behavior of the algorithm with BLER target of 1% and step size of 0.5 dB.

### 9.2.2.4   High Quality Services

High quality services with very low BLER ($<10^{-3}$) are required to be supported by third generation networks. In such services errors are very rare events. If the required BLER $= 10^{-3}$ and the interleaving depth is 40 ms, an error occurs on average every $40/10^{-3}$ ms $= 40$ seconds. If the received quality is estimated based on the errors detected by CRC bits, the adjustments of the $E_b/N_0$ target are very slow and the convergence of the $E_b/N_0$ target to the optimal value takes a long time. Therefore, for high quality services the soft frame reliability information has advantages. Soft information can be obtained from every frame even if there are no errors.

### 9.2.2.5   Limited Power Control Dynamics

At the edge of the coverage area the UE may hit its maximum transmission power. In that case the received BLER can be higher than desired. If we apply directly the outer loop algorithm of Figure 9.10, the uplink SIR target would be increased. The increase of the SIR target does not improve the uplink

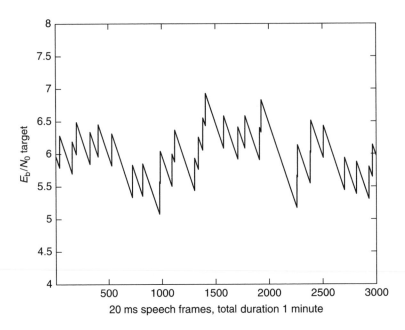

**Figure 9.13** $E_b/N_0$ target in ITU Pedestrian A channel, AMR speech codec, BLER target 1%, step size 0.5 dB, speed 3 km/h

quality if the Node B is already sending only power-up commands to the UE. In that case the $E_b/N_0$ target might become unnecessarily high. When the UE returns closer to the Node B, the quality of the uplink connection is unnecessarily high before the outer loop lowers the $E_b/N_0$ target back to the optimal value. The situation in which the UE hits its maximum transmission power is shown in Figure 9.14. In this example, the AMR speech service with 20 ms interleaving is simulated with the outer loop power control algorithm from Figure 9.12. A BLER target of 1% and an outer loop step size of 0.5 dB are used here. With full power control dynamics an error should occur once every 2 seconds to provide BLER of 1% with 20 ms interleaving. The maximum transmission power of the UE is 125 mW, i.e. 21 dBm.

The same problem could also occur if the UE hits its minimum transmission power. In that case, the $E_b/N_0$ target would become unnecessarily low. The same problems can be observed also in the downlink if the power of the downlink connection is using its maximum or minimum value.

The outer loop problems from limited power control dynamics can be avoided by setting tight limits for the $E_b/N_0$ target or by an intelligent outer loop power control algorithm. Such an algorithm would not increase the $E_b/N_0$ target if the increase did not improve the quality. The 3GPP specifications include this requirement for the UE in [2].

### 9.2.2.6 Multiservice

One of the basic requirements of UMTS is to be able to multiplex several services on a single physical connection. Since all the services have a common fast power control, there can be only one common target for fast power control. This must be selected according to the service requiring the highest target. Fortunately, there should be no large differences between the required targets if unequal rate matching has been applied on Layer 1 to provide the different qualities. The multiservice scenario is shown in Figure 9.15.

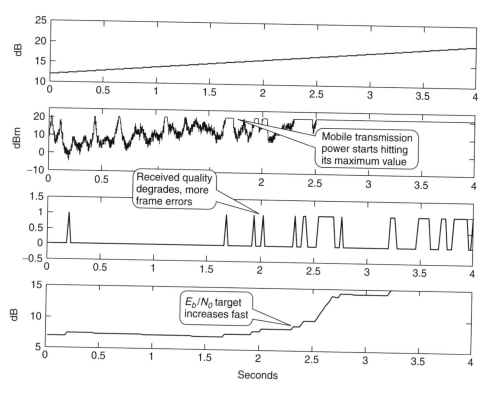

**Figure 9.14** Increase of $E_b/N_0$ target when the UE hits its maximum transmission power. Top: attenuation between UE and Node B; second figure: UE transmission power (dBm); third figure: block errors (1 = error, 0 = no errors); bottom: uplink $E_b/N_0$ target

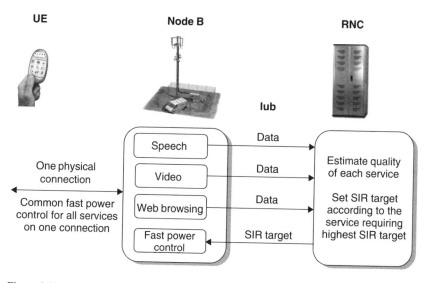

**Figure 9.15** Uplink outer loop power control for multiple services on one physical connection

#### 9.2.2.7  Downlink Outer Loop Power Control

The downlink outer loop power control runs in the UE. The network can effectively control the downlink connections even if it cannot control the downlink outer loop algorithm. First, the network sets the quality target for each downlink connection; that target can be modified during the connection. Second, the Node B does not need to increase the downlink power of that connection even if the UE sends a power-up command. The network can control the quality of the different downlink connections very quickly by not obeying the power control commands from the UE. This approach could be used, for example, during downlink overload to reduce the downlink power of those connections that have a lower priority, such as background-type services (see load control in Section 9.6). This reduction of downlink powers can take place at the frequency of fast power control, i.e. 1.5 kHz.

## 9.3  Handovers

### 9.3.1  Intra-Frequency Handovers

#### 9.3.1.1  Handover Algorithms

The WCDMA soft handover algorithm is introduced in this section, for specifications see [5]. The soft handover uses typically CPICH $E_c/I_0$ (= pilot $E_c/I_0$) as the handover measurement quantity, which is signalled to RNC by using Layer 3 signaling (see Section 7.8). The following terminology is used in the handover description:

- *Active set:* The cells in the active set form a soft handover connection to the UE.
- *Neighbor set/monitored set:* The neighbor set or monitored set is the list of cells that the UE continuously measures, but whose pilot $E_c/I_0$ are not strong enough to be added to the active set.

The soft handover algorithm as described in Figure 9.16 is as follows:

- If $Pilot\_E_c/I_0 > Best\_Pilot\_E_c/I_0 - Reporting\_range + Hysteresis\_event1A$ for a period of $\Delta T$ and the active set is not full, the cell is added to the active set. This event is called Event 1A or Radio Link Addition.
- If $Pilot\_E_c/I_0 < Best\_Pilot\_E_c/I_0 - Reporting\_range - Hysteresis\_event1B$ for a period of $\Delta T$, then the cell is removed from the active set. This event is called Event 1B or Radio Link Removal.
- If the active set is full and $Best\_candidate\_Pilot\_E_c/I_0 > Worst\_Old\_Pilot\_E_c/I_0 + Hysteresis\_event1C$ for a period of $\Delta T$, then the weakest cell in the active set is replaced by the strongest candidate cell (i.e. strongest cell in the monitored set). This event is called Event 1C or Combined Radio Link Addition and Removal. The maximum size of the active set in Figure 9.16 is assumed to be two.

   where:

   – *Reporting_range* is the threshold for soft handover
   – *Hysteresis_event1A* is the addition hysteresis
   – *Hysteresis_event1B* is the removal hysteresis
   – *Hysteresis_event1C* is the replacement hysteresis
   – *Reporting_range* − *Hysteresis_event1A* is also called Window_add
   – *Reporting_range* + *Hysteresis_event1B* is also called Window_drop
   – $\Delta T$ is the time to trigger
   – $Best\_Pilot\_E_c/I_0$ is the strongest measured cell in the active set
   – $Worst\_Old\_Pilot\_E_c/I_0$ is the weakest measured cell in the active set
   – $Best\_candidate\_Pilot\_E_c/I_0$ is the strongest measured cell in the monitored set
   – $Pilot\_E_c/I_0$ is the measured and filtered quantity.

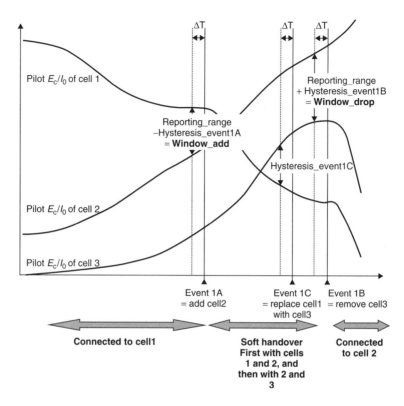

**Figure 9.16**   General scheme of the WCDMA soft handover algorithm

### 9.3.1.2   Handover Measurements

The WCDMA UE continuously scans the other cells on the same frequency when in cell_DCH state. The UE typically uses a matched filter to find the primary synchronization channel, P-SCH, of the neighboring cells. All cells transmit the same synchronization code that the UE seeks. The UE further identifies the cells with secondary synchronization channel, S-SCH and pilot channel, CPICH. The synchronization procedure is described in detail in Chapter 6. After that procedure, the UE is able to make pilot $E_c/I_0$ measurements and identify to which cell that measurement result belongs.

Since the WCDMA Node Bs can be asynchronous, the UE also decodes the system frame number (SFN) from the neighboring cells. SFN indicates the Node B timing with 10 ms frame resolution. SFN is transmitted on the Broadcast channel, BCH, carried on the Primary Common Control Physical Channel, P-CCPCH.

The intra-frequency handover measurement procedure is shown in Figure 9.17.

We evaluate the mobile handover measurement requirements in 3GPP and compare the performance requirements to the typical CPICH $E_c/I_0$ values with the common channel power allocations suggested in Chapter 8.

*Phase (1): Cell Identification*
The cell identification time in phase (1) of Figure 9.17 depends mainly on the number of cells and multipath components that the UE can receive. The UE needs to check every peak in its matched filter; the fewer peaks there are, the faster is the cell identification. The length of the neighbor list has only a marginal effect on the handover measurement performance.

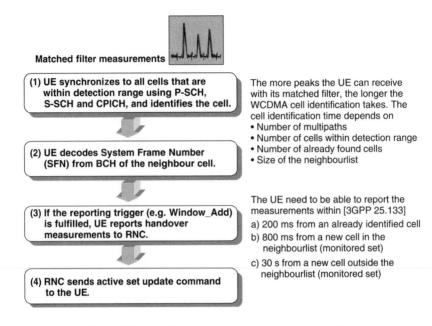

**Figure 9.17**  Intra-frequency handover measurement procedure

The 3GPP handover measurement performance requirements for the UE are given in [6]: with CPICH $E_c/I_0 > -20$ dB and SCH $E_c/I_0 > -20$ dB the UE will be able to report measurements within 200 ms from an already identified cell and within 800 ms from a new cell belonging to the monitored set. We will consider these performance requirements, together with typical common channel power allocations from Chapter 8 and typical handover parameter values. The scenario is illustrated in Figure 9.18 where the UE is connected to cell1 and it needs to identify cell2 that is approaching Window_add value. The resulting $E_c/I_0$ is obtained as follows:

1. We allocate 10% for CPICH and for SCH, giving $E_c/I_0$ r $= -10$ dB.
2. We assume Window_add $= 3$ dB where the UE needs to identify the cell when it is 3 dB below the strongest cell. This gives $E_c/I_0 = -3$ dB.
3. We further assume that the interference from other cells is 3 dB higher than the signal power from the best server. This gives $I_{0c2}/I_{0c1} = 3$ dB.

$$\frac{E_c}{I_0} = \frac{E_c}{I_{0r} + I_{0c1} + I_{0c2}} = \frac{E_c}{I_{0r}(1 + 10^{0.3} + 10^{0.6})} = \frac{E_c}{I_{0r}} = -8.5\,\text{dB} = -18.5\,\text{dB} \qquad (9.1)$$

The $E_c/I_0$ in this scenario is $-18.5$ dB which is better than $-20$ dB given in the performance requirements.

### Phase (2): System Frame Number (SFN) Decoding

In phase (2) of Figure 9.17, the UE decodes the system frame number from BCH that is transmitted on P-CCPCH. If we allocate $-5$ dB power for P-CCPCH compared to CPICH, the resulting $E_c/I_0$ is $-23.5$ dB. The performance requirement for BCH decoding with BLER $= 1\%$ is $-22.0$ dB [2]. As higher BLER levels can be allowed, the planned P-CCPCH power allocation should be adequate.

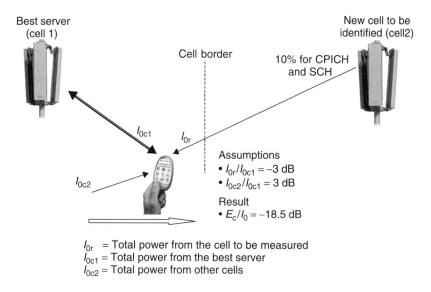

$I_{\text{or}}$ = Total power from the cell to be measured
$I_{\text{oc1}}$ = Total power from the best server
$I_{\text{oc2}}$ = Total power from other cells

**Figure 9.18**   Intra-frequency handover measurement scenario

These calculations of the cell identification and the system frame number decoding show that the assumed common channel powers in Chapter 8 are high enough to guarantee accurate handover measurements. It may even be possible to optimize the common channel powers to be lower than shown in Chapter 8.

Before the pilot $E_c/I_0$ is used by the active set update algorithm in the UE, some filtering is applied to make the results more reliable. The measurement is filtered both on Layer 1 and on Layer 3. The Layer 3 filtering can be controlled by the network. The WCDMA handover measurement filtering is described in Figure 9.19.

The handover measurement reporting from the UE to RNC can be configured to be periodic, as in GSM, or event-triggered. According to [7] the event-triggered reporting provides the same performance as periodic reporting but with less signaling load.

Before we leave the area of handover measurements, we note the size of the neighbor list. The maximum number of intra-frequency cells in the neighbor list is 32. Additionally, there can be 32 inter-frequency and 32 GSM cells on the neighbor list. The inter-frequency and inter-system measurements are covered later in this chapter. The maximum number of cells in the neighbor list is shown in Table 9.5.

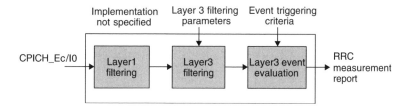

**Figure 9.19**   Handover measurement filtering and reporting

**Table 9.5**   Maximum number of monitored cells

| Intra-frequency | Inter-frequency | Inter-system GSM |
|---|---|---|
| 32 | 32[1] | 32 |

[1]The 32 inter-frequency neighbors can be on two frequencies.

The neighbor list can be defined by the network planning tool or it can also be tuned automatically by network optimization algorithms based on the UE measurements. If there is a neighbor cell missing from the list, it can be noticed based on the UE measurements, since UE is required to identify a new detectable cell that does not belong to the monitored set within 30 s [6].

### 9.3.1.3   Soft Handover Link Gains

The primary purpose of soft handover is to provide seamless handover and added robustness to the system. This is mainly achieved via three types of gain provided by the soft handover mechanism, i.e.,

- *Macro diversity gain*: A diversity gain over slow fading and sudden drops in signal strength due to, e.g., UE movement around a corner.
- *Micro diversity gain*: A diversity gain over fast fading.
- *Downlink load sharing*: A UE in soft handover receives power from multiple Node Bs, which implies that the maximum transmit power to a UE in $X$-way soft handover is multiplied by factor $X$ (i.e. improved coverage).

These three soft handover gains can be mapped to improved coverage and capacity of the WCDMA network. This section presents results of micro diversity soft handover gains that have been obtained by means of link level simulations. The micro diversity gains are presented relative to the ideal hard handover case, where the UE would be connected to the Node B with the highest pilot $E_c/I_0$. The results were presented and discussed in more detail in [3].

Figures 9.20 and 9.21 show the simulation results of 8 kbps speech in an ITU Pedestrian A channel, at 3 km/h, assuming that the UE is in soft handover with two Node Bs. The relative path loss from

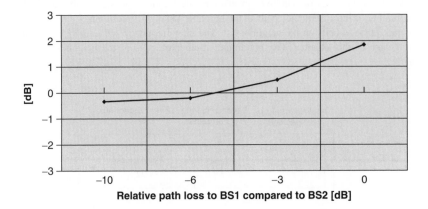

**Figure 9.20**   Soft handover gain in uplink transmission power (positive value = gain, negative value = loss)

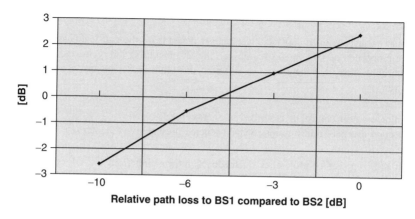

**Figure 9.21**  Soft handover gain in downlink transmission power (positive value = gain, negative value = loss)

the UE to Node B #1 compared to Node B #2 is 0, −3, −6 or −10 dB. The highest gains are obtained when the path loss is the same to both Node Bs, i.e. the relative path loss difference is 0 dB. Figure 9.20 shows the soft handover gain in uplink transmission power with two branch Node B receive antenna diversity. Figure 9.21 shows the corresponding gains in downlink transmission power without transmit or receive antenna diversity. The gains are relative to the single link case where the UE only is connected to the best Node B. It should be noted that ITU Pedestrian A channel has only little multipath diversity, and thus the micro diversity soft handover gains are relatively high. With more multipath diversity, the gains tend to decrease.

In Figure 9.20, the maximum reduction of the UE transmission power due to soft handover is observed to be 1.8 dB if the path loss is the same to both Node Bs. If the path loss difference is very large, the uplink soft handover should – in theory – never increase the UE transmission power since there is no extra transmission but only more Node Bs trying to detect the signal. In practice, if the path loss difference is very large, the soft handover can cause an increase in the UE transmission power. This increase is caused by the signaling errors of the uplink power control commands, which are transmitted in the downlink. But, typically, the Node B would not be in the active set of the UE if the path loss were 3–6 dB larger than the path loss to the strongest Node B in the UE's active set.

In the downlink the maximum soft handover gain is 2.3 dB (Figure 9.21), which is more than in the uplink (Figure 9.20). The reason is that no antenna diversity is assumed in the downlink, and thus in the downlink there is more need for micro diversity soft handover gain.

In the downlink, soft handover causes an increase in the required downlink transmission power if the path loss difference is more than 4–5 dB for the current example. In that case the UE does not experience a gain from the signal transmitted from the Node B with the largest path loss. Hence, the power transmitted from that Node B to the UE will only contribute to the total interference in the network. These simulation results also suggest values for Window_add and for Window_drop. Typical values for those parameters are shown in Table 9.6.

**Table 9.6**  Typical handover parameters

| Window_add | Window_drop |
|---|---|
| 1–3 dB | 2–5 dB |

#### 9.3.1.4 Network Capacity Gains

The potential capacity gain of soft handover mainly depends on the soft handover overhead (i.e. on the relative proportion of UEs in soft handover), the soft handover link gain, and the applied power control algorithm. Note that there are two different downlink power control algorithms for UEs in soft handover:

1. Conventional power control as described in Section 9.2.1.3; and
2. Site selection diversity transmission (SSDT) scheme presented in Chapter 6.

Recall that SSDT power control relies on feedback information from the UE, so only one of the Node Bs in the active set transmits data, while the other Node Bs only transmit physical layer control information. Thus, SSDT is equivalent to selection transmit diversity, while conventional power control of UEs in soft handover may be characterized as equal gain transmit diversity. The potential gain of SSDT comes from the reduced interference in the downlink, which should compensate for the loss of diversity gain in the downlink for the user data. From the conceptual point of view, it is obvious that the gain of SSDT is larger at high data rates where the overhead from the control information is marginal.

The capacity gain of soft handover in combination with SSDT, is on the same order of magnitude as the gain of soft handover and conventional power control. No significant gain of SSDT is observed, and in some cases the gain even turns into a loss. The reason for these observations can be explained as follows. A UE in soft handover periodically sends feedback commands to the Node Bs in the active set, which dictates which of the Node Bs should be transmitting the data. This results in larger power fluctuations at the different Node Bs, because the transmission to UEs is switched on/off on a relatively fast basis, as dictated by the UEs in soft handover. The alternation of Node B transmission towards UEs in soft handover is not within the control of the network, but purely UE controlled. Hence, although the SSDT scheme results in a reduction of the average total transmit power from Node Bs, the variance of the total transmit power also increases. The increased variance of the total transmit power maps to larger required power control headroom, which tends to reduce the potential gain of SSDT. Other aspects to note from a performance point of view include the impact of UE velocity, as with higher velocities the UE feedback is not well synchronized with the actual channel state. At some velocities, resonance problems do occur so the UE constantly asks for the 'wrong' Node B to transmit, via the feedback signaling to the network. This effect does, e.g., become dominant when the fading rate is roughly equal to the feedback rate.

#### 9.3.1.5 Soft Handover Overhead

The soft handover overhead is a common metric, which often is used to quantify the soft handover activity in a network. The soft handover overhead ($\beta$) is defined as

$$\beta = \sum_{n=1}^{N} n P_n - 1 \qquad (9.2)$$

where $N$ is the active set size and $P_n$ is the average probability of a UE being in $n$-way soft handover. In this context one-way soft handover refers to a situation where the UE is connected to one Node B, while two-way soft handover means that the UE is connected to two Node Bs, etc. as shown in Figure 9.22. As each connection between a UE and Node B requires logical baseband resources, reservation of transmission capacity over the Iub, and RNC resources, the soft handover overhead may also be regarded as a measure of the additional hardware/transmission resources required for implementation of soft handover. Radio network planning is responsible for proper handover parameter setting and site planning so that the soft handover overhead is planned to be on the order of 20–40%

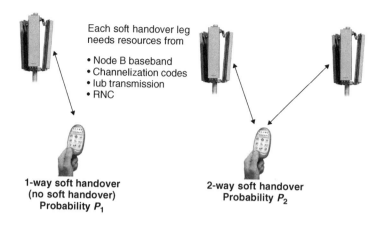

Each soft handover leg
needs resources from

• Node B baseband
• Channelization codes
• Iub transmission
• RNC

**1-way soft handover**
**(no soft handover)**
**Probability $P_1$**

**2-way soft handover**
**Probability $P_2$**

**Figure 9.22**   Soft handover overhead

for a standard hexagonal cell grid with three sector sites. Note that an excessive soft handover overhead could decrease the downlink capacity as shown in Figure 9.22. In the downlink each soft handover connection increases the transmitted interference to the network. When the increased interference exceeds the diversity gain, the soft handover does not provide any gain for system performance.

The soft handover overhead can, to a large extent, be controlled by proper selection of the parameters Window_add, Window_drop, and the active set size. However, there are also factors which influence the soft handover overhead, which are not controllable via the soft handover parameter settings. Some of these are:

• The network topology: How are the sites located relative to each other, how many sectors per site, etc.?
• The Node B antenna radiation patterns.
• The path loss and shadow fading characteristics.
• The average number of Node Bs that a UE can synchronize to.

As an example, the soft handover overhead is plotted in Figure 9.23 for a standard hexagonal cell grid with three sector sites. These results are obtained from a dynamic network level simulator. Results are presented for a cell radii of 666 metres and 2000 metres. A standard 65 degree antenna is assumed for each sector. The deterministic path loss is modelled according to the Okumura–Hata model, while the shadow fading component is assumed to be log-normal distributed with 8 dB standard deviation. The transmit power of the CPICH is fixed at 10% and 20% of the maximum Node B transmit power for the small and large cell radii, respectively. The power of the SCH is −3.0 dB relative to the P-CPICH. The active set size is limited to three.

It is observed that the soft handover overhead increases approximately linearly when Window_add and Window_drop are increased. For the same soft handover parameter settings, the soft handover overhead is typically larger for the scenario with small cells, compared to large cells. This behavior is observed because UEs in the large cell grid can only synchronize to a few Node Bs, while UEs in the small cell grid typically can synchronize to many Node Bs. Assuming that the design goal is to have a soft handover overhead of 20–40%, then the results in Figure 9.23 indicate that appropriate parameter settings are Window_add = 1–3 dB in small cells and slightly larger values in large cells. This conclusion is, however, only valid for a network topology with three sector sites. For the same soft handover parameter settings, the soft handover overhead increases when migrating from three sector sites to six sector sites. As a rule of thumb, the soft handover overhead increases by approximately

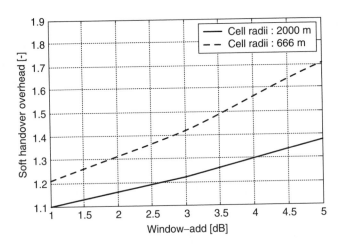

**Figure 9.23** The soft handover overhead versus the soft handover parameter Window_add for a hexagonal cell grid with three sector sites, and two different cell radii. Window_drop = Window_add + 2.0 dB

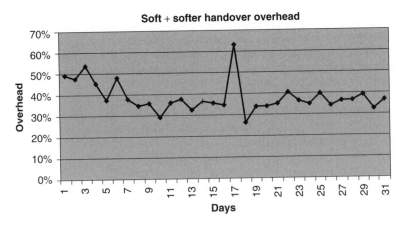

**Figure 9.24** Example soft handover overhead in a live network. Softer handovers are included in this figure

30% when comparing results for three and six sector site configurations. This calls for selection of lower Window_add/Window_drop when increasing the number of sectors.

An example of soft handover overhead in a live WCDMA network in a dense urban area is shown in Figure 9.24. The average overhead, including softer handovers, is 38%. From the transmission point of view the overhead is less than 38% because softer handover combining takes place in Node B and does not require extra transmission resources. The Window_add has been 2–4 dB, the Window_drop 4–6 dB, addition timer 320 ms and drop timer 640 ms.

### 9.3.1.6 Active Set Update Rate

Active set update rate is counted as the time between consecutive active set update commands and includes all the events: addition, removal and replacement. This measure is relevant for RNC dimensioning since active set update signaling causes load to the RNC. Active set update rates are shown in Figure 9.25. The average value in this example is 12 seconds, i.e. 5 active set updates per minute.

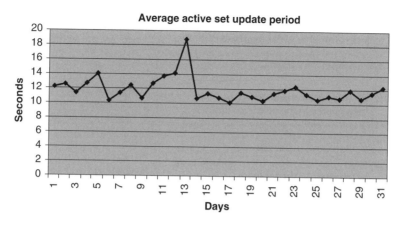

**Figure 9.25**  Example active set update rates in a live network

The value depends on the average mobility of the users, on the cell size, on the network planning and on the handover parameters. Typically, soft handover overhead and active set update rates are related: smaller overhead can be obtained at the expense of a higher active set update rate.

## 9.3.2  Inter-System Handovers between WCDMA and GSM

WCDMA and GSM standards support handovers both ways between WCDMA and GSM. These handovers can be used for coverage or load balancing reasons. At the start of WCDMA deployment, handovers to GSM are needed to provide continuous coverage, and handovers from GSM to WCDMA can be used to lower the loading in GSM cells. This scenario is shown in Figure 9.26. When the traffic in WCDMA networks increases, it is important to have load reason handovers in both directions. The inter-system handovers are triggered in the source RNC/BSC, and from the receiving system point of view, the inter-system handover is similar to inter-RNC or inter-BSC handover. The handover algorithms and triggers are not standardized.

A typical inter-system handover procedure is shown in Figure 9.27. The inter-system measurements are not active all the time but are triggered when there is a need to make inter-system handover. The measurement trigger is an RNC vendor-specific algorithm and could be based, for example, on the quality (block error rate) or on the required transmission power. When the measurements are triggered, the UE measures first the signal powers of the GSM frequencies on the neighbor list. Once those measurements are received by RNC, it commands the UE to decode the BSIC (base station identity code) of the best GSM candidate. When the BSIC is received by RNC, a handover command can be sent to the UE. The measurements can be completed in approximately 2 seconds.

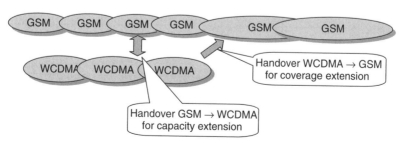

**Figure 9.26**  Inter-system handovers between GSM and WCDMA

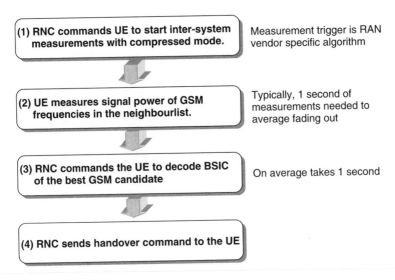

**Figure 9.27**   Inter-system handover procedure

### 9.3.2.1   Compressed Mode

WCDMA uses continuous transmission and reception and cannot make inter-system measurements with a single receiver if there are no gaps generated in the WCDMA signals. Therefore, compressed mode is needed both for inter-frequency and for inter-system measurements. The compressed mode procedure is described in Chapter 6. UE transmission power during compressed mode is illustrated in Figure 9.28.

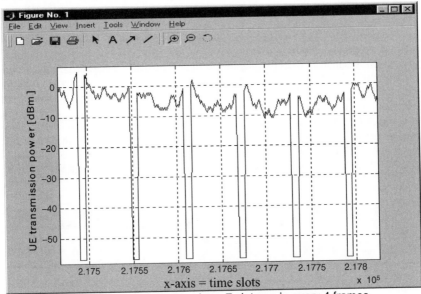

**Figure 9.28**   Measured UE transmission power during compressed mode

**Table 9.7** Effect of compressed mode on the capacity

|  | 100% UEs in compressed mode | 10% UEs in compressed mode |
|---|---|---|
| Every frame compressed | −58% | −5.8% |
| Every 3rd frame compressed | −19% | −1.9% |

During the gaps of the compressed mode, the fast power control cannot be applied and part of the interleaving gain is lost. Therefore, a higher $E_b/N_0$ is needed during compressed frames, leading to a capacity degradation. An example calculation for the capacity effect is shown in Table 9.7. In here it is assumed that the $E_b/N_0$ needs to be 2.0 dB higher during the compressed frames. If all UEs used compressed mode in every frame, the interference levels would increase 58% ($= 10^{\wedge}0.2$), and the capacity would reduce correspondingly.

The measurements capability is typically fast enough if every 3rd frame is compressed. With every 3rd frame compressed, the UE can measure six samples in every gap, i.e. 480 ms/30 ms* 6 = 96 samples per 480 ms [6]. That measurement capability is similar to that in GSM only mobile for intra-GSM measurements.

If we have every 3rd frame compressed, the capacity degradation is still 19%. The results clearly show that compressed mode should only be activated when there is a need to make inter-system and inter-frequency handover. If we assume that 10% of the UEs are simultaneously using the compressed mode, the capacity effect is reduced down to 1.9%. These results show that the effect of the compressed mode to the capacity is negligible if the compressed mode is used only when needed. The RNC algorithms for activating the compressed mode are important to guarantee reliable handovers while maintaining low compressed mode usage.

Compressed mode also affects the uplink coverage area of the real time services where the bit rate cannot be lowered during the compressed mode. Therefore, the coverage reason inter-system handover procedure has to be initiated early enough at the cell edge to avoid any quality degradation during the compressed mode. This situation is shown in Figure 9.29.

The compressed mode affects coverage in two ways:

1. The same amount of data is transmitted in a shorter time. This effect is $10* \log10(15/(15 − 7)) = 2.7$ dB, with a 7-slot gap in a 15-slot frame.
2. The $E_b/N_0$ performance degrades during the compressed mode. The degradation is assumed to be 2.0 dB here.

An example effect to the coverage area of speech is shown in Table 9.8, where the coverage is reduced by 2.4 dB with 20 ms interleaving. AMR voice connection uses 20 ms interleaving. This 2.4 dB

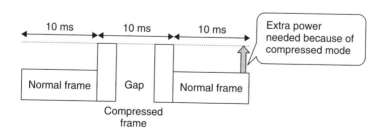

**Figure 9.29** Effect of compressed mode on the coverage

**Table 9.8**  Effect of compressed mode on the coverage

| Interleaving | Reduction in coverage |
|---|---|
| 10 ms | 2.7 dB + 2.0 dB = 4.7 dB |
| 20 ms | (2.7 dB + 2.0 dB)/2 = 2.4 dB |

coverage degradation can be compensated by lowering instantaneously the AMR bit rate during the compressed mode if the UE hits its maximum power, see Section 11.2.1.2.

Inter-system handovers from GSM to WCDMA are initiated in GSM BSC. No compressed mode is needed for making WCDMA measurements from GSM because GSM uses discontinuous transmission and reception.

The service interruption time in the inter-system handovers is 40 ms maximum. The interruption time is the time between the last received transport block on the old frequency and the time the UE starts transmission of the new uplink channel. The total service gap is slightly more than the interruption time because the UE needs to get the dedicated channel running in GSM. The service gap is typically below 80 ms which is similar to that in intra-GSM handovers. Such a service gap does not degrade voice quality. The service gap is illustrated in Figure 9.30.

## 9.3.3   Inter-Frequency Handovers within WCDMA

Most UMTS operators have two or three FDD frequencies available. The operation can be started using one frequency and the second and the third frequency are needed later to enhance the capacity. Several frequencies can be used in two different ways, as shown in Figure 9.31: several frequencies on the same sites for high capacity sites or macro and micro layers using different frequencies. Inter-frequency handovers between WCDMA carriers are needed to support these scenarios.

Compressed mode measurements are used also in the inter-frequency handover in the same way as in the inter-system handovers.

The inter-frequency handover procedure is shown in Figure 9.32. The UE uses the same WCDMA synchronization procedure as for intra-frequency handovers to identify the cells on the target frequency. The synchronization procedure is presented in Chapter 6. The cell identification time depends mainly on the number of cells and multipath components that the UE can receive, in the same way as with intra-frequency handovers. The requirement for the cell identification in 3GPP is 5 seconds with CPICH $E_c/I_0 - 20$ dB [6].

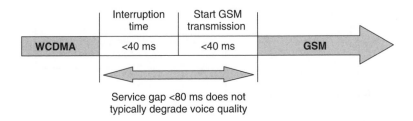

**Figure 9.30**  Service gap in inter-system handover

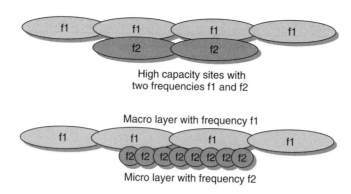

High capacity sites with
two frequencies f1 and f2

Macro layer with frequency f1

Micro layer with frequency f2

**Figure 9.31**  Need for inter-frequency handovers between WCDMA carriers

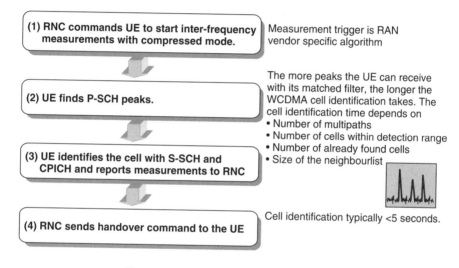

**(1) RNC commands UE to start inter-frequency measurements with compressed mode.**

Measurement trigger is RAN vendor specific algorithm

**(2) UE finds P-SCH peaks.**

The more peaks the UE can receive with its matched filter, the longer the WCDMA cell identification takes. The cell identification time depends on
• Number of multipaths
• Number of cells within detection range
• Number of already found cells
• Size of the neighbourlist

**(3) UE identifies the cell with S-SCH and CPICH and reports measurements to RNC**

**(4) RNC sends handover command to the UE**

Cell identification typically <5 seconds.

**Figure 9.32**  Inter-frequency handover procedure

## 9.3.4  Summary of Handovers

The WCDMA handover types are summarized in Table 9.9. The most typical WCDMA handover is intra-frequency handover that is needed due to the mobility of the UEs. The intra-frequency handover is controlled by those parameters shown in Figure 9.16. The intra-frequency handover reporting is typically event-triggered, and RNC commands the handovers according to the measurement reports. In the case of intra-frequency handover, the UE should be connected to the best Node B(s) to avoid the near–far problem, and RNC does not have any freedom in selecting the target cells.

Inter-frequency and inter-system measurements are typically initiated only when there is a need to make inter-system and inter-frequency handovers. Inter-frequency handovers are needed to balance loading between WCDMA carriers and cell layers, and to extend the coverage area if the other frequency does not have continuous coverage. Inter-system handovers to GSM are needed to extend the WCDMA coverage area, to balance load between systems and to direct services to the most suitable systems.

**Table 9.9**  WCDMA handover types

| Handover type | Handover measurements | Typical handover measurement reporting from UE to RNC | Typical handover reason |
|---|---|---|---|
| WCDMA intra-frequency | Measurements all the time with matched filter | Event-triggered reporting | –Normal mobility |
| WCDMA → GSM inter-system | Measurements started only when needed, compressed mode used | Periodic during compressed mode | –Coverage<br><br>–Load<br>–Service |
| WCDMA inter-frequency | Measurements started only when needed, compressed mode used | Periodic during compressed mode | –Coverage<br><br>–Load |

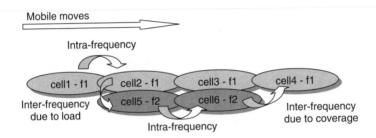

**Figure 9.33**  Example handover scenario

An example handover scenario is shown in Figure 9.33. The UE is first connected to cell1 on frequency f1. When it moves, intra-frequency handover to cell2 is made. The cell2, however, happens to have a high load, and RNC commands load reason inter-frequency handover to cell5 on frequency f2. The UE remains on frequency f2 and continues to cell6. When it runs out of the coverage area of frequency f2, coverage reason inter-frequency handover is made to cell4 on frequency f1.

Handovers are used in Cell_DCH state. In Cell_FACH, Cell_PCH and URA_PCH state the UE makes the cell reselections between frequencies and systems itself according to the idle mode parameters. The states are described in Section 7.8.2.

## 9.4   Measurement of Air Interface Load

If the radio resource management is based on the interference levels in the air interface, the air interface load needs to be measured. The estimation of the uplink load is presented in Section 9.4.1 and the estimation of the downlink load is in Section 9.4.2.

### 9.4.1   Uplink Load

In this section two uplink load measures are presented: load estimation based on wideband received power, and load estimation based on throughput. These are example approaches that could be used in WCDMA networks.

#### 9.4.1.1 Load Estimation Based on Wideband Received Power

The wideband received power level can be used in estimating the uplink load. The received power levels can be measured in the Node B. Based on these measurements, the uplink load factor can be obtained. The calculations are shown below.

The received wideband interference power, $I_{total}$, can be divided into the powers of own-cell ($=$ intra-cell) users, $I_{own}$, other-cell ($=$ inter-cell) users, $I_{oth}$, and background and receiver noise, $P_N$:

$$I_{total} = I_{own} + I_{oth} + P_N \tag{9.3}$$

The uplink noise rise is defined as the ratio of the total received power to the noise power:

$$\text{Noise rise} = \frac{I_{total}}{P_N} = \frac{1}{1 - \eta_{UL}} \tag{9.4}$$

This equation can be rearranged to give the uplink load factor $\eta_{UL}$:

$$\eta_{UL} = 1 - \frac{P_N}{I_{total}} = \frac{\text{Noise rise} - 1}{\text{Noise rise}} \tag{9.5}$$

where $I_{total}$ can be measured by the Node B and $P_N$ is known beforehand.

The uplink load factor $\eta_{UL}$ is normally used as the uplink load indicator. For example, if the uplink load is said to be 60% of the WCDMA pole capacity, this means that the load factor $\eta{UL} = 0.60$.

Load estimation based on the received power level is also presented in [8] and [9].

#### 9.4.1.2 Load Estimation Based on Throughput

The uplink load factor $\eta_{UL}$ can be calculated as the sum of the load factors of the UEs that are connected to this Node B:

$$\eta_{UL} = (1 + i) \cdot \sum_{j=1}^{N} L_j = (1 + i) \cdot \sum_{j=1}^{N} \frac{1}{1 + \dfrac{W}{(E_b/N_0)_j \cdot R_j \cdot \upsilon_j}} \tag{9.6}$$

where $N$ is the number of UEs in the own cell, $W$ is the chip rate, $L_j$ is the load factor of the $j$th UE, $R_j$ is the bit rate of the $j$th UE, $(E_b/N_0)_j$ is $E_b/N_0$ of the $j$th UE, $\upsilon_j$ is the voice activity factor of the $j$th UE, and $i$ is the other-to-own cell interference ratio.

Note that Equation (9.6) is the same as the load factor calculation in radio network dimensioning in Section 8.2.2. In dimensioning, the average number of UEs, $N$, of a cell needs to be estimated, and average values for $E_b/N_0, i$ and $\upsilon$ are used as input parameters. These values are typical for that environment and can be based on the measurements and simulations. In load estimation the instantaneous measured values for $E_b/N_0, i, \upsilon$ and the number of UEs $N$ are used to estimate the instantaneous air interface load.

In throughput-based load estimation, interference from other cells is not directly included in the load but needs to be taken into account with the parameter $i$. Also, the part of own-cell interference that is not captured by the Rake receiver can be taken into account with the parameter $i$. If it is assumed that $i = 0$, then only own-cell interference is taken into account.

#### 9.4.1.3 Comparison of Uplink Load Estimation Methods

Table 9.10 compares the above two load estimation methods. In the wideband power-based approach, interference from the adjacent cells is directly included in the load estimation because the measured wideband power includes all interference that is received in that carrier frequency by the Node B. If the

**Table 9.10** Comparison of uplink load estimation methods

|  | Wideband received power | Throughput | Number of connections |
|---|---|---|---|
| What to measure | Wideband received power $I_{total}$ per cell | Uplink $E_b/N_0$ and bit rates $R$ for each connection | Number of connections |
| What needs to be assumed or measured separately | Thermal noise level (= unloaded interference power) $P_N$ | Other-to-own cell interference ratio, $i$ | Load caused by one connection |
| Other-cell interference | Included in measurement of wideband received power | Assumed explicitly in $i$ | Assumed explicitly when choosing the maximum number of connections |
| Soft capacity | Yes, automatically | Not directly, possible via RNC | No |
| Other interference sources (= adjacent channel) | Reduced capacity | Reduced coverage | Reduced coverage |

loading in the adjacent cells is low, this can be seen in the wideband power-based load measurement, and a higher load can be allowed in this cell, i.e. soft capacity can be obtained. The importance of soft capacity was explained in radio network dimensioning in Section 8.2.3.

The wideband power-based and throughput based load estimations are shown in Figure 9.34. The different curves represent a different loading in the adjacent cells. The larger the value of $i$, the more interference from adjacent cells. The wideband power-based load estimation keeps the coverage within the planned limits and the delivered capacity depends on the loading in the adjacent cells (soft capacity). This approach effectively prevents cell breathing which would exceed the planned values.

The problem with wideband power-based load estimation is that the measured wideband power can include interference from adjacent frequencies. This could originate from another operator's UE located very close to the Node B antenna. Therefore, the interference-based method may overestimate the load of own carrier because of any external interference. The Node B receiver cannot separate the interference from the own carrier and from other carriers by the wideband power measurements.

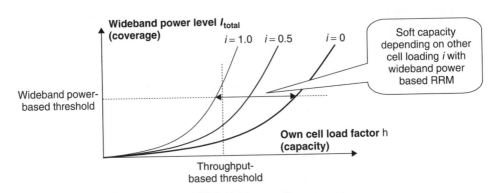

**Figure 9.34** Wideband power-based and throughput-based load estimations

Throughput-based load estimation does not take interference from adjacent cells or adjacent carriers directly into account. If soft capacity is required, information about the adjacent cell loading can be obtained within RNC. The throughput-based RRM keeps the throughput of the cell at the planned level. If the loading in the adjacent cells is high, this affects the coverage area of the cell.

The third load estimation method in Table 9.10, in the right-hand column, is based simply on the number of connections in the Node Bs. This approach can be used in second generation networks where all connections use fairly similar low bit rates and no high bit rate connections are possible. In third generation networks the mix of different bit rates, services and quality requirements prevents the use of this approach. It is unreasonable to assume that the load caused by one 2-Mbps UE is the same as that caused by one speech UE.

## 9.4.2  Downlink Load

### 9.4.2.1  Power-Based Load Estimation

The downlink load of the cell can be determined by the total downlink transmission power, $P_{total}$. The downlink load factor, $\eta_{DL}$, can be defined to be the ratio of the current total transmission power divided by the maximum Node B transmission power $P_{max}$:

$$\eta_{DL} = \frac{P_{total}}{P_{max}} \qquad (9.7)$$

Note that in this load estimation approach the total Node B transmission power $P_{total}$ does not give accurate information concerning how close to the downlink air interface pole capacity the system is operating. In a small cell the same $P_{total}$ corresponds to a higher air interface loading than in a large cell.

### 9.4.2.2  Throughput-Based Load Estimation

In the downlink, throughput-based load estimation can be effected by using the sum of the downlink allocated bit rates as the downlink load factor, $\eta_{DL}$, as follows:

$$\eta_{DL} = \frac{\sum_{j=1}^{N} R_i}{R_{max}} \qquad (9.8)$$

where $N$ is the number of downlink connections, including the common channels, $R_j$ is the bit rate of the $j$th UE, and $R_{max}$ is the maximum allowed throughput of the cell.

It is also possible to weight the UE bit rates with $E_b/N_0$ values as follows:

$$\eta_{DL} = \sum_{j=1}^{N} R_j \cdot \frac{\upsilon_j (E_b/N_0)_j}{W} \cdot [(1 - \bar{\alpha}) + \bar{i}] \qquad (9.9)$$

where $W$ is the chip rate, $(E_b/N_0)_j$ is the $E_b/N_0$ of the $j$th UE, $\upsilon_j$ is the voice activity factor of the $j$th UE, $\bar{\alpha}$ is the average orthogonality of the cell, and $\bar{i}$ is the average downlink other-to-own cell interference ratio of the cell. Note that Equation (9.9) is similar to the downlink radio network dimensioning (see Section 8.2.2.2).

The average downlink orthogonality can be estimated by the Node B based on the multipath propagation in the uplink. The values of $E_b/N_0$ need to be assumed based on the typical values for that environment. The average interference from other cells can be obtained in RNC based on the adjacent cell loading.

## 9.5 Admission Control

### 9.5.1 Admission Control Principle

If the air interface loading is allowed to increase excessively, the coverage area of the cell is reduced below the planned values, and the quality of service of the existing connections cannot be guaranteed. Before admitting a new UE, admission control needs to check that the admittance will not sacrifice the planned coverage area or the quality of the existing connections. Admission control accepts or rejects a request to establish a radio access bearer in the radio access network. The admission control algorithm is executed when a bearer is set up or modified. The admission control functionality is located in RNC where the load information from several cells can be obtained. The admission control algorithm estimates the load increase that the establishment of the bearer would cause in the radio network. This has to be estimated separately for the uplink and downlink directions. The requesting bearer can be admitted only if both uplink and downlink admission control admit it, otherwise it is rejected because of the excessive interference that it would produce in the network. The limits for admission control are set by the radio network planning.

Several admission control schemes have been suggested in [10–15]. In [10, 12 and 13] the use of the total power received by the Node B is supported as the primary uplink admission control decision criterion, relative to the noise level. The ratio between the total received wideband power and the noise level is often referred to as the noise rise. In [10] and [13] a downlink admission control algorithm based on the total downlink transmission power is presented.

### 9.5.2 Wideband Power-Based Admission Control Strategy

In the interference-based admission control strategy the new UE is not admitted by the uplink admission control algorithm if the new resulting total interference level is higher than the threshold value:

$$I_{total\_old} + \Delta I < I_{threshold} \tag{9.10}$$

The threshold value $I_{threshold}$ is the same as the maximum uplink noise rise and can be set by radio network planning. This noise rise must be included in the link budgets as the interference margin: see Section 8.2.1. Wideband power-based admission control is shown in Figure 9.35. The uplink admission control algorithm estimates the load increase by using either of the two methods presented below. The uplink power increase estimation methods take into account the uplink load curve.

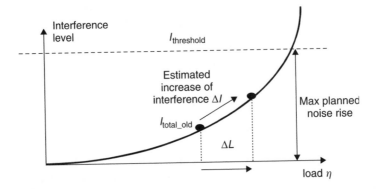

**Figure 9.35**  Uplink load curve and the estimation of the load increase due to a new UE

Two different uplink power increase estimation methods are shown below. They can be used in the interference-based admission control strategy. The idea is to estimate the increase $\Delta I$ of the uplink received wideband interference power $I_{total}$ due to a new UE. The admission of the new UE and the power increase estimation are handled by the admission control functionality.

The first proposed method (the *derivative* method) is presented in Equation (9.13) and the second (the *integral* method) in Equation (9.14). Both take into account the load curve and are based on the derivative of uplink interference with respect to the uplink load factor

$$\frac{d I_{total}}{d\eta} \tag{9.11}$$

which can be calculated as follows

$$\text{Noise rise} = \frac{I_{total}}{P_N} = \frac{1}{1-\eta} \Rightarrow I_{total} = \frac{P_N}{1-\eta} \Rightarrow \frac{d I_{total}}{d\eta} = \frac{P_N}{(1-\eta)^2} \tag{9.12}$$

The change in uplink interference power can be obtained by Equation (9.13). This equation is based on the assumption that the power increase is the derivative of the old uplink interference power with respect to the uplink load factor, multiplied by the load factor of the new UE $\Delta L$:

$$\frac{\Delta I}{\Delta L} \approx \frac{d I_{total}}{d\eta} \Leftrightarrow \Delta I \approx \frac{d I_{total}}{d\eta}\Delta L \Leftrightarrow \Delta I \approx \frac{P_N}{(1-\eta)^2}\Delta L \Leftrightarrow \Delta I \approx \frac{I_{total}}{1-\eta}\Delta L \Leftrightarrow \tag{9.13}$$

The second uplink power increase estimation method is based on the integration method, in which the derivative of interference with respect to the load factor is integrated from the old value of the load factor ($\eta_{old} = \eta$) to the new value of the load factor ($\eta_{new} = \eta + \Delta L$) as follows:

$$\Delta I = \int_{I_{total\_old}}^{I_{total\_old}+\Delta I} d I_{total} \Leftrightarrow$$

$$\Delta I = \int_{\eta}^{\eta+\Delta L} \frac{P_N}{(1-\eta)^2} d\eta \Leftrightarrow$$

$$\Delta I = \frac{P_N}{1-\eta-\Delta L} - \frac{P_N}{1-\eta} \Leftrightarrow \tag{9.14}$$

$$\Delta I = \frac{\Delta L}{1-\eta-\Delta L} \cdot \frac{P_N}{1-\eta} \Leftrightarrow$$

$$\Delta I = \frac{I_{total}}{1-\eta-\Delta L}\Delta L$$

In Equations (9.13) and (9.14) the load factor of the new UE $\Delta L$ is the estimated load factor of the new connection and can be obtained as

$$\Delta L = \frac{1}{1 + \dfrac{W}{\upsilon \cdot E_b/N_0 \cdot R}} \tag{9.15}$$

where $W$ is the chip rate, $R$ is the bit rate of the new UE, $E_b/N_0$ is the assumed $E_b/N_0$ of the new connection and $\upsilon$ is the assumed voice activity of the new connection.

The downlink admission control strategy is the same as in the uplink, i.e. the UE is admitted if the new total downlink transmission power does not exceed the predefined target value:

$$P_{total\_old} + \Delta P_{total} > P_{threshold} \tag{9.16}$$

The threshold value $P_{\text{threshold}}$ is set by radio network planning. Notice that $\Delta P_{\text{total}}$ both includes the power of the new UE requesting capacity and the additional power rise of the existing UEs in the system due to the additional interference contributed by the new UE. The load increase $\Delta P_{\text{total}}$ in the downlink can be estimated based on *a priori* knowledge of the required $E_b/N_0$, the requested bit rate, and the pilot report from the UE. The pilot report implicitly provides information on the path loss towards the new UE as well as the interference level experienced by the UE.

## 9.5.3 Throughput-Based Admission Control Strategy

In the throughput-based admission control strategy, the new requesting UE is admitted into the radio access network if

$$\eta_{\text{UL}} + \Delta L \eta_{\text{UL\_threshold}} \tag{9.17}$$

and the same in downlink:

$$\eta_{\text{DL}} + \Delta L < \eta_{\text{DL\_threshold}} \tag{9.18}$$

where $\eta_{\text{UL}}$ and $\eta_{\text{DL}}$ are the uplink and downlink load factors before the admittance of the new connection and are estimated as shown in Section 9.4. The load factor of the new UE $\Delta L$ is calculated as in Equation (9.15).

Finally, we need to note that different admission control strategies can be used in the uplink and in the downlink.

## 9.6 Load Control (Congestion Control)

One important task of the RRM functionality is to ensure that the system is not overloaded and remains stable. If the system is properly planned, and the admission control and packets scheduler work sufficiently well, overload situations should be exceptional. If overload is encountered, however, the load control functionality returns the system quickly and controllably back to the targeted load, which is defined by the radio network planning.

The possible load control actions in order to reduce load are listed below:

- Downlink fast load control: Deny downlink power-up commands received from the UE.
- Uplink fast load control: Reduce the uplink $E_b/N_0$ target used by the uplink fast power control.
- Reduce the throughput of packet data traffic.
- Handover to another WCDMA carrier
- Handover to GSM.
- Decrease bit rates of real time UEs, e.g. AMR speech codec.
- Drop low priority calls in a controlled fashion.

The first two in this list are fast actions that are carried out within a Node B. These actions can take place within one time slot, i.e. with 1.5 kHz frequency, and provide fast prioritization of the different services. The instantaneous frame error rate of the non-delay-sensitive connections can be allowed to increase in order to maintain the quality of those services that cannot tolerate retransmission. These actions only cause increased delay of packet data services while the quality of the conversational services, such as speech and video telephony, is maintained.

The other load control actions are typically slower. Packet traffic is reduced by the packet scheduler: see Chapter 10. Inter-frequency and inter-system handovers can also be used as load balancing and load control algorithms and they were described in this chapter.

One example of a real-time connection whose bit rate can be decreased by the radio access network is Adaptive Multirate (AMR) speech codec: for further information, see Section 2.2.

# References

[1] Sipilä, K., Laiho-Steffens, J., Wacker, A. and Jäsberg, M., 'Modelling the Impact of the Fast Power Control on the WCDMA Uplink', *Proceedings of VTC'99 Spring*, Houston, TX, 16–19 May 1999, pp. 1266–1270.

[2] 3GPP TS 25.101 UE Radio Transmission and Reception (FDD).

[3] Salonaho, O. and Laakso, J., 'Flexible Power Allocation for Physical Control Channel in Wideband CDMA', *Proceedings of VTC'99 Spring*, Houston, TX, 16–19 May 1999, pp. 1455–1458.

[4] Sampath, A., Kumar, P. and Holtzman, J., 'On Setting Reverse Link Target SIR in a CDMA System', *Proceedings of VTC'97*, Arizona, 4–7 May 1997, Vol. 2, pp. 929–933.

[5] 3GPP, Technical Specification Group RAN, Working Group 2 (WG2), 'Radio Resource Management Strategies', TR 25.922.

[6] 3GPP TS 25.133 'Requirements for Support of Radio Resource Management (FDD)'.

[7] Hiltunen, K., Binucci, N. and Bergström, J., 'Comparison Between the Periodic and Event-Triggered Intra-Frequency Handover Measurement Reporting in WCDMA', *Proceedings of IEEE WCNC 2000*, Chicago, 23–28 September 2000.

[8] Shapira, J. and Padovani, R., 'Spatial Topology and Dynamics in CDMA Cellular Radio', *Proceedings of 42nd IEEE VTS Conference*, Denver, CO, May 1992, pp. 213–216.

[9] Shapira, J., 'Microcell Engineering in CDMA Cellular Networks', *IEEE Transactions on Vehicular Technology*, Vol. 43, No. 4, November 1994, pp. 817–825.

[10] Dahlman, E., Knutsson, J., Ovesjö, F., Persson, M. and Roobol, C., 'WCDMA – The Radio Interface for Future Mobile Multimedia Communications', *IEEE Transactions on Vehicular Technology*, Vol. 47, No. 4, November 1998, pp. 1105–1118.

[11] Huang, C. and Yates, R., 'Call Admission in Power Controlled CDMA Systems', *Proceedings of VTC'96*, Atlanta, GA, May 1996, pp. 1665–1669.

[12] Knutsson, J., Butovitsch, T., Persson, M. and Yates, R., 'Evaluation of Admission Control Algorithms for CDMA System in a Manhattan Environment', *Proceedings of 2nd CDMA International Conference*, CIC '97, Seoul, South Korea, October 1997, pp. 414–418.

[13] Knutsson, J., Butovitsch, P., Persson, M. and Yates, R., 'Downlink Admission Control Strategies for CDMA Systems in a Manhattan Environment', *Proceedings of VTC'98*, Ottawa, Canada, May 1998, pp. 1453–1457.

[14] Liu, Z. and Zarki, M. 'SIR Based Call Admission Control for DS-CDMA Cellular System,' *IEEE Journal on Selected Areas in Communications*, Vol. 12, 1994, pp. 638–644.

[15] Holma, H. and Laakso, J., 'Uplink Admission Control and Soft Capacity with MUD in CDMA', *Proceedings of VTC'99 Fall*, Amsterdam, Netherlands, 19–22 September 1999, pp. 431–435.

# 10

# Packet Scheduling

Jeroen Wigard, Harri Holma, Renaud Cuny, Nina Madsen,
Frank Frederiksen and Martin Kristensson

## 10.1  Introduction

This chapter presents the radio access algorithms for supporting packet switched services and analyses their performance. Such services are, for example, messaging, email, WAP/web browsing, streaming video or Voice over IP. This chapter is organized as follows. Packet data protocols are first discussed in Section 10.2. The network delay aspects in terms of round trip time are analysed in Section 10.3. The transport channels and the user specific packet scheduler are introduced in Section 10.4. The cell specific packet scheduler algorithms are discussed in Section 10.5. The packet data system performance results are presented in Section 10.6 and the application performance results in Section 10.7. The services in general are introduced in Chapter 2.

All four UMTS traffic classes – from background to conversational – can be supported by WCDMA radio networks. See Chapter 2 for the introduction of the traffic classes. Background and interactive traffic classes do not require any guaranteed minimum bit rate and they are transmitted through the packet scheduler. The streaming class requires a minimum guaranteed bit rate but tolerates some delay, and packet scheduling can be utilized for streaming. The conversational class is transmitted without scheduling on dedicated channel DCH. The traffic classes and their mapping to transport channels are shown in Figure 10.1. The main focus in this chapter is on those services where packet scheduling can be applied.

## 10.2  Transmission Control Protocol (TCP)

The characteristics of the incoming data depend on the properties of the transport protocol. Non-real-time traffic is typically carried using Transmission Control Protocol, TCP, and packet-based real-time traffic, like streaming, typically uses the Real-time Protocol, RTP, on top of User Datagram Protocol, UDP. The protocols are shown in Figure 10.2. The characteristics of TCP are described in this section.

The user plane protocol stack for web browsing application is shown in Figure 10.3. It can be seen that the medium access control (MAC), the radio link control (RLC) and the packet data convergence protocol (PDCP) layer are terminated in the RNC, while the Internet Protocol (IP), TCP and hypertext

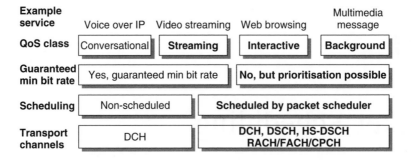

Figure 10.1   Mapping of UMTS traffic classes to scheduling and to transport channels

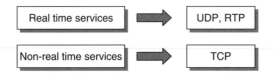

**Figure 10.2**   Typical packet protocols over WCDMA

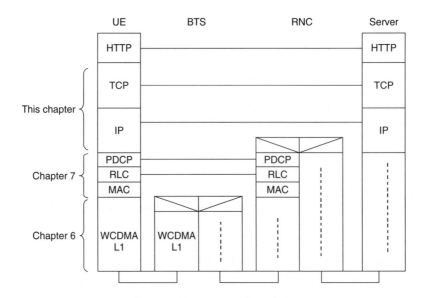

**Figure 10.3**   Data User Plane Protocol Stack for HTTP web browsing

transfer protocol (HTTP) are terminated in a server. MAC, RLC and PDCP are covered in Chapter 7. The protocol employed by Web browsing applications is the HTTP, but other protocols and applications such as file transfer protocol (FTP), and mail protocols, like the Internet mail protocol (IMAP) or Telnet would use a similar architecture over TCP/IP. In the protocol structure each entity of a certain layer uses the services offered by the layer just below to communicate with its peer.

A TCP connection is set up between the UE and a server, or intermediate proxies. This server can be located close to the UE or far away. It is possible for a UE in Europe to set up a TCP connection with a server in the USA. TCP has been designed to provide reliable end-to-end information exchange over an unreliable Internet network since the Internet Protocol does not guarantee any reliability. TCP uses retransmissions and flow control to provide the reliability. TCP is connection oriented between two end points and requires connection establishment and termination. From a wireless performance point of view, one of the most significant characteristics of TCP is the flow control. The receiver end only allows the transmitter to send as much data as the receiver has buffer for. Moreover, the transmitter side adapts the transmission rate to the network capacity/load by means of algorithms such as slow start, and congestion avoidance. The flow control in TCP is accomplished by means of two windows: the congestion window and the offered window. The congestion window is controlled by slow start and congestion avoidance algorithms, whereas the offered window is a window with the size the receiver has buffer for, and the receiver advertises it in every segment transmitted to the server. The offered window is therefore flow control imposed by the receiver. At any time, the minimum of the congestion window and the offered window determines the amount of outstanding data the server can transmit. Therefore, the number of transmitted but unacknowledged data (segments on the fly) can never be larger than the minimum of these two windows. The window, which is the minimum of the two above mentioned windows, is adapted using a sliding window approach:

1. Transmit new segments in the window.
2. Wait for acknowledgements from the receiver.
3. If the acknowledgement is received, slide the window forward and increase the window size, if the offered window is not exceeded.
4. If the acknowledgement is not received before a timer expires, retransmit non-acknowledged segments and decrease the window size.

An example of the progressing of the window, in the case of a positive ACK can be seen in Figure 10.4. In this example the offered window is larger than three segments.

This flow control approach fully utilizes the available bandwidth by adjusting the window size according to the received acknowledgements. The size of the congestion window is adjusted according to the received acknowledgements. Traditionally, the congestion window is initialized to one, two or four times the maximum segment size at the connection establishment. Typical values of maximum segment sizes are 256, 512, 536 or 1460 bytes.

Each time a TCP segment is positively acknowledged, the size of the congestion window is doubled: when the first acknowledgement is received and the initial congestion window was equal to the maximum segment size, the congestion window is increased to two times the maximum segment size, and two segments can be sent. This produces an aggressive exponential increase of the amount of outstanding data the sender can transmit. This approach is called slow start because the starting congestion window is just one maximum segment size. When the congestion window reaches the threshold, called the slow start threshold, it is increased linearly at the reception of acknowledgements.

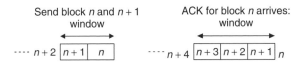

**Figure 10.4**  Example of TCP window progressing

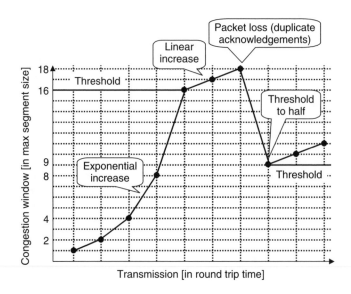

**Figure 10.5**   Example of the congestion window during slow start and fast retransmission

The segment retransmissions in TCP can be triggered by two type of event: reception of several duplicated acknowledgements (usually three duplicated acknowledgements), or timer expiration. The timer is used to detect congestion in the network. Each segment is timed in every cycle, and if its corresponding acknowledgement is not received before the timer expires, the retransmission of non-acknowledged segments is triggered subject to a new window set-up. To avoid further congestion, the slow start threshold is set to half of the current window. The congestion window is set to the maximum segment size and the slow start is started again. This is shown on the right side of Figure 10.5. The Retransmission Time Out (RTO) is calculated from the average and the mean deviation of the round trip time (RTT).

$$RTO = A + 4 \cdot D \tag{10.1}$$

where $A$ represents the smoothed RTT average and $D$ represents the mean deviation of the RTT. The mean deviation is a good approximation to the standard deviation, and it is included in the RTO computation in order to provide better response to wide fluctuations in the RTT. This means that large variations in the delay of TCP packets over the air interface should be avoided in order to limit the number of TCP timeouts. Details of the RTO calculations are given in [1].

Upon the reception of a number of duplicated acknowledgements, i.e. multiple acknowledgements with the same next expected segment number $n$, that exceeds a threshold (usually set to 3), the segment $n$ is assumed to be lost. Note that TCP requires the receiver side to generate an immediate acknowledgement when an out of order segment is received. There are several types of TCP implementation that operate differently in fast retransmission and fast recovery. Here, two of them are mentioned: TCP-Tahoe and TCP-Reno. When the packet loss is detected TCP-Tahoe sets the slow start threshold to half the current window, and the congestion window is set to one maximum segment size which corresponds to what happens at the timer expiration. TCP-Reno sets the slow start threshold and the congestion window to half the current window so the slow start is avoided. While TCP-Reno produces less bursty traffic than TCP-Tahoe, it is much less robust toward phase effects because multiple segment losses of the same congestion episode significantly deteriorate the congestion window size.

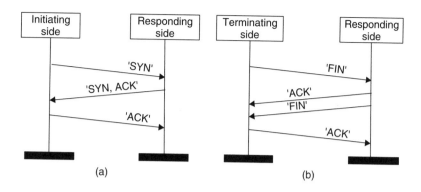

**Figure 10.6**  TCP connection establishment (a) and termination (b)

For small file sizes the slow start will affect the throughput as the radio link capability is not fully utilized during the slow start. Fundamental to TCP's slow start, timeout and retransmission is the measurement of the round trip time (RTT) experienced on a given connection. This is analysed in the next section.

We further note that, during the TCP connection establishment and termination, typically small IP packets of 40 bytes are transmitted, which can be seen in Figure 10.6. The connection establishment consists of the initiating side sending a SYN message, indicating it wants to open a TCP connection. The SYN is used to synchronize the sequence numbers. The responding side simply acknowledges this with an ACK message and sends, at the same time, a SYN message to open the TCP connection in the return direction. As the last part of the connection establishment, the initiating side sends an ACK message when it has received the SYN message correctly. The TCP termination is very similar: the terminating side sends a FIN message, indicating that the connection should be closed. To this the responding side answers with an ACK message and a FIN message in order to also close the TCP connection in the other direction. The connection is completely closed when the terminating side has acknowledged the FIN message. The TCP connection establishment and release both take approximately $1.5 \times$ round trip time since three messages are required.

TCP was originally designed to work over fixed networks. Using TCP over wireless networks brings some challenges: narrow bandwidth, longer delay and larger delay variations than in a fixed network. Third generation networks improve the packet data performance compared to second generation systems and bring it closer to fixed line performance. Some enhancements have been considered to optimize TCP, as described in [2]. Some of those solutions may be beneficial for improving the third generation packet data performance as well. These potential TCP optimization solutions can be classified into standard parameter optimization, non-standard TCP optimization and buffer congestion management. The TCP optimization solutions are briefly described below, but first we list the main TCP parameter optimization solutions:

- *An increase of the initial TCP congestion window.* This is especially effective for the transmission of small amounts of data. These amounts of data are commonly seen in such applications as Internet-enabled mobile wireless devices. For large data transfers, on the other hand, the effect of this mechanism is negligible. An initial congestion window size of two segments or larger is recommended in [3].
- *An appropriate TCP window size that should be set according to the bandwidth delay product.* The bandwidth delay product defines the amount of unacknowledged data that should be transmitted in

order to fully utilize the limiting transmission link. The maximum TCP throughput is limited by the following formula:

$$TCP\_throughput \leq \frac{advertised\_window(bits)}{round\_trip\_time(s)} \qquad (10.2)$$

Typical TCP window sizes are 16–64 kB and they may be adjusted adaptively. The average WCDMA round trip times are shown in Figure 10.12 to be in the order of 200 ms. Additionally, we need some margin for RLC transmissions. If we assume a maximum round trip time of 400 ms, the achievable TCP throughput is 320–1280 kbps. These numbers indicate that the normal advertised TCP window sizes, that are at least equal to 32 kB, are large enough for TCP connections over WCDMA with 384 kbps data rate.

As non-standard TCP optimization examples, we can mention:

- *Redundancy elimination*. This functionality monitors the latest TCP acknowledged sequence number in a router, allowing unnecessary retransmitted TCP segments in that router to be discarded and consequently saving valuable radio capacity. This can be seen in Figure 10.7.
- *Split TCP*. Splitting the TCP connections into two legs (one leg for the wireless domain and one leg for the fixed Internet) allows a freer optimization of the transport protocol in the wireless part. Note, however, that this approach breaks the end-to-end design principle of Internet protocols.

This list is not exhaustive and other TCP optimization methods can be found in the literature. Typically some of these methods (e.g. the split TCP approach), should be used with care and their implications towards end-to-end functionalities such as security should be well understood by network operators. RFC 3135 presents a general IETF view on such network features, underlining the most important issues to consider.

Buffer congestion control mechanisms are used in the case of congestion in shared buffers. The problem arises during buffer overflow since it affects several TCP connections. In the case of user-dedicated buffers no such problems exist. The congestion can generally be avoided with large buffers, but long buffering delays are not desirable and should be avoided whenever possible. Buffer congestion

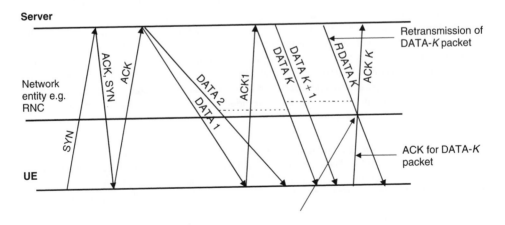

**Figure 10.7**  Redundancy elimination

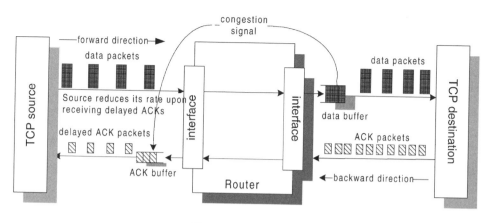

**Figure 10.8**  Fast TCP

most likely happens in the last downlink router buffer before the radio link. Buffer congestion control
mechanisms include:

- *Random Early Detection (RED)*, the best-known standard buffer management method. The principle
  of RED is to increase the packet dropping probability of incoming packets at the same time as
  the buffer occupancy grows. Ultimately, if the buffer length exceeds a predefined threshold, every
  incoming packet will be dropped. The advantage of dropping packets before the buffer is completely
  full is to force some of the TCP sources to reduce their sending rate before congestion actually
  happens. Another advantage is to avoid global synchronization in the routers. Global synchronization
  happens when the buffer load oscillates from almost empty to congested, and is caused mostly by
  the TCP congestion control mechanism of multiple connections acting at the same time.
- *Fast TCP*. An algorithm delays (slightly) TCP acknowledgments if congestion is detected in the
  reverse direction. As a result, TCP senders will send later their next window of data, giving, in
  many cases, enough time for the router buffers to recover from congestion. Furthermore, delaying
  the TCP acknowledgement will slow down the growth of the sending window in the TCP sender,
  which is also beneficial in the case of congestion. Fast TCP is illustrated in Figure 10.8.
- *Window pacing*. This scheme reduces the advertised window of TCP acknowledgements if congestion is detected in the reverse direction. The aim is to force the TCP senders to decrease their
  sending rate. The advantage compared to RED is that packets are actually not dropped in the downlink direction, preventing throughput degradation for the end user. The drawback is that modifying
  the TCP header is a heavier task than just dropping or delaying incoming IP packets. Also, if IP
  security is used, this feature cannot be applied since the TCP headers can no longer be read.

## 10.3   Round Trip Time

The round trip time in WCDMA is defined as the delay of a small packet travelling from UE to a
server behind GGSN and back. The round trip time is illustrated in Figure 10.9. In fixed IP networks,
the round trip time is typically a few tens of milliseconds if the server is close, and up to a few
hundred milliseconds if the server is far away. Mobile networks like UMTS increase the delay and
the round trip time is longer when TCP connection is established over a WCDMA air interface. A
shorter round trip times give a benefit in the response time, which is especially advantageous for TCP
slow start and for interactive services with small packets, like gaming; see more detailed discussion

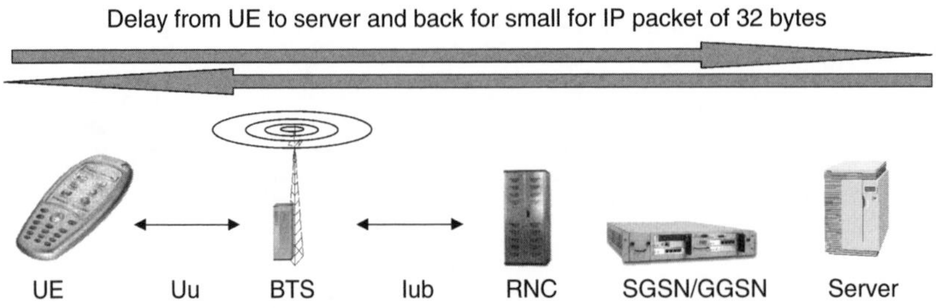

Delay from UE to server and back for small for IP packet of 32 bytes

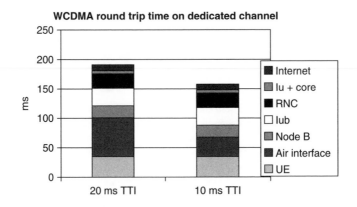

UE          Uu      BTS       lub       RNC      SGSN/GGSN       Server

**Figure 10.9**  Round trip time definition

**WCDMA round trip time on dedicated channel**

Legend:
- Internet
- Iu + core
- RNC
- Iub
- Node B
- Air interface
- UE

(y-axis: ms, 0 to 250; categories: 20 ms TTI, 10 ms TTI)

**Figure 10.10**  WCDMA round trip time for small packets

on the requirements in Chapter 2. In this section we first analyse the WCDMA round trip time delay budget for small packets and then present the effect of the packet size to the round trip time. Small packets are typically transmitted in TCP connection establishment and in games, while larger packets are transmitted during file or web page download.

The round trip time delay budget for small packets is shown in Figure 10.10, and is typically between 150–200 ms depending on the Transmission Time Interval (TTI = interleaving) used in the air interface. The delay components can be found from [4]. We assume UE delay of 35 ms, Node B delay of 20 ms, Iub delay of 30 ms, including frame protocol and AAL2 processing, RNC 25 ms and air interface 33–66 ms, including buffering and retransmissions. Furthermore, we assume that the core network adds 5 ms and fixed Internet 10 ms delay. Example round trip time measurements are shown in Figure 10.11. The round trip time in this example is approximately 160–200 ms, except for when a block error occurs in the air interface and an RLC retransmission takes place, in which case the round trip time is increased to 320 ms. Since WCDMA round trip time is similar to the fixed Internet round trip time that can be seen for inter-continental connections or for dial-up connections, we may expect that those applications that are designed for fixed Internet typically perform at least satisfactorily when used over the WCDMA air interface. WCDMA round trip time will reduce when product platforms are optimized and when HSDPA with shorter TTI is introduced. It is expected that below 100 ms is feasible with HSDPA.

The round trip time above assumes that the packet fits into one TTI. For example, 64 kbps and 10 ms TTI carries 640 bits = 80 bytes in one TTI. If the packet size is large, multiple TTIs are needed

Pinging 213.161.41.37 with 32 bytes of data:

Reply from 213.161.41.37: bytes = 32 time = 201 ms TTL = 255
Reply from 213.161.41.37: bytes = 32 time = 160 ms TTL = 255
Reply from 213.161.41.37: bytes = 32 time = 191 ms TTL = 255
Reply from 213.161.41.37: bytes = 32 time = 190 ms TTL = 255
Reply from 213.161.41.37: bytes = 32 time = 171 ms TTL = 255
Reply from 213.161.41.37: bytes = 32 time = 180 ms TTL = 255
Reply from 213.161.41.37: bytes = 32 time = 321 ms TTL = 255 ⟵ **RLC retransmission**
Reply from 213.161.41.37: bytes = 32 time = 170 ms TTL = 255


Reply from 213.161.41.37: bytes = 32 time = 160 ms TTL = 255
Reply from 213.161.41.37: bytes = 32 time = 180 ms TTL = 255
Reply from 213.161.41.37: bytes = 32 time = 170 ms TTL = 255
Reply from 213.161.41.37: bytes = 32 time = 160 ms TTL = 255

Ping statistics for 213.161.41.37:
    Packets: Sent = 29, Received = 29, Lost = 0 (0 % loss),
Approximate round trip times in milliseconds:
    Minimum = 160 ms, Maximum = 321 ms, Average = 182 ms ⟵ **Average RTT 182 ms**

**Figure 10.11**   Example WCDMA round trip time measurements on dedicated channel

to transmit one packet and the round trip time gets larger. The total average round trip times for a 1500 B downlink packet and a 40 B uplink acknowledgement are shown in Figure 10.12 for different downlink bit rates. It is assumed that for uplink acknowledgements an 8 kbps channel is selected when the downlink bit rate is less than or equal to 64 kbps and a 32 kbps uplink channel is used when

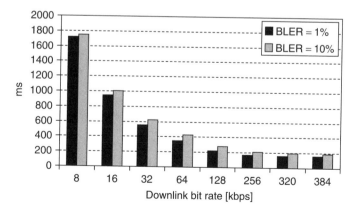

**Figure 10.12**   WCDMA round trip time for a 1500 B packet in downlink and a 40 B acknowledgement in uplink

the downlink bit rate is higher than 64 kbps. The average delay of retransmissions is assumed to be 100 ms. RLC BLERs equal to 1% and 10% are shown. It can be noted that using a lower BLER in the radio transmission can somewhat reduce the average round trip time.

It is not only the average round trip time but also the delay jitter that affects the performance. The delay jitter depends on the BLER, the bit rate and the retransmission delay. For low bit rates, one IP packet is typically transmitted over many TTIs, so the probability of at least one of the TTIs failing is quite large, whereas for large bit rates, one IP packet is sent in a few TTIs. Thus, the size of the delay jitter compared to the round trip time is larger for high bit rates.

This section presented the round trip time on a dedicated channel DCH. The following section discusses the common channel state, its round trip time and the delay of the state transitions.

## 10.4   User-Specific Packet Scheduling

The WCDMA packet scheduler is located in RNC. The base station provides the air interface load measurements and the mobile provides uplink traffic volume measurements for the packet scheduler: see Figure 10.13. The user-specific packet scheduler is presented in this section and the cell-specific scheduler in Section 10.5. The user-specific part controls the utilization of Radio resource control (RRC) states, transport channels and their bit rates according to the traffic volume. The cell-specific part controls the sharing of the radio resources between the simultaneous users.

WCDMA supports three types of transport channel that can be used to transmit packet data: common, dedicated and shared transport channels. These channels are described in Chapter 6 and in this section their properties and feasibility for packet data are discussed.

### 10.4.1   Common Channels (RACH/FACH)

Common channels are the random access channel (RACH) in the uplink and the forward access channel (FACH) in the downlink. Both can carry signaling data but also user data in WCDMA. There

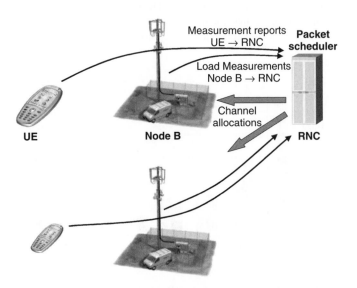

**Figure 10.13**   The WCDMA packet scheduler, located in RNC

Pinging www.nokia.com [147.243.3.73] with 32 bytes of data:

Reply from 147.243.3.73: bytes = 32 time = 301 ms TTL = 242
Reply from 147.243.3.73: bytes = 32 time = 290 ms TTL = 242
Reply from 147.243.3.73: bytes = 32 time = 300 ms TTL = 242
Reply from 147.243.3.73: bytes = 32 time = 241 ms TTL = 242
Reply from 147.243.3.73: bytes = 32 time = 290 ms TTL = 242
Reply from 147.243.3.73: bytes = 32 time = 281 ms TTL = 242
Reply from 147.243.3.73: bytes = 32 time = 260 ms TTL = 242
Reply from 147.243.3.73: bytes = 32 time = 281 ms TTL = 242
Reply from 147.243.3.73: bytes = 32 time = 300 ms TTL = 242
Reply from 147.243.3.73: bytes = 32 time = 311 ms TTL = 242
Reply from 147.243.3.73: bytes = 32 time = 290 ms TTL = 242
Reply from 147.243.3.73: bytes = 32 time = 300 ms TTL = 242
Reply from 147.243.3.73: bytes = 32 time = 261 ms TTL = 242

Ping statistics for 147.243.3.73:
   Packets: Sent = 13, Received = 13, Lost = 0 (0 % loss),
Approximate round trip times in milliseconds:
   Minimum = 241 ms, Maximum = 311 ms, Average = 285 ms  ⇐  **Average RTT 285 ms**

**Figure 10.14**   Example round trip times for 32 B ping on RACH/FACH

are typically only one or a few RACH or FACH channels per sector, for example, 16 kbps RACH with 20 ms TTI and 32 kbps FACH with 10 ms TTI. The advantage of common channels is that no set-up time is needed if the user is in Cell_FACH state. The round trip time on RACH/FACH is illustrated in Figure 10.14: the round trip time is slightly longer than on a dedicated channel but even the first packet goes through as quickly as the other ones.

Common channels do not have a feedback channel, and therefore cannot use fast closed loop power control, but only open loop power control or fixed power. Nor can these channels use soft handover. Therefore, the link level performance of the common channels is not as good as that of the dedicated channels, and more interference is generated than with dedicated channels. The gain of fast power control is analysed in Section 9.2 and the gain of soft handover in Section 9.3. Common channels are most suitable for transmitting small IP packets, for example, during the TCP connection establishment, and for infrequent packets for interactive gaming. Since common channels do not use soft handover but cell reselection, there is a longer delay when the cell reselection takes place. The round trip times on RACH/FACH are shown in Figure 10.15 when cell reselections happen due to mobility. The round trip time is increased from below 300 ms to above 1 s during the cell reselection.

## 10.4.2   Dedicated Channel (DCH)

The dedicated channel is always a bi-directional channel with both uplink and downlink connections. Because of the feedback channel, fast power control and soft handovers can be used. These features improve their radio performance, and consequently less interference is generated than with common channels. No breaks occur due to mobility since soft handover is used. On the other hand, setting up a dedicated channel takes more time than accessing common channels. The dedicated channel set-up time is illustrated in Figure 10.16. The set-up time is the difference in ping delay between the 1st

```
Reply from 130.225.51.16: bytes = 32 time = 290 ms TTL = 239
Reply from 130.225.51.16: bytes = 32 time = 270 ms TTL = 239
Reply from 130.225.51.16: bytes = 32 time = 350 ms TTL = 239
Reply from 130.225.51.16: bytes = 32 time = 301 ms TTL = 239
                            ⋮
Reply from 130.225.51.16: bytes = 32 time = 270 ms TTL = 239
Reply from 130.225.51.16: bytes = 32 time = 1192 ms TTL = 239    ⟸ Cell
Reply from 130.225.51.16: bytes = 32 time = 371 ms TTL = 239       reselection
Reply from 130.225.51.16: bytes = 32 time = 270 ms TTL = 239
                            ⋮
Reply from 130.225.51.16: bytes = 32 time = 281 ms TTL = 239
Reply from 130.225.51.16: bytes = 32 time = 310 ms TTL = 239
Reply from 130.225.51.16: bytes = 32 time = 1042 ms TTL = 239    ⟸ Cell
Reply from 130.225.51.16: bytes = 32 time = 400 ms TTL = 239       reselection
Reply from 130.225.51.16: bytes = 32 time = 230 ms TTL = 239
```

**Figure 10.15**   Round trip times on RACH/FACH when cell reselections take place

Pinging www.nokia.com [147.243.3.73] with 128 bytes of data:

```
Reply from 147.243.3.73: bytes = 128 time = 1121 ms TTL = 242    ⟸ DCH allocation 900 ms
Reply from 147.243.3.73: bytes = 128 time = 221 ms TTL = 242       ( = 1121 – 221 ms)
Reply from 147.243.3.73: bytes = 128 time = 190 ms TTL = 242
Reply from 147.243.3.73: bytes = 128 time = 200 ms TTL = 242
Reply from 147.243.3.73: bytes = 128 time = 190 ms TTL = 242
Reply from 147.243.3.73: bytes = 128 time = 200 ms TTL = 242
```

**Figure 10.16**   Example DCH set-up time from Cell_FACH state

and 2nd packets, which is 900 ms in this example. The signaling flow chart for this RRC state change is illustrated in Section 7.8.2. The DCH set-up time should preferably be minimized to improve the response times and to avoid the potential TCP timeout risk during DCH set-up. A number of proposals have been discussed in 3GPP to reduce the required signaling messages and to minimize the delay.

The dedicated channel can have bit rates from a few kbps up to 384 kbps in the first mobiles, and up to 2 Mbps according to 3GPP. The bit rate can be changed during transmission. When the data transmission is over, the dedicated channel is kept allocated for a few seconds before releasing it and reallocating to another user. During that time the downlink orthogonal code and the network hardware is still dedicated for that connection. Therefore, very bursty traffic on dedicated channels consumes a relatively high number of downlink orthogonal codes and network resources. Also, during TCP slow start the dedicated channel is not fully utilized. The DCH utilization is illustrated in Figure 10.21. The utilization for different files sizes and bit rates is calculated in Figure 10.17. The initial congestion window is assumed to be two MSS and the inactivity timer for all bit rates is set to 2 s. With 100 kB file size the channel utilization is better than 80% if the bit rate is 64 kbps or below, while the utilization with 384 kbps is below 50%. The low utilization is mainly caused by the inactivity timer of 2 s, since the actual download takes less than 3 s. With 1 MB file size the channel utilization is approximately

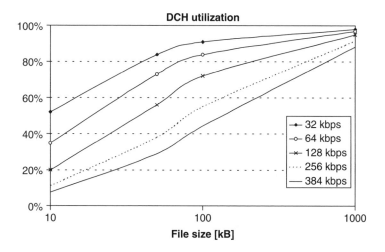

**Figure 10.17** Radio bearer utilization with TCP

90% or better for all channel bit rates. It is shown that allocating high bit rates for downloading small files leads to low utilization factors. This can lead to code shortage and also wastes Iub transmission resources. Therefore, it is beneficial to allocate low bit rates at the start of a download and then increase the bit rate. It can also be noted that the efficiency improves during high load since the data rates are typically decreased from 384 kbps down to 64–128 kbps.

## 10.4.3 Downlink Shared Channel (DSCH)

The downlink shared channel, DSCH, is targeted to transfer bursty packet data. The idea is to share a single physical channel, i.e. orthogonal code, between many users in a time division manner. This approach saves the limited number of downlink orthogonal codes because several users share the code. If the dedicated channel were used instead, the orthogonal code would be reserved according to the maximum bit rate, and the efficiency of code usage would be lower. Shared channels are used in parallel with a lower bit rate dedicated channel. The dedicated channels carry the physical control channel, including the signaling for fast power control. It should be noted also that shared channels cannot use soft handover. The efficient rate adaptation capability of a shared channel is useful with the TCP slow start and during the end of the file download. With DSCH the resources can be immediately allocated to another user before the DCH inactivity timer expires, improving the resource utilization compared to DCH. Shared channel is typically not implemented in the first WCDMA networks, nor in the first terminals. The high speed downlink shared channel HS-DSCH is part of the HSDPA concept that is covered in Chapter 12.

## 10.4.4 Uplink Common Packet Channel (CPCH)

The uplink common packet channel (CPCH) is an extended RACH channel. The CPCH can have fast power control after the access procedure but it cannot use soft handover. The set-up time of the CPCH channel is slightly longer than that of the RACH, but it can be allocated for 64 frames = 640 ms, which means more data can be sent on it than on the RACH. The CPCH is optional for the mobiles. Without wide support of CPCH in mobiles, this channel cannot be efficiently utilized in WCDMA.

## 10.4.5   Selection of Transport Channel

The transport channels for packet data are summarized in Table 10.1. The High Speed Downlink Shared Channel, HS-DSCH is left out of this description, since the High Speed Downlink Packet Access, HSDPA, is covered in Chapter 12.

The state transitions can be controlled by the traffic volume threshold from Cell_FACH to Cell_DCH and by the inactivity timer from Cell_DCH to Cell_FACH. If the amount of data in the mobile uplink buffer or in the RNC downlink buffer exceeds the traffic volume threshold, DCH allocation takes place. When DCH is allocated, either for uplink or for downlink, it must be allocated for the other direction at the same time as well, since DCH is a bi-directional channel. The DCH bit rate can be different in uplink and in downlink depending on which direction triggers the DCH allocation. The following two examples illustrate the state transitions and transport channel utilization. Figure 10.18

**Table 10.1**   Overview of WCDMA transport channels

|  | Dedicated channels DCH | Shared channels DSCH | Common channel | | |
|---|---|---|---|---|---|
|  |  |  | FACH | RACH | CPCH |
| RRC state | Cell_DCH | Cell_DCH | Cell_FACH | Cell_FACH | Cell_FACH |
| Uplink/Downlink | Both | Downlink | Downlink | Uplink | Uplink |
| Code usage | According to maximum bit rate | Code shared between users | Fixed codes per cell | Fixed codes per cell | Fixed codes per cell |
| Fast power control | Yes | Yes | No | No | Yes |
| Soft handover | Yes | No | No | No | No |
| Suited for | Medium or large data amounts | Medium or large data amounts | Small data amounts | Small data amounts | Small or medium data amounts |
| Suited for bursty data | No | Yes | Yes | Yes | Yes |
| Available in first networks and terminals | Yes | No | Yes | Yes | No |

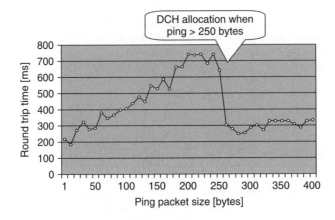

**Figure 10.18**   Measured round trip time with continuous ping with increasing packet size

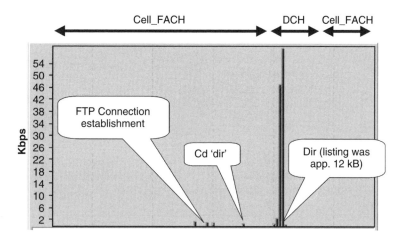

**Figure 10.19**   Measured transport channel utilization example

shows the average round trip time when ping was performed continuously for packet sizes ranging from 1–400 with 10 byte intervals. DCH was allocated in this case when the data amount exceeded 250 bytes, and kept afterwards. The round trip time drops when DCH is allocated, since DCH clearly has a higher bit rate than RACH/FACH. Figure 10.19 shows an example of getting a directory listing. First, an FTP connection is established on RACH/FACH. The request for directory listing ('dir') is sent on RACH. When the directory listing is transmitted in downlink, the 64 kbps DCH is allocated since the data amount is 12 kB. When the data has been transmitted, the DCH is released after the inactivity timer.

When DCH is allocated, its bit rate can be upgraded or downgraded. The bit rate upgrade takes place if the amount of data in the buffer exceeds a higher threshold. When increasing the bit rate, the maximum link powers and the cell capacity need to be checked. The bit rate downgrade takes place if the DCH capacity is under-utilized. The bit rate downgrade may also be needed if the maximum link power is achieved in the weak coverage area. Figure 10.20 illustrates an algorithm for selecting the transport channel and its bit rate. A bit rate allocation example is shown in Figure 10.21 with TCP file download. First, 64 kbps DCH is allocated. When more data arrives in the buffer, the bit rate is upgraded to 384 kbps.

A measurement example of file download is shown in Figure 10.22. The file download is started when there is an on-going voice calls. That particular network supports only 64 kbps with simultaneous voice calls. When the voice call ends, the bit rate is upgraded to 384 kbps. When the second file download starts, 384 kbps is directly allocated.

It is beneficial to allow TCP connection establishment to take place on the common channels. The TCP connection establishment involves transmission of a few small IP packets of 40 bytes. Also, some laptop clients, like VPN clients, tend to send infrequent small or medium size packets. These packets should preferably be transmitted on common channels and no DCH allocation should take place. Suitable traffic volume thresholds are therefore 128–512 bytes.

The DCH inactivity timer depends on the bit rate: the higher the bit rate, the shorter the inactivity timer should be to avoid wasting resources. The resource utilization was presented in Figure 10.17. The inactivity timer should also be relatively short to save the mobile batteries. However, from the end user point of view, a long inactivity timer would be preferred. If more data arrives after the channel is released, the user will experience another DCH set-up time delay. When setting the value for the inactivity timer, one needs to make a trade-off between end user performance, network resource consumption and mobile battery consumption. Typical values for the inactivity timer are 2–10 seconds.

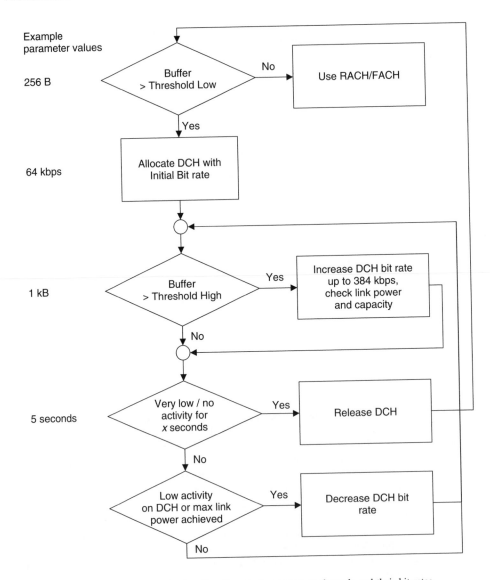

**Figure 10.20**  Example algorithm for selecting transport channels and their bit rates

## 10.4.6  Paging Channel States

The previous section described the utilization of Cell_FACH and Cell_DCH states. This section briefly presents Cell_PCH/URA_PCH states. Both the mobile receiver and the transmitter are active in the Cell_DCH state, causing relatively high power consumption. Only the mobile receiver is active in the Cell_FACH state, reducing power consumption compared to the Cell_DCH state. The Cell_FACH state, however, requires continuous reception, while in the Cell_PCH state, discontinuous reception can be used, providing very low power consumption and long mobile operation times. Cell_PCH/URA_PCH and Cell_FACH states are similar from the air interface capacity point of view: no

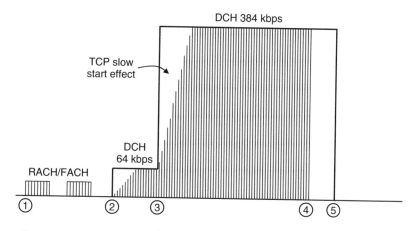

① = TCP connection establishment on RACH/FACH
② = Download starts – DCH allocated with 64 kbps
③ = More data in the buffer – DCH bit rate upgraded to 384 kbps
④ = File transfer over – transmission power to minimum, code still reserved
⑤ = DCH released after inactivity

**Figure 10.21**  Example usage of transport channels in the case of TCP file download

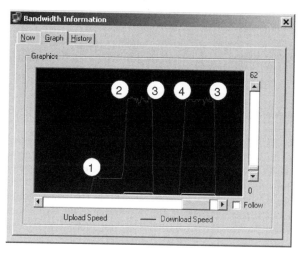

① = File transfer starts, 64 kbps allocated (simultaneous voice call)
② = Bit rate upgraded to 384 kbps due to voice call finish
③ = File transfer over
④ = File transfer starts, 384 kbps allocated

**Figure 10.22**  Example measured bit rate allocations during TCP download

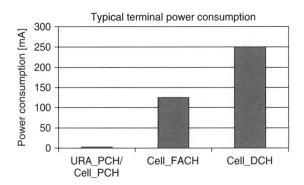

**Figure 10.23**  Typical mobile power consumption in different radio resource control states

interference is caused and no radio resources are reserved. The main difference lies in the mobile power consumption, which is considerably lower in the Cell_PCH state than in the Cell_FACH state. Typical power consumption values are 200–400 mA in Cell_DCH, 100–200 mA in Cell_FACH and <5 mA in Cell_PCH. These values are naturally affected by the network parameters, like paging cycle, and by the required mobile transmission power in the Cell_DCH state. Also, application layer processing, such as video encoding and decoding, will affect the power consumption.

The additional delay when moving from the Cell_PCH to the Cell_FACH state is short if the new capacity request is triggered in uplink. If the new packet arrives in downlink, there is an additional delay caused by the paging cycle to move the mobile from Cell_PCH to Cell_FACH. This extra delay depends on the paging cycle and is, on average, half of the paging period. Paging cycles are normally 320–2560 ms.

The mobile power consumption values are shown in Figure 10.23. These allow roughly four hours of Cell_DCH operation time and hundreds of hours in the Cell_PCH state with a 1000 mAh mobile battery.

## 10.5 Cell-Specific Packet Scheduling

The cell-specific packet scheduler divides the non-real-time capacity between simultaneous users. The cell-specific scheduler operates periodically. This period is a configuration parameter and its value typically ranges from 100 ms to 1 s. If the load exceeds the target, the packet scheduler can decrease the load by decreasing the bit rates of packet bearers; if the load is less than the target, it can increase the load by allocating higher bit rates, as shown in Figure 10.24. The aim of the scheduling is to efficiently use all remaining cell capacity for non-real-time connections but also maintain interference levels within planned values so that real-time connections are not affected. This section presents the main principles of how the cell-specific packet scheduler operates.

The cell-specific packet scheduler uses the following input information:

- Total Node B power. The total load is estimated using power-based load estimation, as described in Section 9.4.
- Capacity used by non-real-time bearers. This capacity can be estimated using throughput-based load estimation.
- Target load level from network planning parameters. This parameter defines the maximum interference level that can be tolerated in the cell without affecting the real-time connection.
- Bit rate upgrade requests from the user-specific packet scheduler.

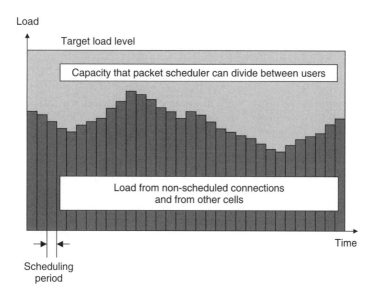

Figure 10.24    The packet scheduler divides the non-real time capacity between non-real time data users

The input information and the calculation principles are illustrated in Figure 10.25. Node B provides periodic total power and link power measurements in radio resource reporting to RNC. The total power measurements normally have a faster reporting cycle than link power measurements. The packet scheduler can estimate the total power used by non-controllable traffic, which consists of real-time connections and inter-cell interference. This part of the interference cannot be affected by the packet scheduler. The remaining capacity can be divided by the packet scheduler between the simultaneous users. The bit rate upgrade information from the user-specific scheduler is taken into account when changing the bit rates.

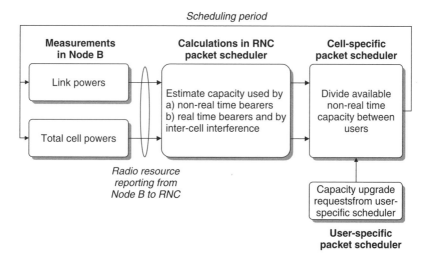

Figure 10.25    Input information and calculation principles of the cell-specific packet scheduler

## 10.5.1  Priorities

The core network provides QoS parameters over the Iu interface to WCDMA RAN; for more details see Chapter 2. The information includes traffic class, allocation/retention priority and traffic handling priority for the interactive class, as shown in Table 10.2.

The aim of these QoS parameters is to share the radio resources more efficiently between the users. When the radio resources are scarce, these parameters help the packet scheduler to decide how to allocate the capacity for the different users. This feature is illustrated in Figure 10.26. The simplest algorithm is one where the higher priority bearers will always get the capacity before the lower priority bearers. Another approach is to give a minimum bit rate, such as 32 kbps, for all users and then allocate the remaining capacity for the high priority users to increase their bit rate.

QoS parameters can also be used to differentiate the services for performance monitoring. If, for example, MMS, WAP and streaming services are allocated different allocation/retention priorities, those services can be separated in the radio network performance counters. This allows the operator to have better visibility of the end user performance of the different services. If the services use the same QoS parameters, their radio performances cannot be separately monitored. QoS parameters can also be used to provide better service quality for operator-hosted services compared to other services.

## 10.5.2  Scheduling Algorithms

The cell-based packet scheduler uses a number of different inputs to select the radio bearer bit rates. An example DCH bit rate allocation is illustrated in Figure 10.27 where new capacity requests keep arriving from user-specific schedulers. The total cell capacity is assumed to be 900 kbps and no real-time connections are present. The packet scheduler can divide that 900 kbps capacity between the users. The first and the second users can be allocated 384 kbps. When there is no more room for the maximum 384 kbps capability, one of the highest bit rate bearers is downgraded to make room for the new request. If one of the bearers is released, some of the other allocations can be upgraded correspondingly. Figure 10.27 shows a simplified case. In real life, the maximum load level is not constant but depends on the user locations, on the simultaneous real-time connections and on the other cell interference.

**Table 10.2**  Main QoS parameters for packet scheduling

| Traffic class | Interactive class | Background class |
| --- | --- | --- |
| Allocation/retention priority | 1, 2, 3 | 1, 2, 3 |
| Traffic handling priority | 1, 2, 3 | – |

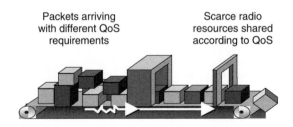

**Figure 10.26**  QoS priorities are targeted for efficient radio utilization

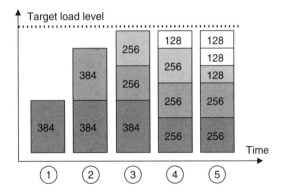

$\begin{array}{l}\textcircled{1}\end{array}$ = 1 capacity request in the queue, 384 kbps allocated

$\begin{array}{l}\textcircled{2}\end{array}$ = 2nd capacity request arrives, room for 2nd 384 kbps

$\begin{array}{l}\textcircled{3}\end{array}$ = 3rd capacity request arrives, no room for 384 kbps, but 256 kbps allocated and one existing connection downgraded to 256 kbps

$\begin{array}{l}\textcircled{4}\end{array}$ = 4th capacity request arrives, 128 kbps allocated and one existing connection downgraded from 384 kbps to 256 kbps

$\begin{array}{l}\textcircled{5}\end{array}$ = 5th capacity request arrives, 128 kbps allocated and one existing connection downgraded from 256 kbps to 128 kbps

**Figure 10.27**   Example DCH bit rate allocation when new requests arrive, no QoS differences

### 10.5.3   Packet Scheduler in Soft Handover

If the mobile is in soft handover, the packet scheduler must take into account the air interface load and the physical resources in all base stations of the active set. The dedicated channels are the only transport channels that can use soft handover. When the mobile is in CELL_DCH state and in soft handover, the packet scheduling can be done for each cell separately. Therefore, the responses (scheduled bit rates) of different packet schedulers may differ. The final selection of the bit rate is made according to the most heavily loaded cell in the active set, which has scheduled the lowest bit rate. The interactions are illustrated in Figure 10.28.

## 10.6   Packet Data System Performance

In this section, packet scheduling performance is analysed. First, link level performance of packet data is discussed in Section 10.6.1, while system level performance is the topic of Section 10.6.2.

### 10.6.1   Link Level Performance

The effect of the block error rate, BLER, and retransmissions on throughput is studied at the link level, and an optimal BLER target level is proposed in this section. Packet data performance is studied in the ITU Vehicular A multipath channel using a mobile speed of 3 km/h in the downlink with 64 kbps DCH. The BLER as a function of $E_b/N_0$ is shown in Figure 10.29.

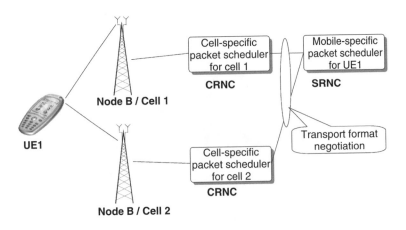

**Figure 10.28**  Packet scheduling in soft handover

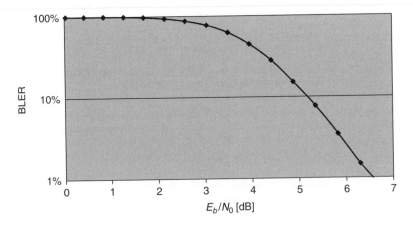

**Figure 10.29**  BLER as a function of $E_b/N_0$ for 64 kbps DCH in the downlink

The higher the BLER, the more retransmissions are needed to deliver error-free data. On the other hand, less power, or lower $E_b/N_0$, is needed for higher BLER levels. What is the optimal BLER operation point that requires the lowest energy per correctly received bit when the retransmissions are taken into account? To find out the optimal BLER point, we use the definition of effective $E_b/N_0$ that should be minimized to maximize the capacity:

$$(E_b/N_0)_{\text{effective}} = \frac{E_b/N_0}{1 - \text{BLER}} \tag{10.3}$$

The BLER vs. effective $E_b/N_0$ is shown in Figure 10.30. The relationship between $E_b/N_0$ and BLER is taken from Figure 10.29. The optimal BLER operation point is around 10%. If BLER is lower, capacity is wasted because the retransmissions are not efficiently utilized to gain from the additional time diversity. The effective $E_b/N_0$ is 0.8 dB higher with BLER 1% than with BLER 10%. If BLER is higher than 10%, there are too many retransmissions, causing additional interference. With higher BLER, the average delay will also be longer due to retransmissions, and the quality of the signaling will be reduced. A higher BLER also consumes more downlink orthogonal codes, Node B hardware

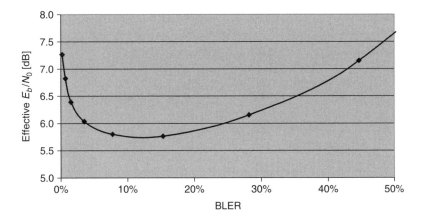

**Figure 10.30** BLER vs. effective $E_b/N_0$ (lowest number = highest capacity)

and Iub resources, because those resources must be reserved for a longer time for the retransmissions. The BLER level of $1-10\%$ is generally assumed for packet data in this book.

## 10.6.2 System Level Performance

In this section, the system level performance of packet scheduling is analysed. The user bit rates and system capacity are shown for different load scenarios. A network with 18 cells in Figure 10.31 is considered in these simulations. The users are distributed uniformly in the network area. Only the

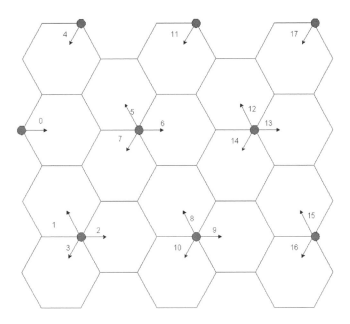

**Figure 10.31** Network layout with 18 cells

statistics from the four cells in the middle are considered in order to avoid border effects. The main parameters can be seen in Table 10.3.

Users arrive according to a Poisson arrival process. Each user performs the downloading of one packet call, where the packet call distribution is given in Figure 10.32. The minimum packet is 500 B, median 30 kB and the maximum equals 1 MB.

The packet scheduler estimates the bit rate that can be given to a user, according to the available air interface capacity and the maximum link power. First, a 32 kbps is allocated, which is upgraded after 4 s if a bit rate upgrade is possible. The maximum possible bit rate will be given to this user and this bit rate will be kept until the file download is finished. The resources will be kept reserved after the download has finished for 2–5 s in order to avoid repeating releasing and setting up of the dedicated channel.

Figure 10.33 shows the cell throughput as a function of the load. It can be seen that the throughput initially increases with the number of users. The reason is that the low number of users do not download enough data to fill the whole cell capacity. The maximum capacity is approximately 800 kbps per cell. When there are more than 300 active UEs in the network, the cell throughput starts decreasing. The

**Table 10.3**  Simulation assumptions

| Parameter | Value |
|---|---|
| BLER target | 5% |
| Packet scheduling period | 200 ms |
| Minimum and initial bit rate | 32 kbps |
| Maximum bit rate | 384 kbps |
| Maximum link power | 2 W |
| Inactivity timer | 2–5 seconds |

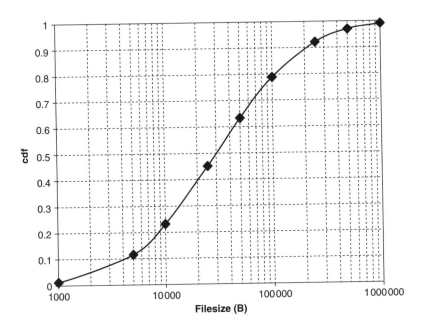

**Figure 10.32**  Cumulative density function of the packet call distribution (simulation input assumption)

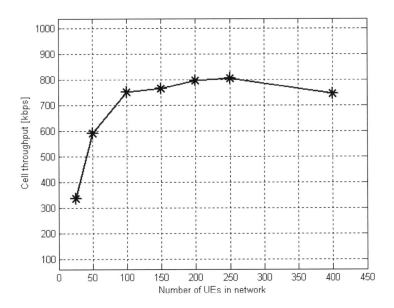

**Figure 10.33**   The cell throughput as a function of the number of UEs in the network (simulation output result)

reason for this is that the users' bit rates decrease with an increasing number of users, while the signaling overhead stays identical per user. Thus, the cell throughput starts to suffer from the large number of signaling channels.

The transmitted bit rate distribution can be seen in Figure 10.34 for three different numbers of users. It can clearly be seen that in the case of a low loaded network, the high bit rates are

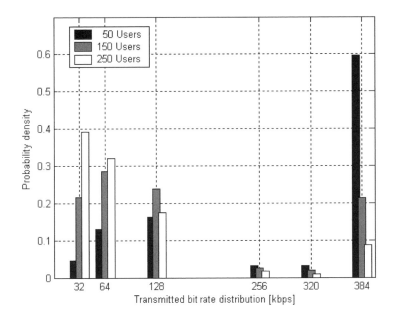

**Figure 10.34**   The probability density of the transmitted user bit rates at three different loads

allocated: nearly 60% of the time the users get 384 kbps, even though they all start out with 32 kbps for the first four seconds. When the load increases, a shift to the lower bit rates can be seen: in the case of 250 users, only 10% get 384 kbps while 40% of the users get 32 kbps.

## 10.7    Packet Data Application Performance

The focus in this chapter has, up to now, been on basic packet data performance items, such as packet round trip times and bit rate throughput on the radio, TCP/IP and UDP/IP layers. What has not been discussed thus far is how the applications on top of the TCP/IP and UDP/IP layers over WCDMA are dependent on the underlying end-to-end performance characteristics. This dependence and end-to-end performance form the topic of the following sections.

### 10.7.1    Introduction to Application Performance

Applications utilize the packet data transport connections with performance properties such as throughput, delay, jitter, residual bit errors and connection set-up times. See Figure 10.35 for an illustration of protocol layers and their corresponding key performance characteristics. One delay component that is usually included in the initial connection set-up time is the activation of the PDP (Packet Data Protocol) context in the GGSN (Gateway GPRS Support Node). The activation or modification of the PDP context is needed as soon as a connection with a new quality of service traffic class is set up to/from the mobile station. Because one application may use multiple quality of service classes, e.g. the interactive traffic class for signaling and the real-time traffic class for user plane data, it is possible that one PDP context is always on and one PDP context is set up on a need basis only.

When designing the applications, both for fixed and wireless networks, it is possible to improve the end user performance by utilizing user and control plane compression and receiver buffering on the application layer. By doing this the perceived end user throughput and outage times are improved from an end user point of view, even though the underlying transport quality is still the same.

The end user performance measures for applications that we look at are service set-up times, response times during service usage and perceived user plane quality of information presentation (throughput, picture quality, outage, availability).

The target is to illustrate the performance with example packet-based applications, i.e. the different performance enhancement methods are not illustrated in more detail. Depending on the application type, the end user performance expectations are different. The applications are therefore grouped into person-to-person applications, content-to-person applications and business connectivity.

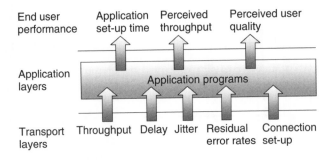

**Figure 10.35**    Application performance dependence on lower level protocol layers

## 10.7.2   Person-to-Person Applications

The performance of the following person-to-person applications is evaluated: push-to-talk over cellular (PoC), real-time video sharing, voice over IP (VoIP) and real-time games.

### 10.7.2.1   Push-to-talk over Cellular (PoC)

The PoC application [5] performance is obviously dependent on the speech quality, but since the speech quality already has been studied extensively for circuit switched connections, this section only includes a study of the response times while using PoC. Two central performance measures related to the response times for the push-to-talk service are the start-to-talk delay in Figure 10.36 and the voice-through delay in Figure 10.37. The start-to-talk delay is the time between user A pressing a button to enable a connection until he/she gets an indication (for example, a beep) that the floor is granted. The voice-through delay is the time from when user A starts speaking until the voice starts playing in user B's terminal.

The delay calculation results presented herein include effects like:

- Typical packet round trip times;
- Representative radio bearer and radio access bearer set-up times;

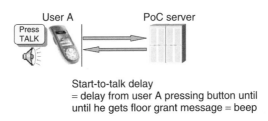

Start-to-talk delay
= delay from user A pressing button until
until he gets floor grant message = beep

**Figure 10.36**   Start-to-talk delay in push-to-talk

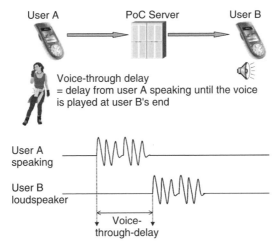

**Figure 10.37**   Voice-through delay in push-to-talk

- Paging delays;
- Voice encoding delays in the mobile station;
- Typical PoC server delays.

The receiving terminal may additionally have a receiver buffer of a few hundred ms to compensate for the jitter delays. The jitter buffer delay is, however, not included in the presented calculations.

The voice-through delay and the start-to-talk delay are both affected by what RRC state the mobile station is in when either pushing the start-to-talk button for the first time or when receiving the first speech burst from its peer. It is assumed that both mobile stations start from the Cell_PCH state and that the floor request messages are small enough for the Cell_FACH state. That is, the transition to the Cell_DCH state takes place first when there are speech packets to send. Start-to-talk I in Figure 10.38 and voice-through delay I in Figure 10.39 refer to the case when the mobiles are initially in Cell_PCH state. Note that once a mobile station has transmitted or received its first speech burst it will be in the Cell_DCH state for some time and subsequent start-to-talk and voice-through delays are hence smaller. The transition back from Cell_DCH to Cell_FACH or Cell_PCH takes place first after an inactivity of several seconds, so it is reasonable to assume that during active use of the application, the mobile stations will stay in the Cell_DCH state. Start-to-talk II and voice through delay II refer to the case when the mobiles are already in Cell_DCH state. The DCH allocation delays and round trip times are assumed to be similar to those presented earlier in this chapter. Note that the start-to-talk delay values are short in all cases, since common channels (RACH/FACH) are used to carry the corresponding packets. The voice-through delay is affected by the RRC state since DCH allocation increases the end user experienced delay. The paging delay does not increase voice-through delay since user B paging is triggered already from the floor request message. For reference: the normal voice-through delay in

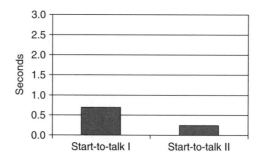

**Figure 10.38**   Start-to-talk delay in push-to-talk application

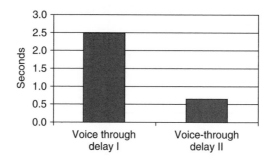

**Figure 10.39**   Voice-through delay in push-to-talk application

circuit switched cellular telephony is approx 200 ms. The higher delay in PoC comes mainly from the fact that the voice packets are larger in PoC than in ordinary voice connections.

The push-to-talk application can be used over GPRS as well. The start-to-talk delay in GPRS is slightly longer than in WCDMA. Also, the GPRS voice-through delay is longer when compared to WCDMA DCH. On the other hand, WCDMA voice-through delay is slightly longer than in GPRS if DCH set-up is required in WCDMA. We may note from the PoC performance calculations, that common channels are handy in WCDMA for sending small packets without DCH allocation, and low round trip time in WCDMA is useful in reducing the voice-through delay. We can also note that minimization of the DCH set-up time is important for optimizing the end user performance.

### 10.7.2.2  Real-Time Video Sharing

Two performance metrics for real-time video sharing are the set-up delay of the streaming connection and the streaming video delay from user A recording to user B display. The streaming video delay is illustrated in Figure 10.40. The two-way voice can be carried on circuit switched channel in this application. It is desirable to minimize the delay of the streaming video so that the video picture and the voice discussion can be kept reasonably well in synchronization.

The set-up delay for real-time video sharing (shown in Figure 10.41) consists of two main components: session initiation and receiver buffering. The session initiation consists of SIP (Session Initiation Protocol) signaling to set up the streaming connection from user A to user B. By looking at the main delays of the SIP signaling the service set-up should not take longer time than setting up a circuit switched connection today between two mobile stations, i.e. less than five to seven seconds. This assumes that efficient SIP compression schemes are used to minimize the SIP message lengths. The

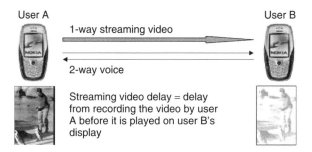

**Figure 10.40**  Streaming video delay in real-time video sharing application

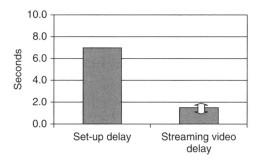

**Figure 10.41**  Typical delays in real-time video sharing

streaming delay is, to some extent, a trade-off of picture quality and delay that is caused by receiver buffers. Because of seamless soft handover, the need to compensate for jitter delays is low in WCDMA. Because of this it should be possible to keep the streaming video delay down to two seconds and below.

### 10.7.2.3   Voice over IP (VoIP)

Voice over IP, video conferencing and other conversational packet services set high requirements on the one-way end-to-end delay: the preferred one-way delay is <150 ms and the limit 400 ms [4]. The short delay is required to enable fluent communication between live end users. The round trip time for WCDMA shown in Figure 10.10 shows that one-way end-to-end delays of approx 200 ms are feasible in WCDMA. This enables good quality conversational packet switched services like, for example, VoIP.

The set-up time is another performance metric that should be minimized to improve end user perceived performance. The estimate of the set-up time consists of RRC state change (DCH allocation), PDP context activation for conversational QoS, paging delay for user B and SIP signaling. The total set-up time is approximately 8 seconds in Figure 10.42. This delay is of the same order as the circuit switched voice call set-up time.

### 10.7.2.4   Multiplayer Games

Real-time network games may require very short delays: action games typically require 200 ms or even shorter round trip time from the UE to the server and back for good quality games. An example action game is Quake II. Real-time strategy games, like Age of Kings, typically require a round trip time of around 1 second. Achievable round trip times in GPRS and WCDMA are shown in Figure 10.43, together with the gaming performance requirements. The current GPRS system allows a number of network games, including real-time strategy games, from the delay point of view. The current WCDMA delay is short enough even for real-time action games with acceptable quality.

The bit rate requirements for real-time games are quite low, mainly <16 kbps. The required data rate is not challenging for WCDMA nor for the current second generation systems. It is mainly the end-to-end delay that is the challenge for real-time gaming over cellular systems.

## 10.7.3   Content-to-Person Applications

The following content-to-person applications are considered: WAP browsing, video streaming and content download.

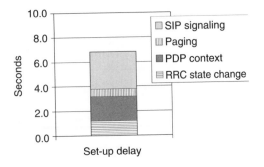

**Figure 10.42**   Set-up time for VoIP connection

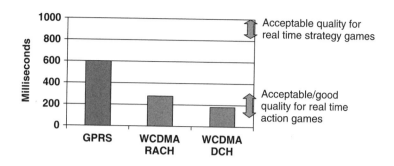

**Figure 10.43**   Round trip time and delay requirements for real-time games

#### 10.7.3.1   WAP Browsing

WAP pages are typically quite small and the download times are fast. The download time of the first WAP page, typically the operator's home page, may be longer. The optimization of first page download time is considered in this section. Before the UE is able to establish a connection to the WAP gateway, it must have RRC connection to RNC, it must be GPRS attached to SGSN and it must have an active PDP context. If all those procedures need to be done before the download, the first page download can take up to 8–10 seconds, even if the actual download of a small WAP page takes only 2–3 seconds. The optimization steps to reduce the download time are illustrated in Figure 10.44. The first step is to keep UEs GPRS attached all the time. The second step is to keep UEs RRC connected to RNC so that RRC connection establishment is not required, but only an RRC state change from PCH state to Cell_DCH state. The RRC connection can be maintained if there is an active PDP context for any application, e.g. for presence. The third optimization step is to use an existing PDP context also

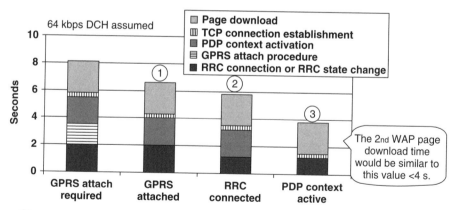

**Figure 10.44**   Download time of first WAP page (home page)

for WAP. In this case, only an RRC state change is required before establishing TCP connection and downloading the WAP page. The download time is reduced from the original 8 seconds to below 4 seconds with these optimization steps. The download time of following WAP pages will be similar to that 4 seconds. If the user already has DCH allocated, the download time is even faster, since no RRC state change is required.

### 10.7.3.2 Video Streaming

Person-to-person real-time video sharing and content-to-person video streaming have a few differences in terms of end user performance. The low delay is important for real-time video sharing, while the delay is not as important for content-to-person streaming. The content-to-person streaming is less challenging also because only a downlink connection is required to carry the data, while in real-time video sharing, both uplink and downlink connections are needed. Most importantly, video streaming requires a high enough bit rate to carry good quality video and voice. The relationship between bit rate and video quality is illustrated in Figure 10.45. The video streaming bit rates of 20–25 kbps are too low for good quality video streaming, while the bit rates of 60–120 kbps provide a clear improvement in quality. Such bit rates are feasible in WCDMA.

### 10.7.3.3 Content Download

The content download times using TCP are shown in Figure 10.46. Three different content sizes are considered: 100 kB file that could be a short video clip or picture, a 300 kB high quality picture and a 4 MB music file. GPRS offers reasonably low response times for file sizes <100 kB which are typical for MMS (Multimedia service). WCDMA offers low download times for high quality photos as well. A 4 MB MP3 music file download time in WCDMA would be less than 100 s. With HSDPA, that music file could be downloaded in less than 1 minute.

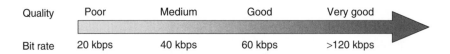

**Figure 10.45**   Relationship between streaming bit rate and video quality in mobile phone display

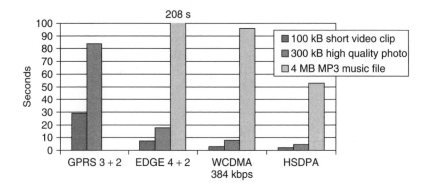

**Figure 10.46**   Content download times

## 10.7.4 Business Connectivity

Business connectivity considers applications running mainly on laptops using a cellular system as a radio modem. The considered applications are Web browsing, Outlook email and Netmeeting. The performance of these applications over WCDMA should preferably be similar to the performance provided by dial-up modems or by broadband DSL and cable modem connections, which set tough requirements on WCDMA performance.

### 10.7.4.1 Browsing

Web page download times are analysed in this section. The web page download times include RRC state change, Domain Name Server (DNS) query, TCP connection establishment and download of text and graphics in one TCP connection using HTTP1.1. The signaling flow chart is shown in Figure 10.47.

Typical average Internet web page sizes are 100–200 kB today and the size of the pages keeps increasing when more pictures and graphics are introduced to the pages. Measured download times are shown in Figure 10.48 and calculated times in Figure 10.49. Most of the download time comes from the

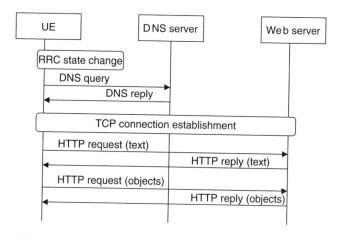

**Figure 10.47**   Flow chart for web page download

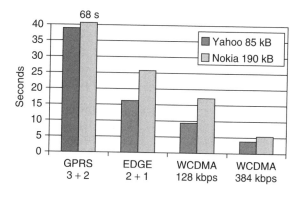

**Figure 10.48**   Measured web page download times. GPRS uses CS-2 and EDGE uses MCS-7

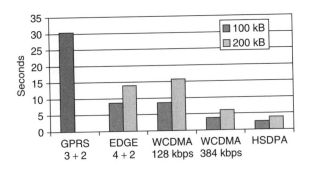

**Figure 10.49**   Estimated web page download times in average loaded networks

HTTP replies, i.e. downloading the content. There are some differences between the measurement set-up and the calculation assumptions, e.g. Virtual private network (VPN) was used in the measurements.

The measurements are done with a three time slot GPRS mobile, such as the Nokia 6600, with a two time slot EDGE mobile, such as the Nokia 6220, and with WCDMA bit rates of 128 kbps and 384 kbps, as in the Nokia 7600. EDGE modulation and coding scheme were fixed to MCS-7, providing approximately 90 kbps throughput with two time slots. The calculations assume a four time slot EDGE mobile, such as the Nokia 6230, with an average 40 kbps per time slot and WCDMA 384 kbps connection. The average assumed data rate for HSDPA is 700 kbps.

EDGE provides a major improvement in web browsing performance compared to GPRS. EDGE is >150% faster than GPRS and faster than using dial-up modem connection. WCDMA download times are 40–60% faster than with EDGE. WCDMA performance in general is similar to low-end DSL connections. HSDPA brings a further 30–40% reduction in download times compared to WCDMA. HSDPA performance is already close to public WLAN performance.

The web browsing response times could be improved by using Performance Enhancement Proxies, PEPs. These proxies are able to reduce the size of the web page, e.g. by reducing the picture resolution.

### 10.7.4.2   Email

Two performance measures are shown for Windows 2000 Outlook: time for receiving an email and time for connecting to a mail server. The measured email reception times for two different sizes of an email, 45 kB and 215 kB, are shown in Figure 10.50. Both EDGE and WCDMA improve the performance compared to GPRS. EDGE is roughly 100% faster than GPRS, and WCDMA is 100%

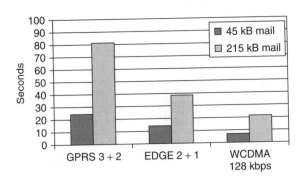

**Figure 10.50**   Measured email reception times

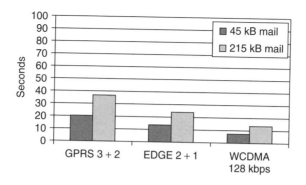

**Figure 10.51**   Measured email reception times with Performance Enhancement Proxy (PEP)

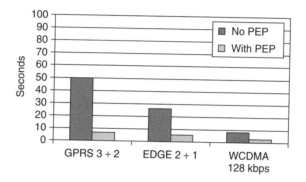

**Figure 10.52**   Measured connection times to mail server

faster than EDGE. The reception times with PEP are shown in Figure 10.51. PEP improves the performance for GPRS by 15% for small emails and up to 50% for large emails. EDGE and WCDMA do not benefit from PEP for small emails but they also gain roughly 30% for large mails.

The measured connection to mail server is illustrated in Figure 10.52. This procedure includes a large number of signaling messages and therefore, a small round trip time is important for the performance. Without PEP, EDGE provides only a minor gain over GPRS, while WCDMA with lower round trip time provides a major improvement. PEP can improve the connection times by a factor of 3–5 and it is important especially for GPRS and EDGE.

### 10.7.4.3   Netmeeting

The delay of sharing a slide from laptop to LAN is shown in Figure 10.53. Sharing a simple slide is relatively fast <5 seconds in GPRS. Sharing a heavy slide takes 11 seconds in WCDMA using 64 kbps uplink connection, and 22–33 seconds in GPRS/EDGE.

## 10.7.5   Conclusions on Application Performance

The performance of the considered person-to-person applications is mainly defined by the delay: a short delay is required for these applications. WCDMA is able to support these applications by providing a lower delay than second generation systems. The content-to-person applications benefit from both

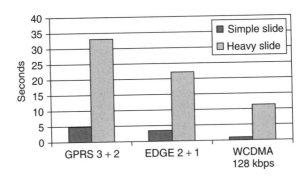

**Figure 10.53**   Measured time of sharing a slide from cellular system to LAN

high bit rate capability and from low delay. The presented business applications are sensitive to the delay because of the application signaling involved. The business applications also benefit from the high bit rate capabilities to provide close to DSL/WLAN levels of performance.

The performance estimates in this chapter show that GPRS is well suited for background downloads without strict delay requirements, and for downloads of small WAP/web pages. GPRS can also be used for narrowband streaming, like audio streaming. EDGE brings a clear improvement in performance for content-to-person and for business applications. EDGE performance is better than dial-up connection and it allows video streaming bit rates. WCDMA enables a number of person-to-person applications with its short delay. WCDMA also improves business applications and download performance. High Speed Downlink Packet Access, HSDPA, brings a further improvement in end user performance for downlink packet data. The application areas are shown in Figure 10.54.

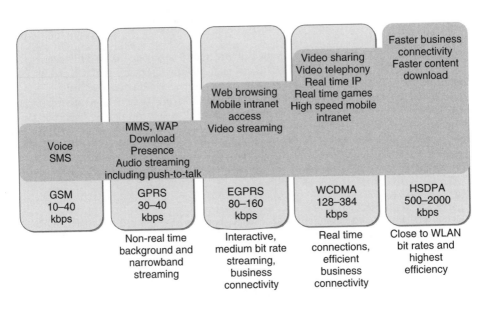

**Figure 10.54**   New applications are enabled by improved network performance

# References

[1] Stevens, W.R., *TCP/IP Illustrated, Volume 1: The Protocols*, Addison-Wesley Professional Computing Series, Reading, MA: Addison-Wesley, 1994.

[2] Inamura, H., Montenegro, G., Ludwig, R., Gurtov, A. and Khafizov, F., '*TCP over 2.5G and 3G Wireless Networks*', internet draft, http://www.ietf.org/internet-drafts/draft-ietf-pilc-2.5g3g-06.txt

[3] Allman, M, Floyd, S. and Partridge, C., 'Increasing TCP's Initial Window', experimental protocol RFC2414, http://www.faqs.org/rfcs/rfc2414.html

[4] 3GPP Technical Specification 25.853 'Delay Budget within the Access Stratum', 2003.

[5] Industry Specification for PoC Submitted to OMA (Open Mobile Alliance) MAG PoC SWG, Doc# OMA-MAG-POC-2003-0007, August 25, 2003.

# 11

# Physical Layer Performance

Harri Holma, Jussi Reunanen, Leo Chan, Preben Mogensen,
Klaus Pedersen, Kari Horneman, Jaakko Vihriälä and Markku Juntti

## 11.1   Introduction

This chapter presents coverage and capacity results and investigates the impact of the radio propagation
environment, advanced base station (Node B) solutions, and WCDMA physical layer parameters.
The advanced base station solutions include both baseband and antenna techniques. Cell coverage
is especially important in terms of the initial network deployment investment, where the offered
traffic load does not set demands for a fine grid of cells; WCDMA network coverage is analysed in
Section 11.2. The importance of capacity will increase after the initial coverage deployment when
the amount of traffic increases; WCDMA capacity is presented in Section 11.3. In this chapter, we
also present the WCDMA air interface capacity, which is limited by interference. We here assume
that there are sufficient baseband hardware resources in Node B, transmission network capacity, and
RNC resources to support the maximum air interface capacity. The results from field capacity trials
are presented in Section 11.4. The 3GPP performance requirements are analysed in Section 11.5, and
Section 11.6 presents the possible performance enhancements that are supported by the 3GPP standard,
including adaptive antenna structures and multiuser detection with advanced baseband processing. The
focus areas of this chapter are illustrated in Figure 11.1.

   The radio network planning and the optimization also have a major impact on coverage and capacity.
Their effects are presented in Chapters 8 and 9. The sharing of the capacity between simultaneous
users by the packet scheduler is discussed in Chapter 10.

## 11.2   Cell Coverage

Cell coverage is important when the offered traffic to a cell is insufficient to fully utilize the operator's
available spectrum; this is typically the case even for urban areas at initial network deployment
phase, and remains the case for rural areas with low traffic density. For conversational class (i.e.
symmetrical bit rate allocation) the macro cell coverage is determined by the uplink direction, because
the transmission power of the UE is much lower than that of the macro cell Node B. The output power
of the UE is typically 21 dBm (125 mW) and that of the macro cell Node B 40–46 dBm (10–40 W)

*WCDMA for UMTS: HSPA Evolution and LTE, Fifth Edition*   Edited by Harri Holma and Antti Toskala
© 2010 John Wiley & Sons, Ltd

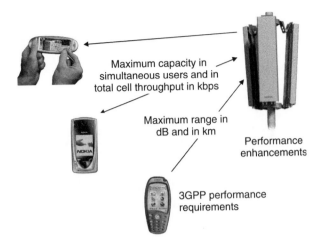

**Figure 11.1** Focus areas of this chapter

per sector (cell). Hence, in this section we focus on uplink coverage. Also, in Section 8.2.1 the macro cell coverage was shown to be limited by uplink link budget.

Let us first consider how the link budget improvement in dB impacts the maximal cell coverage area [km$^2$]. We here assume an improvement in link budget of $\Delta L$ dB, for example, by deploying advanced base station antenna techniques, mast head amplifier, or improved base station sensitivity by other techniques. The effect of the improvements in the link budget, $\Delta L$, on the relative cell radius increase, $\Delta R/R$, can be calculated assuming a macro cell propagation model, for example, the Okumura–Hata model from Section 8.2. $R$ is the default cell range and $\Delta R$ is the relative increase in cell range. In this example, the path loss exponent is set to 3.52, which results in

$$\Delta L = 35.2\log_{10}\left(1 + \frac{\Delta R}{R}\right) \tag{11.1}$$

The relative cell area increase $\Delta A/A$ can be calculated as

$$1 + \frac{\Delta A}{A} = \left(1 + \frac{\Delta R}{R}\right)^2 = \left(10^{\frac{\Delta L}{35.2}}\right)^2 \tag{11.2}$$

The required relative base station site density with a given improvement in the link performance is calculated in Table 11.1. The base station density is inversely proportional to the cell area. For

**Table 11.1** Reduction in the required base station site density with an improved link budget

| Improvement in the link budget | Relative number of sites |
|---|---|
| 0.0 dB = Reference case | 100% |
| 1.0 dB | 88% |
| 2.0 dB | 77% |
| 3.0 dB | 68% |
| 4.0 dB | 59% |
| 5.0 dB | 52% |
| 6.0 dB | 46% |
| 10.0 dB | 27% |

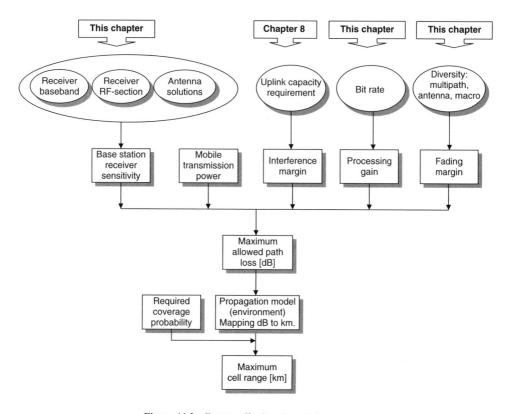

**Figure 11.2**   Factors affecting the uplink coverage

example, with a link performance improvement of 5.3 dB, the base station density can be reduced to approximately 50%.

Typically, the radio access network represents a major part of the total UMTS network investment, and most of the radio access network costs are base station site related costs. Therefore, a reduction in the number of cell sites is important in reducing the required investment for the UMTS network.

The factors affecting the maximum path loss can be seen from the link budget – see Section 8.2 – and are shown in Figure 11.2. The effect of the base station solutions, the bit rate and the diversity is described in this chapter. The relationship between uplink loading and coverage was discussed in Section 8.2.2.

## 11.2.1   Uplink Coverage

In this section we evaluate the effect of the physical layer parameters and the base station solutions on the WCDMA uplink coverage.

### 11.2.1.1   Data Bit Rate

Uplink cell coverage depends on the transmitted bit rate: for higher bit rates, the processing gain is reduced and the coverage range is hence reduced, see the uplink link budgets in the tables in Section 8.2, row $k$. The relationship between bit rate and coverage is quite different for real-time and

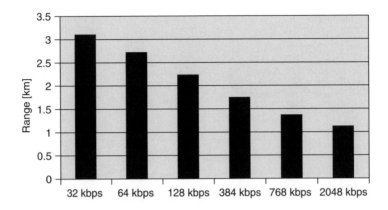

**Figure 11.3**  Uplink range of real-time guaranteed bit rates in suburban area

for non-real-time applications. The relative coverage versus bit rate is first shown in Figure 11.3 for identical outage probability for all bit rates. This plot corresponds to real-time applications demanding a guaranteed bit rate with a high coverage probability, e.g. 95%. The same maximum transmit power of the terminal is assumed for all bit rates and a suburban propagation model is assumed. In this example, the uplink range of 384 Mbps is reduced to 62% of the range of 64 kbps. If the cell layout is planned for 384 kbps uplink coverage instead of 64 kbps, the base station site density must be increased by a factor of $(1/0.62)^2 = 2.6$, which would be quite challenging.

Non-real-time applications do not require a minimum guaranteed bit rate with low outage probability. If we can allow temporarily a lower bit rate, the cell size does not need to be decreased to offer high data rates with high probability. The distribution of the uplink bit rate is shown in Figure 11.4 for a cell that is planned to provide 64 kbps with 95% probability. Such a cell is able to provide 384 kbps with 80% coverage probability. The average bit rate in this case is >330 kbps. The underlying reason for the difference between real-time and non-real-time bit rates is shown in Figure 11.5: the median mobile output power for 64 kbps is 5 dBm and only 8% of the mobiles are transmitting with full power

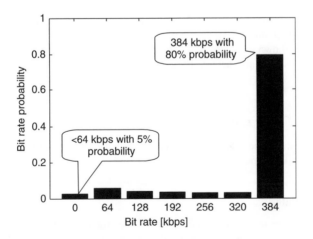

**Figure 11.4**  Distribution of the uplink non-real-time bit rates

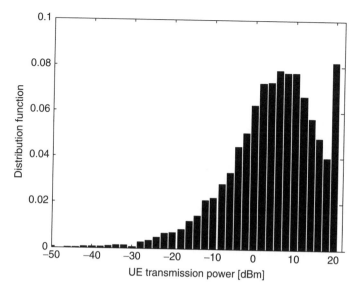

**Figure 11.5** Simulated distribution of the mobile transmission power with 64 kbps (5% outage)

(5% of them are in outage). It is possible for most mobiles to increase their transmission power to obtain a higher data rate for non-real-time services.

There are a number of solutions to improve uplink coverage, for example, beamforming antennas or higher order antenna diversity reception. If we can obtain improvements in the uplink link budget, this will increase the coverage probability of uplink data rates. The uplink bit rate improvements for non-real-time and for guaranteed real-time services are shown in Figure 11.6. The non-real-time bit

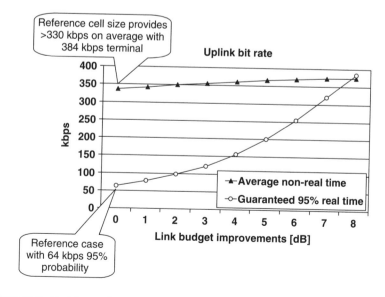

**Figure 11.6** Real-time and non-real-time bit rates as a function of link budget improvements

rate is the average for a 384 kbps capable terminal, while the real-time bit rate is the one that can be provided with 95% probability. The reference case (= 0 dB improvement) provides 64 kbps with 95% probability. The guaranteed bit rate increases 6-fold to 384 kbps when an 8 dB improvement is obtained in the link budget. The average non-real-time bit rate is already >330 kbps in this reference case and hence the improvement in the average bit rates is only moderate with the link budget improvements.

Wide area uplink coverage for high bit rate real-time services will be challenging in UMTS, as shown in Figure 11.3, and providing full 384 kbps or higher data rate uplink coverage requires high base station site density. These results also point out the importance of the solutions that improve uplink coverage in third generation systems. In second generation systems, coverage issues are less challenging since only low bit rate services are offered. The coverage of WCDMA data services is compared to GSM900 and GSM1800 speech coverage in Section 8.4. On the other hand, high average bit rates beyond 300 kbps can be offered for non-real-time services with GSM site densities. Non-real-time applications can benefit from those locations where the mobile has enough transmission power for 384 kbps, but these applications can survive in those locations where only 64 kbps is available. The bit rate adaptation capability of the application is beneficial when running over cellular systems. The reduction of the uplink bit rate is possible for non-real-time packet data services and also for AMR speech service, which supports different bit rates from 4.75 kbps to 12.2 kbps. The coverage of AMR speech service is discussed in the next section.

### 11.2.1.2   Adaptive Multirate Speech Codec

With Adaptive Multirate (AMR) speech codec it is possible to switch to a lower bit rate – and thereby increase the processing gain – if the link budget becomes insufficient to retain a low target BLER due to poor cell coverage. The AMR speech codec is introduced in Chapter 2. The gain in the link budget by reducing the AMR bit rate can be calculated as follows:

$$
\begin{aligned}
\text{Coverage\_gain} &= 10 \cdot \log_{10} \left( \frac{\text{DPDCH(12.2 kbps)} + \text{DPCCH}}{\text{DPDCH(AMR\_bit\_rate[kbps])} + \text{DPCCH}} \right) \\
&= 10 \cdot \log_{10} \left( \frac{12.2 + 12.2 \cdot 10^{\frac{-3\,\text{dB}}{10}}}{\text{AMR\_bit\_rate[kbps]} + 12.2 \cdot 10^{\frac{-3\,\text{dB}}{10}}} \right) \quad (11.3)
\end{aligned}
$$

where the power difference between DPCCH and DPDCH is assumed to be −3.0 dB for 12.2 kbps AMR speech. DPCCH is the physical layer control channel and DPDCH the physical layer data channel. For different AMR bit rates, the DPCCH power is kept the same, while the power of the DPDCH is changed according to the bit rate. The reduction of the total transmission power is calculated in Equation (11.3) and can be used to provide a larger uplink cell range for AMR speech. The coverage gain by reducing the bit rate from 12.2 kbps to 7.95 kbps is 1.1 dB, and the gain by reducing the bit rate from 12.2 kbps to 4.75 kbps is 2.3 dB. An example of cell range values with different AMR bit rates is shown in Figure 11.7.

The link budget gain with lower AMR bit rate can typically not be used in the cell dimensioning to make the cell range larger, since the coverage is often defined and limited by a certain minimum real-time data rate, e.g. by 64 kbps video. The link budget gain can rather be used to guarantee a better voice quality under challenging radio conditions, e.g. indoor environments with a high penetration loss, or during compressed mode measurements. The compressed mode is described in Chapter 9.

### 11.2.1.3   Multipath Diversity

Multipath diversity reduces the fast fading probability of the signal and thus the required fading margin in the link budget. The fast fading margin is on row $p$ in the link budget of Table 8.4. The gain of multipath diversity in terms of coverage is illustrated in Figure 11.8 for the cases of: 1-path, 2-path and

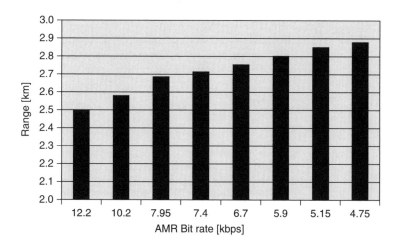

**Figure 11.7**   Example of uplink cell coverage range for different AMR speech codec bit rates

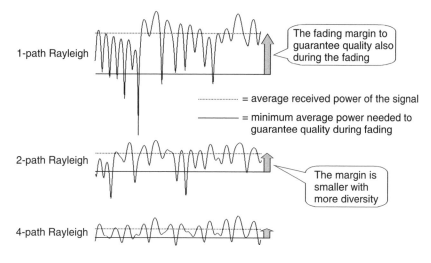

**Figure 11.8**   Diversity reduces the required fading margin

4-path Rayleigh fading. Multipath diversity can be obtained, for example, from Rake reception under time dispersive radio channel conditions or antenna diversity. In this section the effect of the multipath diversity on the uplink coverage is shown in terms of simulation and laboratory measurement results for two different multipath profiles: ITU Pedestrian A – providing only little multipath diversity, and ITU Vehicular A – providing more multipath diversity. The multipath diversity gain is here defined as the reduction of the average mobile transmission power when there is multipath diversity available. Two-branch receive antenna diversity is assumed at the base station.

The multipath diversity gain is shown in Table 11.2 at 3 and 20 km/h [1]. The gain is in the range of 1.0 and 1.6 dB respectively, for both simulation and measurement results. These results are obtained with fast power control active. However, if the terminal is operating at the cell edge and transmitting with constant full power, the fast power control does not compensate for the fast fading and the

**Table 11.2**  Multipath diversity gain for AMR voice with fast power control [1]

|          | Simulations | Laboratory measurements |
|----------|-------------|-------------------------|
| 3 km/h   | 1.0 dB      | 1.3 dB                  |
| 20 km/h  | 1.5 dB      | 1.6 dB                  |

**Table 11.3**  Multipath diversity gain for AMR voice with power control and with constant maximum transmit power

|          | With power control | With constant maximum transmit power |
|----------|--------------------|--------------------------------------|
| 3 km/h   | 1.0 dB             | 2.8 dB                               |

importance of the multipath diversity significantly increases, as shown in Table 11.3. The degree of multipath diversity obtained by a Rake receiver depends on the time dispersion of the radio channel relative to the transmission bandwidth. Hence, the multipath diversity gain is larger for wideband CDMA than for narrowband CDMA for the same radio environment, see Section 3.4.

#### 11.2.1.4  Soft Handover

During soft handover the uplink transmission from the terminal is received by two or more base stations. Since during soft handover there are at least two base stations detecting the UE transmitted signal, the probability of a correctly detected signal increases, and thus a soft handover gain is obtained. The soft handover gain for uplink coverage is shown in Table 11.4 for the case of 3 km/h, two receiving base stations, and AMR speech. We assume here that the fast fading is uncorrelated between the base stations and sectors. Two cases are shown: when the mean path losses to the two base stations are identical, and when there is a 3 dB mean difference in the path loss. These two cases are illustrated on the upper row of Figure 11.9. The first case gives the highest soft handover gain. When the difference in mean path loss becomes large, the soft handover gain vanishes and at a certain mean power difference the terminal will leave soft handover and only remain connected to the strongest base station. A typical value for the window drop is 2–4 dB, see Chapter 9 for more details. The results show that the lower the multipath diversity, the larger the soft handover gain. For equal mean path loss the soft handover gain is 4 dB for an ITU Pedestrian profile and 2.2 dB for an ITU Vehicular A profile.

**Table 11.4**  Soft and softer handover gains against fast fading

|                                                                                  | ITU Pedestrian A | ITU Vehicular A |
|----------------------------------------------------------------------------------|------------------|-----------------|
| Softer handover gain, equal mean path loss to both sectors                       | 5.3 dB           | 3.1 dB          |
| Soft handover gain, equal mean path loss to both base stations                   | 4.0 dB           | 2.2 dB          |
| Soft handover gain, 3 dB higher mean path loss to the worst receiving base stations | 2.7 dB        | 0.8 dB          |

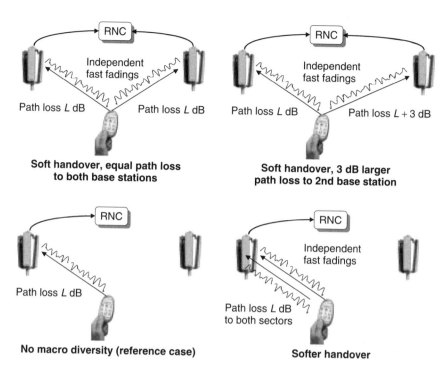

**Figure 11.9** Soft and softer handover cases for the soft handover gain evaluation

Uplink soft handover uses selection combining in RNC based on a CRC check, while in softer handover, the uplink transmission from the mobile is received by two sectors of one Node B. In softer handover the signals from two sectors are maximal ratio combined in the baseband Rake receiver unit of the base station, see Section 3.6. The soft and softer handover gains with equal power to both sectors and base stations are shown in Table 11.4. Softer handover provides 0.9–1.3 dB more gain than soft handover.

### 11.2.1.5 Base Station Receive Antenna Diversity

Ideally, 3 dB coverage gain can be obtained with receive antenna diversity, even if the antenna diversity branches have fully correlated fading. The reason is that the desired signals from two antenna branches can be combined coherently, while the received thermal noises are combined non-coherently. The 3 dB gain assumes ideal channel estimation, but the degradation of non-ideal channel estimation is marginal. Additionally, antenna diversity also provides a significant gain against fast fading for the case of uncorrelated or low correlated antenna branches. Network operators typically select antenna diversity topologies that ensure an envelope correlation of less than 0.7. The Node B receive antenna diversity gains are obtained at the expense of increased or duplicate hardware in the Node B, including RF front-end, baseband hardware, antenna feeders, antennas or antenna ports.

Two different diversity antenna topologies are shown in Figure 11.10. Low correlated antenna branches can be obtained by space or polarization diversity. The advantage of polarization diversity is that the diversity branches do not need separate physical antenna structures, see the left side of Figure 11.10. The performance of polarization diversity in GSM has been presented in [3], [4] and [5], and for WCDMA in [6].

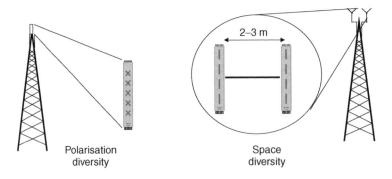

**Figure 11.10**   Polarization and space diversity antennas

Simulated and measured antenna diversity gain results are shown in Table 11.5. It can be observed that the gain is higher at low mobile speeds of 3 km/h and 20 km/h than for 120 km/h. The reason is that for high mobile speeds the link performance benefits from time diversity provided by the interleaving, and hence the additional gain from antenna diversity is reduced. We can also note that the gain is higher when the amount of multipath diversity is small as in the ITU Pedestrian A channel. The antenna diversity gain at low mobile speed is up to 5–6 dB for the ITU Pedestrian A profile and 3–4 dB for ITU Vehicular A profile. For the simulated case, the antenna branches are uncorrelated and for the measured case, the branches are practically uncorrelated.

The performance of uplink diversity reception can be further extended by deploying four-branch antenna reception. The four-branch antenna configuration can be obtained using two antennas with polarization diversity with a separation of 2–3 metres to combine polarization and space diversity, i.e. obtain four low correlated antenna branches. The two antennas can also be placed very close to each other, even in a single radome, to make the visual impact lower. However, in that case the branch correlation between the two polarization antenna structures is expected to be high. The two four-branch antenna options are shown in Figure 11.11.

**Table 11.5**   Antenna diversity gain for AMR speech with fast power control [1]

|          | ITU Pedestrian A | | ITU Vehicular A | |
|          | Simulations | Laboratory measurements | Laboratory | Simulations measurements |
|----------|-------------|-------------------------|------------|--------------------------|
| 3 km/h   | 5.5 dB      | 5.3 dB                  | 3.7 dB     | 3.3 dB                   |
| 20 km/h  | 5.0 dB      | 5.9 dB                  | 3.5 dB     | 3.5 dB                   |
| 120 km/h | 4.0 dB      | 4.4 dB                  | 3.0 dB     | 3.4 dB                   |

Two x-polarized antennas    Single radome solution

**Figure 11.11**   Four-branch receive antenna configurations

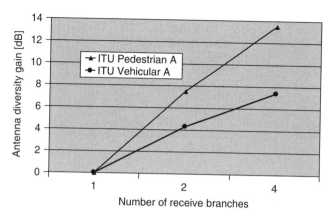

**Figure 11.12**   Antenna diversity gain with one-, two- and four-branch reception for the case of constant maximum transmit power of the mobile

The simulated diversity gains of two- and four-branch diversity are summarized in Figure 11.12. These results assume separate antennas in four-branch reception, i.e. low branch correlation, and constant maximum transmit power of the mobile. Hence, it should be noted that the results cannot be directly compared to the results in Table 11.5. The gain of four-branch diversity over two-branch diversity in ITU Vehicular A is 3.1 dB. The gain of the single radome solution is typically 0.2–0.4 dB lower, due to the higher antenna branch correlation shown in the measurement part.

The more diversity already available, the smaller the diversity gain from an additional diversity feature. This rule applies to antenna diversity and to all different kinds of diversity. Therefore, there is no *a priori* value for any diversity gain, because the gains depend on the degree of diversity from other diversity techniques.

### Field Measurements of Four-branch Receive Antenna Diversity
The field performance of four-branch reception was tested in the WCDMA network in Espoo, Finland. The measurement area is in the middle of Figure 8.18. The measurement environment is of the urban and sub-urban type. The measurement routes are shown in Table 11.6.

In the field measurements the mobile transmission power was recorded slot-by-slot with three different base station antenna configurations:

1. Two-branch reception with one polarization diversity antenna.
2. Four-branch reception with two polarization diversity antennas separated by 1 m.
3. Four-branch reception with two polarization diversity antennas side-by-side (emulates single radome solution).

For each configuration the route was measured several times. The different measurement routes are made comparable using the differential Global Positioning System, GPS. The average transmission

**Table 11.6**   Measurement routes

| | |
|---|---|
| Route A | up to 40 km/h in Leppävaara/Lintuvaara |
| Route B | up to 70 km/h on Ring I |
| Route C | below 10 km/h in Mäkkylä |

**Table 11.7**  Measured logarithmic average mobile transmission powers

| Route | Antenna separation | 2-branch reception | 4-branch reception | 4-branch gain over 2-branch |
|-------|-------------------|-------------------|-------------------|----------------------------|
| Route A | 1 m separation | 6.95 dBm | 4.44 dBm | 2.5 dB |
|         | no separation | 6.95 dBm | 4.83 dBm | 2.1 dB |
| Route B | 1 m separation | 7.90 dBm | 4.59 dBm | 3.3 dB |
|         | no separation | 7.90 dBm | 4.86 dBm | 3.1 dB |
| Route C | 1 m separation | 5.63 dBm | 2.54 dBm | 3.1 dB |

power over the measurement route is calculated from dBm values. These measured mobile transmission powers are shown in Table 11.7.

The multipath propagation in the measured environment is closer to ITU Vehicular A than to ITU Pedestrian A. We therefore compare the measurement results to the simulation results of the ITU Vehicular A profile. The simulated gain of four-branch reception over two-branch reception in Figure 11.12 is 3.1 dB with separate antennas, and the average measured gain with 1 m separation is 3.0 dB in Table 11.7.

The difference between separate antennas and the single radome solution is 0.2–0.4 dB. The impact of antenna branch correlation for the two spaced antenna structures is small because the diversity order is already large: multipath and polarization diversity.

It can be concluded that four-branch receive antenna diversity is an effective technique to increase the uplink coverage area. A 3 dB improvement in the uplink performance reduces the required site density by about 30% according to Table 11.1.

## 11.2.2  Downlink Coverage

The Node B transmit power is typically 20 W (43 dBm), while the mobile transmit power is only 125 mW (21 dBm). With a low number of simultaneous connections, it is possible to allocate a high power per mobile connection in downlink. Hence, better coverage can be given for high bit rate services in downlink than in uplink. The downlink coverage is affected by the maximum link power that is a network planning parameter. The downlink coverage is also affected by the amount of inter-cell interference. In this example the G factor, i.e. own cell to other cell interference ratio, at the cell edge is assumed to be $-2.5$ dB, which corresponds to approximately $-12$ dB CPICH $E_c/I_0$ with medium base station transmission power in large cells. The calculation assumes that CPICH is allocated 2 W and other common channels 1 W. The other cell transmission power is assumed to be 10 W and the maximum path loss at the cell edge 156 dB. The results are shown in Figure 11.13. Thus, 2 W link power provides 384 kbps at 60% of the maximum cell range and 64 kbps with full coverage. 5 W power allocation gives 384 kbps at 80% of the maximum cell range, while 10 W power allocation gives practically full 384 kbps coverage.

## 11.3  Downlink Cell Capacity

The WCDMA downlink air interface capacity has been shown to be less than the uplink capacity [7–9]. The main reason is that better receiver techniques can be used in the Node B than in the mobile. These techniques include receiver antenna diversity and multiuser detection. Additionally, in UMTS, the downlink capacity is expected to be more important than the uplink capacity because of the asymmetric downloading type of traffic. In this section the downlink capacity and its performance enhancements are therefore considered. WCDMA capacity evaluation is studied also in [10].

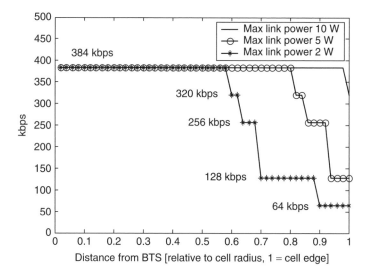

**Figure 11.13**  Downlink coverage with different maximum link powers

The following sections present two aspects that impact upon the downlink capacity, and which are different from the uplink: the issue of orthogonal codes is described in Section 11.3.1 and the performance gain of downlink transmit diversity in Section 11.3.2. Additionally, we discuss the WCDMA voice capacity with AMR codec and Voice over IP (VoIP) in Section 11.3.3.

## 11.3.1  Downlink Orthogonal Codes

### 11.3.1.1  Multipath Diversity Gain in Downlink

The effect of the downlink orthogonal codes on capacity is considered in this section. In downlink, short orthogonal channelization codes are used to separate users in a cell. Within one scrambling code the channelization codes are orthogonal, but only in a one-path channel. In the case of a time dispersive multipath channel, the orthogonality is partly lost, and own-cell users sharing one scrambling code also interfere with each other. The downlink performance in the ITU Vehicular A and ITU Pedestrian A multipath profiles is presented below for the case of 8 kbps, 10 ms interleaving, and 1% BLER. The ITU Pedestrian A channel is close to a single-path channel and does, on one hand, preserve almost full own-cell orthogonality, but does not provide much multipath diversity, while the ITU Vehicular A channel gives a significant degree of multipath diversity but the orthogonality is partly lost. The simulation scenario is shown in Figure 11.14. The required transmission power per speech connection ($= I_c$) as compared to the total base station power ($= I_{or}$) is shown on the vertical axis in Figure 11.15. For example, the value of $-20\,\mathrm{dB}$ means that this connection takes $10^{(-20\,\mathrm{dB}/10)} = 1\%$ of the total base station transmission power. The lower the value on the vertical axis, the better the performance. The horizontal axis shows the total transmitted power from this base station divided by the received interference from the other cells, including thermal noise ($= I_{oc}$). This ratio $I_{or}/I_{oc}$ is also known as the geometry factor, $G$. A high value of $G$ is obtained when the mobile is close to the base station, and a low value, typically $-3\,\mathrm{dB}$, at the cell edge.

We can observe some important issues about downlink performance from Figure 11.15. At the cell edge, i.e. for low values of $G$, the multipath diversity in the ITU Vehicular A channel gives a better

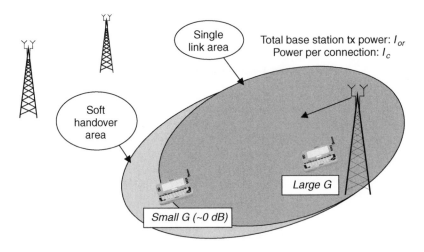

**Figure 11.14**   Simulation scenario for downlink performance evaluation

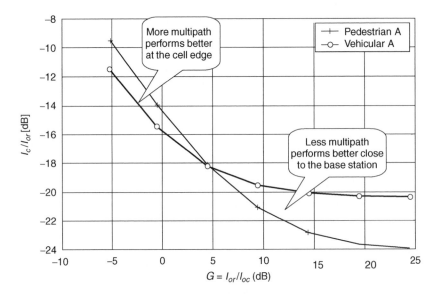

**Figure 11.15**   Effect of multipath propagation

performance compared to less multipath diversity in the ITU Pedestrian A channel. This is because other cell interference dominates over own-cell interference. Close to the base station the performance is better in the ITU Pedestrian A channel because the multipath propagation in the ITU Vehicular A channel reduces the orthogonality of the downlink codes. Furthermore, there is not much need for diversity close to the base station, since the intra-cell interference experiences the same fast fading as the desired user's signal. If signal and interference have the same fading, the signal to interference ratio remains fairly constant despite the fading. The effect of soft handover is not shown in these simulations but it would improve the performance, especially in the ITU Pedestrian A channel at the cell edge by providing extra soft handover diversity – macro diversity. The macro diversity gain is presented in detail in Section 9.3.1.3.

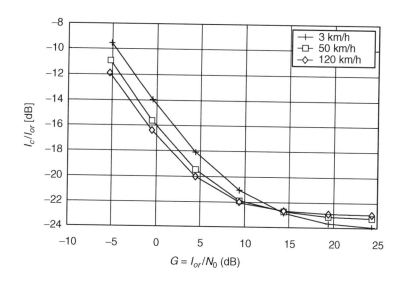

**Figure 11.16**  Effect of mobile speed in the ITU Pedestrian A channel

We note that in the downlink the multipath propagation is not clearly beneficial – it gives diversity gain at the cell edge but at the same time reduces orthogonality close to the Node B. Hence, the multipath propagation does not necessarily improve downlink capacity because of the loss of orthogonality. The loss of the orthogonality in multipath channel could be improved with interference cancellation receivers or equalizers in the mobile. Such receivers are discussed in Chapter 12 for High Speed Downlink Packet Access, HSDPA.

The effect of the mobile speed on downlink performance in the Pedestrian A channel is shown in Figure 11.16. At the cell edge the best performance is obtained for high mobile speeds, while close to the base station, low mobile speeds perform better. This behaviour can be explained by the fact that for high mobile speeds interleaving and channel coding, here convolutional code, provide time diversity and coding gain. In Figure 11.15 it was shown that diversity is important at the cell edge to improve the performance.

### 11.3.1.2  Downlink Capacity in Different Environments

In this section the WCDMA capacity formulas from Section 8.2.2 are used to evaluate the effect of orthogonal codes on the downlink capacity in macro and micro cellular environments. The downlink orthogonal codes make the WCDMA downlink more resistant to intra-cell interference than the uplink direction, and the effect of inter-cell interference from adjacent base stations has a large effect on the downlink capacity. The amount of interference from the adjacent cells depends on the propagation environment and the network planning. Here we assume that the amount of inter-cell interference is lower in micro cells where street corners isolate the cells more strictly than in macro cells. This cell isolation is represented in the formula by the other-to-own cell interference ratio $i$. We also assume that in micro cellular environments there is less multipath propagation, and thus a better orthogonality of the downlink codes. On the other hand, less multipath propagation gives less multipath diversity, and therefore we assume a higher $E_b/N_0$ requirement in the downlink in micro cells than in macro cells.

The assumed loading in uplink is allowed to be 60% and in downlink 80% of WCDMA pole capacity. A lower loading is assumed in uplink than in downlink because the coverage is more challenging in uplink. A higher loading results in smaller coverage, as shown in Section 8.2.2. We assume

**Table 11.8**  Assumptions in the throughput calculations

|                                              | Macro cell | Micro cell |
|----------------------------------------------|------------|------------|
| Downlink orthogonality                       | 0.5        | 0.9        |
| Other-to-own cell interference ratio $I$     | 0.65       | 0.4        |
| Uplink $E_b/N_0$ with 2-branch diversity     | 2.0 dB     | 2.0 dB     |
| Uplink loading                               | 60%        | 60%        |
| Downlink $E_b/N_0$, no transmit diversity    | 5.0 dB     | 6.5 dB     |
| Downlink loading                             | 80%        | 80%        |
| Downlink common channels                     | 15%        | 15%        |
| Block error rate BLER                        | 1%         | 1%         |

**Table 11.9**  Data throughput in macro and micro cell environments per sector per carrier

|          | Macro cell  | Micro cell  |
|----------|-------------|-------------|
| Uplink   | 900 kbps    | 1060 kbps   |
| Downlink | 710 kbps    | 1160 kbps   |

that 15% of the downlink capacity is allocated for downlink common channels, for more information about these channels, see Section 8.2.2. A user bit rate of 64 kbps is assumed in the uplink calculation.

We calculate the example data throughputs in macro and micro cellular environments in both uplink and downlink. The assumptions of the calculations are shown in Table 11.8 and the results in Table 11.9. The capacities in Table 11.9 assume that the users are equally distributed over the cell area and the same bit rate is allocated for all users.

In macro cells the uplink throughput is higher than the downlink throughput, while in micro cells the uplink and downlink throughputs are very similar. We can note that the downlink capacity is more sensitive to the propagation and multipath environment than the uplink capacity. The reason is the application of the orthogonal codes.

The capacity calculations above assume that all cells are fully loaded. If the adjacent cells have lower loading, it is possible to have an even higher cell capacity. The extreme case is an isolated cell without any inter-cell interference. Figure 11.17 shows three different cell capacities with 384 kbps connections.

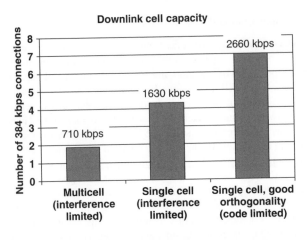

**Figure 11.17**  384 kbps data capacity in multicell and single cell cases

The first is the typical multicell capacity, the second is single cell capacity with orthogonality of 0.5 and the third is single cell capacity with orthogonality close to 1, i.e. single path model. In the third case, the capacity is code limited with a maximum seven simultaneous users of 384 kbps. In the case of favorable orthogonality conditions and low other-cell to own-cell interference ratio, the cell capacity can be clearly higher than in the typical multicell case.

### 11.3.1.3  Number of Orthogonal Codes

The number of downlink orthogonal codes within one scrambling code is limited. With a spreading factor of $SF$, the maximum number of orthogonal codes is $SF$. This code limitation can place an upper limit on the downlink capacity if the propagation environment is favorable and the network planning and hardware support such a high capacity. In this section the achievable downlink capacity with one set of orthogonal codes is estimated. The assumptions in these calculations are shown in Table 11.10 and the results in Table 11.11. Part of the downlink orthogonal codes must be reserved for the common channels and for soft and softer handover overhead. These factors are taken into account in Table 11.10 and Table 11.11. The maximum number of full rate speech channels per sector is 98 with these assumptions, and the maximum data throughput is 2.5 Mbps per sector.

The number of orthogonal codes is not a hard-blocking limitation for the downlink capacity. If this number is not large enough, a second (or more) scrambling code can be taken into use in the downlink, which gives a second set of orthogonal short codes, see Section 6.3. These two sets of orthogonal codes are not orthogonal to each other. Hence, if the second scrambling code is used, the code channels under the second scrambling code cause much more interference to those under the first

**Table 11.10**  Assumptions in the calculation of Table 11.11

| Common channels | 10 codes with SF = 128 |
|---|---|
| Soft handover overhead | 20% |
| Spreading factor (SF) for half rate speech | 256 |
| Spreading factor (SF) for full rate speech | 128 |
| Chip rate | 3.84 Mcps |
| Modulation | QPSK (2 bits per symbol) |
| Average DPCCH overhead for data | 10% |
| Channel coding rate for data | 1/3 with 30% puncturing |

**Table 11.11**  Maximum downlink capacity with one scrambling code per sector

| Speech, full rate (AMR 12.2 kbps and 10.2 kbps) | 128 channels | Number of codes with spreading factor of 128 |
|---|---|---|
| | *(128 − 10)/128 | Common channel overhead |
| | /1.2 | Soft handover overhead |
| | **= 98 channels** | |
| Speech, half rate (AMR ≤ 7.95 kbps) | 2*98 channels | Spreading factor of 256 |
| | **= 196 channels** | |
| Packet data | 3.84e6 | Chip rate |
| | *(128 − 10)/128 | Common channel overhead |
| | /1.2 | Soft handover overhead |
| | *2 | QPSK modulation |
| | *0.9 | DPCCH overhead |
| | /3 | 1/3 rate channel coding |
| | /(1 − 0.3) | 30% puncturing |
| | **= 2.5 Mbps** | |

scrambling code than the other code channels under the first scrambling code. A second scrambling code will be needed with downlink smart antenna solutions, but in this case the scrambling codes can be spatially isolated to reduce the non-orthogonal interference from multiple scrambling codes in a cell, see Figure 11.46.

## 11.3.2   Downlink Transmit Diversity

The downlink capacity could obviously be improved by using receive antenna diversity in the mobile. For small and cheap mobiles it is not, however, feasible to use two antennas and receiver chains. Furthermore, two receiver chains in the mobile will increase power consumption. The WCDMA standard therefore supports the use of Node B transmit diversity. The target of the transmit diversity is to move the complexity of antenna diversity in downlink from the mobile reception to the Node B transmission. The supported downlink transmit diversity modes are described with physical layer procedures in Section 6.6. With transmit diversity, the downlink signal is transmitted via two base station antenna branches. If receive diversity is already deployed in the Node B and we duplex the downlink transmission to the receive antennas, there is no need for extra antennas for downlink diversity. In Figure 11.10 both antennas could be used for reception and for transmission.

In this section we analyse the performance gain from the downlink transmit diversity. The performance gain from transmit diversity can be divided into two parts: (1) coherent combining gain and (2) diversity gain against fast fading. The coherent combining gain can be obtained because the signal is combined coherently, while interference is combined non-coherently. The gain from ideal coherent combining is 3 dB with two antennas. With downlink transmit diversity it is possible to obtain coherent combining in the mobile reception if the phases from the two transmission antennas are adjusted according to the feedback commands (estimated antenna weights) from the mobile in the closed loop transmit diversity. The coherent combining is, however, not perfect because of the discrete values of the antenna weights and delays in the feedback commands. The downlink transmit diversity with feedback is depicted in Figure 11.18.

Both the closed loop and the open loop transmit diversity provide gain against fading because the fast fading is uncorrelated from the two transmit antennas. The gain is larger when there is less multipath diversity. The importance of the diversity is discussed in detail in Section 9.2.1.2. The gains from the downlink transmit diversity are summarized in Table 11.12.

It is important to note the difference between the two sources of diversity in the downlink: multipath and transmit diversity. Multipath diversity reduces the orthogonality of the downlink codes, while transmit diversity keeps the downlink codes orthogonal in flat fading channels. In order to maximize

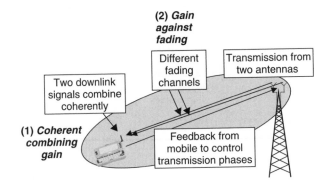

**Figure 11.18**   Downlink transmit diversity with feedback

**Table 11.12**  Comparison of uplink receive and downlink transmit diversity

|                          | Coherent combining gain                                                                                                              | Diversity gain                                      |
| ------------------------ | ------------------------------------------------------------------------------------------------------------------------------------ | --------------------------------------------------- |
| How to obtain gain       | Feedback loop from mobile to base station to control the transmission phases to make received signals to combine coherently in mobile | Uncorrelated fading from the two transmission antennas |
| Non-idealities in obtaining the gain | Discrete steps in feedback loop<br>Delay in feedback loop<br>Multipath propagation                                       | Correlation between transmission antennas           |

the interference-limited downlink capacity, it would be beneficial to avoid multipath propagation to keep the codes orthogonal and to provide the diversity with transmit antenna diversity.

The effect of the downlink transmit diversity gains on downlink capacity and coverage is illustrated in Figure 11.19. The simulation results typically show an average gain of 0.5–1.0 dB in the macro cell environment. A 0.8 dB gain – including coherent combining gain and diversity gain against fading – is assumed here. This gain implies that the average power of each downlink connection can be reduced by 0.8 dB while maintaining the same quality. At the same time the system can support 0.8 dB, i.e. 20% ($=10^{(10.8/10)}$), more users. If we allow, for example, a maximum path loss of 156 dB, the capacity can be increased by 20% from 760 kbps to 910 kbps. The transmit diversity gain can be used alternatively to improve the downlink coverage while keeping the load unchanged. In the example in Figure 11.19, the maximum path loss could be increased by 7 dB, from 156 dB to 163 dB, if the load were kept at 760 kbps. The coverage gain is higher than the capacity gain because of the WCDMA load curve. It may not be possible to utilize the downlink coverage gains and extend the cell size with downlink transmit diversity if the uplink is the limiting direction in coverage. The coverage gain

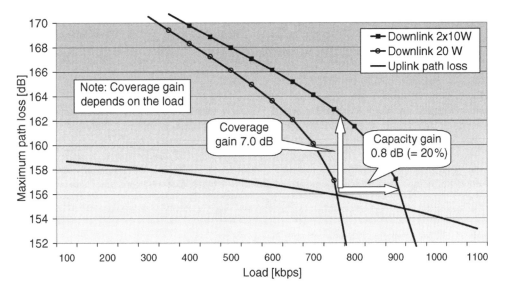

**Figure 11.19**  Downlink capacity and coverage gains with transmit diversity. A 0.8 dB link level gain from transmit diversity is assumed

could be used alternatively to reduce the required base station transmission power. If we keep the load unchanged at 760 kbps and the maximum path loss unchanged at 156 dB, we could reduce the transmission power by 7 dB, from 20 W to 2 × 2.0 W.

Transmit diversity is also supported with Release 5 High-Speed Downlink Packet Access (HSDPA). As HSDPA uses fast scheduling, there is a conflict with benefits from transmit diversity and HSDPA. The fast scheduling with HSDPA benefits from the wider $C/I$ distribution, which is made narrower by the transmit diversity methods. Especially with the open loop transmit diversity, the HSDPA performance can improve in the link level but in the system level there is no clear gain compared to single antenna transmission. With closed loop mode 1 there are some benefits even in the system level due to the feedback.

## 11.3.3 Downlink Voice Capacity

WCDMA voice capacity with AMR voice codec is addressed in this section. Both circuit switched voice and Voice over IP (VoIP) are considered. AMR codec is introduced in Chapter 2. The voice capacity numbers in Chapter 8 refer to the full rate AMR 12.2 kbps. With path loss of 156 dB the maximum number of voice users is 66. The voice capacity can be increased by using a lower bit rate AMR mode. We estimate the capacity of the lower AMR modes with the following equation

$$\text{Voice capacity} = 66 \text{ users} \cdot \frac{12.2 \text{ kbps}}{\text{AMR bit rate[kbps]}} \cdot 10^{\left(E_b/N_{0_{12.2 \text{ kbps}}} - E_b/N_{0_{\text{AMR}}}\right)}/10 \qquad (11.4)$$

where the capacity increases according to the reduction of the bit rate. Additionally, we take into account that the physical layer overhead increases for the lower AMR rates and that is modelled here as a higher $E_b/N_0$. Table 11.13 shows the voice capacity with three AMR modes: 12.2 kbps, 7.95 kbps and 4.75 kbps. The voice capacity approximately doubles when the bit rate is reduced from 12.2 kbps to 4.75 kbps. The AMR bit rate can be controlled by the operator and it allows a trade-off to be made between voice capacity and voice quality. The AMR bit rate can be adjusted dynamically according to the instantaneous network load. The AMR voice capacity can be further increased by 15–20% by using AMR source adaptation, for details, see Chapter 2.

The voice-over IP (VoIP) uses the same dedicated channels in the air interface as the circuit switched voice. The flexibility of the WCDMA air interface allows the introduction of the VoIP service without any modifications to the physical layer standard. The VoIP call from the packet core network includes IP headers that are considerably large compared to the voice payload. In order to save air interface resources, the IP headers are compressed by Packet Data Convergence Protocol, PDCP, in RNC, which is part of the 3GPP Release 4 standard. For more details, see Chapter 7. The compressed IP headers are delivered over the WCDMA air interface. The scenario is shown in Figure 11.20.

The VoIP service will affect the WCDMA voice capacity because of increased overhead, even if IP headers are compressed. That overhead includes compressed IP headers, RLC headers, real-time protocol (RTP) payload headers and real-time control protocol (RTCP). We assume that the average overhead of compressed IP header and other headers is 7 bytes. The 12.2 kbps voice carries 244 bits

**Table 11.13** Voice capacity with different AMR modes

|              | AMR 12.2 kbps | AMR 7.95 kbps | AMR 4.75 kbps |
|--------------|---------------|---------------|---------------|
| $E_b/N_0$    | 7.0 dB        | 7.5 dB        | 8.0 dB        |
| Capacity     | 66 users      | 90 users      | 134 users     |

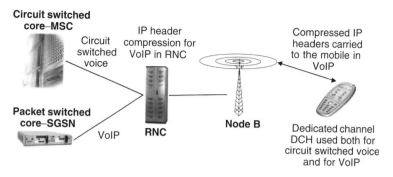

**Figure 11.20** Circuit switched voice and voice over IP with WCDMA

**Table 11.14** Typical circuit switched voice and voice over IP capacity

|  | AMR 12.2 kbps | AMR 7.95 kbps | AMR 4.75 kbps |
| --- | --- | --- | --- |
| Circuit switched voice | 66 users | 90 users | 134 users |
| Voice over IP | 54 users | 70 users | 95 users |

per 20 ms and the overhead can be calculated as

$$10\log_{10}\left(\frac{\text{payload} + \text{IP header}}{\text{payload}}\right) = 10\log_{10}\left(\frac{244 + 7.8}{244}\right) = 0.9\,\text{dB} \qquad (11.5)$$

The overhead of 0.9 dB reduces air interface capacity $1 - 10^{(-0.9/10)} = 19\%$. The typical VoIP capacities are shown in Table 11.14 for three AMR modes. The small loss in air interface capacity with VoIP is compensated by the flexibility of end-to-end IP traffic in rich calls.

## 11.4 Capacity Trials

### 11.4.1 Single Cell Capacity Trials

This section presents the measurement methods and results from capacity trials. The tests are done with a single sector without interference from adjacent sectors or sites. The test environment has usually been the typical suburban case where several test mobiles are located in one or more fixed and stationary locations. The mobiles are located without line-of-sight connection but the coverage is relatively good, i.e. the thermal noise component in the total amount of noise is minimized. In terms of CPICH RSCP good coverage means better than −90 dBm conditions. Also, the base station should be configured so that the hardware and transmission resources are not becoming the limiting factor, however, that is not always possible and therefore some of the pole capacity results shown in this section are extrapolated values.

#### 11.4.1.1 AMR Voice Capacity Uplink

The AMR voice capacity test is carried out with a constant bit rate of 12.2 kbps and has additional conditions such as 100% voice activity factor, 0.8% BLER target and AMR unequal error protection. The received total base station power and the number of simultaneous UEs can be plotted as shown in the example in Figure 11.21. The received power level of approximately $-102$ dBm without any users is the thermal noise level.

From Figure 11.21 the average fractional load per number of connected UEs can be calculated based on the equation:

$$\text{Noise rise} = \frac{P_{\text{rxtotal}}}{P_{\text{noise}}} = \frac{1}{1 - \eta_{\text{UL}}} \tag{11.6}$$

where $\varepsilon_{\text{UL}}$ is the fractional load, $P_{\text{rxtotal}}$ is the total received power, including noise and other cell and own cell users. The noise rise as a function of the AMR users is plotted in Figure 11.22 (a) and the fractional load in Figure 11.22 (b).

The best linear fit, fractional load as a function of number of users, can then be derived as depicted in Figure 11.22, where $y$ is the fractional load $\varepsilon$, $x$ is the number of connected UEs and $A$ is the linear fit slope which is equivalent to the fractional load of a single user. The fractional load can also be expressed according to Equation (8.14) from Chapter 8:

$$\eta_{\text{UL}} = \frac{E_b/N_0}{W/R} \cdot N \cdot \upsilon \cdot (1 + i) \tag{11.7}$$

where $N$ is the number of users, $i$ is the other-to-own cell interference, $W$ is the chip rate, $E_b/N_0$ is the uplink $E_b/N_0$ requirement for a user, $R$ is the bit rate of a user and $\upsilon$ is the voice activity. As all the users are using the same service 12.2 kbps AMR, the other-to-own cell interference is 0 in the single cell scenario and the voice activity is 100%, the equation can be simplified as:

$$\eta_{\text{UL}} = \frac{E_b/N_0}{W/R} \cdot N \tag{11.8}$$

Using the best linear fit slope $A$ from Figure 11.22, the average $E_b/N_0$ can be derived as:

$$A = \frac{E_b/N_0}{W/R} \Rightarrow \frac{E_b}{N_0} = A \frac{W}{R} \tag{11.9}$$

The pole capacity of the service can then be calculated assuming 100% fractional load. The average achieved results for AMR speech capacity testing are presented in Section 11.4.3.

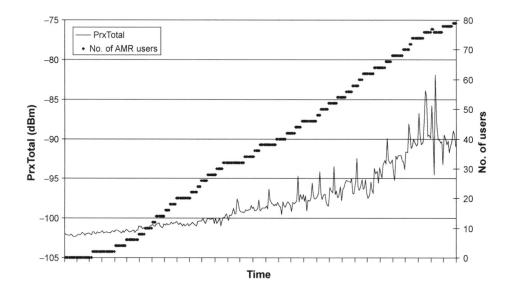

**Figure 11.21**  Uplink received power (Prx) and the number of simultaneous AMR users

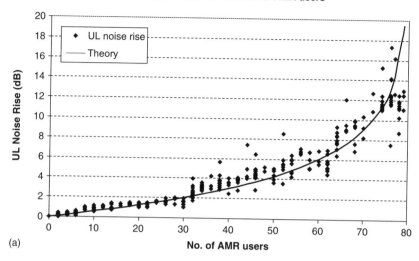

(a)

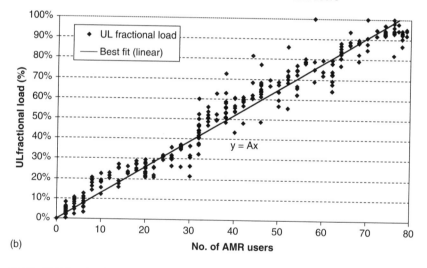

(b)

**Figure 11.22** Uplink noise rise (a) and uplink fractional load (b) as a function of the number of AMR users

### 11.4.1.2 AMR Voice Capacity Downlink

The downlink capacity test has the same assumptions as the uplink test. The orthogonality is assumed to be 0.5. The total transmitted base station power and the number of UEs is presented in Figure 11.23. The transmission power of 35.5 dBm = 3.2 W is caused by the common channel powers.

The downlink analysis starts from the common downlink equation for the connection $E_b/N_0$ requirement as presented in [11].

$$\left(\frac{E_b}{N_0}\right)_i = \frac{W}{R} \cdot \frac{\dfrac{P_i}{L_{m,i}}}{\dfrac{P_{\text{tot},m}}{L_{m,j}}(1-\alpha_i) + \displaystyle\sum_{n=1,n\neq m}^{N} \dfrac{P_{\text{tot},n}}{L_{n,i}} + P_N} \tag{11.10}$$

where $(E_b/N_0)_i$ is the downlink $E_b/N_0$ requirement, $P_i$ is the required transmit power at base station $m$ for the connection $i$, $N$ is the number of interfering base stations, $P_N$ is the noise power at the UE receiver, $R_i$ is the bit rate for the UE $i$, $W$ is the chip rate, $P_{tot,m}$ is the required total transmit power of the serving base station $m$, $P_{tot,n}$ is the required total transmit power of another neighboring base station $n$, $L_{m,i}$ is the path loss from the serving base station $m$ to UE $i$, $L_{n,i}$ is the path loss from another base station $n$ to UE $i$ and $\alpha_i$ is the orthogonality factor for UE $i$. The required transmitted power for connection $i$, $P_i$ can be solved as follows:

$$P_i = \frac{\left(\dfrac{E_b}{N_0}\right)_i \cdot R}{W} \left[ (1 - \alpha_i) \cdot P_{tot,m} + \sum_{n=1,n \neq m}^{N} L_{m,i} L_{n,i} \cdot P_{tot,n} + P_N \cdot L_{m,i} \right] \tag{11.11}$$

In these single cell tests, the other-to-own cell interference can be set to be zero. Also, as the testing is done in good RSCP conditions, the noise power can be assumed to be negligible (i.e. $P_{tot,n}/L_{m,i} \gg P_N$). The two assumptions can be used to simplify the equation to the following format:

$$P_i = \frac{\left(\dfrac{E_b}{N_0}\right)_i \cdot R}{W} \cdot (1 - \alpha_i) \cdot P_{tot,m} \tag{11.12}$$

Then the constant common channel power is added as well as the activity factor and both sides are summed over all the connections under the cell. This gives the total base station downlink transmitted power as:

$$\sum_{i=1}^{I} P_i + P_{CCH} = P_{tot,m} \cdot \sum_{i=1}^{I} \frac{\upsilon \cdot \left(\dfrac{E_b}{N_0}\right)_i \cdot R}{W} \cdot (1 - \alpha_i) + P_{CCH} \tag{11.13}$$

where I is the number of connections in the cell. The power rise over the common channels can be solved as follows

$$P_{tot,m} P_{CCH} = 11 - \sum_{i=1}^{I} \frac{\upsilon \cdot \left(\dfrac{E_b}{N_0}\right)_i \cdot R}{W} \cdot (1 - \alpha_i) \tag{11.14}$$

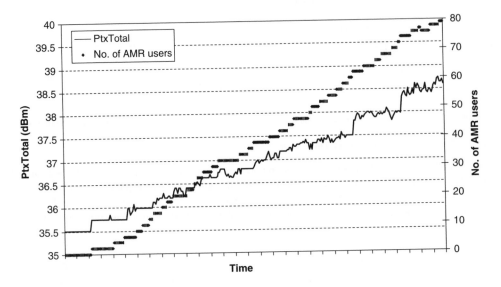

**Figure 11.23**  Downlink total transmitted power (Ptx) as a function of the number of AMR users

Using the downlink load factor, $\varepsilon_{DL}$ we can write

$$\text{Power rise} = \frac{1}{1 - \eta_{DL}} \tag{11.15}$$

The power rise above the common channel power and the fractional load as a function of the number of AMR users can be plotted as in the example charts in Figure 11.24.

The downlink $E_b/N_0$ can be defined from the downlink fractional load by using the best linear fit as:

$$\eta_{DL} = \frac{\left(\dfrac{E_b}{N_0}\right)_i \cdot R}{W} \cdot (1 - \alpha_i) \cdot I = B \cdot I$$

$$B = \frac{\left(\dfrac{E_b}{N_0}\right)_i \cdot R}{W} \cdot (1 - \alpha_i) \tag{11.16}$$

$$\left(\frac{E_b}{N_0}\right)_i = \frac{B \cdot W}{R \cdot (1 - \alpha_i)}$$

The resulting average downlink $E_b/N_0$ values and the calculated pole capacity are shown in Section 11.4.3. The downlink code power can be plotted as well to see the power deviation per connected UE. One example is shown in Figure 11.25.

There is, on average, 6 dB fluctuation on the power per connection. Also, it seems that the fluctuation is increasing as the load increases. The average powers per connection during high load are shown in Figure 11.26. The average downlink transmitted code power per connection varies between 17 dBm and 24 dBm in this example, showing the variation for UEs in different locations experiencing different path losses and multipath conditions as well as different UE models. These average powers and the power fluctuations need to be considered in the network planning when setting the maximum allowed powers per connection.

In poor coverage conditions, the downlink calculation formula shown earlier in this section does not apply any more because the noise power cannot be assumed to be negligible, i.e. $P_{tot,n}/L_{m,i} \gg P_N$ does not apply. In this case, the downlink $E_b/N_0$ can be calculated, based on the average base station transmitted code power, base station total transmitted power and received CPICH $E_c/N_0$, using the definition of geometry factor G:

$$G = \frac{P_i}{L_{m,i}} \Big/ (P_{\text{other}} + P_N) \tag{11.17}$$

Equation (11.10) can now be written

$$\frac{W}{R} \cdot \frac{P_i}{P_{\text{tot},m}} \cdot \frac{1}{(1 - \alpha_i) + \dfrac{1}{G}} = \left(\frac{E_b}{N_0}\right)_i, \quad i = 1, \ldots I \tag{11.18}$$

On the other hand, the geometry factor can be defined using CPICH $E_c/N_0$ as:

$$G = \frac{1}{\dfrac{P_{tx,\text{CPICH}}}{\dfrac{P_{\text{tot},m}}{E_{c,\text{CPICH}}} - 1}} \tag{11.19}$$

This way the noise power can be included in the calculations as well. In the uplink direction, the tests performed in poor coverage conditions tend to show better results than tests performed in good coverage conditions. This is due to the UE in poor coverage conditions not having as many transmitted

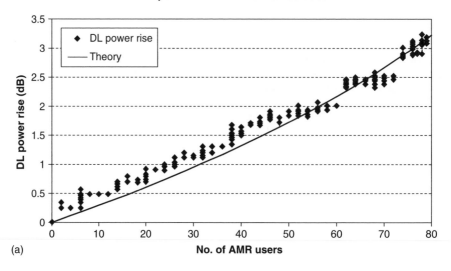

(a)

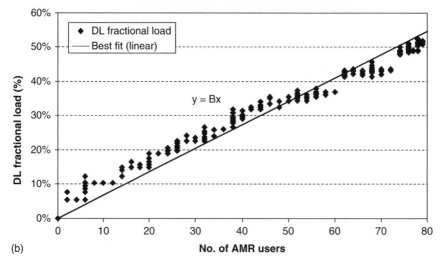

(b)

**Figure 11.24** Downlink power rise (a) and downlink fractional load (b) as a function of the number of AMR users

power resources, therefore, possible power increases due to interference are lower. In the downlink direction, the tests performed in poor coverage conditions usually lead to worse results due to the average transmission power requirement being higher, and the power deviation is higher for the UEs at the coverage border than for the UEs in good coverage conditions. This is shown as a higher $E_b/N_0$ requirement in poor coverage than in good coverage conditions.

As the last topic, we take a look at the CPICH $E_c/N_0$ values as a function of downlink load. Figure 11.27 shows CPICH $E_c/N_0$ as a function of the number of AMR users, and Figure 11.28 as

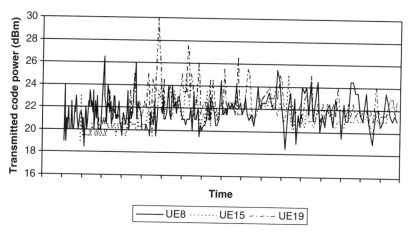

**Figure 11.25** Downlink transmitted code power per AMR connection when load is increased

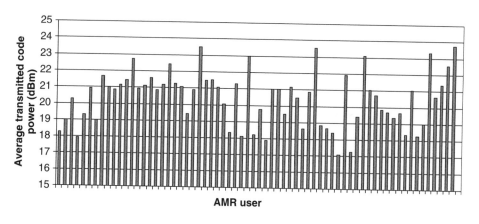

**Figure 11.26** Downlink average transmitted code power per AMR connection

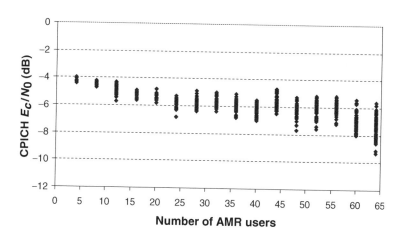

**Figure 11.27** CPICH $E_c/I_0$ as a function of the number of AMR users

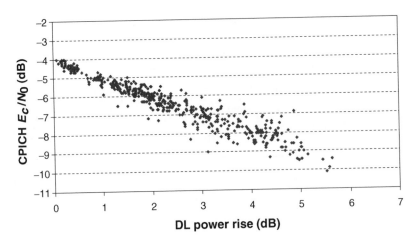

**Figure 11.28** CPICH $E_c/I_0$ as a function of downlink power rise

a function of power rise. When the load of the cell increases, the received CPICH $E_c/N_0$ decreases and deviation becomes larger. As the power rise increases by 5 dB, the CPICH $E_c/N_0$ is decreased correspondingly by 5 dB on average. This scenario leads to the topic of network optimization in high load condition, which is recommended for further study.

### 11.4.1.3 Circuit Switched Video Capacity Uplink

The circuit switched video call capacity test is done with a constant bit rate of 64 kbps and has additional conditions such as: 100% activity factor and 0.5% BLER target. The total received power measurements per amount of connected UEs can be plotted as shown in the example in Figure 11.29.

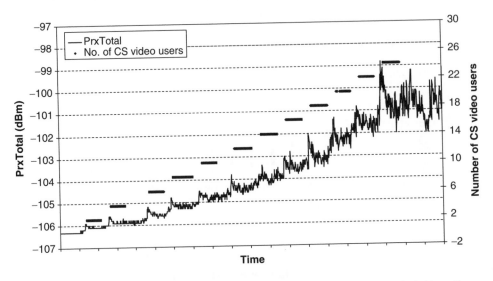

**Figure 11.29** Uplink total received power (Prx) as a function of the number of circuit switched video call users

In a similar way to the AMR speech case, the average fractional load per number of connected UEs can be calculated in good coverage conditions, i.e. assuming $P_{tot,n}/L_{m,i} \gg P_N$, and plotted as in Figure 11.30. The best linear fit, fractional load as a function of number of connections, can then be derived as depicted in Figure 11.30, where $y$ is the fractional load $\varepsilon$, and $x$ is the number of connections. The average achieved results for circuit switched video call capacity testing are presented in Section 11.4.3.

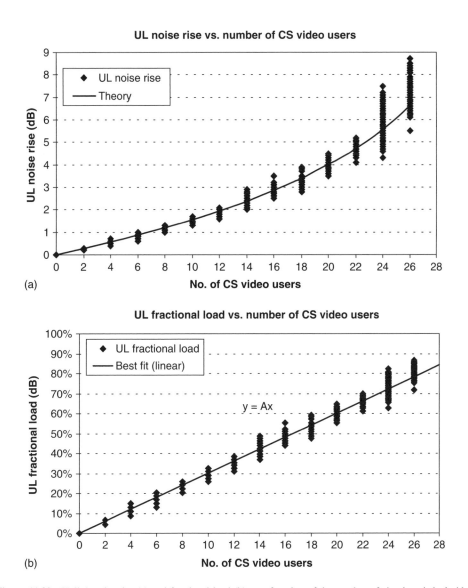

**Figure 11.30** Uplink noise rise (a) and fractional load (b) as a function of the number of circuit switched video call users

#### 11.4.1.4    Circuit Switched Video Capacity Downlink

The downlink capacity test has the same assumptions as the uplink test. It is further assumed that the orthogonality is 0.5. An example of the total transmitted power and the number of connected UEs in the tested cell is shown in Figure 11.31.

Using the same calculation method as for AMR voice, the fractional load can be plotted as depicted in Figure 11.32.

The resulting average downlink $E_b/N_0$ example values are shown in Section 11.4.3. The downlink $E_b/N_0$ for a circuit switched video call is usually lower than for an AMR speech connection in the same conditions. This is due to a higher bit rate having more efficient turbo coding. Assuming 1 dB lower $E_b/N_0$ for circuit switched video, and taking into account the difference in the processing gains between 64 kbps and 12.2 kbps, the expected difference in the downlink connection powers is approximately 6 dB. Example measured downlink code powers are shown in Figure 11.33 and the averaged power in Figure 11.34. The maximum downlink code power fluctuation can be seen to be, on average, 10 to 12 dB, which is higher than for an AMR speech call. This is due to the lower BLER requirement for a circuit switched video call and higher required average transmission power from the base station compared to the AMR speech call. The average downlink power per connection in Figure 11.34 varies between 21 dBm and 30 dBm in this example, showing the variation for UEs in different locations as well as different UE models. The average power difference between a circuit switched video call and an AMR speech call is approx 6–7 dB when comparing Figure 11.26 for AMR voice and Figure 11.34 for circuit switched video. That difference is an expected result based on the processing gain difference.

#### 11.4.1.5    Packet Data Capacity Downlink

The downlink packet data capacity test is done for 384 kbps bit rate and the same analysis methodology is used as for AMR voice and circuit switched video. The BLER target is set to 1%. The application

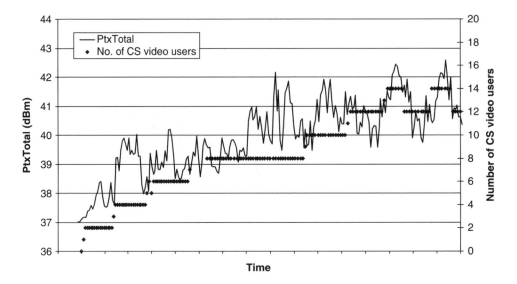

**Figure 11.31**    Downlink total transmitted power (Ptx) as a function of the number of circuit switched video call users

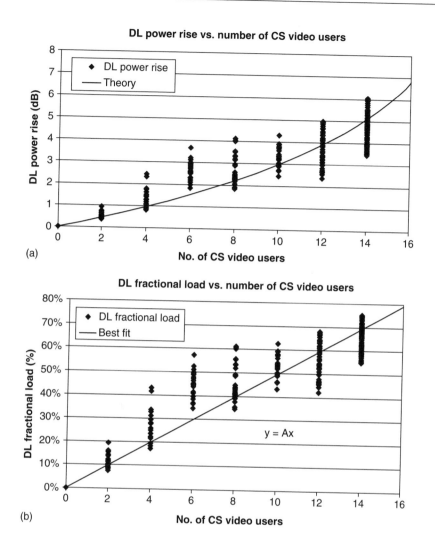

**Figure 11.32** Downlink power rise (a) and downlink fractional load (b) as a function of the number of circuit switched video call users

used in the testing is FTP download, and for every connection the FTP throughput is recorded. The total transmitted power and the number of connected UEs in an example case are shown in Figure 11.35.

The power rise above the common channel powers and the fractional load can be plotted as in Figure 11.36.

The resulting average downlink $E_b/N_0$ example values are shown in Section 11.4.3. The downlink $E_b/N_0$ for a packet switched 384 kbps call is lower than for an AMR speech or for circuit switched video call. This is due to a higher bit rate service having more data to transmit, therefore the Turbo coding and interleaving become more efficient.

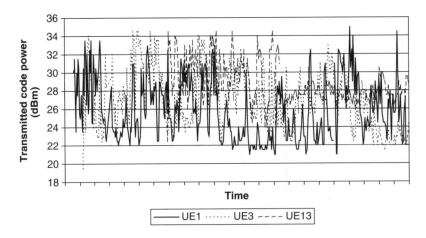

**Figure 11.33**  Downlink transmitted code power per circuit switched video call connection

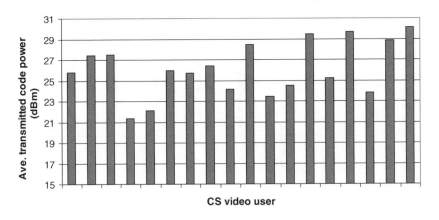

**Figure 11.34**  Downlink average power per circuit switched video call connection

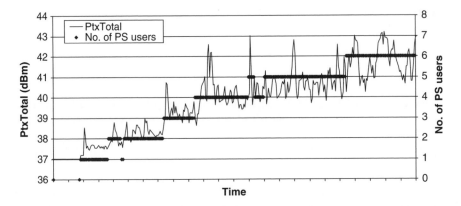

**Figure 11.35**  Downlink total transmitted power (Ptx) as a function of the number of packet switched 384 kbps users

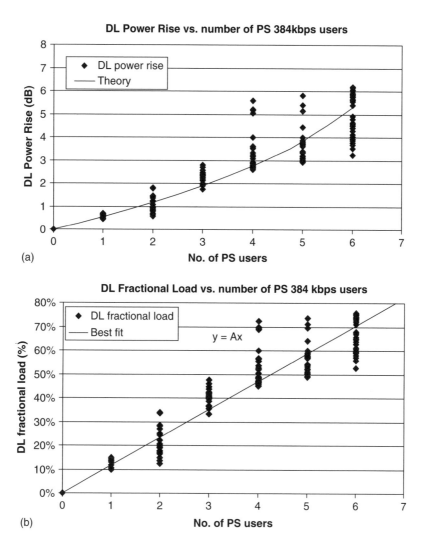

**Figure 11.36** Downlink power rise (a) and downlink fractional load (b) as a function of the number of packet switched 384 kbps users

The downlink code power can be plotted as well to see the power deviation per connected UE. In Figure 11.37 one example is shown. The maximum downlink code power fluctuation can be seen to be in the region of 8 dB, which is slightly higher than for AMR and close that for a circuit switched video call. However, it should be noted that the transmission power in these cases is close to its maximum value, reducing the power variance. This can be seen in the case of UE2 in Figure 11.37. The maximum power per connection was set to 39 dBm.

The average downlink code power per connection is depicted in Figure 11.38, and it can be seen that, also, the average power difference between a packet switched 384 kbps packet call and an AMR speech call is about 12–13 dB, which is an expected result taking into account the difference in the processing gain of 15 dB and the lower $E_b/N_0$ for packet data.

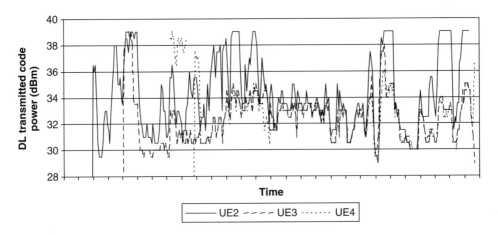

**Figure 11.37** Downlink transmitted code power per packet switched 384 kbps data call connection

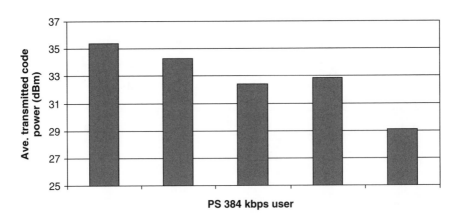

**Figure 11.38** Downlink average transmitted code power per packet switched 384 kbps data call connection

The throughput of each 384 kbps connection is also recorded in the trials. Examples of FTP download throughput are shown in Figure 11.39. The throughput is sampled every second and captured by the application PC connected to each UE. Occasional fluctuation in throughput can be expected in the plots due to the very short sampling period. It can be seen that the throughput performance varies from one UE to another, depending on the location of the UE. Degradation is expected for UE requiring higher downlink transmitted code power under bad radio or high load conditions due to the fact that the maximum downlink transmitted code power is sometimes achieved. The upper part of Figure 11.39 shows some drops in UE2 throughput due to power limitations, as was seen in Figure 11.37. In other cases, little or no impact is seen in the throughput when the downlink transmitted code power is within the downlink power control range.

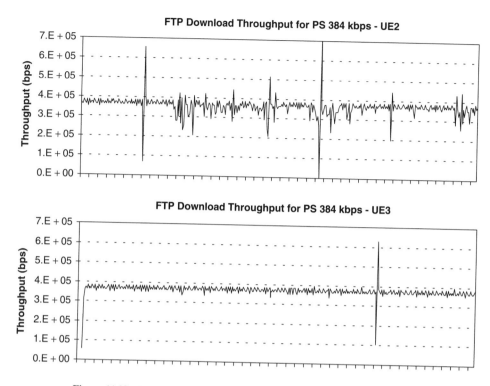

**Figure 11.39** Examples of FTP download throughput during the capacity trial

## 11.4.2   Multicell Capacity Trials

The multicell capacity testing is similar to the single cell capacity testing described in the previous section. The difference is that the other-to-own cell interference $i$ has significant impact and it needs to be included in the calculations.

### 11.4.2.1   Uplink Methodology

The average $E_b/N_0$ can be estimated first by having only one UE driving inside the coverage area. In this way we can define the average $E_b/N_0$ in moving conditions over the whole coverage area. The effect of other-to-own cell interference can be included using Equation (11.7), where the other-to-own cell interference can be determined based on downlink testing, as specified in the following section.

### 11.4.2.2   Downlink Methodology

The downlink calculations start with Equation (11.10). The equation can be simplified by assuming that the total transmitted power of all base stations in the area is the same. This assumption gives slightly pessimistic results, as testing is done having several UEs in the measurement van and the neighbouring base station average total transmitted power is always smaller than the average total transmitted power of the serving base station. Equation (11.10) can be modified by adding the activity factor as well as the common channel power. The total transmitted power, $P_{\text{tot},m}$, can be solved by summing over all the connections $i$ as specified below:

$$\sum_{i=1}^{I} P_i + P_{\text{CCH}} = P_{\text{tot},m} \cdot \sum_{i=1}^{I} \frac{\left(\dfrac{E_b}{N_0}\right)_i \cdot v_i}{W/R} \left[ (1 - \alpha_i) + \sum_{n=1,n\neq m}^{N} \frac{L_{m,i}}{L_{n,i}} \right] + P_N \cdot \sum_{i=1}^{I} \frac{\left(\dfrac{E_b}{N_0}\right)_i \cdot v_i}{W/R}$$

(11.20)

The path loss $L_{x,y}$ in the above equation stands for the air interface path loss, i.e. excluding antenna gain, cable losses and soft handover gain. $P_{\text{tot},m}$ can be solved as:

$$P_{\text{tot},m} = \frac{P_{\text{CCH}} + P_N \cdot \displaystyle\sum_{i=0}^{I} \frac{\left(\dfrac{E_b}{N_0}\right)_i \cdot v_i}{W/R} L_{m,i}}{1 - \displaystyle\sum_{i=1}^{I} \frac{\left(\dfrac{E_b}{N_0}\right)_i \cdot v_i}{W/R} \left[ (1 - \alpha_i) + \displaystyle\sum_{n=1,n\neq m}^{N} \frac{L_{m,i}}{L_{n,i}} \right]}$$

(11.21)

The downlink load factor, $\varepsilon_{\text{DL}}$, can be formulated as the increase in the required base station power in order to maintain the required $I$ connections, taking into account the non-orthogonal inter-cell interference and the non-orthogonal part of the intra-cell interference.

$$\eta_{\text{DL}} = \sum_{i=1}^{I} \frac{\left(\dfrac{E_b}{N_0}\right)_i \cdot v_i}{W/R} \left[ (1 - \alpha_i) + \sum_{n=1,n\neq m}^{N} \frac{L_{m,i}}{L_{n,i}} \right]$$

(11.22)

The average other-to-own cell interference can be calculated in downlink as:

$$\bar{i}_{\text{DL}} = \frac{\displaystyle\sum_{i=1}^{I} \sum_{n=1,n\neq m}^{N} \frac{L_{m,i}}{L_{n,i}}}{I}$$

(11.23)

The path loss can be calculated from the RSCP measurements when the CPICH transmission power is known. The average other-to-own cell interference ratio can be calculated by averaging the values over all measurement points. This other-to-own cell interference can be used also to calculate the uplink load, assuming that the average other-to-own cell interference per cell is the same in uplink and in downlink. This assumption is correct as long as the traffic is uniformly distributed between the cells.

The average $E_b/N_0$ in downlink can be calculated in the same way as described for single cell test cases: by plotting the downlink fractional load as a function of the number of connected UEs and defining the best fit. It should be noted that the $E_b/N_0$ calculated this way also includes the macro diversity gain as the load measurement from the base station cell includes all the served UEs, i.e. UEs in soft handover and UEs not in soft handover.

## 11.4.3 Summary

The results from the capacity trials are summarized in this section. The results show limiting factors in maximum capacity, uplink and downlink pole capacity, uplink and downlink $E_b/N_0$ and the average downlink connection powers. The single cell test results are shown in Table 11.15. This summary includes results also from other single cell tests besides the ones introduced earlier in this chapter.

We can note that for an AMR 12.2 kbps voice service, the pole capacity is always limited by the uplink noise rise. For a 64 kbps circuit switched video service, the pole capacity is usually limited by

**Table 11.15**  Summary of single cell capacity trials

| | AMR 12.2 kbps voice with 100% voice activity | 64 kbps video | 384 kbps packet data |
|---|---|---|---|
| Limiting factor | Uplink noise rise | Total downlink power usually or uplink noise rise | Total downlink power |
| Uplink pole capacity | 80–112 users | 27–38 users | – |
| Downlink pole capacity | 93–125 users | 15–33 users | 5.8–8.9 users (code limit 7 users) |
| Uplink $E_b/N_0$ | 4.5 dB–5.9 dB | 2.1 dB–3.5 dB | – |
| Downlink $E_b/N_0$ | 6.1 dB–7.4 dB | 4.6 dB–7.9 dB | 2.6 dB–4.4 dB |

**Table 11.16**  Comparison of the average downlink transmitted code power for different services

| | AMR 12.2 kbps voice with 100% voice activity | 64 kbps video | 384 kbps packet data |
|---|---|---|---|
| Average downlink code power | 17 dBm – 23 dBm | 21 dBm – 30 dBm | 29 dBm – 35 dBm |
| Relative power compared to AMR voice | Reference | 6 dB – 7 dB higher than AMR | 12 dB – 13 dB higher than AMR |

the maximum downlink power of the cell. However, under good radio conditions, i.e. good coverage with small path loss from the cell, the uplink noise rise could be the limiting factor due to low downlink $E_b/N_0$, as well as the expected low downlink transmitted code power per connection. For a 384 kbps packet switched data service, the pole capacity is always limited by the maximum downlink power of the cell, as the data rate is asymmetrical. When comparing the $E_b/N_0$ values from the field trials to the expected values in Section 11.5, we can note that the values are quite well in line.

The expected downlink connection powers for each service are summarized in Table 11.16. The results are shown as the relative power compared to AMR voice and as absolute powers. The connection powers provide some indications for designing the downlink power control range for different services.

The single cell results are applied to estimate the maximum multicell capacities. The following assumptions are used: voice activity of 50%, downlink orthogonality 0.5 and other-cell to own-cell interference ratio of 0.65. The voice activity of 50% is assumed to lead to physical layer activity factor of 67% in uplink and 58% in downlink. The results in Table 11.17 show that the limiting factor

**Table 11.17**  Estimated multicell pole capacities based on single cell measurements

| | AMR 12.2 kbps voice with 50% voice activity | 64 kbps video | 384 kbps packet data |
|---|---|---|---|
| Limiting factor | Total downlink power | Total downlink power | Total downlink power |
| Uplink pole capacity | 72–101 users | 16–23 users | – |
| Downlink pole capacity | 70–94 users | 7–14 users | 2.5–3.9 users |

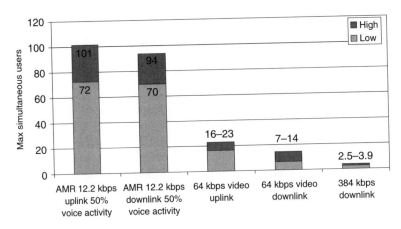

**Figure 11.40**  Estimated multicell pole capacities based on single cell measurements

in the multicell case typically is the total downlink power. The uplink may become a limiting factor for symmetric services if we want to limit the uplink noise rise to a low value to provide maximized coverage. The estimated multicell capacities are summarized in Figure 11.40.

## 11.5   3GPP Performance Requirements

In this section we derive typical $E_b/N_0$ values and receiver noise figures from 3GPP performance requirements [12, 13]. A simplified view of the receiver parts is shown in Figure 11.41. The assumed services are 12.2 kbps AMR voice and packet data 64, 128 and 384 kbps with BLER of 1%.

### 11.5.1   $E_b/N_0$ Performance

The uplink $E_b/N_0$ values are derived as follows:

1. Take the $E_b/N_0$ requirement for static and multipath channels from [12].
2. [12] includes higher layer signaling channel with 100 bits per 40 ms. We can remove that overhead since the signaling channel is active only when there is a signaling need. The overhead is 10 · log 10(user bits per 40 ms + 100)/(user bits per 40 ms)).
3. We assume that the average base station product performs 1.5 dB better than 3GPP requirement.
4. We assume that outer loop power control causes a loss of 0.3 dB, approximately half of a typical power control step size of 0.5 dB.

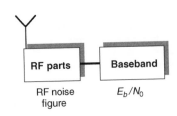

**Figure 11.41**  Simplified view of the receiver and the performance requirements

5. We assume that fast power control degrades performance by 0.3 dB compared to the ideal non-power control case at 120 km/h.
6. We assume that the required $E_b/N_0$ is 1.0 dB lower at 3 km/h than at 120 km/h.

The calculations of the uplink $E_b/N_0$ values are shown in Table 11.18 for a static channel and in Table 11.19 for a multipath fading channel.

The downlink $E_b/N_0$ values are derived as follows:

1. Take the $E_c/I_{or}$ requirement from [13]. Take also the $I_{or}/I_{oc}$ operation point from [13].
2. Calculate $E_b/N_0$ with the following formula

$$\frac{E_b}{N_0} = 10\log_{10}\left(\frac{\dfrac{\text{Chip rate}}{\text{Bit rate}} \cdot \dfrac{E_c}{I_{or}}}{1 - \text{orthogonality} + \dfrac{1}{I_{or}/I_{oc}}}\right) \tag{11.24}$$

We assume an orthogonality of 1.0 for a static channel and 0.5 for a multipath channel.

1. We remove the effect of the signaling channel as in uplink.
2. A slot format of 11 is assumed for voice in [13]. The physical layer overhead can be decreased by 0.8 dB by using a slot format of 8.
3. We assume that the average base station product performs 1.5 dB better than the 3GPP requirement.
4. We assume that outer loop power control causes a loss of 0.3 dB, approximately half of a typical power control step size of 0.5 dB.
5. We assume that fast power control gives the same performance as the ideal non-power control case at 120 km/h.
6. We assume that the required $E_b/N_0$ is 0.3 dB higher at 3 km/h than at 120 km/h because of the power rise.

**Table 11.18** Uplink static channel performance

|  | Voice | Data 64 kbps | Data 128 kbps | Data 384 kbps |
|---|---|---|---|---|
| $E_b/N_0$ [12] | 5.1 dB | 1.7 dB | 0.9 dB | 1.0 dB |
| DCCH overhead | −0.8 dB | −0.2 dB | −0.1 dB | −0.03 dB |
| Product vs. 3GPP | −1.5 dB | −1.5 dB | −1.5 dB | −1.5 dB |
| Outer loop power control | +0.3 dB | +0.3 dB | +0.3 dB | +0.3 dB |
| $E_b/N_0$ | 3.1 dB | 0.3 dB | −0.4 dB | −0.2 dB |

**Table 11.19** Uplink multipath fading channel (Case 3) performance

|  | Voice | Data 64 kbps | Data 128 kbps | Data 384 kbps |
|---|---|---|---|---|
| $E_b/N_0$ [12] | −7.2 dB | 3.8 dB | 3.2 dB | 3.6 dB |
| DCCH overhead | −0.8 dB | −0.2 dB | −0.1 dB | −0.03 dB |
| Product vs. 3GPP | −1.5 dB | −1.5 dB | −1.5 dB | −1.5 dB |
| Outer loop power control | +0.3 dB | +0.3 dB | +0.3 dB | +0.3 dB |
| Fast power control | +0.3 dB | +0.3 dB | +0.3 dB | +0.3 dB |
| $E_b/N_0$ 120 km/h | 5.5 dB | 3.0 dB | 2.5 dB | −3.0 dB |
| 3 km/h vs 120 km/h | −1.0 dB | −1.0 dB | −1.0 dB | −1.0 dB |
| $E_b/N_0$ 3 km/h | 4.5 dB | 2.0 dB | 1.5 dB | 2.0 dB |

**Table 11.20**  Downlink static channel performance

|  | Voice | Data 64 kbps | Data 128 kbps | Data 384 kbps |
|---|---|---|---|---|
| $E_c/I_{or}$ from [13] | −16.6 dB | −12.8 dB | −9.8 dB | −5.5 dB |
| $I_{or}/I_{oc}$ | −1.0 dB | −1.0 dB | −1.0 dB | −1.0 dB |
| $E_b/N_0$ from Eq. (11.24) | 7.4 dB | 3.7 dB | 3.4 dB | 3.4 dB |
| DCCH overhead | −0.8 dB | −0.2 dB | −0.1 dB | −0.03 dB |
| Slot format 8 | −0.8 dB | − | − | − |
| Product vs. 3GPP | −1.5 dB | −1.5 dB | −1.5 dB | −1.5 dB |
| Outer loop power control | +0.3 dB | +0.3 dB | +0.3 dB | +0.3 dB |
| $E_b/N_0$ | 4.6 dB | 2.6 dB | 2.7 dB | 2.3 dB |

**Table 11.21**  Downlink multipath fading channel (Case 3) performance

|  | Voice | Data 64 kbps | Data 128 kbps | Data 384 kbps |
|---|---|---|---|---|
| $E_c/I_{or}$ from [13] | −11.8 dB | −7.4 dB | −8.5 dB | −5.1 dB |
| $I_{or}/I_{oc}$ | −3.0 dB | −3.0 dB | 3.0 dB | 6.0 dB |
| $E_b/N_0$ from Eq.(11.24) | 9.2 dB | 5.7 dB | 5.3 dB | 5.3 dB |
| DCCH overhead | −0.8 dB | −0.2 dB | −0.1 dB | −0.03 dB |
| Slot format 8 | −0.8 dB | − | − | − |
| Product vs. 3GPP | −1.5 dB | −1.5 dB | −1.5 dB | −1.5 dB |
| Outer loop power control | +0.3 dB | +0.3 dB | +0.3 dB | +0.3 dB |
| Fast power control | +0.0 dB | +0.0 dB | +0.0 dB | +0.0 dB |
| $E_b/N_0$ 120 km/h | 6.4 dB | 5.0 dB | 5.0 dB | 4.9 dB |
| 3 km/h vs 120 km/h | +0.3 dB | +0.3 dB | +0.3 dB | +0.3 dB |
| $E_b/N_0$ 3 km/h | 6.7 dB | 5.3 dB | 5.3 dB | 5.2 dB |

The calculations of the downlink $E_b/N_0$ values are shown in Table 11.20 for voice and in Table 11.21 for data.

The $E_b/N_0$ values are summarized in Figure 11.42 for uplink and in Figure 11.43 for downlink. We can note that the uplink $E_b/N_0$ is typically 2–3 dB lower than that for the downlink. The reason is antenna diversity, that is assumed in uplink but not in downlink. The downlink transmit diversity would improve the downlink $E_b/N_0$ values.

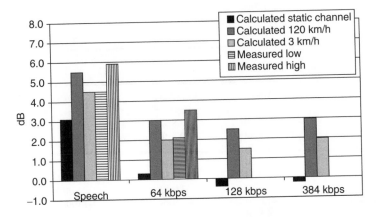

**Figure 11.42**  Summary of uplink $E_b/N_0$

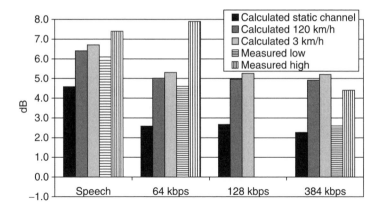

**Figure 11.43**   Summary of downlink $E_b/N_0$

The calculated values can be compared to the values obtained from field measurements. Since stationary mobiles were used in the measurements, we compare the results to the calculated values at 3 km/h. The measured lowest values are similar to those calculated in uplink for voice and 64 kbps, but the measured highest values are 1.5 dB higher than the calculated ones. It is to be expected that the measured worst case $E_b/N_0$ values are higher than the calculated ones. In the case of 64 kbps, part of the difference is caused by the different BLER requirements: calculations are based on packet data BLER 1%, while the measurements are based on BLER 0.5%. The measured lowest voice $E_b/N_0$ values in downlink are 0.6 dB lower, and the measured highest values 0.6 dB higher than the calculated ones. The lowest measured 64 kbps $E_b/N_0$ values are slightly below the calculated ones, but the highest values are clearly above the calculated ones. The difference in BLER target also affects the benchmarking here. The measured 384 kbps values are clearly below the calculated ones. The low measured values are partly explained by the 384 kbps link transmission power hitting its maximum value in the measurements, see Figure 11.37.

This book has typically used the following $E_b/N_0$ values for low mobile speed cases: uplink voice 4.0–5.0 dB, uplink data 1.5–2.0 dB, downlink voice 7.0 dB and downlink data 5.0 dB.

## 11.5.2   RF Noise Figure

The $E_b/N_0$ values represent mainly the baseband performance of the receiver. The receiver RF performance is described with the RF noise figure, which represents the loss of the signal power in the receiver RF parts. The required reference sensitivity level for 12.2 kbps voice for the base station is −121 dBm and for the mobile station −117 dBm [13]. The RF noise figure can be calculated when the baseband $E_b/N_0$ performance is known:

$$\text{RF noise figure} = \text{Reference sensitivity} - (-174\,\text{dBm} + 10\log_{10}(12.2e3) + E_b/N_0) \qquad (11.25)$$

The RF noise figure calculations are shown in Table 11.22. The required base station noise figure is estimated to be 4.5 dB and the mobile station 8.7 dB. The RF noise figure requirement can be more relaxed if the baseband $E_b/N_0$ performance is better than the assumption here. On the other hand, the average product is better than the requirement due to the production variations. In this book we have typically assumed an average base station noise figure of 4.0 dB and an average mobile station noise figure of 7.0 dB.

**Table 11.22**  RF noise figures

|                                                      | Base station                       | Mobile station      |
| ---------------------------------------------------- | ---------------------------------- | ------------------- |
| 3GPP sensitivity requirement for voice               | −121 dBm [13]                      | −117 dBm [1]        |
| 3GPP $E_b/N_0$ requirement in static channel         | 5.1 dB + 2.5 dB  from Table 11.18 | 7.4 dB from [1]     |
| RF noise figure from Equation (11.25)                | 4.5 dB                             | 8.7 dB              |

*Note*: [1]The sensitivity of −121 dBm is without antenna diversity. The 2.5 dB factor is the estimated antenna diversity gain.

## 11.6  Performance Enhancements

WCDMA performance enhancements with advanced antenna structures and with baseband multiuser detection are described in this section. Both solutions are considered here mainly for the base station for improving uplink performance.

### 11.6.1  Smart Antenna Solutions

The use of antenna arrays at the Node B can help increase the capacity of terrestrial cellular systems significantly, or alternatively improve the coverage [14, 15, 16]. The basic principle is to multiply the signals at the different antenna branches with complex weight factors before the signals are transmitted, or before the received signals are summed, as illustrated in Figure 11.44. This set-up can be regarded as a spatial filter, where the signals at the different antenna branches represent spatial samples of the radio channel, and the complex weight factors are filter coefficients. The set of complex weight factors (also known as weight vectors) is typically different for each user.

Antenna array systems can basically be designed to operate in one of two distinct modes: (1) diversity mode or (2) beamforming mode. Diversity techniques help increase the signal-to-interference ratio and reduce the likelihood of deep fades, provided that there is statistical independence between the signals at the antenna elements. Beamforming techniques, on the other hand, are applied when there is coherence between the signals at the antenna elements, so a narrow beam can be created towards

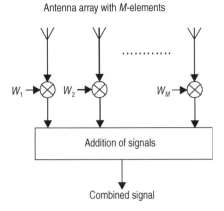

**Figure 11.44**  Basic principle for antenna array processing. The example is for the case where the antenna array is used for reception

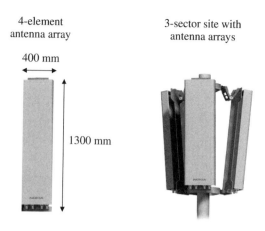

**Figure 11.45** A four-element linear antenna array

the desired user. Such schemes are known to provide an average spatial interference suppression gain, which basically depends on the effective beamwidth and the sidelobe level, assuming a scenario with a large number of users compared to the number of antenna elements. Coherence between the antenna signals in a compact array with half a wavelength element spacing can be expected in environments such as typical urban macro cells, where the azimuth dispersion of the radio channel seen at the antenna array is typically small. As an example, Figure 11.45 illustrates the physical dimensions of a four-element linear antenna array for WCDMA, as well as an example of the use of such arrays in a three-sector site configuration.

During the past two decades, many beamforming algorithms have been proposed and analysed for selection of complex weight vectors; see results for uplink vector Rake receivers in [17] and downlink beamforming techniques in [18, 19, 20], among others. More degrees of freedom are typically available for uplink beamforming, where the radio channel can be more accurately estimated and different beamformer weights can be applied for the different multipath delays. On the contrary, downlink beamforming from the Node B is subject to more constraints, since the downlink radio channel is unknown, only the average direction towards the UEs can be estimated. In the following subsections, we will further address the use of beamforming techniques in UMTS, and shortly discuss how implementation of such techniques impacts on the radio resource management algorithms. Performance results for the gain of beamforming schemes are also presented.

### 11.6.1.1 Beamforming Options

Two different beamforming modes are possible for the downlink of UMTS within one logical cell: (1) user-specific beamforming and (2) fixed beamforming. User-specific beamforming allows for the generation of individual beams to each UE without any restrictions on the selection of weight vectors. However, this prevents the UE from using the primary common pilot channel (P-CPICH) for phase reference, since the P-CPICH must be transmitted on a sector beam which provides coverage in the entire cell area. The phase rotation of the P-CPICH received at the UE is therefore likely to be different from the phase rotation of the signal transmitted under the user-specific beam. UEs receiving signals subject to user-specific beamforming are therefore informed via higher layer protocols to use dedicated pilot symbols for phase reference instead of the P-CPICH. Fixed beamforming refers to the case where a finite set of beams are synthesized at the Node B, so multiple UEs may receive

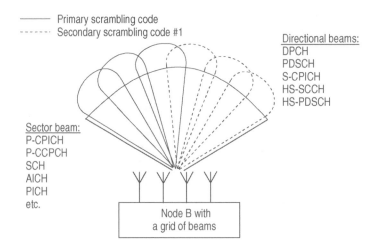

**Figure 11.46** Mapping of physical channels onto a grid of fixed beams for the case where four elements are used to form six directional beams, plus one wide sector beam. For the sake of simplicity, the sidelobes on the directional beams are not illustrated

signals transmitted under the same beam. Even though this beamforming mode is referred to as fixed beamforming, the complex weight factors used to synthesize the beams may be varied over time to facilitate a slow adaptation of the directional beams. Each of the beams is associated with a unique secondary common pilot channel (S-CPICH), which the UEs are informed to use for phase reference via higher layer protocols. Table 6.5 in Chapter 6 shows an overview of the beamforming options for different downlink channels. Notice that beamforming is not allowed on all channels, and some channels support only fixed beamforming, while user-specific beamforming is optional. In this context, optional means that this beamforming mode is not mandatory for the UE, and therefore the network cannot assume that all UEs in the system support, for instance, user-specific beamforming on HS-DSCH. As the HS-DSCH and HS-SCCH belong to the high-speed downlink packet access (HSDPA) concept, the scope of applicability for the combination of user-specific beamforming and HSDPA depends on the UE implementation. Figure 11.46 shows an example with the mapping of physical channels onto a grid of six directional fixed beams plus one sector beam. Notice that, despite the Node B using beamforming, it still needs to offer a sector beam for channels which are not allowed to use beamforming. It is assumed that transmission of transport channels towards one UE from a Node B is conducted via one directional beam per cell only. Reception from two directional beams would involve the use of two S-CPICHs for phase reference at the UE (one for each directional beam), which is not allowed according to the current UMTS specifications. Link level results for both user-specific beamforming and fixed beamforming are available in [18 and 19].

### 11.6.1.2  Higher Order Sectorization via Beamforming

Instead of synthesizing beams within one logical cell, the antenna array can also be exploited to create individual cells where the coverage area of each beam represents a logical cell. This implies that each beam is transmitted under a unique primary scrambling code. In addition, each beam will have its own P-CPICH, BCH, PCH, etc. An example is shown in Figure 11.47, where three uniform linear antenna arrays with four elements each are used to create 18 logical sectors. Provided that the beamforming network is implemented digitally, this solution allows for adaptive cell sectorization to maximize the coverage and capacity in coherence with the traffic distribution in the network. Using

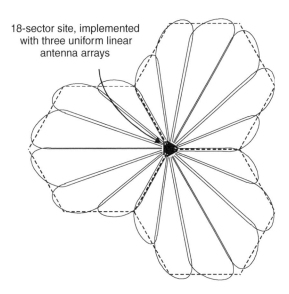

18-sector site, implemented
with three uniform linear
antenna arrays

**Figure 11.47**  An example of an 18-sector site implemented with beamforming on three uniform linear antenna arrays. The figure shows the approximate coverage areas of each beam (sector)

this configuration, the pilot overhead from having an S-CPICH per beam is avoided compared to synthesizing a grid of fixed beams in a logical cell. However, removal of the S-CPICH overhead is obtained at the expense of reduced downlink orthogonality, since each beam is transmitted under a unique primary scrambling code. Finally, higher order sectorization via beamforming allows for flexible allocation of transmit power in the different sectors, compared to the scheme where each sector has its own panel antenna and power amplifier.

### 11.6.1.3  Node B Measurements for Beamforming Support

As discussed in Chapters 9 and 10, the family of radio resource management algorithms such as admission control, packet scheduling, handover control, and congestion control are primarily implemented in the RNC according to the UMTS architecture. As a consequence of this functional split, these algorithms must rely on standardized measurements collected at the Node B and reported via the open Iub interface to the RNC. For this reason the Node B informs the RNC via a configuration message whether it uses fixed beamforming, user-specific beamforming, or no beamforming.

In order to provide measurements of the spatial load distribution in cells with beamforming capabilities, new measurements have been proposed for the Node B per *cell portion*, as well as the corresponding reporting over the Iub interface to the RNC [21]. A cell portion is defined as a geographical part of a cell for which a Node B measurement can be reported to the RNC, and for which the RNC can allocate a phase reference for a UE that is within the cell portion. A cell portion is semi-static, and identical for both the uplink and the downlink. Within a cell, a cell portion is uniquely identified by a cell portion ID [21]. Hence, assuming that fixed beamforming as discussed in Section 11.6.1.1 is applied, the cell portions are typically configured to be identical to the directional beams.

A pseudo uplink direction-of-arrival (DoA) measurement is included in 3GPP Release 5 for Node B to facilitate beam switching in the downlink. The Node B measures the average uplink signal-to-interference ratio (SIR) on the dedicated physical control channel (DPCCH) received from each UE in all the cell portions [21]. The four highest SIR values and the corresponding cell portion IDs are

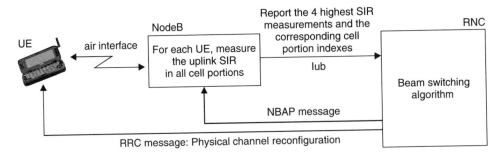

**Figure 11.48**  Simplified sketch of the Node B measurements, and the signalling flow from the RNC to the Node B and the UE during beam switching

sent to the RNC. These measurements are used to implement the beam switching functionality that is required when fixed beamforming is used in the downlink. Based on the cell portion specific SIR measurements, the RNC will typically inform the Node B to transmit the data to a UE under the beam (cell portion) corresponding to the highest uplink measured SIR, as well as informing the UE which S-CPICH it should use for phase reference via a radio resource control (RRC) message, as illustrated in Figure 11.48. Notice that beam switching only includes measurements from the Node B, i.e. no measurements from the UE are used to trigger beam switching. The beam switching operation is discussed in more detail in [20].

For Release 6, measurements of the carrier transmit power and total received wideband power per cell portion are also defined for the Node B. These measurements can be regarded as downlink/uplink measurements of the load per cell portion, i.e. equivalent to directional load measurements. These beamforming-specific measurements make it possible to take full system performance benefits from the beamforming capabilities. As an example, directional power-based radio resource management algorithms are proposed in [22 and 23], to make sure that the spatial interference suppression gain offered by beamforming techniques is mapped into a capacity gain in the network.

### 11.6.1.4   Capacity Results from Dynamic Network Simulations

The results presented in this section are obtained from downlink dynamic network simulations with multiple UEs and 33 cells, with a site-to-site distance of 2.0 km. The simulations include accurate modelling of the radio link performance towards each UE, mobility of the UEs, radio resource management algorithms, traffic models with 64 kbps circuit switched services on dedicated channels, etc. The detailed simulation methodology is described in [20]. The reference configuration is a standard three-sector network topology, where a panel antenna is used for each sector, with a 3 dB beamwidth of 65 degrees, as illustrated in Figure 11.49. The considered beamforming configuration assumes a four-element linear antenna array per sector, which is exploited to synthesize six directional beams with a separate S-CPICH assigned to each beam. Finally, a scenario with higher order sectorization, as illustrated in Figure 11.47, is also considered, where three uniform linear antenna arrays with four elements are used to synthesize a total of 18 sectors per Node B.

Assuming that only one scrambling code is allocated per sector, the relative capacity gain is found be a factor of 1.9 for the beamforming case. For this configuration the capacity gain is severely limited by the shortage of channelization codes, since it happens with a probability of 28% that an incoming call is rejected because there are no available channelization codes. Hence, under these conditions the capacity is hard limited rather than interference/power limited, so the interference suppression gain offered by the use of beamforming techniques cannot be fully exploited. By introducing a secondary

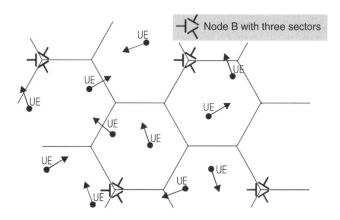

**Figure 11.49** Network topology for the reference configuration

scrambling code in the cell, the capacity gain increases to a factor of 2.4 and the likelihood of channelization code shortage is practically reduced to zero. This indicates that the spatial interference suppression gain from using beamforming techniques is effectively mapped into a capacity gain by using two scrambling codes per cell for the considered scenario. Notice that the capacity gain of 2.4 includes the overhead from the S-CPICH per beam. The use of antenna arrays with more than four elements may require more than two scrambling codes per cell, due to the larger spatial interference suppression gain.

Additional simulation results for the three considered network configurations are summarized in Figure 11.50. The presented results for the net capacity gain and soft handover are normalized with the results for the reference scenario. Two scrambling codes are enabled for the scenario where six

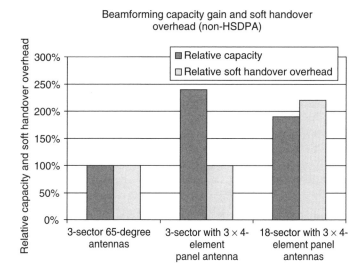

**Figure 11.50** Comparison of the downlink capacity gain for dedicated channel and SHOO for different site configurations

directional beams are formed within each logical cell. The largest capacity gain is achieved for the three-sector configuration with an antenna array in each sector, without increasing the soft handover overhead. The capacity gain for the configuration with 18 sectors is approximately 20% less than this, in spite of the fact that both configurations use three antenna arrays with four elements to create the same number of beams. The configuration with 18 sectors suffers from a slightly worse equivalent downlink orthogonality factor, since different primary scrambling codes are used for each sector (i.e. 18 scrambling codes per site), while the three-sector configuration with beamforming uses only two scrambling codes per sector, or equivalently six scrambling codes per site. In addition, the 18-sector configuration suffers from a large soft handover overhead, which is more than twice as large as the overhead observed for the reference configuration with three-sector sites. The larger overhead not only indicates the need for additional baseband resources and transmission capacity over the Iub, but it also represents potential problems in managing fast updates of the UE's active set for fast moving UEs.

The results above are obtained for the downlink assuming that the traffic is carried on dedicated channels. The equivalent beamforming capacity gain is typically found to be 10–15% higher for the uplink compared to the downlink assuming single antenna uplink reception, as there is no additional pilot overhead for the uplink and different beams can be formed for each multipath delay in the radio channel. However, when compared against Node B configuration two uplink receive antennas and maximal ratio combining, the uplink beamforming capacity gain with a four-element antenna array is found to be of the order of a factor 1.7.

The later results were all obtained for capacity limited scenarios. However, beamforming techniques can also be applied to improve the coverage. Ideally, the additional beamforming gain equals 10 $\log(M)$, where $M$ is the number of antenna elements in the array. However, due to azimuthal dispersion in the radio channel and other imperfections, the actual beamforming gain is typically slightly smaller, i.e. for $M = 4$ antennas, a more realistic estimate of the beamforming antenna gain is of the order of 5 dB, compared to 6 dB in the ideal case. Assuming a path loss exponent of $-3.5$, a beamforming antenna gain of 5 dB is equivalent to approximately a 38% increase in range extension.

### 11.6.1.5  Beamforming for HSDPA

The combination of fixed beamforming and HSDPA is also feasible. This is possible by transmitting to one UE under each beam during each TTI, so the number of served UEs in each TTI equals the number of beams. It is the packet scheduler in the Node B (in the MAC-hs) that determines which UEs should be scheduled under the different beams. As discussed in [24], independent packet schedulers for each of the beams can be used, such as, proportional fair packet scheduling or simple round robin packet scheduling. Results for the HSDPA beamforming cell capacity gain are shown in Figure 11.51 as a function of the number of antenna elements in the array. These results are obtained from dynamic macro cellular network simulations, assuming an ITU Vehicular-A power delay profile and proportional fair packet scheduling [24].

## 11.6.2  Multiuser Detection

The uplink performance improvements with base station multiuser detection (MUD) are discussed in this section. The aim is to give an overview of different multiuser detection algorithms, and references to more detailed information. Another target is to assess capacity and coverage gains of an MUD receiver. Section 11.6.2.1 gives an overview of MUD receivers. Section 11.6.2.2 describes a practical parallel interference canceller (PIC). Section 11.6.2.3 contemplates PIC efficiency and derives equations for capacity and coverage gains that PIC can offer. Section 11.6.2.4 presents numerical results of soft quantization PIC (SQ-PIC) for different propagation channels and bit rates for one, two and four diversity antennas.

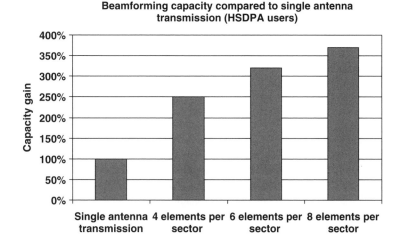

**Figure 11.51** HSDPA cell capacity gain versus the number of antenna elements

### 11.6.2.1 Overview

CDMA systems are inherently interference-limited from both the receiver performance and system capacity points of view [32–34]. From the receiver perspective this means that, if the number of users is large enough, an increase in signal-to-noise ratio yields no improvement in bit or frame error rate. From the system capacity view, it means that the larger the signal-to-interference-plus-noise ratio required for the desired quality of service, the fewer users can be accommodated in the communication channel.

The interference-limited nature of CDMA systems results from the receiver design. In CDMA systems, the core of the receiver is a spreading code matched filter (MF) or correlator [34]. Since the received spreading codes are usually not completely orthogonal, multiple access interference (MAI) is generated in the receiver. If the spreading factor is moderate, the received powers of users are equal (no near–far problem), and the number of interfering users is large (>10), by the central limit theorem the multiple access interference can be modelled as increased background noise with a Gaussian distribution. This approximation has led to the conclusion that the matched filter followed by decoding is the optimal receiver for CDMA systems in additive white Gaussian noise (AWGN) channels. In frequency-selective channels, the Rake receiver [35] can be considered optimal with corresponding reasoning.

Although multiple access interference can be approximated as AWGN, it inherently consists of received signals of CDMA users. Thus, multiple access interference is very structured, and can be taken into consideration in the receiver. This observation led Verdú [36] to analyse the optimal multiuser detectors (MUDs) for multiple access communications. Verdú was able to show that CDMA is not inherently interference-limited, but that is a limitation of the conventional matched filter receiver.

The optimal multiuser detectors [36] can use either maximum *a posteriori* (MAP) detection or maximum likelihood sequence detection (MLSD). In other words, techniques similar to those applied in channels with inter-symbol interference [35] can be used to combat multiple access interference. The drawback of both the MLSD- and MAP-based multiuser detectors is that their implementation complexity is an exponential function of the number of users. Thus, they are not feasible for most practical CDMA receivers. This fact, together with Verdú's observation that CDMA with an MLSD receiver is not interference-limited, has triggered an avalanche of papers on sub-optimal multiuser receivers. A brief summary of the sub-optimal multiuser detection techniques is given below. For a more complete treatment, the reader is referred to the overview paper by Juntti and Glisic [37] and to the book by Verdú [38].

The existing sub-optimal multiuser detection techniques can be categorized in several ways. One way is to classify the detection algorithms as centralized multiuser detection or decentralized single user detection algorithms. The centralized algorithms perform real multiuser joint detection, i.e. they detect jointly each user's data symbols; they can be considered practical in base station receivers. The decentralized algorithms detect the data symbols of a single user based on the received signal observed in a multiuser environment containing multiple access interference; the single user detection algorithms are applicable to both base station and terminal receivers.

In addition to one kind of implementation-based categorization on multiuser and single user detectors, the multiuser detectors can be classified based on the method applied. Two main classes in this category can be identified: linear equalizers and subtractive interference cancellation (IC) receivers. Linear equalizers are linear filters suppressing multiple access interference. The most widely studied equalizers include the zero-forcing (ZF) or decorrelating detector [39–41] and the minimum mean square error (MMSE) detector [42, 43]. The IC receivers attempt to explicitly estimate the multiple access interference component, after which it is subtracted from the received signal. Thus, the decisions become more reliable. Multiple access interference cancellation can be performed in parallel to all users, resulting in parallel interference cancellation (PIC) [44, 45]. Interference cancellation can also be performed in a serial fashion, resulting in serial interference cancellation (SIC) [46, 47].

Both linear equalizers and interference cancellation receivers can be applied in centralized receivers. The linear equalizers can also be implemented adaptively as single-user type decentralized detectors. This is possible if the spreading sequences of the users are periodic over a symbol interval so that multiple access interference becomes cyclostationary. Various adaptive implementations based on training sequences of the MMSE detectors have been studied [48–52]. So-called blind adaptive detectors not requiring training sequences have also been considered [51, 52, 53].

The choice of multiuser detection techniques for WCDMA base station receivers has been studied [55–58]. Both the receiver performance and implementation complexity have been considered. The conclusion of the studies is that a multiuser receiver based on multistage parallel interference cancellation (PIC) is currently the most suitable method to be applied in CDMA systems with a single spreading factor. The PIC receiver principle with one cancellation stage for a two-user CDMA system is illustrated in Figure 11.52. The parallel interference cancellation means that interference is cancelled from all users simultaneously, i.e. in parallel. The cancellation performance can be improved

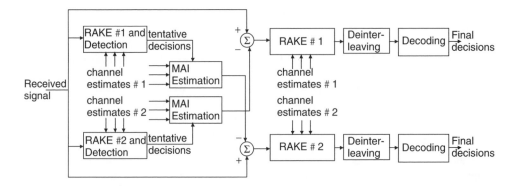

**Figure 11.52**   Parallel interference cancellation receiver for two users

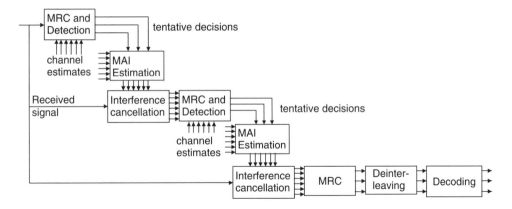

**Figure 11.53** Multistage interference cancellation receiver1

by reusing the decisions made after interference cancellation in a new IC stage. This results in a multistage interference cancellation receiver, which is illustrated in Figure 11.53.

The choice of multiuser detection for multiservice CDMA systems with a variable spreading factor access has been considered by Ojanperä [59]. For such a system, a groupwise serial interference cancellation (GSIC) [60–63] receiver seems attractive. In the GSIC receiver, the users with a certain spreading factor are detected in parallel, after which multiple access interference caused by them is subtracted from the users with other spreading factors. A main reason for the GSIC being efficient is that the power of users depends on the spreading factor. By starting the cancellation with the users with the lowest spreading factor, the highest power users (the most severe interferers) are cancelled first.

The adaptive linear equalizers can be applied only if the spreading sequences of users are periodic over a relatively short time, such as over the symbol interval. Therefore, by using the short scrambling code option in a WCDMA uplink, the adaptive receivers could be utilized therein. In the WCDMA downlink, the spreading codes are periodic over one radio frame, whose duration is 10 ms. The period is so long that conventional adaptive receivers are practically useless. The problem can be partially overcome by introducing chip equalizers [64–69]. The idea here is to equalize the impact of the frequency-selective multipath channel on a chip-interval level. This suppresses inter-path interference (IPI) of the signals and also retains the orthogonality of spreading codes of users within one cell. The latter impact is possible, since synchronous transmission with orthogonal signature waveforms is applied in the downlink. In other words, multiple access interference in the downlink is caused by the multipath propagation, which can now be compensated for by the equalizer. The effect of multipath propagation on downlink performance without any interference suppression receivers is presented in Section 11.3.1.1. For a complete overview of the techniques available and the relevant literature, we refer the reader to [68]. The advanced receiver algorithms can be applied to WCDMA/HSDPA terminals to improve the end user data rates and system capacity. The performance of advanced HSDPA receivers is discussed in Chapter 11.

### 11.6.2.2 Structure of PIC

The PIC algorithm presented here is the so-called residual PIC in Figure 11.54. The first stage is a Rake bank, which makes tentative bit decisions for each user. Also the channel estimation is done at the Rake stage. After the Rake stage, there are one or several cancellation stages. A cancellation stage comprises of a multiple access interference estimator and a Rake receiver with an interference canceller. The tentative decisions and channel estimates are fed to the interference estimator, which

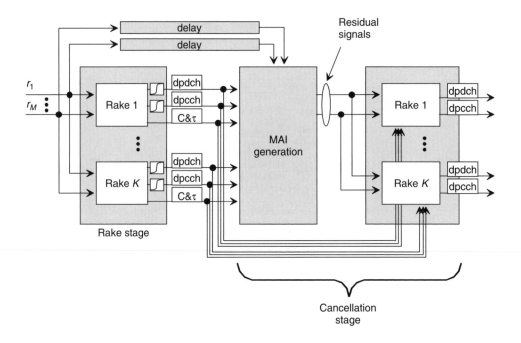

**Figure 11.54**  PIC receiver with $M$ diversity antennas

computes an estimate of the received wideband signal. This estimate is subtracted from the delayed received signal, and the result is called a residual. The residual is then despread and after that, bit decisions from the previous Rake stage, multiplied by channel estimates, are added. After the Rake cancellation stage we get final bit decisions from which interference has been removed.

The reason for using residual PIC lies in the implementation architecture, i.e. only one wideband signal needs to be distributed from the Rake stage to the PIC stage. In conventional PIC there are separate interference estimates for every user. Note that the performance of residual PIC is the same as that of conventional PIC.

A problem of the conventional hard decision PIC is that, when a tentative decision is wrong, the interference from that case is doubled in the cancellation stage. One way to overcome this problem is to use a null zone hard decision device [70] or to use an adaptive decision threshold [71], where the threshold is based on the statistic of matched filter output. The cancellation can also be made only to those signals which are reliable, as in [72]. The reliability of decisions is also covered in [73], where Divsalar *et al.* propose the use of a hyperbolic tangent function as a decision device instead of a sign function. These two functions are depicted in Figure 11.55. A sign function makes hard decisions (either $+1$ or $-1$); a hyperbolic tangent makes soft quantized decisions (anything between $+1$ and $-1$). Ideally, the horizontal axis in Figure 11.55 should be instantaneous signal-to-noise-and-interference ratio (SINR), but in practice, we have to estimate average SINR or assume interference + noise constant and use the power of the symbol. Hence, the reliability of the decision is taken into account. Note that a soft quantized decision is different from a soft decision, where there is no reliability weighting and the values are not limited to $+1$ and $-1$. In the following, a modified version of this receiver is called SQ-PIC.

The 3GPP standard and implementation costs of the receiver set basic requirements for a PIC receiver. The strong channel coding specified in the standard and usually short spreading factor mean

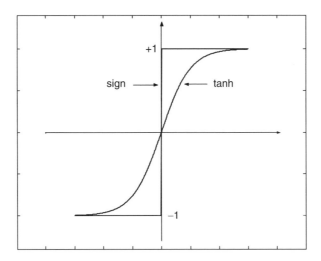

**Figure 11.55** Non-linearities of a PIC receiver. The horizontal axis is the input signal level and the vertical axis, the output signal level

that SINR can be quite low at a receiver, resulting in unreliable tentative decisions and high BER before decoding, up to 15%. This means that the hard decision-based PIC cannot work very well, since for every wrong decision, the corresponding interference is doubled. SQ-PIC tries to overcome this by using a reliability measure to weight the tentative decisions as mentioned. In order to minimize costs, only one PIC stage is suggested as the performance improvement from the second or third stage is not that large.

### 11.6.2.3   PIC Efficiency and Derivation of Network Level Gains

We define PIC efficiency $\beta$ as the amount of own-cell interference it can remove. (It is assumed that PIC cannot remove other inter-cell interference.) We can write $I_{\text{total}}$ for Rake and PIC as

$$I_{\text{total,rake}} = I_{\text{own}} + i I_{\text{own}} + N_0 = (1+i) K_{\text{rake}} P_j + P_N \tag{11.26a}$$

$$I_{\text{total,PIC}} = (1 - \beta) I_{\text{own}} + i I_{\text{own}} + N_0 = (1 + i - \beta) K_{\text{pic}} P_j + P_N \tag{11.26b}$$

where $i$ is the ratio of other-cell interference to own-cell interference, $P_N$ is the thermal noise power and we have assumed that users are homogenous, each having the same received power $P_j \cdot K_{\text{rake}}$ and $K_{\text{pic}}$ are the number of users for Rake and PIC, respectively. $K_{\text{pic}}$ is selected so that $I_{\text{total,rake}}$ and $I_{\text{total,pic}}$ are equal, i.e. noise rises of Rake and PIC are the same. We can solve capacity gain $G_{\text{cap}} = K_{\text{pic}}/K_{\text{rake}}$ from the equations above:

$$G_{\text{cap}} = \frac{K_{\text{pic}}}{K_{\text{rake}}} = \frac{1+i}{1+i-\beta} \tag{11.27}$$

Coverage gain is defined as the ratio of required $E_b/N_0$ s for Rake and PIC when the number of users, $K$, is kept constant:

$$G_{\text{cov}} = \frac{\{E_b/N_0\}_{\text{rake}}}{\{E_b/N_0\}_{\text{pic}}} \tag{11.28}$$

Using the definition of $E_b/N_0$ and using Equations (11.26) we get:

$$\{E_b/N_0\}_K = \frac{W}{R}\frac{P_j}{P_N} = \frac{W}{R}\frac{1}{\frac{1}{L_j} - (1+i-\beta)K}$$

(11.29)

Note that $\beta$ is zero for a Rake receiver. From Equations (11.28) and (11.29) we obtain:

$$G_{\mathrm{cov}} = \frac{\{E_b/N_0\}_{\mathrm{rake}}}{\{E_b/N_0\}_{\mathrm{pic}}} = \frac{\dfrac{W}{R}\dfrac{1}{\frac{1}{L_j} - (1+i)K}}{\dfrac{W}{R}\dfrac{1}{\frac{1}{L_j} - (1+i-\beta)K}} = \frac{\dfrac{1}{L_j} - (1+i-\beta)K}{\dfrac{1}{L_j} - (1+i)K}$$

(11.30)

We can solve $K$ from the noise rise equation, Equation (8.9):

$$K = \frac{\dfrac{1}{L_j} - \dfrac{1}{L_j \Delta I_{\mathrm{rake}}}}{1+i}$$

(11.31)

where $\Delta I_{\mathrm{rake}}$ is the noise rise with a Rake receiver. We can now solve $G_{\mathrm{cov}}$ as a function of $i$, $\beta$ and $\Delta I_{\mathrm{rake}}$:

$$G_{\mathrm{cov}} = 1 + \frac{\beta \cdot (\Delta I_{\mathrm{rake}} - 1)}{1+i}$$

(11.32)

Examples of the capacity gain in Equation (11.27) and the coverage gain in Equation (11.28) as functions of $i$ are depicted in Figure 11.56 and Figure 11.57. In coverage gain examples, the noise rise of rake $\Delta I_{\mathrm{rake}}$ is also a parameter. The gains increase as $i$ decreases, which is expected since PIC cannot cancel other-cell interference. The gains naturally also depend on $\beta$. The dependence is particularly strong in single cell without any other-cell interference, $i = 0$. Coverage gain also depends on the target noise rise with a Rake receiver: the higher the interference level without PIC, the higher the gain from PIC.

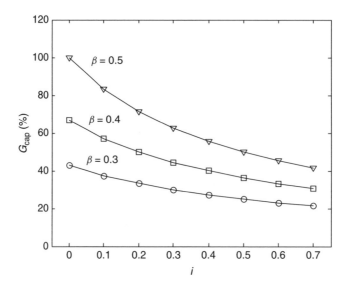

**Figure 11.56**   Capacity gain of PIC as a function of $i$ with PIC efficiency $\beta$ as a parameter

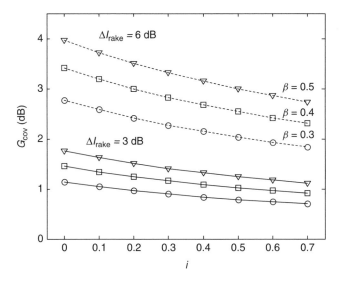

**Figure 11.57** Coverage gain of PIC as a function of inter-cell interference $i$, multiuser efficiency $\beta$ and noise rise $\Delta I_{\mathrm{Rake}}$

### 11.6.2.4 Performance of SQ-PIC

The performance of SQ-PIC was evaluated by Monte Carlo simulations in the link level without inter-cell interference [74]. Two propagation channels were considered, namely a pedestrian type of environment (Case 1 in 3GPP TS 25.141) and a vehicular type of environment (Case 3 in 3GPP TS 25.141). In the pedestrian channel, the UE velocity was 3 km/h, and in the vehicular channel, 120 km/h. The 12.2 kbps speech and 384 kbps data services were studied.

We estimate the PIC efficiency $\beta$ from the simulation results by finding the value of $\beta$ that gives the best fit to the simulation results. The results are shown in Table 11.23. The PIC efficiency is between 24% and 41%. The highest efficiency is obtained at high mobile speed with high data rate 384 kbps. The fast power control cannot keep the received power level exactly constant at high speed and there are larger power differences that can be cancelled by PIC. A high data rate provides better efficiency, since the number of simultaneous users is lower and the number of estimated parameters by PIC is lower, resulting in more accurate estimates. With a data service, the small spreading factor results also

**Table 11.23** PIC efficiency

| Propagation channel | Number of diversity antennas ($M$) | PIC efficiency $\beta$ | |
| --- | --- | --- | --- |
| | | 384 kbps data | 12.2 kbps speech |
| Case 1 $v = 3$ km/h | 1 | 32% | 24% |
| | 2 | 36% | 26% |
| | 4 | 37% | 33% |
| Case 3 $v = 120$ km/h | 1 | 40% | 35% |
| | 2 | 41% | 32% |
| | 4 | 36% | 28% |

**Table 11.24** Capacity and coverage gains with simulated PIC efficiencies (with outer-to-own cell interference ratio $i = 0.55$)

| Propagation channel | M | Capacity gain (%) | | Coverage gain (dB) with 6 dB noise rise | |
|---|---|---|---|---|---|
| | | 384 kbps data | 12.2 kbps speech | 384 kbps data | 12.2 kbps speech |
| Case 1 $v = 3$ km/h | 1 | 26% | 26% | 2.1 dB | 1.6 dB |
| | 2 | 30% | 30% | 2.3 dB | 1.8 dB |
| | 4 | 31% | 31% | 2.3 dB | 2.1 dB |
| Case 3 $v = 120$ km/h | 1 | 35% | 35% | 2.5 dB | 2.2 dB |
| | 2 | 36% | 36% | 2.5 dB | 2.1 dB |
| | 4 | 30% | 30% | 2.3 dB | 1.9 dB |

in high cross-correlation between users, making the performance of Rake poor and hence allowing higher potential gain for SQ-PIC. The lowest gain is obtained at low mobile speed with voice users.

Table 11.24 shows capacity and coverage gains with the estimated efficiencies. The capacity gains are 26–35% in a typical macro cell with $i = 0.55$. The coverage gain is 1.6–2.5 dB. This coverage gain assumes an initial noise rise of 6 dB. If the initial noise rise was 3 dB, the corresponding coverage gain would be 0.6–1.0 dB.

Numerical results for a data service are depicted in Figure 11.58 for channel case 3. The number of diversity antennas was one, two or four for both Rake and SQ-PIC. Rake with diversity antennas can be seen as an alternative to PIC, since increasing the order of diversity also provides substantial gains. The reason for this is Rake's ability to average MAI over diversity antennas, as the interference

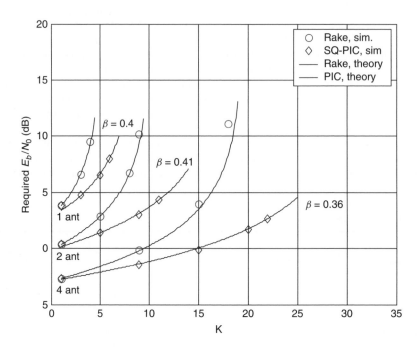

**Figure 11.58** Required $E_b/N_0$ vs. number of users for a 384 kbps data service and BLER target 10% in case 3 channel (120 km/h), the number of diversity antennas 1, 2 and 4

components from different diversity channels are independent. The results show that increasing the number of antennas with a Rake receiver provides higher capacity and better coverage than introducing PIC. In the case of a low number of users, the gain from any interference cancellation is low, while more antennas provide clear coverage benefits. On the other hand, adding more antennas and antenna cables may not be possible from the site solution point of view, while the introduction of interference cancellation as the baseband processing enhancement is easier.

Interference cancellation, namely SQ-PIC, is the most promising method for improving base station receiver performance, as well as system capacity and coverage. Uplink interference cancellation may provide further gains in end user throughput when the uplink peak data rates exceed 1 Mbps in High-Speed Uplink Packet Access, HSUPA, which is a 3GPP study item in Release 6. For more details, see Chapter 13.

# References

[1] Holma, H., Soldani, D. and Sipilä, K. 'Simulated and Measured WCDMA Uplink Performance', *Proceedings VTC 2001* Fall, Atlantic City, NJ, USA, pp. 1148–1152.

[2] UMTS, Selection Procedures for the Choice of Radio Transmission Technologies of the UMTS, ETSI, v.3.1.0, 1997.

[3] Laiho-Steffens, J. and Lempiäinen, J., 'Impact of the Mobile Antenna Inclinations on the Polarization Diversity Gain in DCS1800 Network', *Proceedings of PIMRC'97*, Helsinki, Finland, September 1997, pp. 580–583.

[4] Lempiäinen, J. and Laiho-Steffens, J., 'The Performance of Polarization Diversity Schemes at a Base-Station in Small/Micro Cells at 1800MHz', *IEEE Trans. Vehic. Tech.*, Vol. 47, No. 3, August 1998, pp. 1087–1092.

[5] Sorensen, T.B., Nielsen, A.O., Mogensen, P.E., Tolstrup, M. and Steffensen, K., 'Performance of Two-Branch Polarization Antenna Diversity in an Operational GSM Network', *Proceedings of VTC'98*, Ottawa, Canada, 18–21 May 1998, pp. 741–746.

[6] Grandell, J. and Salonaho, O. 'Macro Cell Measurements with the Nokia WCDMA Experimental System', *IEE Conference on Antennas and Propagation*, UK, 2001, pp. 516–520.

[7] Westman, T. and Holma, H., 'CDMA System for UMTS High Bit Rate Services', *Proceedings of VTC'97*, Phoenix, AZ, May 1997, pp. 825–829.

[8] Pehkonen, K., Holma, H., Keskitalo, I., Nikula, E. and Westman, T., 'A Performance Analysis of TDMA and CDMA Based Air Interface Solutions for UMTS High Bit Rate Services', *Proceedings of PIMRC'97*, Helsinki, Finland, September 1997, pp. 22–26.

[9] Ojanperä, T. and Prasad, R., *Wideband CDMA for Third Generation Mobile Communications*, New York: Artech House, 1998, p. 439.

[10] Holma, H., '*A Study of UMTS Terrestrial Radio Access Performance*', Doctoral thesis, Communications Laboratory, Helsinki University of Technology, Espoo, Finland, 2003.

[11] Sipila, K., Honkasalo, Z.C., Laiho-Steffens, J., Wacker, A. 'Estimation of Capacity and Required Transmission Power of WCDMA Downlink Based on a Downlink Pole Equation', *Proceedings of VTC 2000 Spring*., Tokyo, Japan, May 2000, pp. 1002–1005.

[12] 3GPP Technical Specification 25.104 'UTRA (BS) FDD; Radio transmission and Reception'.

[13] 3GPP Technical Specification 25.101 'UE Radio Transmission and Reception (FDD)'.

[14] Winters, J.H., 'Smart Antennas for Wireless Systems', *IEEE Personal Communications*, Vol. 5, Issue 1, February 1998, pp. 23–27.

[15] Paulraj, A. and Chong-Ng, B. 'Space–Time Modems for Wireless Personal Communications', *IEEE Personal Communications*, February 1998, pp. 36–49.

[16] Anderson, S., Hagerman, B., Dam, H., Forssen, U., Karlsson, J., Kronested, F., Mazur, S. and Molnar, K. 'Adaptive Antennas for GSM and TDMA Systems', *IEEE Personal Communications*, June 1999, pp. 74–86.

[17] Pedersen, K.I. and Mogensen, P.E. 'Evaluation of Vector-Rake Receivers Using Different Antenna Array Configurations and Combining Schemes', *Internal Journal on Wireless Information Networks*, 1999, vol. 6, pp. 181–194.

[18] Osseiran, A., Ericson, M., Barta, J., Goransson, B. and Hagerman, B. 'Downlink Capacity Comparison Between Different Smart Antenna Concepts in a Mixed Service WCDMA System', *IEEE Proc. Vehicular Technology Conference*, September 2001, pp. 1528–1532.

[19] Tiirola, E. and Ylitalo, J. 'Performance Evaluation of Fixed-Beam Beamforming in WCDMA Downlink', *Proc. Vehicular Technology Conference*, Tokyo, Japan, May 2000, pp. 700–704.

[20] Pedersen, K.I., Mogensen, P.E. and Ramiro-Moreno, J. 'Application and Performance of Downlink Beamforming Techniques in UMTS', *IEEE Communications Magazine*, October 2003, pp. 134–143.

[21] 3GPP Technical Specification 25.215 'Physical Layer Measurements', v.5.4.

[22] Pedersen, K.I. and Mogensen, P.E. 'Directional Power Based Admission Control for WCDMA Systems Using Beamforming Antenna Array Systems', *IEEE Trans. on Vehicular Technology*, November 2002, Vol. 51, No. 6, pp. 1294–1303.

[23] Osseiran, A. and Ericson, M. 'On Downlink Admission Control with Fixed Multi-Beam Antennas For WCDMA Systems', *Proc. IEEE Vehicular Technology Conference, VTC-2003-Spring*, April 2003, pp. 1203–1207.

[24] Pedersen, K.I. and Mogensen, P.E. 'Performance of WCDMA HSDPA in a Beamforming Environment under Code Constraints', *Proc. IEEE Vehicular Technology Conference VTC-2003-fall*, October 2003.

[25] Klingenbrunn, T., 'Downlink Capacity Enhancements of UTRA FDD Networks', Ph.D. thesis, Aalborg University, 2001.

[26] Godara, L.C., 'Application of Antenna Arrays to Mobile Communications, Part I: Performance Improvement, Feasibility, and System Considerations', *Proc. IEEE*, Vol. 85, No. 7, 1997, pp. 1031–1060.

[27] Godara, L.C., 'Application of Antenna Arrays to Mobile Communications, Part II: Beamforming and Direction-of-Arrival Considerations', *Proc. IEEE*, Vol. 85, No. 8, 1997, pp. 1195–1245.

[28] Jakes, W.J. (ed.), *Microwave Mobile Communications*, New Jersey: IEEE Press, 1974.

[29] Muszynski, P., 'Interference Rejection Rake-Combining for WCDMA', *Proceedings of WPMC'98*, Yokosuka, Japan, November 1998, pp. 93–97.

[30] Winters, J.H., 'Optimum Combining in Digital Mobile Radio with Co-channel Interference', *IEEE Trans. Vehic. Tech.*, Vol. 33, No. 3, 1984, pp. 144–155.

[31] Monzingo, R.A. and Miller, T.W., *Introduction to Adaptive Arrays*, New York: John Wiley & Sons, 1980.

[32] Pursley, M.B., 'Performance Evaluation for Phase-Coded Spread-Spectrum Multiple-Access Communication – Part I: System Analysis', *IEEE Trans. Commun.*, Vol. 25, No. 8, 1977, pp. 795–799.

[33] Gilhousen, K.S., Jacobs, I.M., Padovani, R., Viterbi, A.J., Weaver, L.A. and Wheatley III, C. E., 'On the Capacity of a Cellular CDMA System', *IEEE Trans. Vehic. Tech.*, Vol. 40, No. 2, 1991, pp. 303–312.

[34] Viterbi, A.J., *CDMA: Principles of Spread Spectrum Communication*, Addison-Wesley Wireless Communications Series, Reading, MA: Addison-Wesley, 1995.

[35] Proakis, J.G., *Digital Communications*, 3rd edn, New York: McGraw-Hill, 1995.

[36] Verdú, S., 'Minimum Probability of Error for Asynchronous Gaussian Multiple-Access Channels', *IEEE Trans. Inform. Th.*, Vol. 32, No. 1, 1986, pp. 85–96.

[37] Juntti, M. and Glisic, S., 'Advanced CDMA for Wireless Communications', in *Wireless Communications: TDMA Versus CDMA*, ed. S. Glisic and P. Leppänen, Chapter 4, pp. 447–490, Dordrecht: Kluwer, 1997.

[38] Verdú, S., *Multiuser Detection*, Cambridge: Cambridge University Press, 1998.

[39] Lupas, R. and Verdú, S., 'Near–Far Resistance of Multiuser Detectors in Asynchronous Channels', *IEEE Trans. Commun.*, Vol. 38, No. 4, 1990, pp. 496–508.

[40] Klein, A. and Baier, P.W., 'Linear Unbiased Data Estimation in Mobile Radio Systems Applying CDMA', *IEEE J. Select. Areas Commun.*, Vol. 12, No. 7, 1999, pp. 1058–1066.

[41] Zvonar, Z., 'Multiuser Detection in Asynchronous CDMA Frequency-Selective Fading Channels', *Wireless Personal Communications*, Kluwer, Vol. 3, No. 3–4, 1996, pp. 373–392.

[42] Xie, Z., Short, R.T. and Rushforth, C.K., 'A Family of Suboptimum Detectors for Coherent Multiuser Communications', *IEEE J. Select. Areas Commun.*, Vol. 8, No. 4, 1990, pp. 683–690.

[43] Klein, A., Kaleh, G.K. and Baier, P.W., 'Zero Forcing and Minimum Mean-Square-Error Equalization for Multiuser Detection in Code-Division Multiple Access Channels', *IEEE Trans. Vehic. Tech.*, Vol. 45, No. 2, 1996, pp. 276–287.

[44] Varanasi, M.K. and Aazhang, B., 'Multistage Detection in Asynchronous Code-Division Multiple-Access Communications', *IEEE Trans. Commun.*, Vol. 38, No. 4, 1990, pp. 509–519.

[45] Kohno, R., Imai, H., Hatori, M. and Pasupathy, S., 'Combination of an Adaptive Array Antenna and a Canceller of Interference for Direct-Sequence Spread-Spectrum Multiple-Access System', *IEEE J. Select. Areas Commun.*, Vol. 8, No. 4, 1990, pp. 675–682.

[46] Viterbi, A. J., 'Very Low Rate Convolutional Codes for Maximum Theoretical Performance of Spread-Spectrum Multiple-Access Channels', *IEEE J. Select. Areas Commun.*, Vol. 8, No. 4, 1990, pp. 641–649.

[47] Patel, P. and Holtzman, J., 'Analysis of a Simple Successive Interference Cancellation Scheme in a DS/CDMA System', *IEEE J. Select. Areas Commun.*, Vol. 12, No. 10, 1994, pp. 796–807.

[48] Madhow, U. and Honig, M.L., 'MMSE Interference Suppression for Direct-Sequence Spread-Spectrum CDMA', *IEEE Trans. Commun.*, Vol. 42, No. 12, 1994, pp. 3178–3188.

[49] Rapajic, P. B. and Vucetic, B. S., 'Linear Adaptive Transmitter–Receiver Structures for Asynchronous CDMA Systems', *European Trans. Telecommun.*, Vol. 6, No. 1, 1995, pp. 21–27.

[50] Miller, S. L., 'An Adaptive Direct-Sequence Code-Division Multiple-Access Receiver for Multiuser Interference Rejection', *IEEE Trans. Commun.*, Vol. 43, No. 2/3/4, 1995, pp. 1746–1755.

[51] Latva-aho, M., 'Advanced Receivers for Wideband CDMA Systems', Vol. C125 of *Acta Universitatis Ouluensis*, Doctoral thesis, University of Oulu Press, Oulu, Finland, 1998.

[52] Latva-aho, M. and Juntti, M., 'Modified LMMSE Receiver for DS-CDMA – Part I: Performance Analysis and Adaptive Implementations', *Proceedings of ISSSTA'98*, Sun City, South Africa, September 1998, pp. 652–657.

[53] Honig, M., Madhow, U. and Verdú, S., 'Blind Adaptive Multiuser Detection', *IEEE Trans. Inform. Th.*, Vol. 41, No. 3, 1995, pp. 944–960.

[54] Latva-aho, M., 'LMMSE Receivers for DS-CDMA Systems in Frequency-Selective Fading Channels', in *CDMA Techniques for 3rd Generation Mobile Systems*, ed. F. Swarts, P. van Rooyen, I. Oppermann and M. Lötter, Chapter 13, Dordrecht: Kluwer, 1998.

[55] Juntti, M. and Latva-aho, M., 'Multiuser Receivers for CDMA Systems in Rayleigh Fading Channels', *IEEE Trans. Vehic. Tech.*, Vol. 49, No. 3, 2000, pp. 885–889.

[56] Correal, N.S., Swanchara, S.F. and Woerner, B.D., 'Implementation Issues for Multiuser DS-CDMA Receivers', *Int. J. Wireless Inform. Networks*, Vol. 5, No. 3, 1998, pp. 257–279.

[57] Juntti, M., 'Multiuser Demodulation for DS-CDMA Systems in Fading Channels', Vol. C106 of *Acta Universitatis Ouluensis*, Doctoral thesis, University of Oulu Press, Oulu, Finland, 1997.

[58] Ojanperä, T., Prasad, R. and Harada, H., 'Qualitative Comparison of Some Multiuser Detector'Algorithms for Wideband CDMA', *Proceedings of VTC'98*, Ottawa, Canada, May 1998, pp. 46–50.

[59] Ojanperä, T., 'Multirate Multiuser Detectors for Wideband CDMA', Ph.D. thesis, Technical University of Delft, Delft, The Netherlands, 1999.

[60] Johansson, A.-L., 'Successive Interference Cancellation in DS-CDMA Systems', Doctoral thesis, Chalmers University of Technology, Göteborg, Sweden, 1998.

[61] Juntti, M., 'Multiuser Detector Performance Comparisons in Multirate CDMA Systems', *Proceedings of VTC'98*, Ottawa, Canada, May 1998, pp. 36–40.

[62] Wijting, C.S., Ojanperä, T., Juntti, M. J., Kansanen, K. and Prasad, R., 'Groupwise Serial Multiuser Detectors for Multirate DS-CDMA', *Proceedings of VTC'99*, Houston, TX, May 1999, pp. 836–840.

[63] Juntti, M., 'Performance of Multiuser Detection in Multirate CDMA Systems', *Wireless Pers. Commun.*, Kluwer, Vol. 11, No. 3, 1999, pp. 293–311.

[64] Werner, S. and Lillberg, J., 'Downlink Channel Decorrelation in CDMA Systems with Long Codes', *Proceedings of VTC'99*, Houston, TX, May 1999, pp. 836–840.

[65] Hooli, K., Latva-aho, M. and Juntti, M., 'Multiple Access Interference Suppression with Linear Chip Equalizers in WCDMA Downlink Receivers', *Proceedings of Globecom'99*, Rio de Janeiro, Brazil, December 1999, pp. 467–471.

[66] Hooli, K., Juntti, M. and Latva-aho, M., 'Performance Evaluation of Adaptive Chip-Level Channel Equalizers in {WCDMA} Downlink', *Proceedings of ICC'01*, Helsinki, Finland, June 2001, pp. 1974–1979.

[67] Komulainen, P. and Heikkilä, M., 'Adaptive Channel Equalization Based on Chip Separation for CDMA Downlink', *Proceedings of PIMRC'99*, Osaka, Japan, September 1999, pp. 1114–1118.

[68] Hooli, K. 'Equalization in WCDMA Terminals'. Doctoral thesis. Acta Universitatis Ouluensis C 192, University of Oulu Press, Oulu, Finland, 2003.

[69] Grant, P.M., Spangenberg, S.M., Cruickshank, G.M., McLaughlin, S. and Mulgrew, B., 'New Adaptive Multiuser Detection Technique for CDMA Mobile Receivers', *Proceedings of PIMRC'99*, Osaka, Japan, September 1999, pp. 52–54.

[70] Divsalar, D. and Simon, M. K., 'Improved CDMA Performance Using Parallel Interference Cancellation', *Proc. IEEE MILCOM'94*, Fort Monmouth, N. J. USA, Oct. 2–5, 1994, pp. 911–917.

[71] Cho, Bong Youl and Lee, Jae Hong, 'Nonlinear Parallel Interference Cancellation with Partial Cancellation for a DS-CDMA System', *IEICE Trans. on Communications*, Vol. E83-B, September 2000.

[72] Bae, J., Song, I. and Won, D. H., 'A Selective and Adaptive Interference Cancellation Scheme for Code Division Multiple Access Systems', *Signal Processing*, Vol. 83, No. 2, February 2003, Elsevier Science B.V.

[73] Divsalar, D., Simon, M.K., and Raphaeli, D., 'Improved Parallel Interference Cancellation for CDMA', *IEEE Trans. Commun.*, Vol. 46, No. 2, February 1998, pp. 258–268.

[74] Vihriälä, J. and Horneman, K., 'Impacts of SQ-PIC to Capacity and Coverage in WCDMA Uplink', ISSSTA'04, August 30–September 2 2004.

# 12

# High-Speed Downlink Packet Access

Antti Toskala, Harri Holma, Troels Kolding, Preben Mogensen,
Klaus Pedersen and Jussi Reunanen

## 12.1  Introduction

This chapter presents High-Speed Downlink Packet Access (HSDPA) for Wideband Code Division
Multiple Access (WCDMA), the key new feature included in the Release 5 specifications. The HSDPA
concept has been designed to increase downlink packet data throughput by means of fast physical
layer (L1) retransmission and transmission combining, as well as fast link adaptation controlled by the
Node B (Base Transceiver Station (BTS)). This chapter is organized as follows: first, HSDPA key
aspects are presented and a comparison with Release 99 downlink packet access possibilities is made.
Next, the impact of HSDPA on the terminal uplink (user equipment (UE)) capability classes is summarized and an HSDPA performance analysis is presented, including a comparison with Release 99
packet data capabilities and performance in the case of a shared carrier between HSDPA and non-
HSDPA traffic. Then the link budget for both HSDPA and High-Speed Uplink Packet Access (HSUPA)
is presented and followed by the Iub dimensioning example for HSDPA. The chapter concludes with
a description of the Release 6 enhancements for HSDPA.

## 12.2  Release 99 WCDMA Downlink Packet Data Capabilities

Various methods for packet data transmission in WCDMA downlink already exist in Release 99. As
described in Chapter 10, the three different channels in Release 99/Release 4 WCDMA specifications
that can be used for downlink packet data are:

- Dedicated Channel (DCH)
- Downlink-shared Channel (DSCH)
- Forward Access Channel (FACH).

*WCDMA for UMTS: HSPA Evolution and LTE, Fifth Edition*   Edited by Harri Holma and Antti Toskala
© 2010 John Wiley & Sons, Ltd

The DCH can be used basically for any type of service, and it has a fixed spreading factor (SF) in the downlink. Thus, it reserves the code space capacity according to the peak data rate for the connection. For example, with Adaptive Multirate (AMR) speech service and packet data, the DCH capacity reserved is equal to the sum of the highest rate used for the AMR speech and the highest rate allowed to be sent simultaneously with full rate AMR. This can be used even up to 2 Mbps, but reserving the code tree for a very high peak rate with low actual duty cycle is obviously not a very efficient use of code resources. The DCH is power controlled and may be operated in soft handover as well. Further details of the downlink DCH can be found in Section 6.4.5.

The DSCH, in contrast to DCH (or FACH), has a dynamically varying SF informed on a 10 ms frame-by-frame basis with the Transport Format Combination Indicator (TFCI) signaling carried on the associated DCH. The DSCH code resources can be shared between several users and the channel may employ either single-code or multi-code transmission. The DSCH may be fast power-controlled with the associated DCH but does not support soft handover. The associated DCH can be in soft handover, e.g. speech is provided on DCH if present with packet data. The DSCH operation is described further in Chapter 6. However, the 3rd Generation Partnership Project (3GPP) recognized that HSDPA was such a major step that motivation for DSCH was no longer there, so it was agreed to remove DSCH from the 3GPP specifications from Release 5 onwards.

The FACH, carried on the secondary common control physical channel (S-CCPCH), can be used to downlink packet data as well. The FACH is operated normally on its own, and it is sent with a fixed SF and typically at rather high-power level to reach all users in the cell owing to the lack of physical layer feedback in the uplink. There is no fast power control or soft handover for FACH. The S-CCPCH physical layer properties are described in Section 6.5.4. FACH cannot be used in cases in which a simultaneous speech and packet data service is required.

## 12.3  The HSDPA Concept

The key idea of the HSDPA concept is to increase packet data throughput with methods known already from Global System for Mobile Communications (GSM)/Enhanced data rates for global evolution (EDGE) standards, including link adaptation and fast physical layer (L1) retransmission combining. The physical layer retransmission handling has been discussed earlier, but the inherent large delays of the existing Radio Network Controller (RNC)-based Automatic Repeat reQuest (ARQ) architecture would result in unrealistic amounts of memory on the terminal side. Thus, architectural changes are needed to arrive at feasible memory requirements, as well as to bring the control for link adaptation closer to the air interface. The transport channel carrying the user data with HSDPA operation is denoted as the High-Speed Downlink-Shared Channel (HS-DSCH). A comparison of the basic properties and components of HS-DSCH and DCH is conducted in Table 12.1.

A simple illustration of the general functionality of HSDPA is provided in Figure 12.1. The Node B estimates the channel quality of each active HSDPA user on the basis of, for instance, power control,

**Table 12.1**  Comparison of fundamental properties of DCH and HS-DSCH

| Feature | DCH | HS-DSCH |
|---|---|---|
| Soft handover | Yes | No |
| Fast power control | Yes | No |
| AMC | No | Yes |
| Multi-code operation | Yes | Yes, extended |
| Fast L1 Hybrid ARQ (HARQ) | No | Yes |
| BTS scheduling | No | Yes |

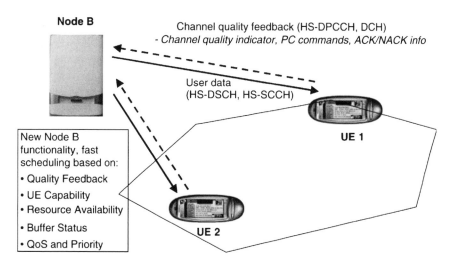

**Figure 12.1**   General operation principle of HSDPA and associated channels

acknowledgement/negative acknowledgement (ACK/NACK) ratio and HSDPA-specific user feedback. Scheduling and link adaptation are then conducted at a fast pace depending on the active scheduling algorithm and the user prioritization scheme. The channels needed to carry data and downlink/uplink control signaling are described later in this chapter.

With HSDPA, two of the most fundamental features of WCDMA, variable SF and fast power control, are disabled and replaced by means of *adaptive modulation and coding* (AMC), extensive multi-code operation and a fast and spectrally efficient retransmission strategy. In the downlink, WCDMA power control dynamics is in the order of 20 dB, compared with the uplink power control dynamics of 70 dB. The downlink dynamics is limited by the intra-cell interference (interference between users on parallel code channels) and by the Node B implementation. This means that, for a user close to the Node B, the power control cannot reduce power maximally; on the other hand, reducing the power to beyond 20 dB dynamics would have only marginal impact on the capacity. With HSDPA, this property is now utilized by the link adaptation function and AMC to select a coding and modulation combination that requires higher $E_c/I_{or}$, which is available for the user close to the Node B (or with good interference/channel conditions in a short-term sense). This leads to additional user throughput, basically for free. To enable a large dynamic range of the HSDPA link adaptation and to maintain a good spectral efficiency, a user may simultaneously utilize up to 15 multi-codes in parallel. The use of more robust coding, fast HARQ and multi-code operation removes the need for variable SF.

To allow the system to benefit from the short-term variations, the scheduling decisions are done in the Node B. The idea in HSDPA is to enable a scheduling such that, if desired, most of the cell capacity may be allocated to one user for a very short time, when conditions are favorable. In the optimum scenario, the scheduling is able to track the fast fading of the users.

The physical layer packet combining basically means that the terminal stores the received data packets in soft memory and, if decoding has failed, the new transmission is combined with the old one before channel decoding. The retransmission can be either identical to the first transmission or contain different bits compared with the channel encoder output that was received during the last transmission. With this incremental redundancy strategy, one can achieve a diversity gain as well as improved decoding efficiency.

## 12.4   HSDPA Impact on Radio Access Network Architecture

All Release 99 transport channels presented earlier in this book are terminated at the RNC. Hence, the retransmission procedure for the packet data is located in the Serving RNC (SRNC), which also handles the connection for the particular user to the core network. With the introduction of HS-DSCH, additional intelligence in the form of an HSDPA Medium Access Control (MAC) layer is installed in the Node B. This way, retransmissions can be controlled directly by the Node B, leading to faster retransmission and thus shorter delay with packet data operation when retransmissions are needed. Figure 12.2 presents the difference between retransmission handling with HSDPA and Release 99 when the serving and controlling RNCs are the same. If no relocation procedure is used in the network, the actual termination point could be several RNCs further into the network. With HSDPA, the Iub interface between Node B and RNC requires a flow-control mechanism to ensure that Node B buffers are used properly and that there is no data loss due to Node B buffer overflow.

The MAC layer protocol in the architecture of HSDPA can be seen in Figure 12.3, showing the different protocol layers for the HS-DSCH. The RNC still retains the functionalities of the Radio

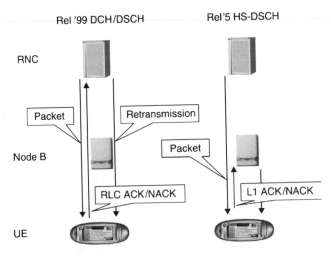

**Figure 12.2**   Release 99 and Release 5 HSDPA retransmission control in the network

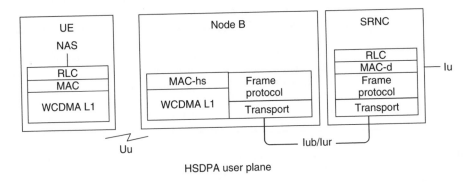

**Figure 12.3**   HSDPA protocol architecture

Link Control (RLC), such as taking care of the retransmission in case the HS-DSCH transmission from the Node B would fail after, for instance, exceeding the maximum number of physical layer retransmissions. Although there is a new MAC functionality added in the Node B, the RNC still retains the Release 99/Release 4 functionalities. The key functionality of the new Node B MAC functionality (MAC-hs) is to handle the ARQ functionality and scheduling, as well as priority handling. Ciphering is done in any case in the RLC layer to ensure that the ciphering mask stays identical for each retransmission to enable physical layer combining of retransmissions.

The type of scheduling to be done in Node B is not defined in 3GPP standardization, only some parameters such as discard timer or scheduling priority indication that can be used by RNC to control the handling of an individual user. As the scheduler type has a big impact to the resulting performance and Quality of Service (QoS), example packet scheduler types are presented in connection with the performance in Section 12.9.

## 12.5   Release 4 HSDPA Feasibility Study Phase

During Release 4 work, an extensive feasibility study was performed on the HSDPA feature to investigate the gains achievable with different methods and the resulting complexity of various alternatives. The items of particular interest were obviously the relative capacity improvement and the resulting increases in the terminal complexity with physical layer ARQ processing, as well as backwards compatibility and coexistence with Release 99 terminals and infrastructure. The results presented in [1] compared the HSDPA cell packet data throughput against Release 99 DSCH performance as presented, and the conclusions drawn were that HSDPA increased the cell throughput up to 100% compared with the Release 99.

The evaluation was conducted for a one-path Rayleigh fading channel environment using carrier/interference (C/I) scheduling. The results from the feasibility study phase were produced for relative comparison purposes only. The HSDPA performance with more elaborate analysis is discussed later in Section 12.9.

## 12.6   HSDPA Physical Layer Structure

The HSDPA is operated similar to DSCH together with DCH, which carries the services with tighter delay constraints, such as AMR speech. To implement the HSDPA feature, three new channels are introduced in the physical layer specifications [2]:

- HS-DSCH carries the user data in the downlink direction, with the peak rate reaching up to the 10 Mbps region with 16 quadrature amplitude modulation (QAM).
- High-Speed Shared Control Channel (HS-SCCH) carries the necessary physical layer control information to enable decoding of the data on HS-DSCH and to perform the possible physical layer combining of the data sent on HS-DSCH in case of retransmission or an erroneous packet.
- Uplink High-Speed Dedicated Physical Control Channel (HS-DPCCH) carries the necessary control information in the uplink, namely ARQ acknowledgements (both positive and negative ones) and downlink quality feedback information.

These three channel types are discussed in the following sections.

### 12.6.1   High-Speed Downlink Shared Channel (HS-DSCH)

The HS-DSCH has specific characteristics in many ways compared with existing Release 99 channels. The Transmission Time Interval (TTI) or interleaving period has been defined to be 2 ms (three slots) to achieve a short round trip delay for the operation between the terminal and Node B for retransmissions.

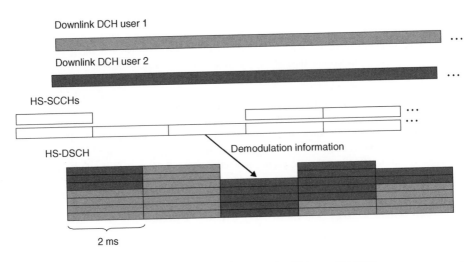

Downlink DCH user 1

Downlink DCH user 2

HS-SCCHs

HS-DSCH

Demodulation information

2 ms

**Figure 12.4**   Code multiplexing example with two active users

The HS-DSCH 2 ms TTI is short compared with the 10, 20, 40 or 80 ms TTI sizes supported in Release 99. Adding a higher-order modulation scheme, 16 QAM, and lower encoding redundancy has increased the instantaneous peak data rate. In the code domain perspective, the SF is fixed; it is always 16, and multi-code transmission as well as code multiplexing of different users can take place. The maximum number of codes that can be allocated is 15, but depending on the terminal (UE) capability, individual terminals may receive a maximum of 5, 10 or 15 codes. The total number of channelization codes with SF 16 is respectively 16 (under the same scrambling code), but as there is need to have code space available for common channels, HS-SCCHs and for the associated DCH, the maximum usable number of codes was thus set to be 15. A simple scenario is illustrated in Figure 12.4, where two users are using the same HS-DSCH. Both users check the information from the HS-SCCHs to determine which HS-DSCH codes to despread, as well as other parameters necessary for correct detection.

### 12.6.1.1   HS-DSCH Modulation

As stated earlier, 16 QAM was introduced in addition to Release 99 Quadrature Phase Shift Keying (QPSK) modulation. Even during the feasibility study phase, 8 PSK and 64 QAM were considered, but eventually these schemes were discarded for performance and complexity reasons. With the constellation example shown in Figure 12.5, 16 QAM doubles the peak data rate compared with QPSK and allows up to 10 Mbps peak data rate with 15 codes of SF 16. However, the use of higher-order

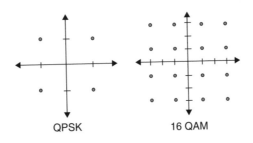

QPSK                          16 QAM

**Figure 12.5**   QPSK and 16 QAM constellations

modulation is not without cost in the mobile radio environment. With Release 99 channels, only a phase estimate is necessary for the demodulation process. Even when 16 QAM is used, amplitude estimation is required to separate the constellation points. Further, more accurate phase information is needed, since constellation points have smaller differences in the phase domain compared with QPSK. The HS-DSCH-capable terminal needs to obtain an estimate of the relative amplitude ratio of the DSCH power level compared with the pilot power level, and this requires that Node B should not adjust the HS-DSCH power between slots if 16 QAM is used in the frame. Otherwise, the performance is degraded, as the validity of an amplitude estimate obtained from Common Pilot Channel (CPICH) and estimated power difference between CPICH and HS-DSCH would no longer be valid.

### 12.6.1.2  HS-DSCH Channel Coding

The HS-DSCH channel coding has some simplifications when compared with Release 99. As there is only one transport channel active on the HS-DSCH, the blocks related to the channel multiplexing for the same users can be left out. Further, the interleaving only spans over a single 2 ms period and there is no separate intra-frame or inter-frame interleaving. Finally, turbo coding is the only coding scheme used. However, by varying the transport block size, the modulation scheme and the number of multi-codes, rates other than 1/3 become available. In this manner, the effective code rate can vary from 1/4 to 3/4. By varying the code rate, the number of bits per code can be increased at the expense of reduced coding gain. The major difference is the addition of the HARQ functionality, as shown in Figure 12.6. When using QPSK, the Release 99 channel interleaver is used, and two parallel (identical) channel interleavers are applied when using 16 QAM. As discussed earlier, the HSDPA-capable Node B has the responsibility of selecting the transport format to be used along with the modulation and number of codes on the basis of the information available at the Node B scheduler.

The HARQ functionality is implemented by means of a two-stage rate-matching functionality, with the principle illustrated in Figure 12.7. The principle shown in Figure 12.7 contains a buffer between the rate-matching stages to allow tuning of the redundancy settings for different retransmissions between the rate-matching stages. The buffer shown should be considered only as virtual buffer, as the obvious practical rate-matching implementation would consist of a single rate-matching block without buffering any blocks after the first rate-matching stage. The HARQ functionality is basically operated in two different ways. It is possible to send identical retransmissions, which is often referred

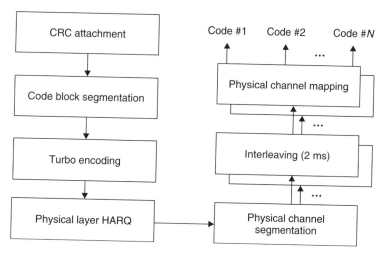

**Figure 12.6**  HS-DSCH channel coding chain

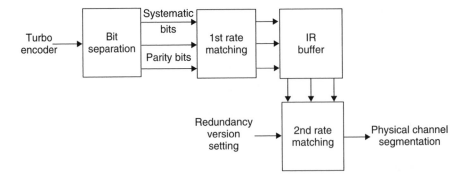

**Figure 12.7** HARQ function principle

to as chase or soft combining. With different parameters, the transmissions will not be identical and then the principle of incremental redundancy is used. In this case, for example, the first transmission could consist of systematic bits, while the second transmission would consist of only parity bits. The latter method has a slightly better performance, but it also needs more memory in the receiver, as the individual retransmissions cannot just be added.

The terminal default memory requirements are done on the basis of soft combining and at maximum data rate (supported by the terminal). Hence, at the highest data rate, only soft combining may be used, whereas incremental redundancy can also be used with lower data rates.

With a 16 QAM constellation, the different bits mapped to the 16 QAM symbols have different reliabilities. This is compensated in connection with the ARQ process with a method called *constellation rearrangement*. With constellation rearrangement, the different retransmissions use slightly different mapping of the bits to 16 QAM symbols to improve the performance. Further details on the HS-DSCH channel coding can be found in [3].

### 12.6.1.3 HS-DSCH versus Other Downlink Channel Types for Packet Data

Table 12.2, presents a comparison of different channel types with respect to the key physical layer properties. In all cases except for the DCH, the packet data do not operate in soft handover. The

**Table 12.2** Comparison of different channel types

| Channel | HS-DSCH | Downlink DCH | FACH |
|---|---|---|---|
| SF | Fixed, 16 | Fixed, (512−4) | Fixed (256−4) |
| Modulation | QPSK/16QAM | QPSK | QPSK |
| Power control | Fixed/slow power setting | Fast with 1500 kHz | Fixed/slow power setting |
| HARQ | Packet combining at L1 | RLC level | RLC level |
| Interleaving | 2 ms | 10−80 ms | 10−80 ms |
| Channel coding schemes | Turbo coding | Turbo and convolutional coding | Turbo and convolutional coding |
| Transport channel multiplexing | No | Yes | Yes |
| Soft handover | For associated DCH | Yes | No |
| Inclusion in specification | Release 5 | Release 99 | Release 99 |

HARQ operation with HS-DSCH will also be employed at the RLC level if the physical layer ARQ timers or the maximum number of retransmissions are exceeded.

## 12.6.2 High-Speed Shared Control Channel (HS-SCCH)

The HS-SCCH carries the key information necessary for HS-DSCH demodulation. The Universal Mobile Telecommunication Services (UMTS) Terrestrial Radio Access Network (RAN) (UTRAN) needs to allocate a number of HS-SCCHs that correspond to the maximum number of users that will be code-multiplexed. If there are no data on the HS-DSCH, then there is no need to transmit the HS-SCCH either. From the network point of view, there may be a high number of HS-SCCHs allocated, but each terminal will only need to consider a maximum of four HS-SCCHs at a given time. The HS-SCCHs that are to be considered are signalled to the terminal by the network. In reality, the need for more than four HS-SCCHs is very unlikely. However, more than one HS-SCCH may be needed to match the available codes better to the terminals with limited HSDPA capability.

Each HS-SCCH block has a three-slot duration that is divided into two functional parts. The first slot (first part) carries the time-critical information that is needed to start the demodulation process in due time to avoid chip-level buffering. The next two slots (second part) contain less time-critical parameters, including a Cyclic Redundancy Check (CRC) to check the validity of the HS-SCCH information and HARQ process information. For protection, both HS-SCCH parts employ terminal-specific masking to allow the terminal to decide whether the control channel detected is actually intended for the particular terminal.

The HS-SCCH uses SF 128, which can accommodate 40 bits per slot (after channel encoding) because there are no pilot or Transmit Power Control (TPC) bits on HS-SCCH. The HS-SCCH used half-rate convolution coding with both parts encoded separately from each other because the time-critical information is required to be available immediately after the first slot and thus cannot be interleaved together with Part 2.

The HS-SCCH Part 1 parameters indicate the following:

- Codes to despread. This also relates to the terminal capability in which each terminal category indicates whether the current terminal can despread a maximum of 5, 10 or 15 codes.
- Modulation to indicate if QPSK or 16 QAM is used.

The HS-SCCH Part 2 parameters indicate the following:

- Redundancy version information to allow proper decoding and combining with the possible earlier transmissions.
- ARQ process number to show which ARQ process the data belongs to.
- First transmission or retransmission indicator to indicate whether the transmission is to be combined with the existing data in the buffer (if not successfully decoded earlier) or whether the buffer should be flushed and filled with new data.

Parameters such as actual channel coding rate are not signaled, but can be derived from the transport block size and other transport format parameters.

As illustrated in Figure 12.8, the terminal has a single slot duration to determine which codes to despread from the HS-DSCH. The use of terminal-specific masking allows the terminal to check whether data were intended for it. The total number of HS-SCCHs that a single terminal monitors (the Part 1 of each channel) is at a maximum of four, but if there are data for the terminal in consecutive TTIs, then the HS-SCCH will be the same for that terminal between TTIs to increase signaling reliability. This kind of approach is necessary not only to avoid the terminal having to

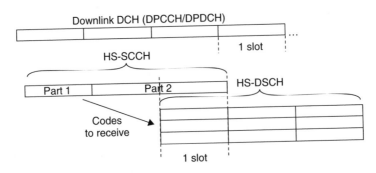

**Figure 12.8**  HS-SCCH and HS-DSCH timing relationship

buffer data not necessarily intended for it, but also as there could be more codes in use than supported by the terminal capability. The downlink DCH timing is not tied to the HS-SCCH (or consequently HS-DSCH) timing.

## 12.6.3  Uplink High-Speed Dedicated Physical Control Channel (HS-DPCCH)

The uplink direction has to carry both ACK/NACK information for the physical layer retransmissions and the quality feedback information to be used in the Node B scheduler to determine to which terminal to transmit and at which data rate. It was required to ensure operation in soft handover in the case that not all Node Bs have been upgraded to support HSDPA. Thus, it was decided to leave the existing uplink channel structure unchanged and add the new information elements needed on a parallel code channel that is termed the *Uplink HS-DPCCH*. The HS-DPCCH is divided into two parts, as shown in Figure 12.9, and carries the following information:

- ACK/NACK transmission, to reflect the results of the CRC check after the packet decoding and combining.
- Downlink Channel Quality Indicator (CQI) to indicate which estimated transport block size, modulation type and number of parallel codes could be received correctly (with reasonable *block error rate* (BLER)) in the downlink direction.

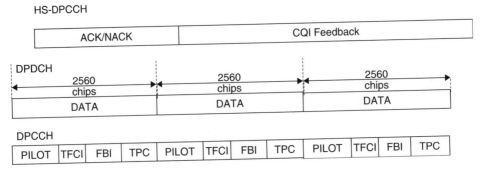

**Figure 12.9**  HS-DPCCH structure

In 3GPP standardization, there was a lively discussion on this aspect, as it is not a trivial issue to define a feedback method that (1) takes into account different receiver implementations and so forth and (2) simultaneously, is easy to convert to suitable scheduler information in the Node B side. In any case, the feedback information consists of 5 bits that carry quality-related information. One signaling state is reserved for the state 'do not bother to transmit' and other states represent what is the transmission that the terminal can receive at the current time. Hence, these states range in quality from single-code QPSK transmission up to 15 codes 16 QAM transmission (including various coding rates). Obviously, the terminal capability restrictions need to be taken into account in addition to the feedback signaling; thus, the terminals that do not support the certain number of codes' part of the CQI feedback table will signal the value for power-reduction factor related to the most demanding combination supported from the CQI table. The CQI table consists of roughly evenly spaced reference transport block size, number of codes and modulation combination that also define the resulting coding rate.

The HS-DPCCH needs some part of the uplink transmission power, which has an impact on the link budget for the uplink. The resulting uplink coverage impact is discussed later in connection with the performance in Section 12.9.

## 12.6.4 HSDPA Physical Layer Operation Procedure

The HSDPA physical layer operation goes through the following steps:

1. The scheduler in the Node B evaluates for different users what are the channel conditions, how much data are pending in the buffer for each user, how much time has elapsed since a particular user was last served, for which users retransmissions are pending, and so forth. The exact criteria that have to be taken into account in the scheduler are naturally a vendor-specific implementation issue.
2. Once a terminal has been determined to be served in a particular TTI, the Node B identifies the necessary HS-DSCH parameters. For instance, how many codes are available or can be filled? Can 16 QAM be used? What are the terminal capability limitations? The terminal soft memory capability also defines which kind of HARQ can be used.
3. The Node B starts to transmit the HS-SCCH two slots before the corresponding HS-DSCH TTI to inform the terminal of the necessary parameters. The HS-SCCH selection is free (from the set of maximum four channels) assuming there were no data for the terminal in the previous HS-DSCH frame.
4. The terminal monitors the HS-SCCHs given by the network and, once the terminal has decoded Part 1 from an HS-SCCH intended for that terminal, it will start to decode the rest of that HS-SCCH and will buffer the necessary codes from the HS-DSCH.
5. Upon having the HS-SCCH parameters decoded from Part 2, the terminal can determine to which ARQ process the data belong and whether it needs to be combined with data already in the soft buffer.
6. Upon decoding the potentially combined data, the terminal sends in the uplink direction an ACK/NACK indicator, depending on the outcome of the CRC check conducted on the HS-DSCH data.
7. If the network continues to transmit data for the same terminal in consecutive TTIs, then the terminal will stay on the same HS-SCCH that was used during the previous TTI.

The HSDPA operation procedure has strictly specified timing values for the terminal operation from the HS-SSCH reception via HS-DSCH decoding to the uplink ACK/NACK transmission. The key timing value from the terminal point of view is the 7.5 slots from the end of the HS-DSCH TTI to the start of the ACK/NACK transmission in the HS-DPCCH in the uplink. The timing relationship between downlink and uplink is illustrated in Figure 12.10. The network side is asynchronous in

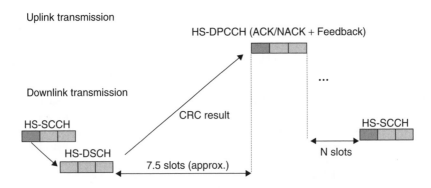

Figure 12.10   Terminal timing with respect to one HARQ process

terms of when to send a retransmission in the downlink. Therefore, depending on the implementation, different amounts of time can be spent on the scheduling process in the network side.

Terminal capabilities do not influence the timing of an individual TTI transmission but do define how often one can transmit to the terminal. The capabilities include information of the minimum inter-TTIinterval that tells whether consecutive TTIs may be used or not. Value 1 indicates that consecutive TTIs may be used, while values 2 and 3 correspond to leaving a minimum of one or two empty TTIs between packet transmissions.

Since downlink DCH and, consecutively, uplink DCH are not slot-aligned to the HSDPA transport channels, the uplink HS-DPCCH may start in the middle of the uplink slot as well, and this needs to be taken into account in the uplink power setting process. The uplink timing is thus quantized to 256 chips (symbol aligned) and minimum values as 7.5 slots −128 chip, 7.5 slots +128 chips. This is illustrated in Figure 12.11.

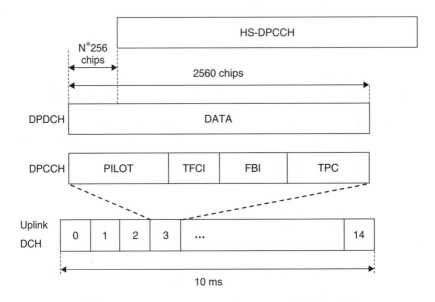

Figure 12.11   Uplink DPCH and HS-SCCH timing relationship

## 12.7 HSDPA Terminal Capability and Achievable Data Rates

The HSDPA feature is optional for terminals in Release 5 with a total of 12 different categories of terminals (from the physical layer point of view) with resulting maximum data rates ranging between 0.9 to 14.4 Mbps. The HSDPA capability is otherwise independent from Release 99-based capabilities; but, if HS-DSCH has been configured for the terminal, then DCH capability in the downlink is limited to the value given by the terminal. A terminal can indicate 32, 64, 128 or 384 kbps DCH capability, as described in Chapter 6.

The terminal capability classes are shown in Table 12.3. The first 10 HSDPA terminal capability categories need to support 16 QAM, but the last two, categories 11 and 12, support only QPSK modulation. The differences between classes lie in the maximum number of parallel codes that must be supported and whether the reception in every 2 ms TTI is required. The highest HSDPA class supports 10 Mbps. Besides the values indicated in Table 12.3, there is the soft buffer capability with two principles used for determining the value for soft buffer capability. The specifications indicate the absolute values, which should be understood in the way that a higher value means support for incremental redundancy at maximum data rate and a lower value permits only soft combining at full rate. While determining when incremental redundancy can also be applied, one needs to observe the memory partitioning per ARQ process defined by the SRNC. There is a maximum of eight ARQ processes per terminal.

Category number 10 is intended to allow the theoretical maximum data rate of 14.4 Mbps, permitting basically the data rate that is achievable with rate 1/3 turbo coding and significant puncturing resulting in the code rate close to 1. For category 9, the maximum turbo-encoding block size (from Release 99) has been taken into account when calculating the values, and thus resulting in the 10.2 Mbps peak user data rate value with four turbo-encoding blocks. It should be noted that, for HSDPA operation, the terminal will not report individual values, only the category. The classes shown in Table 12.3 are as included in [4] with 12 distinct terminal classes. From a Layer 2/3 point of view, the important terminal capability parameter to note is the RLC reordering buffer size that basically determines the window length of the packets that can be 'in the pipeline' to ensure in-sequence delivery of data to higher layers in the terminal. The minimum values range from 50 to 150 kB depending on the UE category. In the first phase of HSDPA market the devices offered had 1.8 Mbps capability but then the market quickly moved to 3.6 Mbps capable devices. Currently the most advanced devices provide

**Table 12.3** HSDPA terminal capability categories

| Category | Max. no. of parallel codes HS-DSCH | Min. inter-TTI | Transport channel bits per TTI | ARQ type at max. data rate | Achievable max. data rate (Mbps) |
|----------|------------|------------|----------|------|------|
| 1  | 5  | 3 | 7 298  | Soft | 1.2  |
| 2  | 5  | 3 | 7 298  | IR   | 1.2  |
| 3  | 5  | 2 | 7 298  | Soft | 1.8  |
| 4  | 5  | 2 | 7 298  | IR   | 1.8  |
| 5  | 5  | 1 | 7 298  | Soft | 3.6  |
| 6  | 5  | 1 | 7 298  | IR   | 3.6  |
| 7  | 10 | 1 | 14 411 | Soft | 7.2  |
| 8  | 10 | 1 | 14 411 | IR   | 7.2  |
| 9  | 15 | 1 | 20 251 | Soft | 10.2 |
| 10 | 15 | 1 | 27 952 | IR   | 14.4 |
| 11 | 5  | 2 | 3 630  | Soft | 0.9  |
| 12 | 5  | 1 | 3 630  | Soft | 1.8  |

**Table 12.4** Theoretical bit rates with 15 multi-codes for different TFRCs

| TFRC | Modulation | Effective code rate | Max. throughput (Mbps) |
|------|------------|---------------------|------------------------|
| 1 | QPSK | $\frac{1}{4}$ | 1.8 |
| 2 | QPSK | 2/4 | 3.6 |
| 3 | QPSK | $\frac{3}{4}$ | 5.3 |
| 4 | 16QAM | 2/4 | 7.2 |
| 5 | 16QAM | $\frac{3}{4}$ | 10.7 |

10.2 Mbps or even 14.4 Mbps. The next step is towards support of Release 7 and 8 features to reach higher data rates with 64QAM and dual-cell HSDPA and also to enable mapping also CS services on top of HSDPA as described in Chapter 15.

Besides the parameters part of the UE capability, the terminal data rate can be largely varied by changing the coding rate as well. Table 12.4 shows the achievable data rates when keeping the number of codes constant (15) and changing the coding rate as well as the modulation. Table 12.4 shows some example bit rates without overhead considerations for different transport format and resource combinations (TFRCs).

These theoretical data rates can be allocated to a single user or divided between several users. This way, the network can match the allocated power/code resources to the terminal capabilities and data requirements of the active terminals. In contrast to Release 99 operation, it is worth noting that the data rate negotiated with the core network is typically smaller than the peak data rate used in the air interface. Thus, even if the maximum data rate negotiated with the core network were, for example, 1 Mbps or 2 Mbps, the physical layer would use (if conditions permit) a peak data rate of, for example, 3.6 Mbps.

## 12.8   Mobility with HSDPA

The mobility procedures for HSDPA users are affected by the fact that transmission of the HS-PDSCH and the HS-SCCH to a user belongs to only one of the radio links assigned to the UE, namely the serving HS-DSCH cell. UTRAN determines the serving HS-DSCH cell for an HSDPA-capable UE, just as it is the UTRAN that selects the cells in a certain user's active set for DCH transmission/reception. Synchronized change of the serving HS-DSCH cell is supported between UTRAN and the UE, so that connectivity on HSDPA is achieved if the UE moves from one cell to another, so that start and stop of transmission and reception of the HS-PDSCH and the HS-SCCH are done at a certain time dictated by the UTRAN. This allows implementation of HSDPA with full mobility and coverage to fully exploit the advantages of this scheme over Release 99 channels. The serving HS-DSCH cell may be changed without updating the user's active set for the Release 99 dedicated channels or in combination with establishment, release, or reconfiguration of the dedicated channels. In order to enable such procedures, a new measurement event from the user is included in Release 5 to inform UTRAN of the best serving HS-DSCH cell.

In the following subsections we will briefly discuss the new UE measurement event for support of mobility for HSDPA users, as well as outline the procedures for intra- and inter-Node B HS-DSCH to HS-DSCH handover. Finally, in Section 12.8.4 we address handover from HS-DSCH to DCH. To narrow the scope further, we only address intra-frequency handovers for HSDPA users, even though inter-frequency handovers are also applicable for HSDPA users triggered by, for instance, compressed mode measurements from the user, as discussed in Chapter 9.

## 12.8.1 Measurement Event for Best Serving HS-DSCH Cell

As discussed in Section 9.3, it is the user's SRNC that determines the cells that should belong to the user's active set for transmission of dedicated channels. The SRNC typically bases its decisions on requests received from the user that are triggered by measurements on the P-CPICH from the cells in the user's candidate set. Similarly, for HSDPA, a measurement event 1d has been defined, which is called the measurement event for the best serving HS-DSCH cell [5]. This measurement basically reports the best serving HS-DSCH cell to the SRNC based on a measurement of the P-CPICH $E_c/I_0$ or the P-CPICH received signal code power (RSCP) measurements for the potential candidate cells for serving HS-DSCH cell, as illustrated in Figure 12.12. It is possible to configure this measurement event so that all cells in the user's candidate set are taken into account, or to restrict the measurement event so that only the current cells in the user's active set for dedicated channels are considered. Usage of a hysteresis margin to avoid fast change of the serving HS-DSCH cell is also possible for this measurement event, as well as specification of a cell individual offset to favor certain cells, i.e. to extend their HSDPA coverage area, for instance.

## 12.8.2 Intra-Node B HS-DSCH to HS-DSCH Handover

Once the SRNC decides to make an intra-Node B handover from a source HS-DSCH cell to a new target HS-DSCH cell under the same Node B as illustrated on Figure 12.13, the SRNC sends a synchronized radio link reconfiguration prepare message to the Node B, as well as a radio resource control (RRC) physical channel reconfiguration message to the user. At a specified time index where the handover from the source cell to the new target cell is carried out, the source cell stops transmitting to the user, and the MAC-hs packet scheduler in the target cell is thereafter allowed to control transmission to the user. Similarly, the terminal starts to listen to the HS-SCCH (or several HS-SCCHs depending on the MAC-hs configuration) from the new target cell, i.e. the new serving HS-DSCH cell. This also implies that the CQI reports from the user are measured from the channel quality corresponding to the new target cell. It is typically recommended that the MAC-hs in the target cell does not start transmitting to the user until it has received the first CQI report that is measured from the target cell.

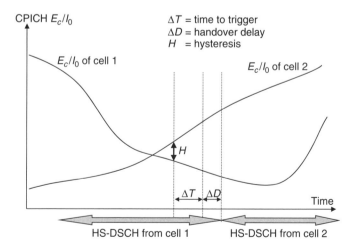

**Figure 12.12** Best serving HS-DSCH cell measurement

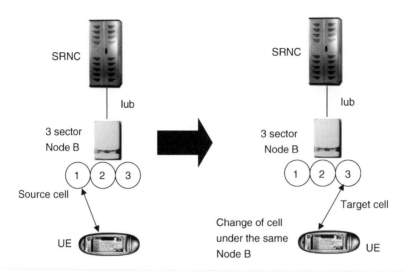

**Figure 12.13**   Example of intra-Node B HS-DSCH to HS-DSCH handover

Prior to the HS-DSCH handover from the source cell to the new target cell, there are likely to be several protocol data units (PDUs) buffered in the source cell's MAC-hs for the user, both PDUs that have never been transmitted to the user and pending PDUs in the HARQ manager that are either awaiting ACK/NACK on the uplink HS-DPCCH or PDUs that are waiting to be retransmitted to the user. Assuming that the Node B supports MAC-hs preservation, all the PDUs for the user are moved from the MAC-hs in the source cell to the MAC-hs in the target cell during the HS-DSCH handover. This means that the status of the HARQ manager is also preserved without triggering any higher layer retransmission, such as RLC retransmissions during intra-Node B HS-DSCH to HS-DSCH handover. If the Node B does not support the MAC-hs preservation, then handling of the PDU not completed is the same as in inter-Node B handover case.

During intra-Node B HS-DSCH to HS-DSCH handover, it is highly likely that the user's associated DPCH is potentially in a two-way softer handover. Under such conditions the uplink HS-DPCCH may also be regarded as being in a two-way softer handover, so Rake fingers for demodulation of the HS-DPCCH are allocated to both cells in the user's active set. This implies that uplink coverage of the HS-DPCCH is improved for users in softer handover and no power control problems are expected.

## 12.8.3   Inter-Node–Node B HS-DSCH to HS-DSCH Handover

Inter-Node B HS-DSCH to HS-DSCH handover is also supported by the 3GPP specifications, where the serving HS-DSCH source cell is under one Node B while the new target cell is under another Node B, and potentially also under another RNC, as illustrated in Figure 12.14. Once the SRNC decides to initiate such a handover, a synchronized radio link reconfiguration prepare message is sent to the drifting RNC and the Node B that controls the target cell, as well as an RRC physical channel reconfiguration message to the user. At the time where the cell change is implemented, the MAC-hs for the user in the source cell is reset, which basically means that all buffered PDUs for the user are deleted, including the pending PDUs in the HARQ manager. At the same time index, the flow control unit in the MAC-hs in the target cell starts to request PDUs from the MAC-d in the SRNC, so that it can start to transmit data on the HS-DSCH to the user.

As the PDUs that were buffered in the source cell prior to the handover are deleted, these PDUs must be recovered by higher layer retransmissions, such as RLC retransmissions. When the RLC

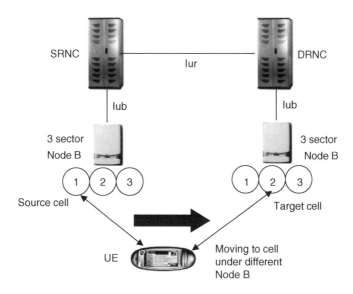

**Figure 12.14**   Example of inter-Node B HS-DSCH to HS-DSCH handover

protocol realizes that the PDUs it has originally forwarded to the source cell are not acknowledged, it will initiate retransmissions, which basically implies forwarding the same PDUs to the new target cell that was deleted in the source cell. In order to reduce the potential PDU transmission delays during this recovery phase, the RLC protocol at the user can be configured to send an RLC status report to the UTRAN at the first time instant after the serving HS-DSCH cell has been changed [6]. This implies that the RLC protocol in the RNC can immediately start to forward the PDUs that were deleted in the source cell prior to the HS-DSCH cell change.

For user applications that do not include any higher layer retransmission mechanisms, such as applications running over User Datagram Protocol and RLC transparent or unacknowledged mode, the PDUs that are deleted in the source cell's MAC-hs prior to the handover are lost forever. For such applications, therefore, having large data amounts (many PDUs) buffered in the MAC-hs should be avoided, as these may be lost if an inter-Node B HS-DSCH to HS-DSCH handover is suddenly initiated.

### 12.8.4   HS-DSCH to DCH Handover

Handover from an HS-DSCH to DCH may potentially be needed for HSDPA users that are moving from a cell with HSDPA to a cell without HSDPA (Release 99-compliant only cell), as illustrated in Figure 12.15. Once the SRNC decides to initiate such a handover, a synchronized radio link reconfiguration prepare message is sent to the Node Bs involved, as well as an RRC physical channel reconfiguration message to the user. Similarly, for the inter-Node B HS-DSCH to HS-DSCH handover, the HS-DSCH to DCH handover results in a reset of the PDUs in MAC-hs in the source cell, which subsequently requires recovery via higher layer retransmissions, such as RLC retransmissions.

The Release 5 specifications also support implementation of handover from DCH to HS-DSCH. This handover type may, for instance, be used if a user is moving from a non-HSDPA-capable cell into an HSDPA-capable cell, or to optimize the load balance in between HSDPA and DCH use in a cell.

Table 12.5 presents a summary of the different handover modes and their characteristics. Notice that the handover delay is estimated to be on the order of 300 to 500 ms, which indicates that the

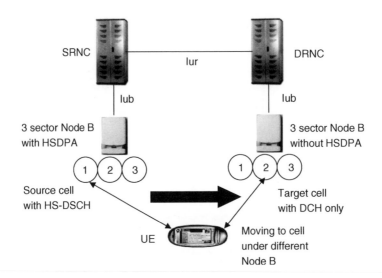

**Figure 12.15**   Example of HS-DSCH to DCH handover

**Table 12.5**   Summary of HSDPA handover types and their characteristics

|  | Intra-Node BHS-DSCH to HS-DSCH | Inter-Node BHS-DSCH to HS-DSCH | HS-DSCH to DCH |
| --- | --- | --- | --- |
| Handover measurement |  | By UE |  |
| Handover decision |  | By SRNC |  |
| Packet retransmissions | Packets forwarded from source MAC-hs to target MAC-hs | Packets not forwarded. RLC retransmissions used from SRNC | RLC retransmissions used from SRNC |
| Packet losses | No | No, when RLC acknowledged mode used | No, when RLC acknowledged mode used |
| Delay (ms) | 300–500 | 300–500 | 300–500 |
| Uplink HS-DPCCH | Softer handover can be used for HS-DPCCH | HS-DPCCH received by one cell |  |

activation time for the synchronized handover should be 300 to 500 ms from the time where the SRNC decides to make the handover. The actual handover delay will, in practice, depend on the RNC implementation and the size of the RRC message that is sent to the user during the handover phase and the data rate on the Layer 3 signaling channel on the associated DPCH.

## 12.9   HSDPA Performance

In this section, different performance aspects related to HSDPA are discussed. Since the two most basic features of WCDMA, fast power control and variable SF, have been disabled, a performance evaluation of HSDPA involves considerations that differ somewhat from the general WCDMA analysis. For packet data traffic, HSDPA offers a significant gain over the existing Release 99 DCH and

DSCH bearers. It facilitates very fast per-2-ms switching among users, which gives high trunking efficiency and code utilization for bursty packet services. Further, with the introduction of higher order modulation and reduced channel encoding, even very high radio quality conditions can be mapped into increased user throughput and cell capacity. Finally, advanced packet scheduling, which considers the user's instantaneous radio channel conditions, can produce a very high cell capacity while maintaining tight end-to-end QoS control. In the following subsections, single-user and multi-user issues are discussed separately. After this description, some examples of HSDPA system performance are given, looking first at the system performance in the 'all HSDPA users' scenario and then looking at the situation when operating the system in emigration phase, where large numbers of terminals do not yet have HSDPA capability.

## 12.9.1 Factors Governing Performance

The HSDPA mode of operation encounters a change in environment and channel performance by fast adaptation of modulation, coding and code resource settings. The performance of HSDPA depends on a number of factors, including the following:

- *Channel conditions*. Time dispersion, cell environment, terminal velocity as well as the ratio of experienced own-cell interference with other cell interference ($I_{or}/I_{oc}$). Compared with the DCHs, the average $I_{or}/I_{oc}$ ratio at the cell edge is reduced for HSDPA owing to the lack of soft handover gain. Macrocell network measurements indicate typical values down to -5 dB compared with approximately -2 to 0 dB for DCH.
- *Terminal performance*. Basic detector performance (e.g. sensitivity and interference suppression capability) and HSDPA capability level, including supported peak data rates and number of multi-codes.
- *Nature and accuracy of radio resource management*. Power and code resources allocated to the HSDPA channel and accuracy/philosophy of Signal-to-Interference (SIR) power ratio estimation and packet-scheduling algorithms.

For a terminal with high detection performance, some experienced SIR would potentially map into a higher throughput performance experienced directly by the HSDPA user.

## 12.9.2 Spectral Efficiency, Code Efficiency and Dynamic Range

In WCDMA, both spectral efficiency and code efficiency are important optimization criteria to accommodate code-limited and power-limited system states. In this respect, HSDPA provides some important improvements over Release 99 DCH and DSCH:

- Spectral efficiency is improved at lower SIR ranges (medium to long distance from Node B) by introducing more efficient coding and fast HARQ with redundancy combining. HARQ combines each packet retransmission with earlier transmissions, such that no transmissions are wasted. Further, extensive multi-code operations offer high spectral efficiency, similar to variable SF but with higher resolution. At very good SIR conditions (vicinity of Node B), HSDPA offers higher peak data rates and, thus, better channel utilization and spectral efficiency.
- Code efficiency is obtained by offering more user bits per symbol and, thus, more data per channelization code. This is obtained through higher-order modulation and reduced coding. Further, the use of time multiplexing and shared channels generally leads to better code utilization for bursty traffic, as described in Chapter 10.

The principle of HSDPA is to adapt to the current channel conditions by selecting the most suitable modulation and coding scheme, leading to the highest throughput level.

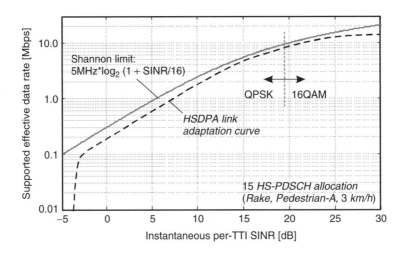

**Figure 12.16**   SINR to throughput mapping table with a single HS-PDSCH

In reality, the available data rate range may be slightly limited in both ends due to reasons of packet header overhead and practical detection limitations. The maximum peak data rate is, thus, often described to be on the order of 11–12 Mbps. The key measure for describing the link performance is the narrowband *signal-to-interference-and-noise ratio* (SINR) as experienced by the UE detector (e.g. the received $E_s/N_0$). In hostile environments, the availability of high SINR is limited, which reduces the link and cell throughput capabilities.

An example SINR-to-throughput mapping function is illustrated in Figure 12.16 for a Pedestrian-A profile with a Rake receiver moving at 3 km/h. The curve includes the first transmission BLER and, thus, considers the basic HARQ mechanism. The HARQ mechanism provides some additional data rate coverage in the lower end and provides a smoother transmission between the different transport block size settings. On the curve, the operating regions for the two modulation options are also illustrated. As QPSK requires less power per user bit to be received correctly, the available options of higher code rate and multiple HS-PDSCHs are used before switching to 16 QAM. Measured in the SINR domain, the total link adaptation dynamic range is on the order of 30–35 dB. It is comparable to the dynamic range of power control with variable SF but is shifted in order to work at higher SINR and throughput values. When only a single HS-PDSCH code is employed, the transition curve saturates earlier at a maximum peak data rate value around 900 kbps. For reference, Figure 12.16 also illustrates the theoretical Shannon capacity for a 5 MHz bandwidth. There is a 1–2 dB difference, due mainly to decoder limitations, receiver estimation inaccuracies, and a relatively low chip rate to channel bandwidth ratio.

The single-user link adaptation performance depends on other issues, such as CQI measuring, transmission, and processing delays. This adds to the inherent delay associated with the two-slot time difference between the HS-SCCH frame and the corresponding HS-DSCH packet. The minimum total delay is around 6 ms between the time of estimation of the CQI report and the time when the first packet based on this report can be received by the UE. If the UE employs CQI repetition to gain in uplink coverage, this delay increases further. As mentioned earlier, the target BLER for the CQI report is 10%, but even higher spectral efficiency can be achieved by operating the system at a first transmission BLER level of 15–40%. However, operation of the system at a lower target BLER may be attractive from delay and hardware utilization considerations; thus, in the simulation in this chapter, 10% is chosen as the target value for first transmission BLER. The link adaptation performance when

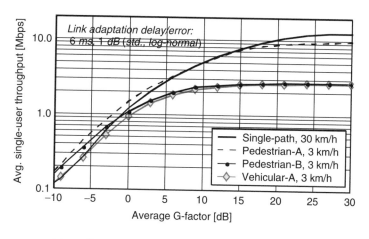

**Figure 12.17** Link adaptation performance versus G-factor

only a single user is being scheduled with a certain average G-factor is depicted in Figure 12.17 for the 15 code case as well (G-factor is the ratio between wideband received own-cell power and other cell interference plus noise). Figure 12.17 assumes the use of non-identical retransmissions and 75% power allocation for HSDPA use. For a typical macrocell environment, the G-factor near the cell edge is approximately -3 dB, whereas the median G-factor is around 2 dB. For users in good conditions, the G-factor may be on the order of 12–15 dB. A Rake receiver is assumed and it is seen that this receiver type is limited at low interference levels by the lack of orthogonality in the Pedestrian-B and Vehicular-A environments.

While the HS-DSCH offers high spectral efficiency, it should be noted that at least one (non-power-controlled) HS-SCCH is needed to operate the system. This also implies that the data rate carried on the HS-DSCH should be sufficient to compensate for the interference due to the relative HS-SCCH overhead. As mentioned previously, code multiplexing can be used to send HSDPA data to several users within the same TTI by sharing the HS-PDSCH code set between them. Code multiplexing is useful when a single user cannot utilize the total power and/or code resources due to lack of buffered data or due to the network being able to transmit more codes than the UE supporting. Considering the overhead of having multiple parallel HS-SCCHs and all UEs will support a minimum of five codes, it is not expected that more than three users need to be code multiplexed in practice even if all the cell traffic were to use HSDPA. In general, HSDPA offers the best potential for large packet sizes and bit rates. Therefore, services resulting in small packet sizes at low data rates, e.g. gaming applications, may be best served using other channel types.

The dependence between the average user throughput per code and the code power is shown in Figure 12.18 for different $I_{or}/I_{oc}$ conditions and different channel profiles using HARQ with soft combining. Owing to the code efficiency inherent in the higher-order TFRCs, HS-DSCH supports higher data rates when more power is allocated to the code. However, by noting that the slopes of the curves in Figure 12.18 generally decrease, it is clear that the spectral efficiency degrades as the power is increased. However, if only limited code resources are available, then the available power can be utilized better compared with, for instance, DSCH, which is hard-limited to 128 kbps per code at an SF 16 level. To achieve 384 kbps with DSCH, the code resources must be doubled (SF 8). Comparing the difference between the Pedestrian-A and Vehicular-A channel profiles, it is evident that the gain achieved by increasing the power is higher when the terminal is limited by time dispersion. At low values of $I_{or}/I_{oc}$, the terminal is mainly interference limited and the two cases become similar.

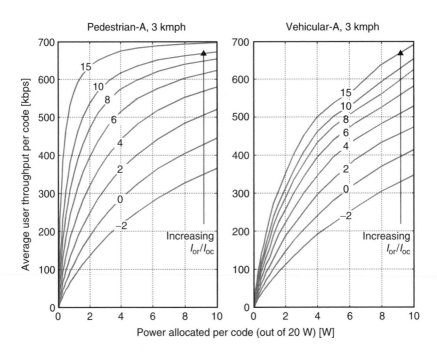

**Figure 12.18**  Average user throughput per code versus code power allocation

## 12.9.3  User Scheduling, Cell Throughput and Coverage

The HSDPA cell throughput depends significantly on the interference distribution across the cell, the time dispersion and the multi-code and power resources allocated to HSDPA. In Figure 12.19, the *cumulative distribution function* (CDF) of instantaneous user throughput for both macrocell outdoor and microcell outdoor–indoor scenarios is considered. The CDFs shown correspond to the case in which fair time scheduling is employed. Fair time scheduling means that the same power is allocated to all users such that users with better channel conditions experience a higher throughput. Figure 12.19 assumes that the available capacity of the cell is allocated to the user studied and that other cells are fully loaded. Note that, in the microcell case, 30% of the users have sufficient channel quality to support peak data rates exceeding 10 Mbps due to limited time dispersion and high cell isolation. The mean bit rate that can be obtained is more than 5 Mbps. For the macrocell case, the presence of time dispersion and high levels of other cell interference widely limits the available peak data rates. Nevertheless, peak data rates of more than 512 kbps are supported 70% of the time and the mean bit rate is more than 1 Mbps. For users located in the vicinity of the Node B, time dispersion limits the maximum peak data rate to around 6 to 7 Mbps. As discussed earlier, with 16 QAM the channel estimation is more challenging than with QPSK and, thus, it is not usable in all cell locations. With macrocell environment (with Vehicular-A channel model) the probability for using 16 QAM is between 5 and 10%, assuming the terminal has a normal Rake receiver. When the delay profile is more favorable and cell isolation is higher with a microcell environment, then the probability increases to approximately 25% with the Pedestrian-A environment. The value of 30% for the cell area in Figure 12.19 is lacking some imperfections, such as the interference between code channels due to hardware imperfections, which shows more in the Pedestrian-A-type environment, where the orthogonality is well preserved by the channel itself.

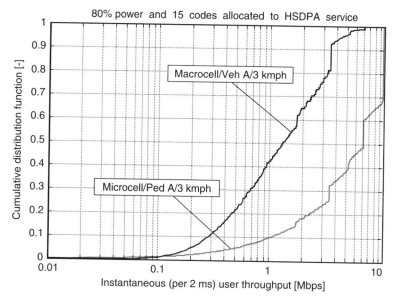

**Figure 12.19** Instantaneous user throughput CDF for microcell and macro cell scenarios

The packet-scheduling method chosen has a significant impact on the overall cell throughput and the end user's perceived QoS. This aspect is related to the gain by *multi-user diversity*. With fast scheduling and multiple users it is possible at any given time to pick the 'best' user in the cell, e.g. a selection diversity mechanism that may be of very high order. The concept of multi-user diversity is illustrated in Figure 12.20a. Such scheduling is denoted as advanced or opportunistic packet scheduling, as opposed to the blind packet scheduling methods that do not consider the radio conditions. Examples of the latter types are the round robin in time and fair throughput packet schedulers. Probably the most straightforward and aggressive advanced packet scheduler is the maximum-throughput or maximum-C/I packet scheduler, which always schedules the user with the best instantaneous channel quality. Its principle is depicted in Figure 12.20b. The main drawbacks of this scheduler are mainly its inherent unfairness and coverage limitations. Several publications list different HSDPA packet scheduler options, including [7, 8].

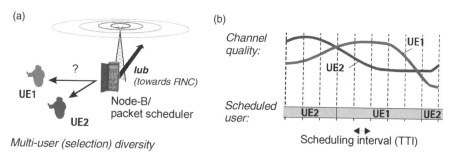

**Figure 12.20** (a) Multi-user diversity principle and (b) scheduling to user with highest instantaneous radio channel quality

One of the fast scheduling methods often referred to is the *proportional fair* algorithm [9, 10], which offers an attractive trade-off between user fairness and cell capacity. The absolute performance of proportional fair scheduling with WCDMA/HSDPA has been studied by several groups, e.g. [11–13]. The proportional fair scheduling idea is to schedule users only when they experience good instantaneous channel conditions (e.g. experience constructive fading), thereby improving both the user throughput and cell throughput for time-shared channels. To identify the best user for scheduling, a relative CQI is calculated, which is calculated for each user as the ratio:

$$\text{Scheduling metric} = \frac{\text{User's instantaneously supported data rate}}{\text{User's average served throughput}} \tag{12.1}$$

Thus, a user is prioritized if either (1) they have good instantaneous conditions compared with the average level or (2) the user has been served with little throughput over the past. The latter ensures scheduling robustness, such that users with static channel conditions are also supported. To compute this scheduling metric, the packet scheduler utilizes the CQI information and the information from the previous transmissions. In deploying the proportional fair packet scheduler, the averaging function must be designed to take the service requirements into account to establish the right trade-off between delays and convergence of the algorithm [11].

The proportional fair scheduling method results in all users getting approximately an equal probability of becoming active even though they may experience very different average channel quality [14]. The performance of advanced scheduling can be modified to meet applicable QoS requirements, e.g. see [8, 15, 16]. While the proportional fair method in Equation (12.1) gives high emphasis on the users near the cell edge, it still offers a non-uniform data distribution across the cell.

The HSDPA bearer capacity gain for proportional fair scheduling over simple round robin in time scheduling is on the order of 40–60% for macrocell environments, and can be theoretically higher for certain environments and operating conditions [17]. These gain numbers assume that the user selection diversity order is higher than, for example, 6–10. If users have low service activity cycles, then this means that the physical number of users needs to be larger for high scheduling gain. Another fundamental requirement for the proportional fair method to give a significant system gain is that the channel variations must be slow enough such that the scheduler can track the channel conditions when considering inherent link adaptation and packet scheduling delays. Previous studies indicate that significant performance is achieved as long as the UE velocity is less that around 25–30 km/h [12]. Beyond this point, the proportional fair scheduler gives a performance similar to the traditional round-robin in time scheduler. Further, the user's channel conditions should be changing fast enough that packet delay requirements do not prevent the scheduler from waiting for the following constructive fade for the user.

Figure 12.21 presents the relative performance between Release 99 and HSDPA in two different environments. As seen from the numbers, HSDPA increases the cell throughput more than 100% compared with Release 99 in the macrocell case, and in the microcell case the gain of HSDPA exceeds 200% (even up to 300%) owing to the availability of very high user peak data rates. However, for the most extreme cases, the practical imperfections associated with the terminal and Node B hardware, link adaptation and packet scheduling may limit the achievable cell throughput in practice. Further, it is assumed that, in favorable conditions, a user will always utilize the available throughput. The application level impacts with HSDPA are contained in Chapter 10.

Another important area for observation is the coverage with HSDPA. The downlink coverage is of interest in terms of what kind of data rate can be offered at the cell edge. As such, the downlink data rate will adapt automatically to the coverage situation based on the CQI feedback from the UE.

As the HS-DSCH does not employ a fast power control, the coverage is defined as the area over which the average user throughput is of some value. The average user data rate coverage follows the $I_{or}/I_{oc}$ distribution of the cell and the amount of time dispersion. The user data rate downlink coverage for a macrocell scenario including significant AMC errors is illustrated in Figure 12.22. Compared

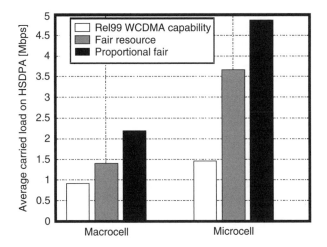

**Figure 12.21**    HSDPA and Release 99 performance comparison

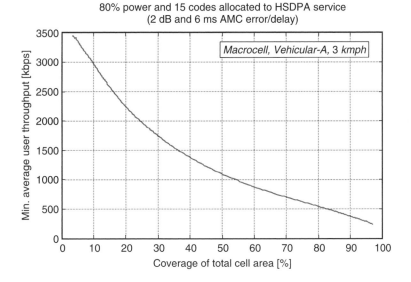

**Figure 12.22**    Minimum average user throughput versus cell coverage

with cell throughput capacity, the single-user data rate coverage is significantly lower, since there is no gain of switching between users with favorable channel conditions; however, the total cell capacity can still benefit from the operation in a soft handover area assuming reasonable scheduling and not too tight timing constraints for the scheduler operation. The flexible support for different handover types for HSDPA-capable users makes it possible to obtain full coverage and mobility for HSDPA users receiving data on the HS-DSCH within an area that is covered by HSDPA-capable Node Bs. This implies that even users with an active set size larger than one (i.e. the associated DPCH is in soft handover) can receive data on the HS-DSCH and thereby benefit from the higher data rate supported for this channel type compared with DCH. The HS-DSCH is still more spectrally efficient than the

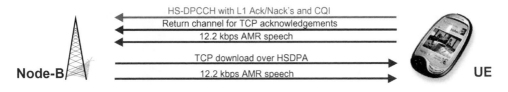

**Figure 12.23**  Different data flows between the UE and Node B

DCH in soft handover (benefits from soft handover macro- and micro-diversity gain), since it benefits from fast link adaptation, effective time diversity and soft combining from the Layer-1 HARQ scheme, and multi-user diversity from using fast Node B scheduling.

The uplink data coverage, as such, is not directly impacted by HSDPA operation, but there needs to be a sufficient power margin available in the uplink for the signaling in the HS-DPCCH as well as for the associated DCH. In the case of Transport Control Protocol (TCP)/Internet Protocol (IP)-based traffic, e.g. web browsing, the uplink traffic consists in addition to the application data, and of the TCP/IP acknowledgements. The resulting uplink data rate will vary depending on the TCP/IP block size, e.g. from 16 kbps onwards for the 500 kbps downlink data rate. These acknowledgements need to be carried by the uplink as well as the minimum necessary layer 2/3 signaling (e.g. handover-related measurements); thus, uplink planning should have coverage roughly equal to 64 kbps data rate. This ensures that downlink throughput is not compromised by the poor uplink performance due to missing or delayed TCP/IP acknowledgements. On top of this, possible service multiplexing, e.g. with AMR speech service, as shown in Figure 12.23, may also need to be accommodated. Note that Layer 2/3 signaling or the HS-SCCH are not shown. The exact value to be used in cell planning will depend on many parameters, including the power offsets and repetition factors for the HS-DPCCH ACK/NACK and CQI fields. With the 3.6 Mbps or higher peak rates there is obviously going to be a need for more TCP/IP acknowledgements, but those data rates are not expected to be available at the cell edge in any case.

Besides the transmission itself, the addition of a new code channel will increase the peak-to-average ratio of the terminal transmission when HS-DPCCH is present. This causes the terminal to use more back-off to ensure maintaining the required spectrum mask for the transmission. The specifications are expected (the topic is recently still being addressed in 3GPP) to allow the terminal transmission power to be reduced by at most 1 dB in cases when the DPCCH/DPDCH power ratio is reasonable, e.g. with the user data rates around 32 kbps. With the high-power DPDCH with higher data rates the transmission power is not reduced at all. This allows configurations with a DPDCH (user) data rate in the order of 64 kbps or higher to have no additional impact on the link budget, except for the actual power needed for HS-DPCCH and for DCH transmission. For the very low data rates, such as 16 or 32 kbps with a 1 dB reduction, the uplink connection will not suffer range problems if the network is dimensioned to enable uplink transmission rate of 64 kbps or more in the whole network.

### 12.9.4  HSDPA Network Performance with Mixed Non-HSDPA and HSDPA Terminals

Typically, the WCDMA networks starts using HSDPA when there is a large existing user base in place. Thus, it becomes essential to understand the HSDPA performance in case of mixed non-HSDPA mobiles and HSDPA mobiles. System-level simulation results are studied for the example case where traffic is carried on both Release 99-dedicated channels and over Release 5 HSDPA on the same carrier and five HS-PDSCH codes are allocated for an HSDPA user with a single HS-SCCH. Release 99 channels can use the remaining code resources. This provides a maximum peak data rate of 3.6 Mbps with 16 QAM and allows one user to be scheduled at the time.

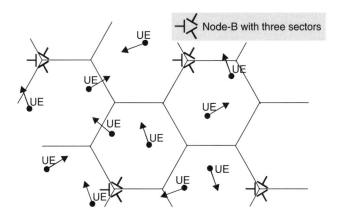

**Figure 12.24** Network topology for the reference simulation set-up

The simulation results are obtained from dynamic cellular network simulations, where users are moving within an area covered by many three-sector Node Bs, as illustrated in Figure 12.24. Dynamic models for user mobility, traffic models, variations of the radio propagation conditions, etc., are used. The ITU Pedestrian-A delay profile is used, whereas otherwise the setting is closer to a macro cell environment.

All traffic on DCHs uses a constant data rate of 64 kbps, with power control. The data rate on the HS-DSCH is adjusted for every TTI as a function of the CQI received. The proportional fair scheduler, as discussed in earlier, is used with HARQ assuming soft combining. The available Node B transmit power for HS-PDSCH and HS-SCCH codes is fixed in each simulation to a value in the range from 3 to 9 W with 20 W total power.

Let us first consider the total average cell capacity that can be achieved with such a system configuration, assuming that the traffic offered in the network is sufficiently high so that the HS-DSCH is utilized in every TTI, and the average power allocated to transmission of DCH is used. Figure 12.25 shows the average cell throughput for the Release 99 DCH and Release 5 HS-DSCH as a function of the power that is allocated to HSDPA transmission. The total cell throughput (i.e. the sum of the Release 99 DCH and Release 5 HS-DSCH throughput) is also plotted. It is observed that the HS-DSCH throughput increases when the HSDPA power is increased, while the DCH throughput decreases as less and less power becomes available for transmission of such channels.

At 7 W HSDPA power, we can achieve an average cell throughput of 1.4 Mbps on the HS-DSCH and an average cell throughput of approximately 440 kbps on the Release 99 DCHs. With only non-HSDPA terminals active in the cell and no power/codes reserved for HS-PDSCH/HS-SCCH transmission, we are able to achieve an average cell throughput of 1.0 Mbps. This implies that with HSDPA enabled the cell throughput is increased by a factor of 1.7, which is basically equivalent to an average gain in cell throughput of 70%. The capacity gain is mainly achieved due to the multi-user diversity gain offered by the fast MAC-hs proportional fair scheduler and the higher spectral efficiency on the HS-DSCH by using fast link adaptation with AMC, as well as the improved Layer-1 HARQ scheme with soft combining of retransmissions. If the radio channel power delay profile is more challenging, such as Vehicular A, then there is also a similar gain observed, though the absolute values are lower.

The average throughput that the HSDPA users experience depends on the number of simultaneous users that are sharing the HS-DSCH channel and their relative signal quality experienced, i.e. symbol energy to noise plus interference ratio ($E_s N_0$).

Figure 12.26 shows the CDF of the average experienced throughput per HSDPA user, depending on the number of simultaneous active users sharing the HS-DSCH. These results are obtained for the

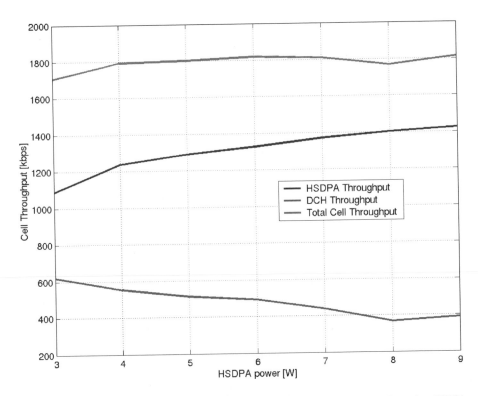

**Figure 12.25**  Average DCH and HSDPA cell throughput as a function of the power allocated to HSDPA

case where 7 W and five HS-PDSCH codes are allocated to HSDPA transmission. For the case with only one active HSDPA user in the cell, the average experienced per user throughput is 800 kbps at the median, and with a 10% probability it is higher than 1.3 Mbps (typically observed for those users that are close to the Node B). When increasing the number of simultaneously active HSDPA users to four, the median per user throughput is decreased to approximately 400 kbps because more users have to share the available capacity on the HS-DSCH. However, notice that the median throughput only is decreased by a factor of two when increasing the number of users from one to four. This behavior is observed because HSDPA benefits from fast scheduling multi-user diversity gain when four users are present, whereas there is, of course, no such gain available for the single user scenario. For eight simultaneously active HSDPA users, the achievable median per user throughput is on the order of 220 kbps. Hence, the per user throughput experienced depends strongly on the number of simultaneously active HSDPA users that are sharing the HS-DSCH.

## 12.10  HSPA Link Budget

HSDPA and HSUPA link budget calculations are presented in this section. The link budgets are used in the network-dimensioning phase together with suitable propagation models to estimate the required number of sites. The relative link budgets can also be used to define the feasibility of base station site reuse, e.g. by studying the relative link budgets between GSM and HSPA.

The uplink HSUPA link budget is presented in Table 12.6. The link budget is calculated for 64 kbps data rate at the coverage edge. The terminal transmission power is assumed 24 dBm and no body loss

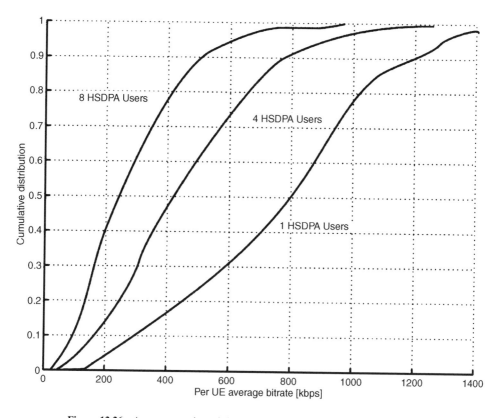

**Figure 12.26** Average experienced throughput as a function of active users per cell

is included for the data connection. The base station receiver assumes a radio-frequency (RF) noise figure of 2.0 dB and the receiver noise floor, therefore, is $-106.2$ dBm. The receiver sensitivity becomes $-123.9$ dBm without interference by assuming $E_b/N_0$ of 0.0 dB for BLER of 10% and by including the processing gain of 17.8 dB. We assume 50% loading, making the interference margin 3.0 dB. We further assume that the base station cable loss is compensated with the masthead amplifier. We reserve 2.0 dB margin for the fast power control. This margin is also called the fast fade margin. The soft handover gain is assumed 2.0 dB. The three-sector macrocell antenna gain is 18 dBi for 65° antennas. The maximum allowed path loss between mobile and base station antenna then becomes 162.9 dB.

The path loss can be measured with the received power level of the pilot channel, CPICH RSCP. The path loss of 162.9 dB corresponds to a CPICH RSCP level of $-113.9$ dBm:

$$RSCP[dBm] = CPICH\_tx[dBm] - Cable\_loss[dB] + Antenna\_gain[dB] - Path\_loss[dB]$$

$$= 33\,dBm - 2\,db + 18\,dBi - 162.9\,dB$$

$$= -113.9\,dBm$$

The downlink link budget is calculated in Table 12.7. The calculation is done for 512 kbps data rate for a single user on HS-DSCH channel. The base station power is assumed 40 W and 80% of the power is allocated for HS-DSCH, making the HS-DSCH output power 45 dBm. The cable loss reduces the output power by 2 dB to the antenna. The average mobile RF noise figure is assumed to be 7 dB, which is approximately 2 dB better than the minimum 3GPP requirement. The SINR requirement can

**Table 12.6**   Uplink/HSUPA link budget for 64 kbps

|   | Data rate (kbps) | 64 | |
|---|---|---|---|
| | *Transmitter – UE* | | |
| a | Max. TX power (dBm) | 24.0 | |
| b | TX antenna gain (dBi) | 0.0 | |
| c | Body loss (dB) | 0.0 | |
| d | EIRP[1] (dBm) | **24.0** | = a + b + c |
| | *Receiver – Node B* | | |
| e | Node B noise figure (dB) | 2.0 | |
| f | Thermal noise (dBm) | −108.2 | = k (Boltzmann) × T (290 K) × B (3.84 Mcps) |
| g | Receiver noise floor (dBm) | −106.2 | = f + e |
| i | $E_b/N_0$ (dB) | 0.0 | From simulations with BLER = 10% |
| k | Processing gain (dBm) | 17.8 | = $10\log_{10}$(3.84 Mcps/data rate) |
| l | Receiver sensitivity (dBm) | −123.9 | = i + j + k |
| g | Load factor (%) | 50 | |
| h | Interference margin (dB) | 3.0 | = $10\log_{10}[1/(1 − g)]$ |
| m | RX antenna gain (dBi) | 18.0 | |
| n | Cable loss (dB) | 2.0 | |
| o | MHA gain (dB) | 2.0 | |
| p | Fast fade margin (dB) | 2.0 | |
| q | Soft handover gain (dB) | 2.0 | |
| | *Maximum path loss* | **162.9** | = d − l − h + m − n + o − p + q |

[1]Equivalent isotropic radiated power.

be obtained from the link simulations [18]. The processing gain in HSDPA is fixed at 16, which equals 12 dB. The load factor is assumed 70%, which corresponds to a relatively high other cell loading.

The SINR value of 6.0 dB assumes a single antenna terminal. If the terminal had a receive diversity, then the data rate can be approximately doubled to 1 Mbps with the same SINR value; see further discussion on the enhanced terminals in Chapter 15.

That path loss of 163 dB is similar to GSM or WCDMA voice path loss, making it possible to provide 0.5 Mbps wireless broadband service using existing sites if the same frequency band is used for GSM and HSPA.

## 12.11   HSDPA Iub Dimensioning

HSDPA is pushing the radio data rates and capacities higher. In order to take full benefit of the enhanced radio capability, the Iub transport capacity needs to be dimensioned accordingly. The Iub transport can be organized in a number of different ways by using leased E1 (T1) lines, each 2 Mbps (1.5 Mbps) capacity, by using Ethernet or by using microwave radio links. If Ethernet connection is available, it can provide large Iub bandwidth. Microwave radio capacity can typically be easily extended beyond 10 Mbps. The leased line case is the most challenging one for HSDPA, since the cost of a single 2 Mbps connection can be even up to 500 EUR per month and multiple E1s are needed for a high-capacity HSDPA site. Therefore, the Iub capacity dimensioning and optimization is most relevant for leased line case. This section discusses the Iub throughput with E1 leased-line transport.

The following overhead factors need to be considered in Iub dimensioning: common channels for the first E1, ATM overhead, AAL2 overhead, Frame protocol overhead and RLC overhead. The E1 user plane capacity is assumed to be 32 × 64 kbps = 1920 kbps and the common channel allocation

**Table 12.7**  Downlink/HSDPA link budget for 512 kbps with single antenna 3.6 Mbps terminal

|  |  | HS-DSCH | HS-SCCH |  |
|---|---|---|---|---|
|  | Data rate (kbps) | 512 |  |  |
| *Transmitter – Node B* |  |  |  |  |
| a | HS-DSCH power (dBm) | 45.0 | 31.7 | 46 dBm BTS and 80% power for HS-DSCH |
| b | TX antenna gain (dBi) | 18.0 | 18.0 |  |
| c | Cable loss (dB) | 2.0 | 2.0 |  |
| d | EIRP (dBm) | **61.0** | **47.7** | = a + b + (c) |
| *Receiver – UE* |  |  |  |  |
| e | UE noise figure (dB) | 7.0 | 7.0 |  |
| f | Thermal noise (dBm) | −108.2 | −108.2 | = $k$ (Boltzmann) × $T$ (290 K) −102 × B (3.84 Mcps) |
| i | Receiver noise floor (dBm) | −101.2 | −101.2 | = f + e |
| j | SINR (dB) | 6.0 | 1.5 | From simulations |
| k | Processing gain (dB) | 12.0 | 21.0 | HS-DSCH SF = 16, HS-SCCH SF = 128 |
| l | Receiver sensitivity (dBm) | −107.2 | −120.7 | = i + j + k |
| g | Load factor (%) | 70 | 70 |  |
| h | Interference margin (dB) | 5.2 | 5.2 | = $10\log_{10}[1/(1 - g)]$ |
| m | RX antenna gain (dBi) | 0.0 | 0.0 |  |
| n | Body loss (dB) | 0.0 | 0.0 |  |
| o | Fast fade margin (dB) | 0.0 | 0.0 |  |
| p | Soft handover gain (dB) | 0.0 | 0.0 |  |
| *Maximum path loss* |  | **162.9** | **163.1** | = d − l − h + m − n − o + p |

300 kbps. The first E1 connection can then provide 1.6 Mbps data capacity. The total overhead is assumed to be 35% together for ATM, AAL2, Frame protocol and RLC.

Figure 12.27 shows the maximum throughout from the radio and from the Iub point of view. If we want to enable the maximum throughput for Category 6 terminal with five codes and 16 QAM, we need minimum 3 × E1. For Category 8 and 9, the respective Iub capacities are 5 × E1 and 7 × E1.

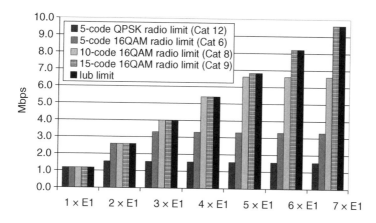

**Figure 12.27**  HSDPA throughput with limited Iub capacity

In order to achieve the high peak rates with HSDPA requires higher Iub capacity. On the other hand, HSDPA improves Iub efficiency considerably compared with WCDMA Release 99. The improvement is due to two factors:

- HSDPA does not require soft handover, whereas WCDMA uses soft handover;
- HSDPA has shared Iub flow control, whereas WCDMA uses dedicated bit pipes over Iub.

The soft handover overhead in WCDMA is typically 30–50%. That overhead can be saved in HSDPA. The effect of shared flow control can be even higher. The dedicated WCDMA bit pipe is not fully utilized during TCP slow start or other application protocol limitations. The WCDMA dedicated channel is also reserved for some inactivity timer after the file download is over. The Iub capacity is not used during the inactivity timer. When these two factors are calculated together, HSDPA can improve Iub efficiency even by four to five times compared with WCDMA Release 99 [19].

## 12.12   HSPA Round Trip Time

HSPA improves end-user performance by providing higher data rates. HSPA also brings improvements in the latency, which further boosts the practical application performance. The importance of round trip time (RTT) for the packet applications is discussed in Chapter 10.

The RTT is the latency from the mobile throughput the UMTS network to the server and back. The RTT of a WCDMA Release 99-dedicated channel is typically 100–200 ms. Commercial HSDPA deployments have shown that the RTT can be pushed below 70 ms, and with the first HSUPA deployments with 10 ms TTI below 50 ms.

The 3GPP specifications set a minimum value for the RTT due to the air interface frame structure. The minimum TTI is 2 ms, both in uplink and in downlink. On average, the packet needs to wait for half the transmission time. This value is called TTI alignment. The downlink SCCH transmission starts two time slots, or 1.3 ms, before the HS-DSCH data. Therefore, the minimum RTT would be 7.3 ms if there were no delay in any network elements or in the terminal.

Assuming a 2 ms processing delay in the Node-B receiver, the Node-B transmitter and in the RNC and packet core together, brings the RTT to 13 ms. Additionally, we need to include the processing time in the terminal. It is expected that HSPA with 2 ms TTI and delay-optimized implementation enables end-to-end RTT below 30 ms, including all the network and terminal delays.

Figure 12.28 illustrates the HSPA round trip time components.

## 12.13   Terminal Receiver Aspects

The terminal receiver aspects were discussed in Section 12.6.1.1, since one of the new challenges is the need for amplitude estimates for the 16 QAM detection. However, there are other challenges coming from the use of 16 QAM as well. A good quality voice call in WCDMA typically requires a C/I of $-20$ dB compared with 10 dB for GSM. Since the interference, including the inter-symbol interference, can be 20 dB above the signal level, the WCDMA voice signal is very robust against interference and does not benefit significantly from equalizers. However, for the high peak data rates provided with an HSDPA service, higher C/I ($E_b/N_0$) values above 0 dB are required and, consequently, the signal becomes less robust against inter-symbol interference.

Hence, the HSDPA concept with 16 QAM transmission potentially benefits from equalizer concepts that reduce the interference from multi-path components. The multi-path interference cancellation receiver shown in Figure 12.29 was discussed and analysed in [1]. The same receiver front-end as employed in the Rake receiver is used as a pre-stage to provide draft symbol estimates. Those estimates are then used to remove the multi-path interference from the received signal, and new symbol estimates

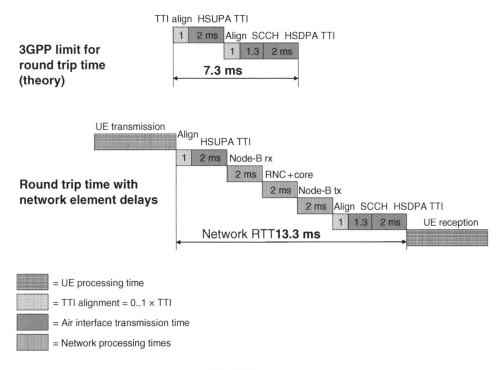

**Figure 12.28** HSPA round trip time

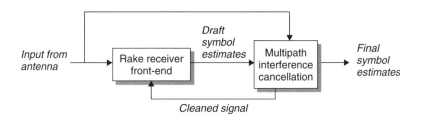

**Figure 12.29** Example multi-path interference cancellation

can be obtained with the same matched filter. After a few iterations, the final symbol estimates are calculated. Another type of advanced receiver is a linear equalizer. The advanced receiver algorithms (with uplink focus) are discussed in more detail in Section 11.6.

Advanced receivers make it possible to provide higher bit rates in multi-path channels compared with what is achievable with normal Rake receivers. On the other hand, the complexity of such receivers is significantly higher than for the standard Rake receiver. In 3GPP standardization there is no intention to specify any receiver solutions, just performance requirements in particular cases. The performance requirements, as such, are always derived using a common baseline received in the simulations so that multiple companies can verify the results using different simulation platforms.

During 2004, work on the improved HSDPA performance requirements was started in 3GPP, with the focus being on two technology directions: advanced receivers and RX diversity. The RX diversity can improve the performance of HSDPA when diversity is small, but, as such, the link level improvements with RX diversity (or additional diversity in general) are not necessarily additive with the gains from the scheduling. The advanced receiver battles the inter-symbol interference and, thus, especially makes the 16 QAM usable more often by enabling higher data rates, especially in the Vehicular-type of environment. The 3GPP specifications now contain the following additional receivers for HSDPA, and these receivers will provide further HSDPA capacity improvements than the results presented in this chapter:

- Type 1 receiver (dual antenna Rake-based requirements);
- Type 2 receiver (single antenna equalizer-based requirements);
- Type 3 receiver (dual antenna equalizer-based requirements). This is found in the Release 7 version of the specifications.
- There is a further version of the Type 3 receiver, known as Type 3i that is operating as aware of the interference situation. The potential benefit is due to very low geometry factors though typically the UE is then moved to be connected to a better cell for better overall network performance.

The use of these receivers does not require any updates on the network side and can be implemented as release-independent improvements in the UEs. Thus one can implement a Release 5 device that meets, for example, the Type 3 receiver performance requirements part of Release 7 specifications. The Uplink CQI signaling will automatically take the receiver characteristics into account and, thus, benefits from the potentially improved throughput on the network side.

## 12.14   Evolution in Release 6

As described previously, the HSDPA concept of Release 5 is able to provide a clear increase in the WCDMA downlink packet data throughput. It is obvious that further enhancements on top of the HSDPA feature can be considered for increased user bit rates and cell throughput. Possible techniques raised previously include further improvements in the downlink with advanced antenna techniques and also applying similar techniques to HSDPA for the uplink direction. This is know as HSUPA in Release 6 specifications and is covered in Chapter 13. HSUPA is intended to be operated with HSDPA, and for that reason HSDPA support is mandatory for those devices that support HSUPA. Thus, in Release 6 the devices could be categorized as follows:

- devices with DCH only support;
- devices with DCH and HSDPA support;
- devices with DCH, HSDPA and HSUPA support.

The second edition of this book also contained fast cell selection (FCS), which was then determined in 3GPP as not worth adding to the specifications. Release 6 contains the following new HSDPA-related features, which are related to range, capacity or RTT reduction:

- fractional DPCH (FDPCH);
- HS-DPCCH pre/post-amble.

FDPCH aims to reduce the need to book the downlink code space for DCHs. FPDCH is intended to be used when only packet-switched services are in use and when in such a case everything can be mapped to HSDPA. As Release 6 also supports mapping the signaling radio bearer to HS-DSCH, then DCH can be replaced by FDPCH carrying only power control commands. A single F-DPCH with SF 256 can carry power control commands for 10 users. There are, however, some timing constraints for practical FDPCH operation, and those have been then addressed in the Release 7 specification. The use of FPDCH is illustrated in Figure 12.30, indicating that, when using different timing offsets, the TPC command stream for different users can be time multiplexed on the same channelization code.

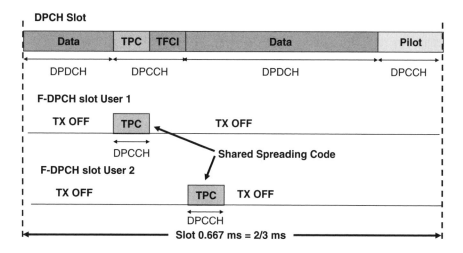

**Figure 12.30**  Use of FDPCH

The HS-DPCCH preamble/post-amble was included in the Release 6 specifications to increase the uplink range. The simple principle was to avoid the need to detect between ACK/NACK and DTX levels in the BTS receiver. This was avoided by sending at the start of the data burst a preamble from UE upon detection of the HS-SCCH; thus, after that, the BTS would only need to detect between ACK and NACK positions. With this method the power offset needed for signaling on HS-DPCCH can be reduced and, thus, more power is left for actual user data. The preamble/post-amble method is illustrated in Figure 12.31.

The performance evaluation in [20] has shown up to 6 dB reduction of HS-DPCCH transmission power in soft handover areas, thus contributing around 1 dB to the link budget with 64 kbps data rate. The FPDCH studies are also covered in [20], though the actual solution was adjusted before entering the specifications and thus details in [20] do not present the FDPCH in the specifications.

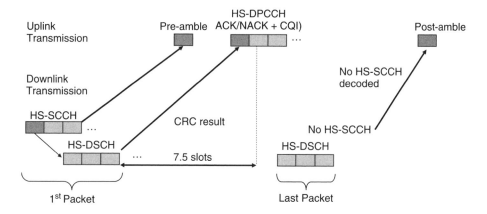

**Figure 12.31**  Pre-/post-method for uplink range improvement

## 12.15 Conclusion

The HSDPA concept has been introduced and its performance considered. The main aspects discussed can be summarized as follows:

- The HSDPA concept utilizes a distributed architecture in which the processing is closer to the air interface at Node B for low-delay link adaptation.
- The HSDPA concept provides a 50–100% higher cell throughput than the Release 99 DCH/DSCH in macrocell scenarios and more than a 100% gain in microcell scenarios. For microcells, the HS-DSCH can support up to 5 Mbps per sector per carrier, i.e. 1 bit/s/Hz/cell.
- The HSDPA concept offers more than 100% higher peak user bit rates than Release 99, and the difference is even larger if observing the maximum downlink DCH data rate supported by the networks being 384 kbps. HS-DSCH bit rates are comparable to Digital Subsriber Line modem bit rates. The mean user bit rates in a large macrocell environment can exceed 1 Mbps and in small microcells 5 Mbps.
- The HSDPA concept is able to efficiently support not only non-real-time UMTS QoS classes, but also real-time streaming UMTS QoS class with guaranteed bit rates.
- The use of HSDPA also provides significant benefits in cases of mixed terminal deployment with cell code and power resources shared between HSDPA and non-HSDPA users.
- The applicability of HSDPA techniques for uplink direction was investigated in 3GPP, and Release 6 specifications contain, in addition to the HSDPA-related improvements, HSUPA for improved uplink packet access, as covered in Chapter 13.

## References

[1] 3GPP Technical Report 25.848, Physical layer aspects of UTRA High Speed Downlink Packet Access, version 4.0.0, March 2001.
[2] 3GPP Technical Specification 25.211, Physical Channels and Mapping of Transport Channels onto Physical Channels (FDD), version 5.0.0, March 2002.
[3] 3GPP Technical Specification 25.212, Multiplexing and Channel Coding (FDD), version 5.0.0, March 2002.
[4] 3GPP Technical Specification 25.306, UE Radio Access Capabilities, version 5.1.0, June 2002.
[5] 3GPP Technical Specification 25.331, Radio Resource Control (RRC), Release 5, December 2003.
[6] 3GPP Technical Specification 25.322, Radio Link Control (RLC), December 2003.
[7] Elliot, R. C. and Krzymien, W. A., 'Scheduling Algorithms for the cdma2000 Packet Data Evolution', *Proceedings of the IEEE Vehicular Technology Conference (VTC)*, Vancouver, Canada, September 2002, Vol. 1, 2002, pp. 304–310,
[8] Ameigeiras, P., 'Packet Scheduling and Quality of Service in HSDPA', PhD thesis, Department of Communication Technology, Aalborg University, Denmark, October 2003.
[9] Kelly, F., 'Charging and Rate Control for Elastic Traffic', *European Transactions on Telecommunications*, Vol. 8, 1997, pp. 33–37.
[10] Jalali, A., Padovani, R. and Pankaj, R., 'Data Throughput of CDMA-HDR High Efficiency-High Data Rate Personal Communication Wireless System', *Proceedings of Vehicular Technology Conference (VTC)*, May 2003, Tokyo, Japan, Vol. 3, 2000, pp. 1854–1858.
[11] Kolding, T. E., 'Link and System Performance Aspects of Proportional Fair Scheduling in WCDMA/HSDPA', *Proceedings of 58th IEEE Vehicular Technology Conference (VTC)*, Florida, USA, October 2003, Vol. 2, 2003, pp. 1454–1458.
[12] Ramiro-Moreno, J., Pedersen, K. I. and Mogensen, P. E., 'Network Performance of Transmit and Receive Antenna Diversity in HSDPA under Different Packet Scheduling Strategies', *Proceedings of 57th IEEE Vehicular Technology Conference (VTC)*, Jeju, South Korea, April 2003.
[13] Parkvall, S., Dahlman, E., Frenger, P., Beming, P. and Persson, M., 'The High Speed Packet Data Evolution of WCDMA', *Proceedings of the 12th IEEE Symposium of Personal, Indoor, and Mobile Radio Communications (PIMRC)*, San Diego, California, USA, September 2001, Vol. 2, 2001, pp. G27–G31.

[14] Holtzman, J. M., 'Asymptotic Analysis of Proportional Fair Algorithm', *IEEE Proc. Personal Indoor Mobile Radio Communications (PIMRC)*, September, 2001, pp. F33–F37.

[15] Andrews, M., Kumaran, K., Ramanan, K., Stolyar, A. and Whiting, P., 'Providing Quality Of Service over a Shared Wireless Link', *IEEE Communications Magazine*, Vol. 39, 2001, pp. 150–154,

[16] Kolding, T. E., Pedersen, K. I., Wigard, J., Frederiksen, F. and Mogensen, P. E., 'High Speed Downlink Packet Access: WCDMA Evolution', *IEEE Vehicular Technology Soceity (VTS) News*, Vol. 50, 2003, pp. 4–10.

[17] Hosein, P. A., 'QoS Control for WCDMA High Speed Packet Data', *International Workshop on Mobile and Wireless Communications Networks*, Stockholm, Sweden, September 2002, pp. 169–173.

[18] Holma, H. and Toskala, A. (eds), *HSDPA/HSUPA for UMTS*, New York: John Wiley & Sons, Ltd, 2006, Chapter 7.

[19] Toskala, A., Holma, H., Metsala, E., Pedersen, K. and Steele, P., 'Iub Efficiency of HSDPA', WPMC-05, Aalborg, Denmark, September 2005.

[20] 3GPP technical Report, 25.899, HSDPA Enhancements, version 6.0.0. June 2004.

# 13

# High-Speed Uplink Packet Access

Antti Toskala, Harri Holma and Karri Ranta-aho

## 13.1   Introduction

This chapter presents High-Speed Uplink Packet Access (HSUPA) for Wideband Code Division Multiple Access (WCDMA), the key new feature included in Release 6 specifications. The HSUPA solution has been designed to deliver similar benefits for the uplink as did the High-Speed Downlink Packet Access (HSDPA) in Release 5, covered in Chapter 12, for the downlink. The technologies applied with HSUPA are to improve uplink packet data performance by means of fast physical layer (L1) retransmission and transmission combining, as well as fast Node B (Base Transceiver Station (BTS)) controlled scheduling. The chapter is organized as follows: first, HSUPA key aspects are presented and then a comparison with Release 99 uplink packet access possibilities is made. Next, the impact of HSUPA on the terminal (or User Equipment (UE) in 3rd Generation Partnership Project (3GPP) terms) capability classes is summarized and HSUPA performance analysis is presented, including a comparison with Release 99 uplink packet data capabilities.

## 13.2   Release 99 WCDMA Downlink Packet Data Capabilities

In Release 99, various methods exist for packet data transmission in WCDMA uplink. As described in Chapters 6 and 10, the three different channels in Release 99 and Release 4 WCDMA specifications that can be used for uplink packet data are:

- Dedicated Channel (DCH);
- Common Packet Channel (CPCH);
- Random Access Channel (RACH).

From Release 5 onwards, however, only DCH and RACH will be retained, as 3GPP finished in June 2006 with the removal of a set of features (including CPCH) not implemented nor in the plans for introduction in the market place by operators and equipment manufacturers. Thus, CPCH is not discussed further in this chapter, but the principles of CPCH can be found in Chapter 6. In any case, the relevant case for comparison with HSUPA is the DCH, as RACH (and CPCH) is only useful in the Cell_FACH state for sending limited amounts of data and not in the Cell_DCH state.

The DCH can basically be used for any type of service, and it has a dynamically variable spreading factor (SF) in the uplink, with an adjustment period of 10 to 40 ms. The momentary data rate can thus vary every interleaving period, which is between 10 and 40 ms. The issue of code space occupancy is not a real concern in the uplink direction, as each user has a user-specific scrambling code and, thus, can use the full core tree if needed. Rather, the uplink DCH consumes both noise rise budget and network resources according to the peak data rate configured for the connection. The theoretical data rate with Release 99 runs up to 2 Mbps, but in practice the devices and networks have implemented typically 384 kbps as the maximum uplink capability. Higher numbers would easily mean reserving the whole cell/sector capacity for a single user regardless of the actual data rate being used. The DCH uses power control and may be operated in soft handover, as described in Chapter 6.

## 13.3   The HSUPA Concept

The main idea of the HSUPA concept is to increase uplink packet data throughput with methods similar to HSDPA, base station scheduling and fast physical layer (L1) retransmission combining. While the telecoms industry uses the term HSUPA widely, it is not used in 3GPP specifications. In the specifications, the term Enhanced DCH (E-DCH) is applied to the transport channel carrying the user data with HSUPA. A comparison of the basic properties and components of E-DCH and DCH is conducted in Table 13.1.

The general functionality of HSUPA is illustrated in a simple fashion in Figure 13.1. The Node B estimates the data rate transmission needs of each active HSUPA user based on the device-specific feedback. The scheduler in Node B then provides instruction to devices on the uplink data rate to be used at a fast pace depending on the feedback received, the scheduling algorithm and the user prioritization scheme. Further, the retransmissions are initiated by the Node B feedback. The channels needed to carry data and downlink/uplink control signaling are described in Section 13.7.

Whereas in Chapter 12 it was explained that HSDPA no longer uses power control, the same does not hold with HSUPA. HSUPA retains the uplink power control with a 70 dB or more dynamic range (exact range depends on the power class and terminal minimum power level). Thus, with HSUPA the signal never arrives at too high a symbol energy level, which is the case with HSDPA, and thus a justification for the use of higher-order modulation with HSDPA. Thus, the key thing for increased data rate is extensive multi-code operation together with base-station-based scheduling and retransmission handling. The Release 99 uplink feature of variable SF is retained; the range of SFs is only slightly changed.

As the control of the scheduling is now in the base station, i.e. the receiving side of the radio link, there is added delay in the operation. This is in contrast to the HSDPA operation, where the

**Table 13.1**   Comparison of fundamental properties of DCH and E-DCH

| Feature | DCH | E-DCH |
|---|---|---|
| Variable SF | Yes | Yes |
| Fast power control | Yes | Yes |
| Adaptive modulation | No | No |
| Multi-code operation | Yes (in specs, not used) | Yes, extended |
| Fast L1 HARQ | No | Yes |
| Soft handover | Yes | Yes |
| Fast BTS scheduler | No | Yes |

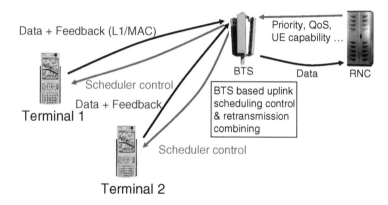

**Figure 13.1** General operating principles of HSUPA

**Table 13.2** Comparison of fundamental properties of DCH and E-DCH

| Feature | HSUPA | HSDPA |
|---|---|---|
| Variable spreading factor | Yes | No |
| Fast power control | Yes | No |
| Adaptive modulation | No | Yes |
| Scheduling | Multipoint to Point | Point to Multipoint |
| Fast L1 HARQ | Yes | Yes |
| Soft handover | Yes | No |
| Non-scheduled transmissions | Yes | No |

base station scheduler resides in the transmitting side of the radio link. Thus, tracking the fast fading of the user for scheduling the uplink is not necessarily that beneficial. Rather, the key idea is to enable the scheduling to track the instantaneous transmission needs and capabilities of each device and then allocate such a data rate when really needed by the device. The fast scheduling allows dynamic sharing not only of the interference budget, but also of network resources, such as baseband processing capacity and Iub transmission resources.

The physical layer retransmission combining is similar to HSDPA: now, it is just the base station that stores the received data packets in soft memory and, if decoding has failed, it combines the new transmission attempt with the old one. The key functionalities between HSDPA and HSUPA are compared in Table 13.2.

## 13.4  HSUPA Impact on Radio Access Network Architecture

As with HSDPA, the additional retransmission procedure is now also handled in the base station and, thus, new Medium Access Control (MAC) layer functionality is added to the base station to cover that and the intelligence for the uplink scheduling functionality. The Radio Link Control (RLC) layer retransmission is still kept in case the physical layer retransmission fails for some reason, but in most cases there is no need for RLC layer retransmissions. Figure 13.2 presents the retransmission handling with HSUPA. The physical layer packet combining now takes place in the base station where the soft buffers are located. The additional element is the MAC layer packet re-ordering, which needs to take place in the Radio Network Controller (RNC).

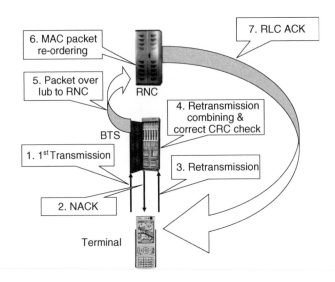

**Figure 13.2**   HSUPA retransmission control in the network

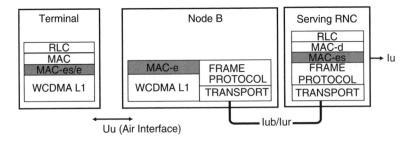

**Figure 13.3**   HSUPA protocol architecture

The MAC layer protocol in the architecture for HSUPA can be seen in Figure 13.3. The aim of the new Node B MAC functionality (MAC-e) is to handle the Automatic Repeat reQuest (ARQ) functionality and scheduling, as well as the priority handling. Now, both the UE and the RNC also have an additional MAC functionality. On the UE side, the new functionality represents the uplink scheduling and retransmission handling (being controlled by the MAC-e in Node B). The MAC-es functionality in the RNC is to cover for packet re-ordering to avoid changes to the layers above. This reordering is needed due to the uplink soft handover operation, as covered in Section 13.7, which may cause the packets to arrive out of sequence from different base stations.

## 13.4.1   HSUPA Iub Operation

The use of HSUPA requires parameterization from RNC to the Node B over the Iub interface similar to HSDPA. The Node B needs to obtain key parameters and terminal-specific Quality of Service (QoS) information from RNC, as shown in Figure 13.1. HSUPA also impacts the Iub efficiency, as now the Iub data rate dimensioning is not required to be the sum of peak rates (unless data loss is allowed); rather, RNC can give guidance to the scheduler on whether the uplink data rates should be

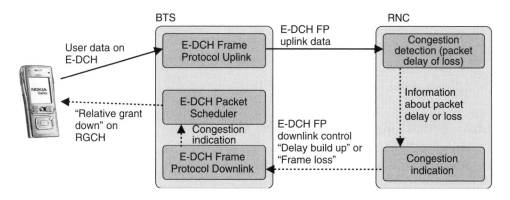

**Figure 13.4**  Iub congestion control

reduced from the Iub congestion point of view. This is illustrated in Figure 13.4, where the RNC can command the Node B either to increase or decrease the allowed total data rate. This allows improved Iub efficiency compared with Release 99 operation.

## 13.5   HSUPA Feasibility Study Phase

After having completed HSDPA specifications, 3GPP started a feasibility study to investigate how the methods known from the HSDPA feature could be applied to the uplink direction and what the resulting benefits would be from doing so. The key difference from HSDPA was the finding that, now, there was no capacity benefit from the higher-order modulation due to the previously mentioned power control possibility with large dynamic range and the resulting higher $E_b/N_0$ from the use of the higher-order modulation. The use of higher-order modulation was considered more from the point of view of whether one could obtain a transmission with better envelope properties when comparing the use of multicode binary phase shift keying (BPSK) operation with a single code 8 PSK case. The Node B-based scheduling and physical layer retransmissions were found beneficial. The results presented in the feasibility study report [1] showed a 50–70% increase in the uplink packet data throughput with HSUPA from Release 99 DCH operation. Thus, in March 2004, 3GPP decided to close the study and to start the actual specification work, which was finalized then for the end of 2004. A more comprehensive capacity comparison based on the actual specification details can be found in Section 13.10.

## 13.6   HSUPA Physical Layer Structure

The transport channel, E-DCH, is sent in the uplink together with the Release 99 DCH, and at least the control part of DCH (Dedicated Physical Control Channel (DPCCH), as covered in Chapter 6) is always present to carry the pilot bits and downlink power control commands in the uplink direction. The presence of the Dedicated Physical Data Channel (DPDCH) for user data depends on whether there is e.g. AMR speech call operated in parallel to uplink packet data transmission. The following new physical channels are introduced in the 3GPP Release 6 specifications to enable HSUPA operation [2]:

- Enhanced DPDCH (E-DPDCH) carries the user data in the uplink direction and reaches the physical layer peak rate of 5.76 Mbps when up to four parallel code channels are in use.
- Enhanced DPCCH (E-DPCCH) carries in the uplink the E-DPDCH-related rate information, retransmission information and the information to be used by the base station for scheduling control.

- E-DCH Hybrid ARQ (HARQ) Indicator Channel (E-HICH) carries information in the downlink direction on whether a particular base station has received the uplink packet correctly or not.
- E-DCH Absolute Grant Channel (E-AGCH) and E-DCH Relative Grant Channel (E-RGCH) carry the Node B scheduling control information to control the uplink transmission rate.

The following sections look in more detail at the channels listed above.

## 13.7   E-DCH and Related Control Channels

The E-DCH transport channel consists of two channels, as discussed above: E-DPDCH to carry the user data and E-DPCCH to carry the physical layer uplink control information.

### 13.7.1   E-DPDCH

The Release 99 DPDCH may use a 10, 20 or 40 ms Transmission Time Interval (TTI), whereas with E-DPDCH 10 ms and 2 ms are available. The reason for not having only one value like with HSDPA is the uplink range. While the round-trip time can be made shorter with a 2 ms TTI, the resulting control signaling is too much for cell edge operation. Thus, the 3GPP specifications adopted both the 10 ms and the 2 ms solutions. The modulation is unchanged from Release 99 and is based on the BPSK modulation; Release 7 also introduces the possibility to use 16 QAM, as discussed in Chapter 15.

Since the modulation is unchanged, the increased uplink peak data rates have been achieved with the more extensive use of uplink multi-code transmission. As described in Chapter 6, the Release 99 smallest spreading factor is 4 and, in reality, devices have not implemented more than a single code in the uplink, resulting in a peak user data rate of 384 kbps. With E-DPDCH, the data rates are created with a combination of multi-codes and with the introduction of SF 2 as well. Thus, a single code channel has increased (uncoded) the bit rate from 960 kbps to 1920 kbps with the use of SF 2. When the single code capacity with SF 4 is exceeded, another E-DPDCH is added, with the same SF, as shown in Figure 13.5. This obviously assumes that the Node B scheduler allows the increase of data rate.

Table 13.3 presents the different steps from the single code case onwards to the highest data rate of 5.76 (uncoded) being implemented with two parallel codes of SF 2 and two parallel codes with SF 4.

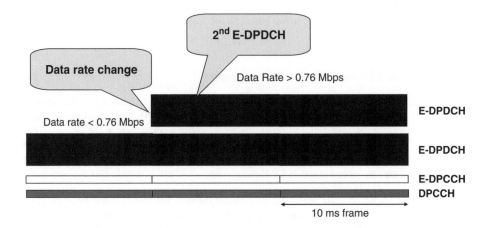

**Figure 13.5**   Adding another E-DPDCH when exceeding single code capacity

**Table 13.3** Different data rate steps with code combinations

| Number of codes | Data rate without channel coding (kbps) |
| --- | --- |
| One code with SF 4 | 960 |
| Two codes with SF 4 | 1920 |
| Two codes with SF 2 | 3840 |
| Two codes with SF 4 and two codes with SF 2 | 5760 |

Another key difference from HSDPA is that, with HSUPA, users are not expected to be totally silent when not being scheduled a high data rate. Instead, while some users transmit at a lower power level, others may user higher transmission power. Also, in the uplink direction, it is not the code tree that the users are sharing; rather, it is the interference budget of the uplink. Each user has a user-specific scrambling code; thus, it does not matter which channelization code one is using in the uplink direction. Compared with the HSDPA case, where a single BTS transmitter with 20 to 40 W (or even more) was behind the transmission, it does not make sense to try to have only one device, with a maximum of 250 mW, transmitting alone. Thus, the uplink has been designed to ramp the user data rates step by step up or down depending on the need of the devices, as illustrated in Figure 13.6.

As different users are like noise to each other, one can combine Release 99 users and HSUPA users on the same carrier in the uplink. In such a case, the BTS cannot obviously control the data rate variations of Release 99 users and has to leave some margin for those. When an overload situation is detected, the Release 99 users can be effected only by sending a measurement report of uplink interference level to RNC which may then, with Radio Resource Control (RRC) signaling, restrict the data rates being used by Release 99 users. This is obviously a much slower process than with BTS scheduling done locally with fast L1 signaling.

The channel coding on E-DPDCH is the same turbo coding as in Release 99; convolutional coding is not available for E-DPDCH, as it was mainly targeting a circuit-switched voice service. There is no need to deal with changes in the number of bits due to modulation changes as there is no adaptive modulation in use. Also, a single transport channel at a time is being transmitted, and a compressed mode is applied in such a way that the coding rate would not need to be adjusted. Instead, with 2 ms TTI the whole TTI is simply skipped if overlapping with the uplink transmission gap, while with

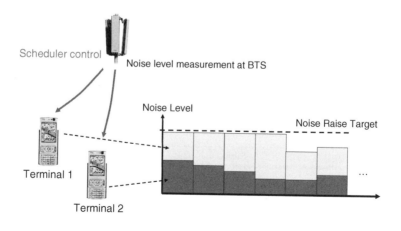

**Figure 13.6** Uplink resource sharing example with two users active

10 ms TTI only those slots that are not transmitted are overlapping with the uplink compressed mode gap, but the momentary data rate is not increased during the active slots.

The E-DPDCH does not carry any other information than user data and it is dependent on the DPCCH to carry the pilot symbols for channel estimation and on the E-DPCCH to carry the HSUPA-related physical layer control information.

The HARQ operation, as illustrated in Figure 13.2, is more or less a reversed operation of that of HSDPA described in Chapter 12. Both chase and soft combining methods are available and now just the soft buffer burden is on the BTS side. The key difference results from the uplink soft handover, as now several base stations are receiving the transmission and may send feedback as covered in the E-HICH description.

## 13.7.2   E-DPCCH

The E-DPCCH carries three different types of information:

- E-DPDCH-related rate information, which uses 7 bits.
- Two bits of retransmission information to indicate whether the packet is new or a retransmission of a previous transmission attempt, as well as the redundancy version of the transmission.
- One bit of information on whether the device could increase data rate or not. This is sent in the form of a happy bit, which defines whether the device could use a higher uplink data rate or not. If not, the bit is set to the 'happy' position and, thus, there is no need for the scheduler to increase the uplink data rate.

The E-DPCCH structure is shown in Figure 13.7. The 10 information bits are coded, resulting in a total of 30 bits spanning over three consecutive slots, i.e. 10 channel bits in each slot. The E-DPCCH will follow the TTI length of E-DPDCH. With the 10 ms TTI the contents in the three slots is simply repeated five times (five times identical three slots transmitted). This allows a lower power level for E-DPCCH with 10 ms TTI and to keep the link budget from the cell edge as well. E-DPCCH is only sent with E-DPDCH. If there are no data being transmitted on E-DCH then E-DPCCH will not be transmitted either. Note that DPCCH is always present, whereas DPDCH is only needed when there are services not mapped on E-DPDCH actively transmitting data.

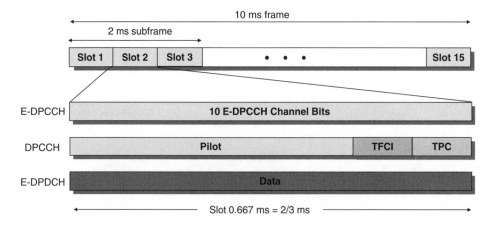

**Figure 13.7**   E-DPCCH with 2 ms TTI

### 13.7.3   E-HICH

The E-HICH has the simple task of indicating in the downlink direction whether a packet has been correctly received in the uplink by the BTS. In order to save downlink code space resources, one code channel is shared by multiple users and each user is allocated one orthogonal signature out of 40 available as the E-HICH, and another one as the E-RGCH (see Section 13.7.4). This allows the accommodation of up to 20 users with each having a dedicated E-HICH and E-RGCH on the single downlink code channel with SF 128. The orthogonal signatures are one slot long and are extended to cover 2 ms by applying three signatures in a sequence over three consecutive slots.

As typically in the active set there is one dominant base station, the serving E-DCH cell, the other cells are only sending E-HICH if the packet was correctly decoded, as indicated in Figure 13.8. Otherwise they do not send anything, which is assumed to mean that the packet was not decoded correctly or not detected to be present at all. This helps to keep the additional downlink interference to a minimum. The E-DCH serving cell is always the same as the HSDPA serving cell.

With a 2 ms E-DCH, a TTI of 2 ms is used with E-HICH, whereas with a 10 ms TTI on E-DCH, the three-slot signature structure is repeated four times, resulting in an 8 ms length. The remaining 2 ms is used for BTS/UE processing time.

Dealing with multiple cells is also the reason why in the uplink direction there are 2 bits for retransmission information. When one base station has acknowledged the packet, another base station might miss a few rounds of control information and it could not then tell whether to combine the packet or not if only a 1-bit new data indicator was used.

### 13.7.4   E-RGCH

As discussed in the previous section, the relative grant channel is sharing the same code channel as E-HICH to save code space. The function of the E-RGCH channel is either to increase or decrease the uplink transmission rate based on the scheduler decisions. The relative grants transmitted effectively control the gain factors to be used, which then map in practice to a particular data rate or rates allowed for the device. A similar principle to E-HICH is used for 2 ms and 8 ms durations of E-RGCH with 2 ms and 10 ms TTIs respectively. The non-serving E-DCH cell will use E-RGCH only as an overload control method and will normally send nothing, but it can in an overload situation send a rate down command that is typically common for all the UEs to which the cell is a non-serving cell.

### 13.7.5   E-AGCH

The E-AGCH is operated as an independent shared channel and all users in the cell monitor one E-AGCH, although multiple could be configured for the cell. The absolute grants sent on the channel will

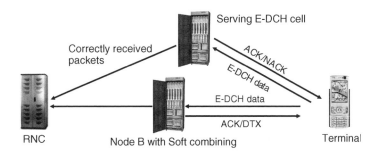

**Figure 13.8**   E-HICH transmission from multiple cells

allow movement, in principle, from a minimum data rate to the maximum one or vice versa, as well as any smaller data rate change in between the two extremes. This also means that signaling has to be more reliable than with the relative grants, as jumping by accident from 16 kbps to 5 Mbps would result in major problems in the network. For this reason, absolute grants are sent with convolutional coding and accompanied by a user-specific 16-bit cyclic redundancy check that is also used for identifying the UE that the E-AGCH transmission was intended for.

## 13.8   HSUPA Physical Layer Operation Procedure

The HSUPA Node B scheduler operation can be described as follows:

- The scheduler in the Node B measures, for example, the noise level at the base station receiver to decide whether additional traffic can be allocated or whether some users should have smaller data rates.
- The scheduler also monitors the uplink feedback, the happy bits, on E-DPCCHs from different users sent in every TTI. This tells which users could transmit at a higher data rate both from the buffer status and the transmission power availability point of view. There is also more detailed information in the MAC-e header on the buffer occupancy and uplink power headroom availability. The latter states how much reserve transmission power the terminal still has, and the former gives the Node B scheduler the information on whether the UE would actually benefit from having a higher data rate or whether it could be downgraded to a lower one.
- Depending on possible user priorities given by the RNC, the scheduler chooses a particular user or users for data rate adjustment. The respective relative or absolute rate commands are then send on the E-RGCH or E-AGCH.

Thus, in the uplink direction, the possible user data rate restriction needs to be informed to the base station from RNC so that the maximum data rate for the service subscribed is not exceeded. The RNC may also give different priorities for the different services for the same users, based on the MAC flow identifiers. The general scheduler operation with the control channels is shown in Figure 13.9. In addition to the scheduled traffic there may be also non-scheduled transmissions, such as Signaling Radio Bearer (SRB) or, for example, a Voice-over-IP connection. Both types of data have a limited delay or delay variance budget and low data rate. Thus, scheduling them would not add much value and would, in the worst case, just degrade the system operation because of, for example, delayed measurement reports; thus, such services are given a permanent grant by the RNC that the Node B scheduler cannot influence. Thus, the non-scheduled transmission operates similar to the Release 99 DCH, but only taking advantage of the physical layer retransmission procedure.

The HSUPA physical layer retransmissions operate as follows:

- Depending on the TTI, four or eight HARQ processes are in use.
- The terminal will send a packet in line with the allowed data rate.
- After the packet has been transmitted, E-HICH is monitored from all the cells in the active set with E-DCH activated. A maximum of four cells can have E-DCH allocated from the maximum active set of six cells.
- If any of the cells indicates a positive acknowledgement (ACK), then the terminal will proceed for a new packet, otherwise retransmission occurs.

The HSUPA operation procedure has strictly specified timing values for the terminal operation as well as for the ACK/negative acknowledgement (NACK) timing response from the base stations, i.e. the whole procedure is synchronized starting from the initial transmission by the UE and ending with the positive ACK reception from the Node B covering the potential retransmissions in between. This

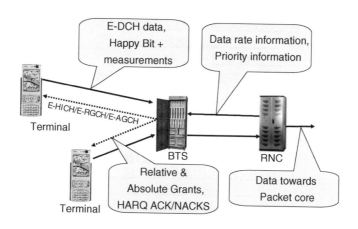

**Figure 13.9**   HSUPA scheduling procedure and relevant signaling

removes the need to signal separately any HARQ process numbers, as the transmission timing always tells for which HARQ channel the retransmission or HARQ feedback is for.

The timing and number of HARQ channels now depend only on the TTI in use; there are no HARQ channels to configure, unlike with HSDPA. With a 10 ms TTI the number of HARQ channels is four and with a 2 ms TTI the number of channels is eight. The resulting timing is given in Figure 13.10 for the 10 ms case, where the delay between the end of the transmission and start of the retransmission is 30 ms, as there is 3 × 10 ms TTIs in between before the same HARQ process is transmitted again.

## *13.8.1   HSUPA and HSDPA Simultaneous Operation*

The operation of HSUPA is typically expected to occur simultaneously with the HSDPA operation. The physical layer channels are defined in such a way that HSDPA operation is not needed simultaneously if the network were for some reason to use Release 99 as the downlink solution. The resulting performance is best when operated simultaneously with HSDPA, especially from the delay and delay variance point of view, as when both directions use L1 retransmissions then also RLC layer retransmission is very seldom needed; and even when needed, then they will happen faster. Also, the SRB will benefit from the use of HSDPA as supported in Release 6, cutting signaling delay then in both directions.

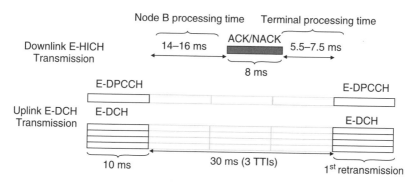

**Figure 13.10**   Timing with 10 ms case for HSUPA operation

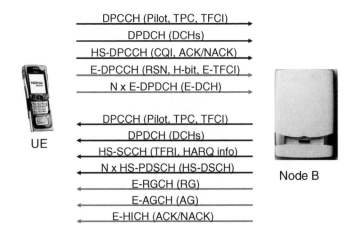

**Figure 13.11**   L1 channels in use for simultaneous HSDPA, HSUPA and DCH operation

The number of channels with simultaneous operation is rather high, though several channels (all downlink control channels) serve multiple users, as covered previously for HS-SCCH in Chapter 12. Figure 13.11 presents the physical channels in use in the case of simultaneous HSDPA and HSUPA operation. The configuration in Figure 13.11 assumes the use of Release 99 DCH as well, e.g. for a speech call. Were this not the case, then in the uplink direction one could omit DPDCH if signaling is also carried on HSUPA; and in the downlink direction, DPCCH/DPDCH could be replaced with Fractional DPCH. Fractional DPCH is covered in Chapter 12.

## 13.9   HSUPA Terminal Capability

The HSUPA feature is optional for terminals in Release 6 with six different categories of terminals allowed by the standard, with resulting maximum physical layer data rates ranging between 0.72 Mbps and 5.76 Mbps (Table 13.4). If a terminal supports HSUPA, then it is also mandatory to support HSDPA. Thus, it is possible in Release 6 to have three kinds of device, with the main classifications as:

- DCH-only device;
- DCH and HSDPA-capable device;
- DCH, HSDPA- and HSUPA-capable device.

**Table 13.4**   HSUPA terminal capability categories

| Category | Max no. of parallel codes for E-DPDCH | TTIs supported (ms) | Smallest E-DPDCH spreading factor | Max. L1 data rate with 10 ms TTI (Mbps) | Max. L1 data rate with 2 ms TTI (Mbps) |
|---|---|---|---|---|---|
| 1 | 1 | 10 | 4 | 072 | N/A |
| 2 | 2 | 2, 10 | 4 | 1.45 | 1.45 |
| 3 | 2 | 10 | 4 | 1.45 | N/A |
| 4 | 2 | 2, 10 | 2 | 2 | 2.91 |
| 5 | 2 | 10 | 2 | 2 | N/A |
| 6 | 4 (with 2 SF4 and 2 SF2) | 2, 10 | 2 | 2 | 5.76 |

**Table 13.5** HSUPA terminal RLC data-rate capability categories

| Category | Maximum RLC data rate with 10 ms TTI (Mbps) | Maximum RLC data rate with 2 ms TTI (Mbps) |
|---|---|---|
| 1 | 0.67 | N/A |
| 2 | 1.38 | 1.28 |
| 3 | 1.38 | N/A |
| 4 | 1.88 | 2.72 |
| 5 | 1.88 | N/A |
| 6 | 1.88 | 5.44 |

**Figure 13.12** Example of HSUPA-capable UEs

For the HSDPA and HSUPA, a terminal can obviously choose the category defining the maximum data rate supported, and especially with HSUPA whether the optional 2 ms TTI is supported or not.

For the data rates at the RLC layer, the overhead for MAC/RLC headers takes some of the data rate, with the resulting values shown in Table 13.5. In the marketplace the categories actually implemented have been category 5 (2 Mbps and 10 ms only TTI support) as in devices such as the Nokia E72 shown in Figure 13.12 or in the products intended for a direct laptop connection. The products more recently on the market in USB stick format typically support category 6 (5.76 Mbps and also 2 ms TTI support), such as the Nokia Internet Stick CS-18 also visible in Figure 13.12.

## 13.10 HSUPA Performance

In this section, different performance aspects related to HSUPA are discussed. The two fundamental HSUPA features, i.e. physical layer packet combining and Node B scheduling, are considered in the analysis for performance improvement from Release 99. With HSUPA, the fast power control is retained and the modulation is unchanged; thus, the fundamental operation is not that far from Release 99 operation.

The HSUPA data rate increase and improved capacity fundamentally come from the following key elements:

- Physical layer retransmission combining, which enables higher initial BLER compared with Release 99, thus resulting in a smaller energy for the first transmission for a packet. Thus, in order to get benefit from this, a higher initial BLER needs to be used, with the expected change to be from 1% BLER on DCH to 10% BLER on E-DCH.
- Fast reaction to data rate/load variations with the Node B scheduler allows one to fill the capacity reserved for transmission better. With Release 99, a user with 384 kbps has all the resources occupied for the allowed peak rate even though the data rate varies. This reservation can only be changed with rate restrictions and reconfigurations from RNC, which are slow in reacting to changes (RRC signaling). Thus, for each Release 99 DCH user, one basically needs to assume that a user may at any time use the maximum data rate allocated, which will lead to a large variance in the resulting interference level and, thus, a large difference between the actual average interference and the desired limit for maximum noise rise.
- The increased data rate is now achieved with more extensive use of multi-code transmission, whereas uplink modulation is unchanged. Additionally, for the higher data rates, an SF as small as 2 is being used.

The following sections will look more into the details of the issues mentioned above.

## 13.10.1   Increased Data Rates

From the performance point of view, the increase in data rates is just done by allowing the use of an SF as small as 2. This, together with four parallel codes (two with SF 2 and two with SF 4) allows 5.76 Mbps to be reached. This kind of peak data rate is, of course, rather theoretical from a range perspective, and also requires quite easy channel conditions to survive the multi-path impact due to low processing gain.

## 13.10.2   Physical Layer Retransmission Combining

The physical layer retransmissions allow, as discussed, a smaller $E_b/N_0$ for the initial transmission. The BLER for the first transmission is also a trade-off between delay, capacity and resulting baseband resource usage. With too low a BLER, there are too few retransmissions and, for example, only 1% of the packets are transmitted more than once, leaving the benefits of fast retransmissions rather marginal. Thus, in the order of 10% of the packets should fail for the first transmission. This provides a bit more than 1 dB reduction for the transmission power needed for a given data rate when comparing with a 1% initial BLER target. Too high a BLER will not eventually boost capacity, it will just use up more resources in the network. If every packet gets transmitted two times on average, then supporting a 1 Mbps data stream requires resources equal to 2 Mbps and then also the maximum single user data rate reduces as so many packets need to be retransmitted.

## 13.10.3   Node B-Based Scheduling

As discussed earlier, the Node B-based scheduling allows faster reaction to the transmission needs from the terminal. This means that the air interface capacity is better utilized, resulting in higher capacity. Also, one can then allocate more high bit-rate users simultaneously, as now they are not all going to be allowed to use the maximum bit rate simultaneously, which means better availability of a high bit-rate uplink service. So with the varying uplink data rate the improvement can be seen in

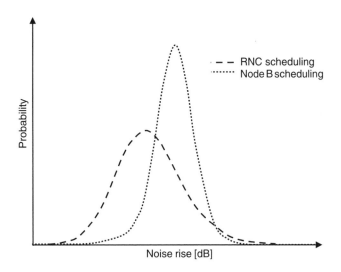

**Figure 13.13**   Noise rise variance distributions with Node B and RNC scheduling cases

Figure 13.13, where the more narrowly distributed curve represents the noise rise variance with fast scheduling and the curve with wider variance corresponds to the case of Release 99 uplink scheduling where control is coming from RNC. In the RNC case, one clearly needs a higher margin between the average and maximum desired noise rise.

The issue could be addressed from the load-level perspective. Faster scheduling can also be considered as taking care of uplink congestion control: the marginal load area can be made narrower and the Prx$_t$arget can be moved closer to Prx$_t$hreshold, which is considered an overload limit, as shown in Figure 13.14.

The dynamic scheduling operation is important for the practical system performance when considering the operation with HSDPA. When a user is downloading a file with HSDPA, the TCP/IP acknowledgements in the uplink will generate an increase in the uplink data rate as well. The correlation depends on whether the actual application will also send something and which TCP/IP version is

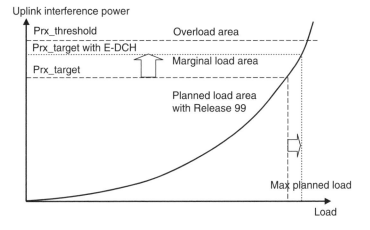

**Figure 13.14**   Increasing the load area

being used. But the use of a 3.6 Mbps peak rate, for example, would be expected to generate at least a 100 kbps uplink transmission need. Thus, in order to get the full benefit from HSDPA operation, there is a need to have a higher than just 64 kbps uplink available. By being able to respond dynamically to the uplink transmission needs, the network also ensures good HSDPA user experience.

### 13.10.4 HSUPA Link Budget Impact

The biggest impact for the link budget comes from the reduced $E_b/N_0$ due to the smaller initial BLER. This extends the cell range for a given data rate compared with Release 99. With some data rates the increase in peak to average power ratio, however, causes the terminal to reduce the transmission power from the maximum power level, similar to the corresponding behaviour with HSDPA. Typically, this occurs only with small data rates and does not impact the coverage of higher data rates at 128 kbps or above or then at the data rates needing more than two parallel E-DPDCHs. Figure 13.15 shows an example of HSUPA uplink data rates achievable as a function of signal strength. The detailed uplink link budget example for HSUPA operation is presented in Chapter 12 together with the HSDPA link budget example.

### 13.10.5 Delay and QoS

The key improvement for the service from the end-user point of view, in addition to the better high bit-rate availability, is the reduced delay variations. The use of fast retransmissions provides a high-reliability physical layer operation that needs RLC layer retransmissions rather seldom. For services with a high delay budget, there may be a need for a smaller amount of allowed retransmissions, such as a maximum of two retransmissions with conversation-type services. Such a service could not use any RLC-level retransmission; thus, better quality can be achieved, compared with Release 99, in the case of non-acknowledged mode operation.

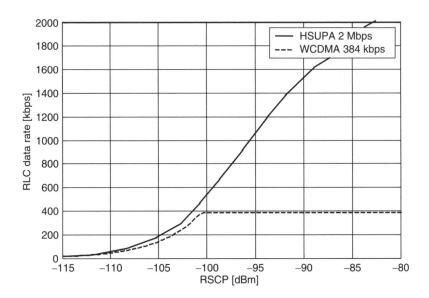

**Figure 13.15**   HSUPA link budget compared with 384 kbps with Release 99

### 13.10.6  Overall Capacity

The overall capacity is thus the outcome from the scheduling impact and chosen initial BLER strategy. The resulting capacity above the physical layer is shown in Figure 13.16. The case studied, as presented originally in [3], is a Vehicular-A environment with a 2800 m cell site distance. The TargetRNS describes the number of retransmissions after which the BLER target is evaluated. For example, a TargetRNS = 0 with a BLER target of 50% means that, after initial transmission, 50% of the packets are erroneous and require a retransmission; and, for example, a TargetRNS = 1 a with BLER target of 10% means that, after one retransmission, 10% of the packets are still erroneous and require further retransmissions. The overall capacity improvement over Release 99 is in the order of 30–50%, depending on the scenario.

The impact of initial BLER for the power needed for the first transmission is presented in Figure 13.17. In order to reach good system performance, the first transmission thus needs to be sent

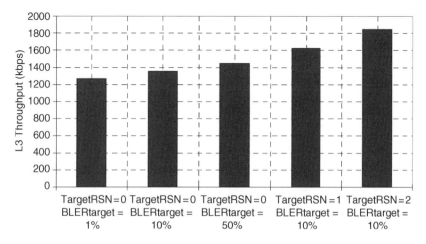

**Figure 13.16**  HSUPA capacity for different BLER and number of transmission targets

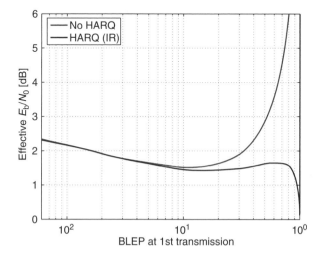

**Figure 13.17**  Initial BLER impact on the first transmission $E_b/N_0$

at a low enough power level and the resulting BLER level needs to be such that the physical channel resources do not become blocked by retransmissions. Thus, operating at 50% BLER is typically not sensible, as it will end up limiting the data rate due to a large amount of retransmissions.

## 13.11   Conclusion

This chapter covered the Release 6 HSUPA principles and the resulting performance. HSUPA represents a major improvement not only in the uplink capacity, but also in the efficiency of baseband and transmission resource utilization. The use of HSUPA fits well with the use of HSDPA as the downlink solution by enabling dynamic uplink data rate allocation. HSUPA was rolled out for many networks from 2007 onwards and represents the natural second major WCDMA network evolution step following HSDPA introduction to complement high data rates in the downlink direction as well as facilitating low latency. All the latest WCDMA USB data modems have support for HSUPA as a standard feature with typically 5.76 Mbps peak physical layer data rate support. Further 3GPP developments on top of HSUPA are presented in Chapter 15.

## References

[1] 3GPP Technical Report, TR 25.896, Feasibility Study for Enhanced Uplink for UTRA FDD (Release 6), Version 6.0.0, March 2003.
[2] 3GPP Technical Specification, TS 25.211, Physical channels and mapping of transport channels onto physical channels (FDD), Version 6.7.0 December 2005.
[3] Wigard, J., Boussif, M., Madsen, N. H., Brix, M., Corneliussen, S. and Laursen, E. A. 'High Speed Uplink Packet Access Evaluation by Dynamic Network Simulations', *PIMRC 2006*, Helsinki, Finland, September 2006.

# 14

# Multimedia Broadcast Multicast Service (MBMS)

Harri Holma, Martin Kristensson and Jorma Kaikkonen

## 14.1 Introduction

WCDMA Release 99 and HSDPA Release 5 are used in commercial networks to deliver multimedia content such as mobile TV and streaming videos. The point-to-point data delivery in WCDMA Release 99 and HSDPA Release 5 allows full flexibility when delivering user specific and on-demand content to many users at any time. The point-to-point delivery is also spectrum efficient when the number of users is not too large. When the number of users increases and they would like to receive the same content at the same time, it is more efficient to use point-to-multipoint delivery. The 3GPP Release 6 Multimedia Broadcast Multicast Service (MBMS) specification introduces point-to-multipoint support. Cellular networks can with MBMS provide the same content efficiently to a large number of subscribers at the same time. This chapter introduces the MBMS concept, 3GPP MBMS specifications and the expected MBMS performance. The terminal capability and MBMS use cases are presented as well.

MBMS reuses the radio and core network elements in the legacy 3GPP architecture to make the introduction of MBMS cost efficient, see Figure 14.1. In addition to the legacy elements, MBMS introduces the Multimedia Broadcast and Multicast Service center (BM-SC). The new BM-SC element supports a number of functions, e.g., service scheduling, service authentication, content encryption, key management and charging. In the legacy 3GPP elements, MBMS requires additional software functionality and possibly additional hardware due to capacity reasons. Adding MBMS functionality to the UMTS terminal platforms is similarly straightforward since, except for minor additions, existing transport channels and physical layer channels are re-used and the MBMS data rates are low compared to WCDMA or HSDPA peak data rates. It is therefore expected that MBMS functionality can be part of almost all 3G terminals in the future as long as the terminal form factor allows using the services introduced with MBMS. A promising key strength of MBMS is hence a large terminal base and a large number of MBMS capable networks using the same standard globally. Having a large terminal base is a prerequisite for all broadcast technologies.

Since MBMS is designed on top of the UMTS network, it can automatically benefit from the existing WCDMA point-to-point channels to carry any potential service level feedback information,

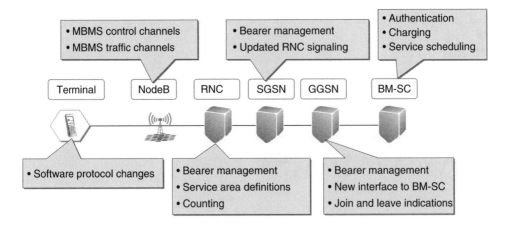

**Figure 14.1**   MBMS uses the existing UMTS network

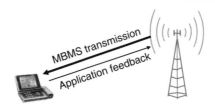

**Figure 14.2**   Uplink feedback with MBMS

see Figure 14.2. The feedback information can be used for chatting, to track how many users are watching a certain content, by the content provider for interactive applications, etc.

MBMS user plane bearers support both point-to-point and point-to-multipoint connections (Figure 14.3). This is because if the number of users per cell is low, it is most efficient to deliver the MBMS user plane data over ordinary point-to-point connections. If, on the other hand, the number of users increases and they all would like to receive the same content at the same time, it is more efficient to use point-to-multipoint connections. The point-to-point channels are more

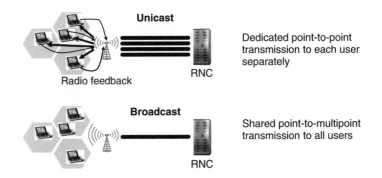

**Figure 14.3**   Concept of broadcast compared to unicast

efficient for a low number of users because with point-to-point connections it is possible to utilize feedback techniques including closed loop power control, link adaptation and retransmissions. Those techniques reduce the required transmission power and interference and they cannot be efficiently applied to point-to-multipoint broadcast transmission. The MBMS standard allows the radio network to dynamically select between point-to-point and point-to-multipoint transmission depending on the number of simultaneous users receiving the same content. The most popular services could be carried over point-to-multipoint bearers for efficient mass delivery and other less popular services could use point-to-point delivery to provide a large selection of different channels. The point-to-multipoint delivery is also beneficial from the terminal power consumption point of view since no uplink transmission is required for the radio feedback.

MBMS contains two modes of operation: broadcast and multicast. In case of broadcast the 3GPP core network (GGSN and SGSN elements) are mainly unaware of when and how individual users view certain content and channels. The broadcast mode supports both point-to-point and point-to-multipoint connections and the transmission in the radio cell is on or off depending on the number of users listening to a certain content channel in this cell. Note, however, that as long as the broadcast bearer is active, the user data is always transmitted from the BM-SC all the way to all RNCs, i.e., in the core network the transmission is always on from the BM-SC to all RNCs as long as the MBMS session is active. The key difference between the multicast mode and the broadcast mode is that in the multicast mode a user needs to register for each service that the user is interested in receiving, i.e., the user needs to join a multicast group. This is done by the mobile station transmitting a joining command to the network. Hence, in this case the core network knows which mobiles are actually listening or are interested in listening to each service. Because the core network is aware of the users in case of multicast, it is possible to activate the MBMS bearers only to those SGSNs and RNCs where there are users listening to the content. In addition, the knowledge of individual users can allow additional charging possibilities as well as the introduction of group-oriented services. From a radio resource point of view the multicast mode works similar to the broadcast mode. The differences between broadcast and multicast modes are illustrated in Figure 14.4. The procedures are described in Section 14.3.

The dedicated broadcast solutions, like traditional TV and radio broadcasting, or DVB-T and DVB-H (Digital Video Broadcasting – Terrestrial and Handheld), use high power transmission from high masts. The same content is transmitted over a large area. There may be just a few tens of large broadcast towers per country. The large broadcast towers are then typically complemented with smaller sites to

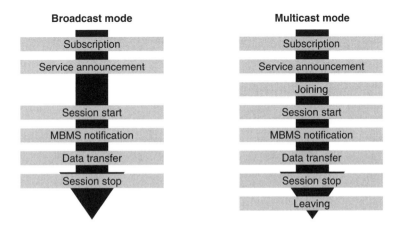

**Figure 14.4**   Concept of multicast compared to broadcast

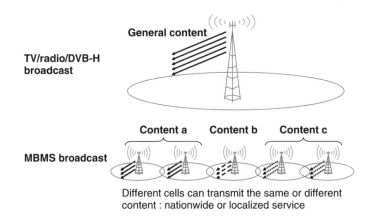

**Figure 14.5**  Localized service in MBMS

provide coverage throughout the continuous area also indoors. The dedicated broadcast solutions do not need to have an uplink connection from the user equipment to the base station. Hence, the link budget for a broadcast solution can be improved by increasing the output power in the base station without worrying about the limited power available in the mobile station in the uplink direction. This enables dedicated broadcast solutions to have both high capacity and coverage.

The MBMS radio transmission uses existing cellular sites and equipment. There can be thousands of base station sites per operator per country. The large number of sites and the limited broadcast capacity can make the MBMS activation more costly than the high power broadcast solutions. On the other hand, the base station sites in many cases are already available and with software upgrades the cost may be limited as long as extra capacity is not needed. MBMS also allows localized content to be delivered. When it comes to deployment costs, then MBMS can compete well with dedicated broadcast solutions as long as the usage is not too large, while the high capacity offered by the broadcast solutions makes them more cost efficient for heavy usage. The broadcast service with MBMS can combine nationwide channels with local information like traffic or weather data. The difference between high power broadcasting and MBMS transmission is illustrated in Figure 14.5.

MBMS can be used to offer and support, e.g., mobile TV, news updates via file downloads, operator home pages pre-downloaded to mobiles, local content distribution, traffic information, software downloads to terminals and distribution of emergency information to all terminals. MBMS is part of 3GPP Release 6. The most relevant 3GPP specifications and technical reports are collected in the references [1–15].

## 14.2   MBMS Impact on Network Architecture

The starting point for the MBMS architecture [3] is the UTRAN and packet core network architectures as originally defined in 3GPP Release 99. The MBMS architecture is hence designed to re-use 3GPP Release 99 network elements and protocols as much as possible. The re-use of existing structures provides backwards compatibility both in the networks and in the terminals. It enables receiving MBMS data in parallel to other services and signaling; this includes paging and circuit switched voice calls.

The MBMS network architecture is illustrated in Figure 14.6. The MBMS architecture introduces one new network element: the BM-SC. The BM-SC element resides between the server that provides

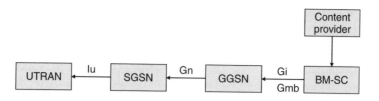

**Figure 14.6**  MBMS network architecture

the content and the GGSN. The BM-SC provides partial or full support for the following MBMS functions:

- *User authentication and authorization*. User authentication and authorization for MBMS services are implemented according to the generic authentication architecture [14]. Together with encryption, the authentication and authorization mechanisms in MBMS can control which services an end user is allowed to access.
- *User, service and traffic key delivery, and encryption*. Encryption can be used in MBMS to ensure that users are only able to listen to content that they have subscribed to. In ordinary point-to-point connections in WCDMA Release 99 and Release 5 there is encryption on the bearer level. In MBMS the radio access bearer is not encrypted as such. Instead, the encryption takes place in the BM-SC closer to the application level. To implement the encryption and the restricted access to content delivered over MBMS, the BM-SC is in charge of key distribution and content encryption [15].
- *Service announcements*. The service announcements are used to inform the end users about the available MBMS services. The service announcement contains information such as, for example, IP multicast addresses [3, 10]. The service announcement can take place via, e.g., SMS, HTTP (home page) and also via information delivered via an MBMS broadcast bearer.
- *Service/session scheduling (sending of session start/stop messages)*. The BM-SC sends the session start message when it is ready to send data and this message triggers reservation of resources for the MBMS bearer in the network [3]. The session start message contains information of, for example, the areas where the MBMS content should be delivered (service areas) and quality of service parameters.
- *Streaming data and downloading files*. The BM-SC sends the user plane data towards the GGSN. In the case of file download, the BM-SC uses the FLUTE (File Delivery over Unidirectional Transport) protocol and in case of streaming the BM-SC uses the RTP (Real Time Protocol) protocol [10] to deliver the data.
- *Forward error correction (FEC) of user plane content*. In point-to-point connections according to 3GPP Release 5, it is possible for the terminal to ask for retransmissions if packets get corrupted over the air interface. However, in the case of broadcast, the terminals cannot ask the radio network for retransmission. To compensate for the lack of re-transmissions in the radio interface, additional forward error correction has been included as an additional option in the BM-SC [10].
- *End user and bearer level charging*. The BM-SC provides support for bearer and user level charging and for this purpose it collects information about both bearers and end users [13].

For each MBMS bearer, the BM-SC keeps a corresponding MBMS bearer context. The MBMS bearer context contains information about the MBMS bearer like its IP multicast address, the temporary mobile group identity (TMGI), quality of service to be applied and service areas where the content should be transmitted. In the case of multicast services where the mobile stations join to the service, the BM-SC also keeps a UE context for each terminal that has joined the MBMS service. The bearer and UE contexts are similar to the PDP contexts of ordinary point-to-point bearers in 3GPP Release 99 and Release 5.

The BM-SC is connected to the GGSN via the Gmb (signaling) and Gi (user plane) interfaces. The signaling interface between the BM-SC and the GGSN includes message support for the items in the above bullet list such as session start/stop messages, authentication, authorization, and joining/leaving sessions. The BM-SC transmits the data that it receives from content providers to the GGSN. The GGSN maps the user plane data in the form of IP multicast packets to the correct GTP (GPRS tunneling protocol) tunnels. For each MBMS bearer the GGSN and the SGSN keep, like the BM-SC, an MBMS bearer context. In addition, in the case of the multicast mode, the GGSN and SGSN also keep UE contexts.

The MBMS specifications include QoS support for streaming and background traffic classes. Hence, in comparison to traditional point-to-point bearers, the interactive traffic class is missing from the MBMS specifications. The BM-SC provides information in the session start message on what QoS class that the MBMS bearer should use. This QoS information is thereafter stored in the bearer concept in the network elements. It is up to the network (GGSN, SGSN and RAN) to determine how to treat the MBMS bearers when reserving resources. MBMS bearers may in the radio interface be both point-to-point and point-to-multipoint. The QoS concepts for MBMS should hence support both point-to-point and point-to-multipoint connections.

The MBMS specifications also include the concept of service areas. The service areas are defined in the radio access network operations support systems. From a standards point of view, each cell in the radio access network can belong to multiple services areas and a service area can include multiple cells in different parts of the networks. There is no relation between service areas and routing/location areas. The definition of service areas is hence very flexible and made on a cell-by-cell basis.

## 14.3  High Level MBMS Procedures

This section introduces the high-level procedures for the broadcast mode in MBMS including subscription, service announcement, authentication, key delivery, session start, and session stop. A more detailed description of the over the air level procedures are provided in the next section. The procedures are illustrated in Figure 14.7 and commented on below:

1. *Subscription.* To be able to use the MBMS services it may be necessary for the end user to subscribe to the MBMS services. The user subscription information is needed in the BM-SC. The subscription to the service happens once.

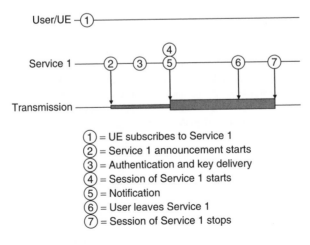

**Figure 14.7**  Example MBMS procedures

2. *Service announcement*. The BM-SC is in charge of delivering the service announcements to the terminals. In this way the mobile station becomes aware of the services that are offered via MBMS bearers and potentially also about other services. The service announcement may take place via, for example, an HTTP Web page. The announcement includes information such as, for example, the service start time and the IP multicast address.

3. *Authentication and key delivery*. Without encryption in the BM-SC; the content on the MBMS bearers is available to all users in the network regardless of whether a user has subscribed to the service or not. If the BM-SC uses encryption, then it is possible to ensure that only subscribers that have paid for the service can access the service. If a user has subscribed to the MBMS service, then the user can, via a point-to-point connection, obtain user and service keys to open up the content carried on the MBMS bearers.

4. *Session start*. When the MBMS session is about to start, the BM-SC sends the session start message to the GGSN. The session start message contains, e.g., the QoS parameters of the bearer, the service areas where the session should be transmitted, and the session duration. The session start message is then further distributed to the SGSN and the RNC. All network elements then allocate resources for the MBMS bearer guided by the parameters in the session start message.

5. *Notification*. At the end of the set-up the bearer is advertised in the radio interface. The notification channel is used to inform the mobile stations about changes in the MBMS bearer set-up. After a while the BM-SC can start to transmit data to the MBMS bearer.

6. *Terminal listens and decodes and finally leaves*. The terminal finds out from the common channels in the radio interface that the MBMS bearer is set up. The terminal can then start listening to the MBMS bearer and decrypt the content by using the keys it has obtained via the point-to-point connection and the appended keys provided in the traffic stream in the MBMS bearer.

7. *Session stop*. When the session is over, the BM-SC sends the session stop message to the GGSN, which then starts tearing down the MBMS bearer by contacting the SGSN.

## 14.4   MBMS Radio Interface Channel Structure

The logical, transport and physical channels for MBMS are described in this section. A design criterion when developing the MBMS solution has been to re-use the already available channels as much as possible from the earlier 3GPP Releases to facilitate an easy introduction in the radio network.

### 14.4.1   Logical Channels

MBMS introduces the following new logical channels: MBMS point-to-multipoint Control Channel (MCCH), MBMS point-to-multipoint Traffic Channel (MTCH) and MBMS point-to-multipoint Scheduling Channel (MSCH).

MCCH is used for control plane information and there is only one MCCH per cell. The MCCH provides information about the MBMS configuration in the particular cell and in the network. The messages that are transmitted on the MCCH control channel provide information about the currently active MBMS services and their identities, where and how to find the MTCH channels in the cell, configurations and timing in neighboring cells to enable combining, and how MBMS services are provided on different carrier frequencies. MTCH is the channel that carries the user plane data to the terminals. It is possible to time multiplex multiple MBMS services in the same physical layer channel. This time multiplexing can bring gains in, for example, terminal power consumption. If multiple services are multiplexed into one physical layer channel, the time multiplexing information for the physical channel is included in the MSCH.

In the case of point-to-point MBMS transmission Release'99 dedicated traffic and control channels, DTCH and DCCH are used.

## 14.4.2   Transport Channels

The already available forward access channel (FACH) is used as the transport channel to carry the MBMS logical channels. The MCCH is mapped to a pre-defined FACH and the mapping is indicated on the Broadcast control channel (BCCH). Also MTCH and MSCH are mapped to pre-defined FACH channels and their mappings are indicated on MCCH. The MTCH channel carries user plane data, while the MSCH channel carries information on how multiple MTCH channels are multiplexed in one physical channel. Because MTCH carries user data and MSCH carries control information, they have different requirements when it comes to target block error rates. Therefore, MTCH and MSCH are mapped to different FACH to enable different block error rate targets.

## 14.4.3   Physical Channels

The FACH transport channel is mapped to the Secondary Common Control Physical Channel (S-CCPCH). The MCCH can be mapped to an already existing S-CCPCH used also for non-MBMS purposes, or to an S-CCPCH dedicated to MBMS. The MCCH and MTCH are mapped to different S-CCPCH in case of soft combining because the MCCH is a control channel that contains information for how to access the MBMS channels in the network and also information about neighboring cells. Hence, the terminals must be able to decode the MCCH without knowing about neighboring cells and without using combining. For MTCH, soft combining can be applied and information about how to do the combining of neighboring cells is found from the MCCH. Soft combining is explained in Section 14.5.1. The transmission time intervals (TTI) can range from 10 ms up to 80 ms.

MBMS also introduces a new physical channel: the MBMS Notification Indicator Channel (MICH), which is an MBMS specific Paging Indicator Channel (PICH). MICH informs UEs of an upcoming change in the MCCH information. The MBMS channel mapping is illustrated in Figure 14.8. Note that, according to the discussion earlier of mappings, the logical channels may be mapped to different FACHs.

MCCH information can be changed at the beginning of a modification period. MICH is transmitted in the previous modification period before the MCCH content is changed. The modification period is long enough so that the terminals can receive MICH during normal paging occasions. The timing relationship is shown in Figure 14.9. During a modification period the MCCH content may be repeated several times. This can be used by a terminal to improve the MCCH reception performance.

## 14.4.4   Point-to-Point and Point-to-Multipoint Connections

As was briefly discussed in Section 14.1, point-to-point connections are more efficient in the radio interface if the number of users is low, while delivery over point-to-multipoint channels is more efficient if there are many users. The threshold when point-to-multipoint connections start becoming

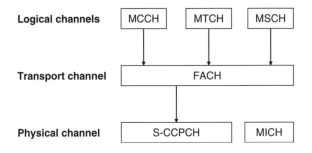

**Figure 14.8**   MBMS channel mapping

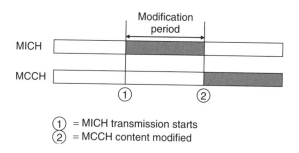

**Figure 14.9**  Timing between MICH transmission and MCCH content change

more efficient is around three to six users. To minimize the required radio investment, the radio network should dynamically select point-to-point connections when the number of users is limited and point-to-multipoint connections when the number of users is large. This procedure is supported in MBMS and it is often denoted briefly as switching.

To support switching, the radio network (RNC) must be able to estimate the number of users that are listening to the MBMS transmission. The so-called counting procedure allows the RNC to determine the number of mobiles that are interested in listening to an MBMS service. Counting is MBMS service-specific and this procedure can be used to determine the optimal delivery method. Counting is indicated by RNC on MICH and UEs respond by sending MBMS Counting Response message. The UEs in idle mode establish RRC connection, UEs in Cell_FACH state use signaling on FACH and the UEs in URA_PCH and in Cell_PCH state make a cell update. If the number of terminals in the same MBMS group is large, counting may cause a large number of simultaneous RRC connection establishment. That can be avoided by setting a probability factor.

If the operator has deployed multiple WCDMA frequencies, it is preferable to have all those UEs on the same carrier that are interested in receiving the MBMS content. That can be achieved by UTRAN providing UE-specific cell reselection parameters for MBMS subscribers.

## 14.4.5  Example Radio Interface Procedure during MBMS Session Start

An example MBMS session start is described below. The MBMS session start procedure is initialized by the BM-SC when it sends the MBMS session start message to the GGSN. After the BM-SC has sent the session start message, it waits until a timer expires and then it starts sending user plane data towards the GGSN. In the radio network the following needs to take place before the user data can be delivered to the mobile stations:

1. The SGSN sends the MBMS session start message to the RNCs. This informs the RNC that soon MBMS data is to be delivered over the air interface.
2. The RNC reserves transport and radio resources to set up the MBMS Radio Access Bearer (RAB).
3. The RNC updates the MICH to prepare for updated MCCH information. The updated MCCH information explains the set-up of the new MBMS RAB.
4. The RNC informs the terminals about the MBMS service start using the MCCH. If there are terminals in the Cell_DCH state, this information is sent on DCCH.
5. The RNC counts the terminals to decide the MBMS delivery method: point-to-point or point-to-multipoint bearer.
6. Terminals establish mobility management connections in response to the counting request. If the probability factor is used, then only a fraction of the mobiles will answer the counting request.

7. The RNC establishes the S-CCPCH and FACH to carry the MTCH if the RNC decides to use point-to-multipoint delivery of the MBMS user plane data.
8. The RNC informs UEs about MBMS transmission using MCCH.

## 14.5 MBMS Terminal Capability

The minimum terminal requirements are defined in [6]. The MBMS terminals support minimum 256 kbps data rate with TTI (Transmission Time Interval) between 10 and 80 ms. That is, all terminals must be able to support the maximum bit rate, 256 kbps, that is supported in MBMS. The terminals are required to support both soft and selective combining. The combining solutions are described in the following section since the combining is an important factor in defining the MBMS performance.

### 14.5.1 Selective Combining and Soft Combining

With selective combining or soft combining UE at the cell edge can combine the MBMS transmissions of same service from the adjacent cells. The mobile station handles the MBMS combining autonomously. The combining allows reduction of the MBMS transmission power levels while maintaining good quality also at the cell edge. The combining also improves MBMS performance for mobile users making frequency cell reselections. The soft combining principle is introduced in Figure 14.10. The use of soft combining to receive MBMS service requires that identical physical layer bits are transmitted from the adjacent cells and the relative delay between the radio links to be combined is short enough. UE does not need to read MCCH from the neighboring cells since the network sends MBMS neighboring cell information that contains MTCH configurations. The UE determines the neighboring cell suitable for selective or soft combining based on the CPICH $E_c/N_0$ threshold. The UE cannot perform soft and selective combining simultaneously.

The implementation of soft combining in the terminal is illustrated in Figure 14.11. The soft combining is similar to the macro diversity combining in normal soft handover. The Rake receiver processes both signals from the base stations. The Rake output consists of soft-valued quantities, which are then combined before feeding them to the channel decoding.

Selective combining is implemented at the RLC (Radio Link Control) layer, where the UE attempts to recover the information by combining packets received from different cells. Hence the receiver

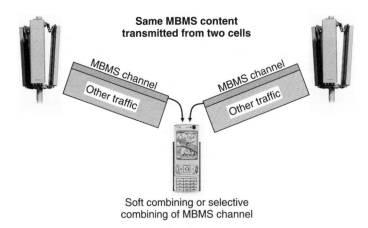

**Figure 14.10** Combining of MBMS transmission from two cells

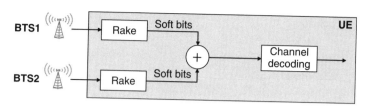

**Figure 14.11** Principle of soft combining

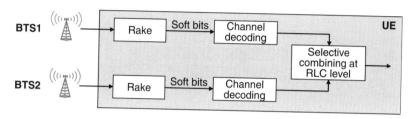

**Figure 14.12** Principle of selective combining

simultaneously decodes transport blocks from different cells delivering the correctly received packets to the RLC layer where the packets from different sources are reordered and combined to construct a complete RLC SDU (Service Data Unit) packet. This is illustrated in Figure 14.12. Selective combining requires the transport block content to be the same across the different cells and is therefore only applied only to MTCH. However, part of the selective combining mechanism can also be used on MCCH. The repetitions of the MCCH messages over the modification period of a single cell can be reordered on the RLC, thus providing a more reliable MCCH reception.

The maximum timing difference between base stations must be limited in order to enable the combining in the terminal. If a large timing difference were allowed, the UE memory requirements would increase for soft combining. The relative base stations synchronization requirement is defined to be one TTI plus one time slot, which is 40.67 ms for 40 ms TTI and 80.67 ms for 80 ms TTI. If the timing difference happens to be larger, the terminal is not able to combine the MBMS transmissions. In the case of selective combining the timing requirements are set so that the transmissions between neighboring cells should be synchronized to less than 64 RLC packet data units. Since one RLC packet covers one TTI, the synchronization accuracy with 40-ms TTI equals $64 \times 40\,\text{ms} = 2.56\,\text{s}$, so essentially the base station synchronization requirements are very relaxed in the case of selective combining, see Figure 14.13.

## 14.6 MBMS Performance

This section first presents 3GPP performance requirements and shows cell capacity results with soft combining and also with multi-antenna terminals.

### 14.6.1 3GPP Performance Requirements

The 3GPP performance requirements in [5] define the maximum RLC SDU error rate for the given scenario and for the given S-CCPCH $E_c/I_{or}$. The assumed multipath channel is Vehicular A 3 km/h and the amount of other-cell interference is $I_{or}/I_{oc} = -3\,\text{dB}$. The Ioc models noise and other interfering

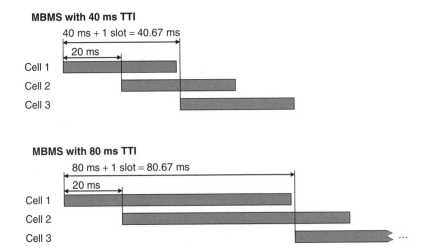

**Figure 14.13** Base station synchronization requirements for soft combining

cells not actually present in the simulations. However, the total interference seen by the UE includes also the other radio links modelled in the simulations resulting $I_{or}/I_{oc} = I_{or1}/(I_{or2} + I_{or3} + \hat{I}_{oc}) = -6$ dB with three cells. In other words, these requirements can be considered to correspond to an MBMS service which can provide good quality reception also at the cell edge. The cell capacity can be estimated from the performance requirements as follows by assuming the common channel overhead of 15%.

$$\text{MBMS cell capacity} = \frac{\text{User data rate}}{10^{\frac{E_c/I_{or}[\text{dB}]}{10}}} \cdot (1 - \text{Common channels}) \qquad (14.1)$$

Table 14.1 summarizes the different performance requirement scenarios and capacities estimated based on those test cases. The requirements are defined both for 1-antenna UEs and for 2-antenna UEs. Cases 2 and 3 show MBMS cell capacity of 770–790 kbps with 1-antenna UE and 1540–1610 kbps with

**Table 14.1** 3GPP performance requirements for MBMS UEs [5]

| Test case number | Bit rate [kbps] | TTI [ms] | Combining | $E_c/I_{or}$ [dB] | $E_c/I_{or}$ [dB] | Estimated capacity [kbps] | Estimated capacity [kbps] |
|---|---|---|---|---|---|---|---|
| | | | | 1-ant | 2-ant | 1-antenna | 2-antenna |
| 1 | 128 | 40 | Selective combining as the link delays are 160 ms and 1240 ms | −4.9 | −7.7 | 340 | 640 |
| 2 | 256 | 40 | Soft combining of three radio links with max delay 40.67 ms | −5.6 | −8.7 | 790 | 1610 |
| 3 | 128 | 80 | Soft combining of three radio links with max delay 80.67 ms | −8.5 | −11.5 | 770 | 1540 |

2-antenna UEs. In these cases the cells are sufficiently well synchronized to allow soft combining to be used. In case 1 the estimated capacity is clearly lower as the delays between cells are larger and therefore only selective combining can be assumed. The MBMS capacity must typically be designed assuming 1-antenna UEs. Using the 2-antenna capacity values requires that all MBMS UEs in the network have a 2-antenna receiver.

## 14.6.2   Simulated MBMS Cell Capacity

Macro cellular simulation results for MBMS capacity are presented in Figure 14.14. The multipath channel used during the simulation is Vehicular A. The y-axis shows the throughput for the dedicated MBMS carrier with and without soft combining and with one and two-antenna terminals.

The capacity without any combining method is very limited – below 300 kbps per carrier. The reason is that the terminal must be able receive the MBMS transmission correctly also at the cell edge with high other cell interference. The cell reselection hysteresis need to be considered as well: the mobile cannot be always connected to the best cell due to the cell reselection hysteresis. The required transmission power needs to be high to guarantee MBMS reception quality. The soft combining can drastically improve the MBMS efficiency to 600–1200 kbps depending on the terminal window size. Soft combining with single antenna terminals provides higher capacity than dual antenna terminals without combining. It is difficult to get the benefit of the dual antenna terminals unless all or most MBMS terminals have two antennas, which is unlikely to happen in the near future. Therefore we can conclude that soft combining is important for MBMS cell capacity.

Assume an MBMS capacity of around 1 Mbps with single antenna terminals and soft combining. This capacity is equivalent to 16 channels á 64 kbps, or 8 channels á 128 kbps. The MBMS point-to-multipoint capacity is independent of the number of simultaneous users receiving the transmission. The HSDPA point-to-point capacity, on the other hand, is directly dependent on the number of simultaneous users, but it is not dependent on the number of offered channels. The HSDPA capacity is typically estimated at 2 Mbps per cell [HSDPA chapter]. The point-to-point HSDPA transmission can gain the benefit of the feedback techniques enhancing capacity including Hybrid ARQ retransmissions and link adaptation.

The HSDPA transmission allows independent offerings in different cells. The point-to-multipoint transmission can also be cell-specific, but then it is not possible to gain the full benefit of the combining.

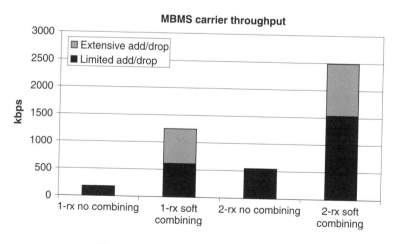

**Figure 14.14**   MBMS cell capacity [16, 17]

The mobile power consumption will also differ in these delivery options: point-to-multipoint transmission needs no uplink transmission while point-to-point operation requires continuous uplink feedback. The differences between point-to-point and point-to-multipoint transmission are summarized in Table 14.2.

The optimal switching point between point-to-point and point-to-multipoint depends on the number of users and offered channels. For a low number of users, point-to-point delivery is more efficient while for a high number of users point-to-multipoint is preferred. For a low number of channels point-to-multipoint is preferred while for a high number of channels point-to-point is preferred. The trade-offs from the air interface point of view are illustrated in Figure 14.15. The air interface capacity is assumed to be 1 Mbps for MBMS and 2 Mbps for HSDPA. The channel data rate is assumed to be 128 kbps. If the operator would like to offer four channels, point-to-multipoint is more efficient with more than 8 users. With 8 different channels, the switching point is at 16 users.

When switching the delivery method, cells cannot be considered separately. If point-to-multipoint delivery is activated just in a single cell, it is not possible to take benefit of the combining. It may be necessary to consider several cells when switching the delivery method.

**Table 14.2**  Benchmarking of point-to-point and point-to-multipath

|  | HSDPA point-to-point | MBMS point-to-multipoint |
| --- | --- | --- |
| Number of offered channels | No impact to capacity | Capacity requirements increase linearly when more channels are added to multicast transmission |
| Number of simultaneous users | Capacity requirements increase linearly when more users are added | No impact to capacity |
| Cell specific transmission | Different cells can have independent content | If different cells use different content, it is not possible to optimize the capacity by using combining solutions |
| Mobile power consumption | Similar to normal HSDPA operation | Lower than HSDPA since mobile transmitter is not running, only receiver. |

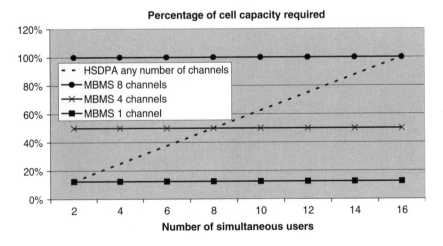

**Figure 14.15**  Capacity benchmarking of point-to-point HSDPA and point-to-multipoint MBMS

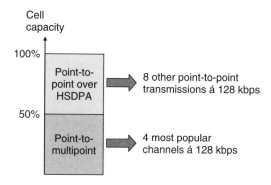

**Figure 14.16**   Combination of point-to-point and point-to-multipoint delivery

One option is to use MBMS point-to-multipoint for the most popular channels and the less popular channels are transmitted using point-to-point delivery. Such an approach is illustrated in Figure 14.16. Some 50% of the cell capacity is allocated to point-to-multipoint providing four channels at 128 kbps. The remaining 50% capacity is used for point-to-point transmission for up to 8 users. Those 8 users can receive any channels.

## *14.6.3   Iub Transport Capacity*

The section so far has considered the air interface capacity. The MBMS solution brings also benefit in Iub capacity requirements compared to point-to-point transmission. The point-to-point solution requires dedicated Iub capacity for each user while MBMS point-to-multipoint solution only requires single Iub capacity allocation. 3GPP Release 6 uses single Iub allocation per sector and Release 7 single Iub allocation per base station. The Iub capacity allocation with MBMS is illustrated in Figure 14.17.

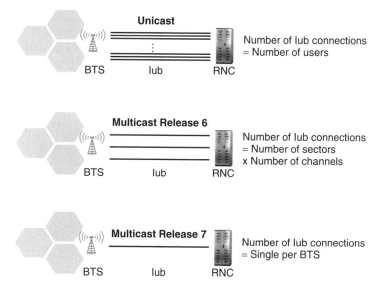

**Figure 14.17**   Iub capacity requirements

## 14.7   MBMS Deployment and Use Cases

A number of different services and applications can be provided over the MBMS solution. The following type of service classes can be delivered with MBMS:

- *Real-time distribution of multimedia content*. The MBMS servers, core network equipment and radio interface solutions provide support for real-time services such as TV and radio. See example in Figure 14.18.
- *Non-real-time distribution of content*. File download via the point-to-multipoint system is enabled by protocol support in the BM-SC and also with support for background quality of service in the network for the radio access bearer. The file download can be used for many different services, e.g., downloading the latest news updates to the terminals in the background including text and streaming clips, and software downloads to terminals.
- *Carousel service*. A carousel service provides content that is repeated at certain intervals. The content may be both real-time TV as well as background files. An example of an application utilizing the Carousel service is a 'ticker-tape' type service in which the data is provided to the user repetitively and updated at certain times to reflect changing circumstances. The required data rate for file download and carousel service can be lower than for the real-time streaming service.

From the resource consumption point of view, real-time content such as on-line mobile TV or radio requires a certain amount of capacity from the radio interface during peak hours to guarantee the service delivery. For non-real-time distribution of content, the capacity requirement could be significantly lower because the content may not need to be delivered during the peak capacity hours of the network. Instead, the non-real-time content could be delivered during off-peak times during both day and night. For example, a news service may supply the most popular streaming clips to the mobiles early in the morning and then later during the day update this content on the memory cards whenever there is network capacity available. The non-real-time distribution can hence provide a cost-efficient multimedia content delivery solution to a large terminal base.

For file download services, one MBMS radio access bearer of 256 kbps can deliver in total 2.7 GB of data during 24 hours to all the terminals in the network. In other words, it can potentially provide 115 MB of data every hour to all the terminals in the network. This data amount could, for example, be represented by one hour streaming clips of 256 kbps content. An application in the mobile station can compile the received content and automatically update with the latest provided streaming clips. From a battery point of view, terminals cannot listen to the MBMS carrier all the time so the services provided to a mobile station must be limited in time even if the service is provided to the mobile station in background mode.

**Figure 14.18**   Real-time TV distribution

## 14.8   Benchmarking of MBMS with DVB-H

DVB-H (Digital Video Broadcasting – Handheld) is a dedicated broadcast solution for mobile devices [18, 19]. The DVB-H solution provides both streaming and downloading support. DVB-H has an OFDM (Orthogonal Frequency Division Multiplexing) air interface designed particularly for broadcasting and the standard has full mobility support. The combining of different DVB-H transmitters in the mobile station is taken care of automatically when using single frequency network (SFN) functionality and OFDM modulation. A single frequency network uses the same frequency to deliver the same content throughout an area. This means that all the transmitters in an area transmit the same signal on the same frequency. From a terminal point of view, the received signal looks as if there only were one base station. DVB-H can use high broadcast masts with high power transmission allowing SFN cell sizes of up to tens of kilometers. Repeaters are typically used when building the DVB-H coverage. DVB-H may use lower frequency bands than MBMS in many countries to further boost the coverage. DVB-H is able to offer higher data rates and more simultaneous channels than MBMS. Because DVB-H is a pure broadcast solution, it relies on the cellular network support for on demand services.

3GPP MBMS is compared to DVB-H in Table 14.3. DVB-H can provide a larger selection of channels and higher data rates for larger displays while MBMS functionality is, on the other hand, simple to integrate into 3G terminals and the number of MBMS-capable terminals is therefore expected to be high once the first software implementations are available.

**Table 14.3**   Benchmarking of MBMS and DVB-H

|  | MBMS point-to-point over HSDPA | MBMS point-to-multipoint | DVB-H |
|---|---|---|---|
| User data rate | In theory >1 Mbps, but <512 kbps for good coverage | Max 256 kbps, typical 64–128 kbps | Typical 128–384 kbps |
| Number of offered channels á 128 kbps | No limits – but the number of users is limited | 5-10 channels per 5 MHz | Tens of channels |
| Support of on-demand services | Yes | Provided via point-to-point channels over HSDPA | Provided via a cellular network |
| Number of users per channel with á 128 kbps | Approx 16 per 5 MHz | No limits | No limits |
| Frequency | UMTS frequencies[1] | UMTS frequencies[1] | UHF band 474–746 MHz (ch 21–69) |
| Network solution | UMTS network | UMTS network | High power, high masts, and repeaters |
| Cell size | UMTS cell sizes: 1–2 km in urban and 2–10 km in suburban and >10 km in rural. | UMTS cell sizes: 1–2 km in urban and 2–10 km in suburban and >10 km in rural. | Up to 50 km |
| Receivers | UMTS terminals | UMTS terminals | Dedicated DVB-H receiver required |

*Note*: [1]Any of different 3GPP defined UMTS bands. Currently more than 10 bands.

## 14.9   3GPP MBMS Evolution in Release 7

3GPP Release 6 can use soft combining of the MBMS transmission from the adjacent cells. The soft combining clearly improves MBMS performance at the cell edge compared to receiving the signal from a single cell only. Release 6 MBMS provides therefore a very good starting point for broadcast services from the performance point of view.

Even if the soft combining can be utilized in Release 6, the other cell signals still cause interference to the MBMS reception since the adjacent cells are not orthogonal due to different scrambling codes. If the same scrambling code were used in all cells and the cells were synchronized, the other cell interference would turn into multipath propagation. The received signal looks as if there were only single base station transmitting but with more multipath components. Therefore, this approach is called single frequency network (SFN). That is similar approach as used in DVB-H. The MBMS multipath propagation could be tackled by the terminal equalizer. The MBMS capacity could be increased by this approach. The enhanced MBMS concept is illustrated in Figure 14.19. This concept is specified as part of 3GPP Release 7.

The SFN approach can be implemented with relatively minor modifications to the radio network while providing a major gain in the broadcast data rates. On the other hand, the SFN transmission requires that the whole 5-MHz carrier is allocated for MBMS usage only, which requires that the total amount of spectrum for the operator is large enough. The terminal receivers must include equalizers, similar to those presented in Chapter 11.

Since all the cells use the same scrambling code, the terminal cannot distinguish between the cells. The mobility management must be designed using the unicast carrier. The interworking between the unicast and multicast carriers need to be defined. The working assumption in the 3GPP work item is to use separate receivers for unicast and multicast carriers.

## 14.10   Why Did MBMS Fail?

MBMS was fully standardized in 3GPP, it was included in product roadmaps, it was tested, but as yet it had not been deployed commercially by the publication of this fifth edition. There are a few reasons for the low interest in the MBMS solution:

- Unicast video delivery over HSPA is more efficient for a low number of simultaneous users than MBMS. Unicast delivery with HSPA does not require any changes to the radio network nor to

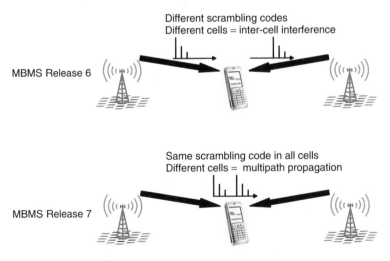

**Figure 14.19**   Dedicated MBMS carrier in 3GPP Release 7

the core network nor to the terminal. The efficiency of unicast delivery further improves with new HSPA+ features.

- Video and TV consumption is moving from linear TV to on-demand viewing. Linear TV refers to the standard television service where the person receives the scheduled program at a pre-defined time. On-demand viewing refers to the flexibility for the end user to watch any content at any time. One example is that streaming video usage – like YouTube – is growing fast in internet traffic. MBMS represents linear TV model while unicast delivery can serve the on-demand usage.
- Operators have a limited spectrum for 3G systems. The typical spectrum allocation is 15 MHz per operator for downlink and for uplink. The HSPA usage has been growing fast with HSPA USB modems and with HSPA smart phones. It would be difficult to allocate part of the paired spectrum and system capacity for MBMS services.
- One example of low interest for MBMS was shown also in 3GPP LTE standardization where MBMS was not included in Release 8, but only later in Release 9.

## 14.11  Integrated Mobile Broadcast (IMB) in Release 8

The TDD spectrum at 1900–1920 MHz was auctioned in many countries together with the FDD spectrum with uplink 1920–1980 MHz and downlink 2110–2170 MHz. The FDD spectrum has been heavily utilized while the TDD band has been sitting empty. 3GPP Release 8 defined the Integrated Mobile Broadcast (IMB) solution, that is the MBMS solution for that TDD band. 3GPP uses the term Mobile Broadcast Single Frequency Network (MBSFN). IMB provides a broadcast service without sacrificing any existing FDD spectrum. There are some differences in the physical layer between MBMS and IMB because IMB is a combination of WCDMA and TD-CDMA physical layers.

The design of IMB needs to consider the potential interference between IMB transmission and the FDD uplink reception on the adjacent frequency. The interference scenario is shown in Figure 14.20. The interference needs to be considered at the base station and at the terminal side. The worst case analysis shows relatively high isolation requirement between IMB transmission and WCDMA FDD uplink reception. The WCDMA base station blocking requirement is $-40$ dBm against IMB transmission at 1900–1920 MHz when there is a minimum 10 MHz separation between IMB transmission and FDD reception [20]. If the IMB base station transmission power is 43 dBm, it would lead to $43 - (-40) = 83$ dB isolation requirement. The isolation requirement is illustrated in Figure 14.21. In practice, the WCDMA base station blocking performance is better especially when using the lower IMB frequencies at 1900–1910 and when using the upper part of FDD spectrum with sub-banded filters. The co-siting of IMB and FDD base stations is typically possible but separate antennas are needed and the interference needs to be considered in the filtering and in the site configurations. On the terminal side, the IMB reception may need to be interrupted during the unicast usage due to interference.

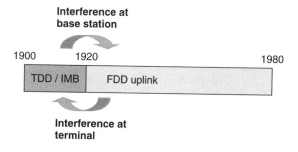

**Figure 14.20**  Potential interference between IMB transmission and FDD uplink reception

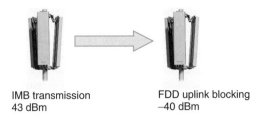

IMB transmission                 FDD uplink blocking
43 dBm                           −40 dBm

**Figure 14.21**   FDD blocking requirements

**Table 14.4**   Summary MBMS and IMB solutions

|                              | MBMS on shared carrier (no-SFN)        | IMB (MBSFN)                                        |
| ---------------------------- | -------------------------------------- | ------------------------------------------------- |
| Spectrum                     | FDD spectrum                           | TDD spectrum 1900–1920 MHz                         |
| Carrier sharing with unicast | Yes, takes part of unicast capacity    | No                                                |
| Antennas and RF modules      | Existing antennas and RF modules used  | New antennas and RF modules required              |
| Base station synchronization | No need for synchronization            | Yes, synchronization with GPS for single frequency network |
| Capacity                     | 1 Mbps per 5 MHz                       | 2–3 Mbps per 5 MHz                                |

While IMB has the benefit of using the TDD spectrum, it also implies that new RF modules and new antennas are required at the base station. The single frequency network requires GPS at the base station for synchronization, but it can also provide clearly higher spectral efficiency compared to MBMS without synchronization. One deployment option is to build a shared IMB network by multiple operators. The total IMB spectrum would then be more than 5 MHz. If we assume 15 MHz of spectrum and 3 Mbps capacity per 5 MHz, the total IMB sector throughput would be 9 Mbps, which is equal to 35 channels at 256 kbps.

If all operators work together with IMB, it would also make it easier to manage the site solutions to minimize the interference. The IMB solution is compared to MBMS in Table 14.4.

## 14.12   Conclusion

3GPP Release 6 has defined MBMS functionality including broadcast and multicast modes. The MBMS solution can use the existing WCDMA/HSPA network with the addition of Broadcast Multicast Service Center (BM-SC). MBMS uses the existing WCDMA transport and physical channels FACH and S-CCPCH. The integration of MBMS functionality to the 3G terminals is also expected to be simple, making it possible to have a large volume of MBMS terminals in the market.

MBMS can provide a maximum data rate of 256 kbps per channel. The limiting factor, however, is typically the total cell capacity limiting the maximum number of channels. The simulations show that 5 MHz MBMS carrier can typically provide 1 Mbps cell capacity, which is equal to 8 channels at 128 kbps. If the MBMS carrier is shared with WCDMA/HSPA point-to-point traffic, the MBMS capacity is correspondingly reduced.

MBMS multicast transmission enables the network to count the number of mobiles receiving each channel. Based on the counting procedure, the network can decide the optimal delivery method: point-to-multipoint over FACH for a large number of users and point-to-point over HSDPA for a low number of users.

MBMS can be used to provide streaming services including mobile TV but also to deliver content downloads or static pictures or video clips efficiently to large number of receivers.

Even if the MBMS technology were standardized, implemented and tested, it has not been used in commercial networks. The general trend of TV usage is moving from linear TV to on-demand viewing. Unicast HSPA has been able to serve the mobile TV market.

3GPP Release 8 defined Integrated Mobile Broadcast (IMB) which is similar to MBMS but defined on the TDD spectrum especially for 1900–1920 MHz. The benefit of IMB is that it does not use the scarce FDD spectrum for broadcast services. The dedicated IMB carrier can also utilize single frequency transmission that boots the network capacity. A shared IMB network between multiple operators could provide a high number of broadcast channels. IMB deployment needs to consider the potential interference from IMB transmission at 1900–1920 MHz to WCDMA uplink reception at 1920–1980 MHz.

# References

[1] 3GPP Technical Specification 22.146, Multimedia Broadcast/Multicast Service, Stage 1.

[2] 3GPP Technical Specification 22.246, Multimedia Broadcast/Multicast Service (MBMS) User Services; Stage 1.

[3] 3GPP Technical Specification 23.246, Multimedia Broadcast/Multicast Service (MBMS); Architecture and Functional Description.

[4] 3GPP Technical Report 23.846, Multimedia Broadcast/Multicast Service (MBMS); Architecture and Functional Description.

[5] 3GPP Technical Specification 25.101, User Equipment (UE) Radio Transmission and Reception (FDD).

[6] 3GPP Technical Specification 25.306, UE Radio Access Capabilities.

[7] 3GPP Technical Specification 25.346, Introduction of the Multimedia Broadcast Multicast Service (MBMS) in the Radio Access Network (RAN); Stage 2.

[8] 3GPP Technical Report 25.803, S-CCPCH Performance for MBMS.

[9] 3GPP Technical Report 25.992, Multimedia Broadcast Multicast Service (MBMS); UTRAN/GERAN Requirements.

[10] 3GPP Technical Specification 26.346, Multimedia Broadcast/Multicast Service (MBMS); Protocols and Codecs.

[11] 3GPP Technical Report 26.946, Multimedia Broadcast/Multicast Service (MBMS); User Service Guidelines.

[12] 3GPP Technical Report 29.846, Multimedia Broadcast/Multicast Service (MBMS); CN1 Procedure Description.

[13] 3GPP Technical Specification 32.273, Multimedia Broadcast/Multicast Service (MBMS) Charging.

[14] 3GPP Technical Specification 33.220, Generic Authentication Architecture (GAA); Generic Bootstrapping Architecture.

[15] 3GPP Technical Specification 33.246, Security of Multimedia Broadcast/Multicast Service.

[16] Aho, K., Ristaniemi, T., Kurjenniemi, J. and Haikola, V. 'System Level Performance of Multimedia Broadcast Multicast Service (MBMS) with Macro Diversity', *IEEE Wireless Communications and Networking Conference (WCNC)*, Hong Kong, 1–15 March 2007.

[17] Aho, K., Ristaniemi, T., Kurjenniemi, J. and Haikola, V. 'Performance Enhancements of Multimedia Broadcast Multicast Service (MBMS) with Receive Diversity', *International Conference on Networking (ICN)*, Sainte-Luce, Martinique, 22–28 April 2007.

[18] *Digital Video Broadcasting (DVB); Transmission System for Handheld Terminals (DVB-H)*; ETSI EN 302 304 V1.1.1 (2004-11), European Telecommunications Standards Institute.

[19] Faria, G., Henriksson, J. A., Stare, E. and Talmola, P. 'DVB-H: Digital Broadcast Services to Handheld Devices', *Proc. IEEE*, Vol. 94, 2006, pp. 194–209.

[20] 3GPP Technical Specification 25.104, Base Station (BS) Radio Transmission and Reception (FDD).

# 15

# HSPA Evolution

Harri Holma, Karri Ranta-aho and Antti Toskala

## 15.1 Introduction

High speed packet access (HSPA) was included in the Third Generation Partnership Project (3GPP) Releases 5 and 6 for downlink and for uplink. The 3GPP Releases 7, 8 and 9 have brought a number of HSPA enhancements providing major improvements to the end user performance and to network efficiency. The work continues further in Release 10. The HSPA evolution work has progressed in parallel to LTE work in 3GPP. Many of the technical solutions in HSPA evolution and LTE are also similar. The overview of the roles of the HSPA evolution and LTE is illustrated in Figure 15.1. HSPA evolution is optimized for co-existence with WCDMA/HSPA supporting legacy Release 99 UEs on the same carrier and designed for simple upgrade on top of HSPA. The HSPA evolution aims to improve the end user performance by lower latency, lower power consumption and higher data rates. HSPA evolution includes also interworking with LTE which enables both packet handovers and voice handovers from LTE voice-over IP (VoIP) to HSPA circuit switched (CS) voice.

The HSPA evolution features are introduced in this chapter. The HSPA evolution is also known as HSPA+. This chapter covers other parts of HSPA evolution except for Dual cell and multicarrier HSPA, covered in Chapter 16.

## 15.2 Discontinuous Transmission and Reception (DTX/DRX)

The technology evolution in general helps to decrease the terminal power consumption. Also, the fast and accurate power control in WCDMA helps to minimize the transmitted power levels. The challenge in 3GPP from Release 99 to Release 6 is still the continuous reception and transmission when the mobile terminal is using HSDPA/HSUPA in the Cell_DCH state. HSPA evolution introduces a few improvements that help to reduce the power consumption for CS voice calls and for all packet services.

3GPP Release 6 UE keeps transmitting the physical control channel even if there is no data channel transmission. The control channel transmission and reception continue until the network commands the UE to Cell_FACH or Cell_PCH state. The Release 7 UE can cut off the control channel transmission as soon as there is no data channel transmission allowing it to shut down the transmitter completely. This

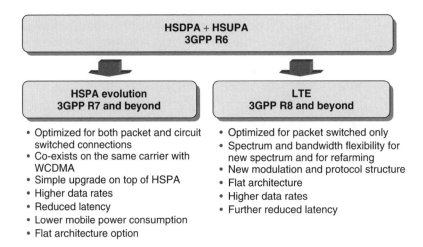

**Figure 15.1**  Overview of HSPA evolution and LTE roles

solution is called the discontinuous uplink transmission and it brings clear savings in the transmitter power consumption.

A similar concept is introduced also in the downlink where during the inactive phases in downlink data transmissions the UE needs to wake up only occasionally to check if the downlink data transmission is starting again. The UE can use the power-saving mode during other parts of the DRX cycle if there is no data to be received. This solution is called downlink discontinuous reception. The discontinuous transmission concept is illustrated in Figure 15.2 for web browsing. As soon as the web page is downloaded, the connection enters discontinuous transmission and reception. Notably, perhaps contrary to traditional wisdom, the WCDMA/HSPA receiver with continuous transmission and reception consumes significant power even if the transmit powers are far below the maximum levels, making the power savings due to DRX an attractive addition to the savings available with DTX. The DTX/DRX feature is also known as Continuous Packet Connectivity (CPC) in 3GPP. Detailed DTX/DRX timing diagrams and the power consumption calculations are presented in Chapter 20.

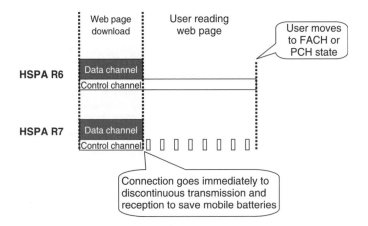

**Figure 15.2**  Discontinuous transmission and reception concept

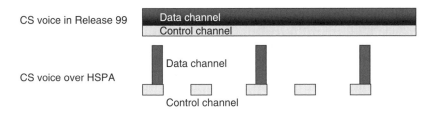

**Figure 15.3**   Discontinuous uplink transmission for circuit switched (CS) voice

The DTX/DRX features can be used also for the voice service with CS voice-over HSPA, see Figure 15.3. Better talk time is one of the main benefits of running CS voice-over HSPA. The concept is described in more detail in the next section.

The Release 99 FACH solution requires continuous reception by the UE. That is challenging from the UE power consumption point of view especially for always-on applications transmitting frequent keep-alive messages. Every time a keep-alive packet needs to be sent, the UE is forced to move to Cell_FACH state and stay there until the network inactivity timer expires. The discontinuous reception is introduced also for Cell_FACH state with Enhanced FACH state in HSPA evolution helping stand-by times when these types of applications are used. In addition, a feature named Fast Dormancy was specifically developed for managing the power consumption of keep-alive messages by minimizing the time the UE needs to be in Cell_FACH or Cell_DCH state. Fast Dormancy is introduced later in this chapter.

## 15.3   Circuit Switched Voice on HSPA

Voice has remained an important service for the mobile operators. WCDMA Release 99 supports Circuit switched (CS) voice on Dedicated channel (DCH) with quite high spectral efficiency. Efficient Voice-over IP (VoIP) capability on top of HSPA was defined in Release 7, but VoIP mass market has not yet started. Therefore, CS voice-over HSPA was also defined in HSPA evolution. CS voice-over HSPA is part of Release 8 but because of the capability indication for the UE support of the feature was introduced to Release 7, it is possible to implement this feature before other Release 8 features. There are two main benefits when running voice on HSPA: UE power consumption is reduced because of DTX/DRX and the spectral efficiency is improved with HSPA features.

The different 3G voice options are illustrated in Figure 15.4. CS voice-over DCH is used currently in commercial networks. Dedicated Release 99 channel is used in Layer 1 and Transparent mode RLC in Layer 2. From the radio point of view, CS voice-over HSPA and VoIP over HSPA use exactly the same Layer 1 and Layer 2 including unacknowledged mode RLC. PDCP layer IP header compression is not needed for CS voice. From the core network point of view, there is again no difference between CS voice-over DCH and CS voice-over HSPA. In fact, CS core network is not aware whether the radio maps CS voice on DCH or on HSPA. CS voice-over HSPA could be described as CS voice from the core point of view and VoIP from the radio point of view.

CS voice-over HSPA was a relatively simple feature from the 3GPP standards' point of view. The following changes were implemented to 3GPP specifications:

- Iu interface = no changes.
- Physical Layer = no changes.
- MAC layer = no changes.
- RLC layer = forwarding RLC-UM sequence numbers to upper layers.

- PDCP Layer = modification of header to identify and timestamp the CS AMR frames + interfacing /inclusion of Jitter Buffer Management.
- New de-jitter buffer = UE/RNC implementation dependent entity that absorbs the radio jitter created by HSPA scheduling and HARQ operation so that the CS AMR frames can be delivered in a timely constant fashion to upper layers, either to the voice decoder in the UE for downlink, or to the CS core network in the RNC for the uplink. The algorithm for the de-jitter buffer is not standardized in 3GPP.

CS voice-over HSPA concept is presented in Figure 15.5. The CS voice connection can be mapped on DCH or on HSPA depending on UE capability and RNC algorithms. The AMR data rate adaptation can be controlled by RNC depending on the system loading. When CS voice-over HSPA is used,

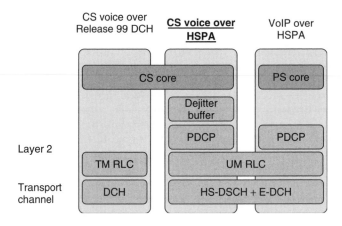

**Figure 15.4**  Voice options in WCDMA/HSPA

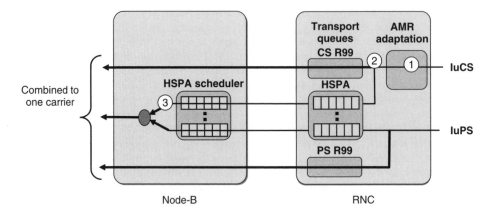

① = AMR bit rate adaptation according to the system load
② = CS voice can be mapped on DCH or HSPA depending on UE capability
③ = HSPA scheduler prioritizes voice packets

**Figure 15.5**  CS voice over HSPA overview

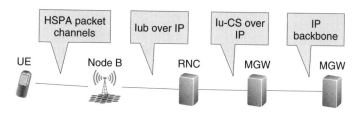

**Figure 15.6** CS voice in different interfaces

there is a clear need for QoS differentiation algorithms in HSPA scheduling to guarantee low delays for voice packets also during the high packet data traffic load. Since the packet scheduling and the prioritization are similar for VoIP and for CS voice-over HSPA, it will be simple from the radio perspective to add VoIP support later on top of CS voice-over HSPA. Therefore, CS voice-over HSPA is paving the way for future VoIP introduction. Similar radio solutions also make the handover simpler between VoIP and Circuit switched domains.

CS voice on HSPA can benefit from IP and packet transmissions in all interfaces: the air interface carried by HSPA, Iub over IP, Iu-CS over IP and the backbone between Media Gateways (MGW) using IP. The call control signaling is still based on Circuit switched protocols [1]. Use of the interfaces is shown in Figure 15.6. CS voice-over HSPA can benefit from the packet network performance and cost while maintaining the existing end-to-end protocols and ecosystem. No changes need to be made to charging, emergency calls or to roaming. The voice codec with CS over HSPA can be Narrowband AMR (Adaptive multirate codec) or Wideband AMR.

CS voice-over HSPA also increases the spectral efficiency compared to voice-over DCH because voice can also take the benefit of the HSPA physical layer enhancements:

- The UE equalizer increases downlink capacity. Equalizer is included in practice in all HSPA terminals.
- Optimized L1 control channel in HSPA reduces control channel overhead. The downlink solution is Fractional DPCH with Discontinuous Reception and the uplink solution is Discontinuous Transmission.
- L1 retransmissions can be used also for voice on HSPA since the retransmission delay is only 12–16 ms.
- HSDPA optimized scheduling improves the capacity even if the tough delay requirements for voice limit the scheduling freedom compared to best effort data.

The gain in spectral efficiency with CS voice-over HSPA is estimated at 50–100% compared to CS voice-over DCH. The voice capacity is illustrated in Figure 15.7.

The evolution of the voice spectral efficiency from GSM to WCDMA and to HSPA is illustrated in Figure 15.8. Most of the global mobile voice traffic is carried with GSM EFR or AMR coding. The GSM spectral efficiency can typically be measured with Effective Frequency Load (EFL) which represents the percentage of the time slots of full rate users that can be occupied in case of frequency reuse one. For example, EFL = 8% corresponds to 8%*8 slots/200 kHz = 3.2 users/MHz. The simulations and the network measurements show that GSM EFR can achieve EFL 8% and GSM AMR can achieve 20% assuming all terminals are AMR capable and the network is optimized. The GSM spectral efficiency can be further pushed by the network feature Dynamic Frequency and Channel Allocation (DFCA) and by using Single Antenna Interference Cancellation (SAIC) known also as Downlink Advanced Receiver Performance (DARP). We assume up to EFL 35% with those enhancements. For further information, see [2, 3].

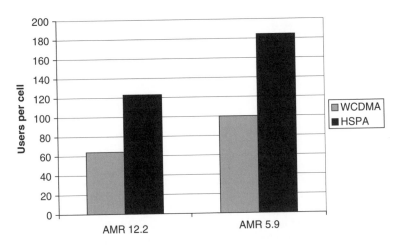

**Figure 15.7**  Circuit switched voice spectral efficiency with WCDMA and HSPA

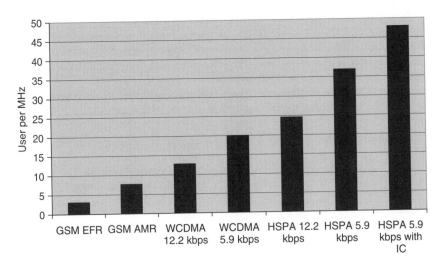

**Figure 15.8**  Evolution of voice spectral efficiency (IC = Interference Cancellation)

The overhead from GSM Broadcast control channel (BCCH) is not included in the calculations. BCCH requires higher reuse than the hopping carriers.

WCDMA voice capacity is assumed to have 64 users with AMR12.2 kbps and 100 users with AMR5.9 kbps on 5 MHz carrier. HSPA voice capacity is assumed to have 123 users with AMR12.2 kbps and 184 users with AMR5.9 kbps. We further assume that the Node B Interference Cancellation (IC) can improve the voice capacity by +30%. Also an HS-SCCH less HSDPA transmission, introduced as a part of Continuous Packet Connectivity feature package, can push the HSDPA voice capacity 4–7% higher by eliminating the HS-SCCH overhead from the first transmission attempts of small packets [4].

HSDPA Release 5 uses the Dedicated Physical Channel (DPCH) to carry the Signaling Radio Bearer (SRB). DPCH can utilize the soft handover also in downlink. When SRB is mapped on HSDPA in

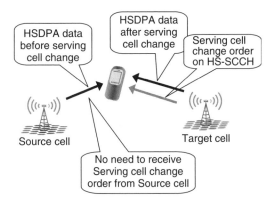

HSDPA data before serving cell change

HSDPA data after serving cell change

Serving cell change order on HS-SCCH

Source cell

Target cell

No need to receive Serving cell change order from Source cell

**Figure 15.9** Enhanced HS-DSCH serving cell change [5]

Release 6, there is no soft handover gain available for SRB. If the downlink radio link gets weak very rapidly, UE may not receive Active set update message and the handover fails. 3GPP Release 8 introduced an enhancement to the HS-DSCH serving cell change to reduce the latency and increase robustness, when SRB is mapped on HSPA channels, see Figure 15.9. The network can preconfigure the UE with a potential target cell's HSDPA configuration. If later the preconfigured target becomes better than the current serving cell, the UE sends a measurement report to the RNC and start monitoring the target cell's HS-SCCH for possible serving cell change order. The RNC can then prepare the target cell and instruct it to transmit the order using HS-SCCH. The benefit of this approach is that if the source cell is rapidly fading out the reliability of the transmissions from the target can be expected to be better. Notably only the cells in the active set are preconfigurable, i.e. the enhanced HS-DSCH serving cell change can only take place between cells in the UE's active set. Uplink transmissions are thus received by both the source and the target cell before and after the procedure, thus maintaining good uplink reliability.

Another, simpler option for coping with the cases where the source cell link fades too rapidly is to let the radio link fail and then recover from the failure by using the RRC re-establishment procedure where the UE transmit a cell update to the target cell, the RNC sets up radio link to the target cell and the voice call continues. The re-establishment procedure has been successfully used also with WCDMA Release 99. The radio link timeout and the re-establishment procedure causes a small break in the voice audio.

## 15.4 Enhanced FACH and Enhanced RACH

WCDMA network data rate and latency are improved with the introduction of Release 5 HSDPA and Release 6 HSUPA. The end user performance can be further improved by minimizing the channel allocation time. Once the packet call has been established, user data can flow on HSDPA/HSUPA in the Cell_DCH (Dedicated Channel) state. When the data transmission is inactive for a few seconds, the UE is moved to Cell_PCH (Paging Channel) state to minimize the UE power consumption. When there is more data to be sent or received, the UE is moved from Cell_PCH to Cell_FACH (Forward Access Channel) and to Cell_DCH state. Release 99 RACH and FACH can be used for signaling and for small amounts of user data. The RACH data rate is very low, typically below 10 kbps, limiting the use of the common channels. Release 5 or Release 6 do not provide any improvements in RACH or FACH performance. The idea in Release 7 Enhanced FACH and Release 8 Enhanced RACH is to utilize the Release 5 and Release 6 HSPA transport and physical channels also in the Cell_FACH state

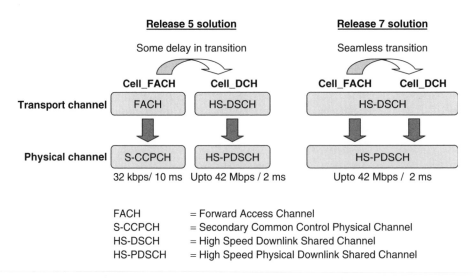

FACH        = Forward Access Channel
S-CCPCH     = Secondary Common Control Physical Channel
HS-DSCH     = High Speed Downlink Shared Channel
HS-PDSCH    = High Speed Physical Downlink Shared Channel

**Figure 15.10**   Enhanced FACH concept in downlink [11]

to improve the end user performance and system efficiency. The concept is illustrated in Figure 15.10. Enhanced FACH and RACH bring a few performance benefits:

- RACH and FACH data rates can be increased beyond 1 Mbps. The end user can get immediate access to relatively high data rates without the latency of channel allocation.
- The state transition from Cell_FACH to Cell_DCH will be practically seamless. Once the network resources for the channel allocation are available, a fast transition can take place to Cell_DCH since the physical channel is not changed.
- Unnecessary state transitions to Cell_DCH can be avoided when more data can be transmitted in Cell_FACH state. Many applications create some background traffic that is today carried on Cell_DCH but can later be carried on the Cell_FACH.
- Discontinuous reception can be used in the Cell_FACH to reduce the power consumption. The discontinuous reception can be implemented since Enhanced FACH uses short 2 ms Transmission time interval of HSDPA instead of Release 99 10 ms. The discontinuous reception in the Cell_FACH state is introduced in the 3GPP Release 8 specifications together with Enhanced RACH.

Since the existing physical channels are utilized in Enhanced FACH, there are only minor changes in Layer 1 specifications. Enhanced FACH can co-exist with Release 99 and with HSDPA/HSUPA on the same carrier. No new power allocation is required for Enhanced FACH since the same HSDPA power allocation is used as for the existing HSDPA.

Similar to using HS-DSCH in the Cell_FACH state for downlink, the Enhanced RACH concept introduced a method for using E-DCH for uplink transmissions, see Figure 15.11. The E-DCH resources and initial data rate allocation to be used in the Cell_FACH state are broadcast to the cell on BCH, see Figure 15.12. At the point of PRACH preamble acquisition, the Node B points the UE to a specific E-DCH resource with AICH. In the initial phase the UE includes its identity to MAC header and if it does not have its identity echoed back on E-AGCH within a specified time limit, it will stop transmitting, release E-DCH resources, consider the random access procedure to have failed and retry. This contention resolution phase is used to resolve the rare cases where two UEs end up sending the same PRACH preamble at the same time and both think that the AICH resource indication is meant for it.

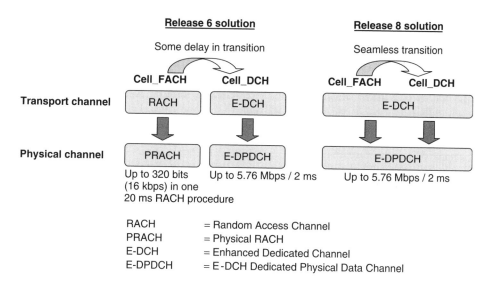

**Figure 15.11** Enhanced RACH concept in uplink

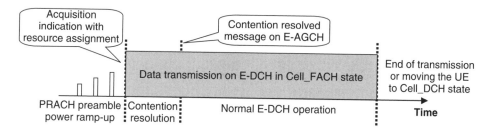

**Figure 15.12** Data transmission on E-DCH in Cell_FACH state

## 15.5 Latency

Most applications are not sensitive to the very high data rate but are sensitive to the low latency. The end user performance can be enhanced by minimizing the latency even if the data rate is not increased. For example, interactive applications such as browsing or gaming benefit from low latency. The latency in this section is defined as the round trip time of a small IP packet through the UMTS radio and core network. The definition of latency is shown in Figure 15.13.

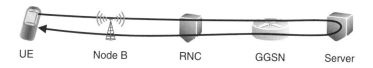

**Figure 15.13** Definition of latency

The aim of radio evolution is to make latency so short that the radio is not the limiting factor in the end-to-end latency. The latency evolution is shown in Figure 15.14. The latency goes down mainly with the shorter frame size and faster retransmissions in the air interface. WCDMA uses 20-ms and 10-ms TTIs, HSDPA uses 2-ms TTI, HSUPA uses 10-ms and 2-ms TTI and LTE uses 1-ms TTI. When the frame size gets shorter, the air interface transmission time is faster and also the processing time in the terminals and in the base station must be faster. The end-to-end round trip time with HSPA can be below 30 ms and with LTE below 15 ms. For reference, 30 ms latency is similar to the two-way propagation time in the optical fiber over 3000 km distance and 15 ms over 1500 km distance.

Example 32-byte ping measurements in live HSPA network are shown in Figure 15.15. The best ping values are below 20 ms and the average values below 30 ms. Some individual ping values are larger due to the retransmissions.

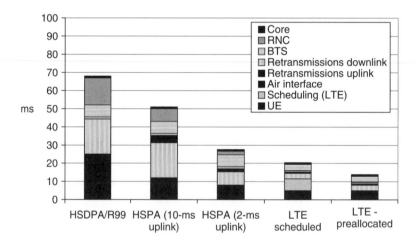

**Figure 15.14**   Latency evolution with HSPA and LTE

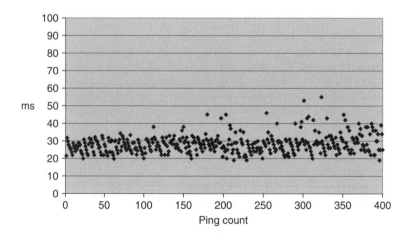

**Figure 15.15**   Example latency measurement in the field with HSPA

## 15.6  Fast Dormancy

Smart phones have an increasing number of applications that send only a small amount of data, but the transmission frequency of the packets is relatively high. Such always-on applications include, for example, email, instant messaging and widgets. It is essential to keep the UE power consumption low while having frequent transmissions. The original idea in 3GPP was that the Radio Resource Control (RRC) states include the Cell_PCH and URA_PCH states that allow very low UE power consumption. When the data transmission is over, UE should move to the PCH state quickly. There is no need to move UE to the idle state for low power consumption. The idle state should be avoided because the next packet will then cause packet connection (PS RAB) set-up leading to increased latencies and increasing signaling traffic in the network. The basic principles of RRC state transitions are illustrated in Figure 15.16. The RRC state transitions are controlled by the RNC.

The problem in practice has been that many networks are configured with relatively long inactivity timers for Cell_DCH and for Cell_FACH states and PCH states have not been activated. Therefore, transmissions of even a small packet may lead to high battery drain. In order to keep UE power consumption low, UEs have implemented proprietary (that is, a functionality not defined by 3GPP standards) fast dormancy. When using such a fast dormancy, the mobile application informs the radio layers when the data transmission is over, and the UE can then send Signaling Connection Release Indication (SCRI) to the RNC simulating a failure in the signaling connection. Consequently, the UE releases RRC connection and moves to idle state. This approach keeps the UE power consumption low, but it causes frequent set-ups of packet connections unnecessarily increasing the signaling load. The worst case UEs caused up to 25 packet data connections per 5 minute measurement period in a commercial HSPA network causing quite a lot of signaling traffic. There were also differences between UE vendors in how the fast dormancy functionality was implemented. In addition to the high signaling load, the network counters indicate a large number of signaling connection failures as this battery-saving method cannot be distinguished from a genuine signaling connection failure in the network.

3GPP Release 8 specified a Fast Dormancy functionality clarifying the UE behavior and providing the network with information of what the UE actually wants to do, but leaving the network in charge of the UE RRC state. That is, the UE is not allowed to release RRC connection and move to idle state on its own without network control. When the RNC receives SCRI from a UE with a special newly introduced cause value indicating a packet data session end, the RNC can command the UE to use the PCH state instead of releasing the RRC connection and dropping the UE into the idle state. This approach avoids unnecessary PS RAB set-ups and enables networks to separate signaling connection failures from fast dormancy-related signaling connection release indications. The Release 8 Fast dormancy feature helps to save the UE battery when the radio layers get information directly from the application layer when the data transmission is over. The fast dormancy functionality is illustrated in Figure 15.17.

The network broadcasts an inhibit timer (T323), setting a minimum delay between two SCRI messages for the fast dormancy a UE is allowed to send. This is to prevent a UE from sending a constant flow of SCRI messages if for some reason the network is temporarily unable to move

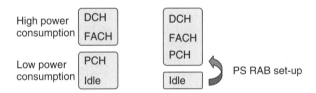

**Figure 15.16**  RRC states and power consumption

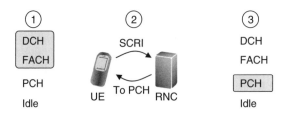

(1) = UE is in high power consumption state = DCH or FACH
(2) = UE sends Signalling Connection Release Indication (SCRI) to RNC
(3) = RNC moves UE to PCH state for low power consumption. No need to
      move UE to idle.

**Figure 15.17**   Fast dormancy functionality

the UE to a battery-saving state. The presence of T323 in the broadcast can also be considered an indication of network support of the Fast Dormancy functionality triggering the UE to use the SCRI messages with the cause value indicating the PS data session end. The 3GPP specification allows UE to send SCRI also from Cell_PCH and URA_PCH states but in practice there should be no reason for UE to do so.

Notably the 3GPP standard-based Fast Dormancy was defined to be early implementable, i.e. even though the behavior and changes in the signaling are defined in the Release 8 specifications, it is possible to build these extensions to a UE and network that is not Release 8 compatible. Such an approach enables the fast introduction of a feature that has only a very limited set of modifications in the radio protocols.

## 15.7   Downlink 64QAM

The downlink peak data rate with Release 6 HSDPA is 10.8 Mbps with $^3/_4$ coding and 14.4 Mbps without any channel coding. There are a number of ways in theory to push the peak data rate higher: larger bandwidth, higher order modulation or multi-antenna transmission with Multiple Input Multiple Output (MIMO). All these solutions are part of HSPA evolution. MIMO and higher order modulation are included in HSPA evolution in Release 7, Dual carrier HSDPA in Release 8 and up to 4 carrier HSDPA is on its way to becoming part of Release 10.

Higher order modulation allows higher peak bit rate without increasing the transmission bandwidth. Release 6 supported QPSK (Quadrature Phase Shift Keying) and 16QAM (Quadrature Amplitude Modulation) transmission in the downlink and dual-BPSK (Binary Phase Shift Keying) in the uplink. Dual-channel BPSK modulation is similar to QPSK, but the data streams are first split into two I/Q multiplexed BPSK modulated bit-streams instead of running a QPSK modulator to a pair of bits of a single bit-stream. The resulting signal looks exactly the same in both cases. The main difference being that if only one bit-stream is needed to support the required data rate, then the dual-BPSK waveform falls naturally back to normal BPSK modulation. The Release 7 introduced 64QAM transmission for the downlink and 16QAM for the uplink. Again strictly speaking the 16QAM modulation used in HSUPA is dual-4PAM, where two I/Q multiplexed amplitude modulated signals form a waveform just like 16QAM.

16QAM can double the bit rate compared to QPSK by transmitting 4 bits instead of 2 bits per symbol. 64QAM can increase the peak bit rate by 50% compared to 16QAM since 64QAM transmits

6 bits with a single symbol. On the other hand, the constellation points are closer to each other for the higher order modulation and the required signal-to-noise ratio for correct reception is higher. The difference in the required signal-to-noise ratio is approximately 6 dB between 16QAM and QPSK and also between 64QAM and 16QAM. Therefore, downlink 64QAM and uplink 16QAM can be utilized only when the channel conditions are favorable.

[7] defines CQI reporting from UE to Node B. The new CQI table uses 64QAM with CQI = 26 or higher assuming 10 HS-DSCH codes, however the Node B link adaptation algorithms are not forced to follow that 3GPP table. The optimized table should utilize as many codes as possible and use lower order modulation as long as possible. Therefore, 64QAM should be used only with CQI 27 or 28 if there are more than 10 codes available. The HSDPA networks based on Release 5 and 6 UEs usually show very low probability (<5%) for CQI >26. It is expected that the UE performance will improve with 64QAM in the high CQI area in terms of RF performance. The 64QAM data rates require good RF performance including cancellation of the interference from the synchronization channel (SCH). SCH is not transmitted under the scrambling code because UE must be able to find SCH before UE can identify the scrambling code. Therefore, SCH is not orthogonal and UE needs to cancel the non-orthogonal SCH interference in order to be able to achieve very high data rates with 64QAM. The probability of obtaining CQI of 27–28 in macro cells with optimized two-antenna 64QAM UEs is expected up to 25%.

The link level performance with 64QAM is illustrated in Figure 15.18. 64QAM gives benefit when the data rate is more than 10 Mbps. UE Category 10 can maximally give 14 Mbps but it is better to utilize 64QAM and more turbo coding above 10 Mbps than use 16QAM with less coding.

The probability of obtaining 64QAM in the networks in practice depends on the availability of high signal-to-noise ratios. The system simulation results with 64QAM in three-sector macro cells are illustrated in Figure 15.19. Assuming two Rx-antenna advanced received in the UE, the 64QAM modulation is available with 10-25% probability depending on the packet scheduling (RR = round

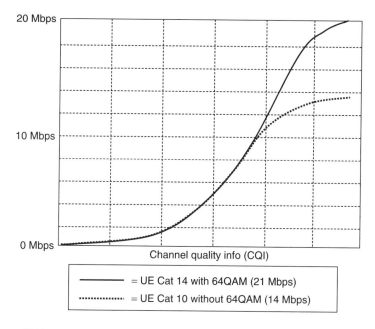

**Figure 15.18**  64QAM link performance, 64QAM only used for data rates beyond 10 Mbps

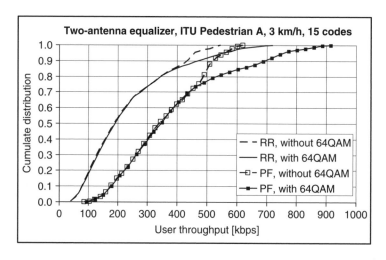

**Figure 15.19** Macro cell data rates per user with 64QAM with 20 active users in a cell [6]

robin, PF = proportional fair). The rest of the time the channel conditions are not good enough to enable the reception of 64QAM modulation. The capacity gain from 64QAM in most cases is 5–10%.

## 15.8 Downlink MIMO

The 3GPP MIMO concept employs two transmit antennas in Node B and two receive antennas in UE and uses a closed loop feedback from the UE to adjust the transmit antenna weighting. The MIMO transmission is shown in Figure 15.20. MIMO allows double the data rate by transmitting dual data streams in good channel conditions. MIMO fallback mode uses single stream transmission (HS-DSCH$_2$ of Figure 15.20 is absent) to get the benefit from the transmit diversity when a dual stream is not beneficial due to channel conditions. The adaptation between single stream and dual stream is based on the UE feedback and it is part of the fast link adaptation.

UE feedback to Node B includes Channel Quality Information (CQI) and Precoding Control Information (PCI). CQI indicates the maximum transport block size while PCI gives the precoding vector

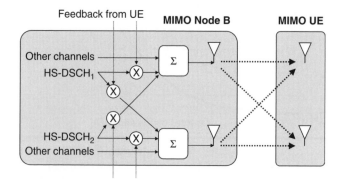

**Figure 15.20** 2 × 2 MIMO transmission concept

information. The aim of the single stream precoding vector is to make the signals from the two antennas combine coherently in the UE receiver. A MIMO UE uses a periodic Type A CQI feedback consisting of two 4-bit CQIs ($CQI_1$ and $CQI_2$ in Equation 15.1, construction of a Type A CQI report) when two streams are preferred, or a single 5-bit CQI ($CQI_s$ in Equation 15.1) when a single stream transmission is preferred.

$$CQI = \begin{cases} 15 \times CQI_1 + CQI_2 + 31 & \text{when 2 transport blocks are preferred by the UE} \\ CQI_S & \text{when 1 transport block is preferred by the UE} \end{cases} \quad (15.1)$$

The UE may also be configured to provide a less frequent, periodic Type B CQI, which corresponds to a single stream Type A CQI report. The motivation for this is that even if the UE preferred a dual stream transmission, the Node B may not have sufficient data to transmit it with two streams and thus would need to know the single stream CQI. Type A CQI feedback is 8 bits and 3 bits go unused when the UE prefers single stream. Type B CQI feedback is 5 bits exactly as is the case with no MIMO. PCI feedback is two bits. The feedback information is summarized in Figure 15.21.

The transmission of non-MIMO channels, like common channels, and the transmission to non-MIMO UEs were assumed to be done using open loop Space Time Transmit Diversity (STTD) at the time of standardizing the feature. The original Release 7 MIMO, however, has a few performance issues:

- Equalizer UEs suffer when STTD is activated. The equalizer receiver estimates the multipath channel and removes the multipath interference. When the transmission happens with STTD, there are two times more multipath components and the equalization becomes more difficult. Also, equalization of two channels simultaneously is more challenging.
- If the data to non-MIMO UEs is transmitted from one antenna only without STTD, the power of second power amplifier in Node B is not utilized for non-MIMO terminals.
- The code multiplexing degrades performance when the precoding vectors are different for code multiplexed UEs because different precoding vectors make the signal structure suboptimal.

Due to these MIMO issues, the first MIMO deployments based on Release 7 are implemented on dedicated frequency without any non-MIMO UEs. That solution clearly is not an efficient approach

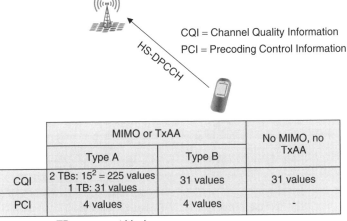

CQI = Channel Quality Information
PCI = Precoding Control Information

| | MIMO or TxAA | | No MIMO, no TxAA |
| --- | --- | --- | --- |
| | Type A | Type B | |
| CQI | 2 TBs: $15^2$ = 225 values 1 TB: 31 values | 31 values | 31 values |
| PCI | 4 values | 4 values | - |

TB = transport block

**Figure 15.21** MIMO feedback information from UE to Node B

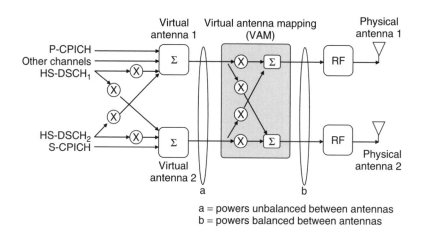

a = powers unbalanced between antennas
b = powers balanced between antennas

**Figure 15.22**  Virtual antenna mapping (VAM)

from a spectrum usage point of view. More efficient MIMO performance can be obtained by transmitting to non-MIMO UEs without transmit diversity but by using Virtual Antenna Mapping (VAM) for power balancing. The VAM solution is illustrated in Figure 15.22. The non-MIMO channels are transmitted only via one branch – virtual antenna 1. The transmission powers are not balanced at point a. The VAM solution takes the signal from one virtual antenna, rotates the signal and sums to the other antenna. The resulting transmission powers in the physical antennas (point b) are now balanced and the two RF power amplifiers can be fully utilized also for non-MIMO and non-transmit diversity transmissions.

The original Release 7 MIMO had four precoding vectors. That causes problems with VAM solution since some combinations of the precoding feedback weights make the powers unbalanced in the two power amplifiers when used in conjunction with Virtual Antenna Mapping. Therefore, there is a need to limit the number of precoding vectors from four to two as shown in Figure 15.23. The simulations show that limiting the precoding vectors to two has typically less than 5% impact to the throughput. The network informs UE how many precoding vectors UE can use in the feedback reporting. This feature was a late inclusion to Release 7 specifications during 2010. Even if UE uses four feedback vectors, Node B can still limit the vectors to two, but that causes some performance penalty.

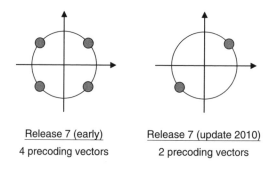

Release 7 (early)            Release 7 (update 2010)
4 precoding vectors          2 precoding vectors

**Figure 15.23**  Limiting the number of precoding vectors to two with MIMO

**Table 15.1** MIMO workaround solution summary

|  | Release 7 Original MIMO | Release 7 Work-around MIMO |
|---|---|---|
| 2nd antenna CPICH | Diversity Primary CPICH | Secondary CPICH |
| Transmission to non-MIMO UEs | Space Time Transmit Diversity (STTD) | Virtual antenna mapping (VAM) |
| Precoding feedback weights | 4 | 2 |

**Table 15.2** MIMO data rate capabilities in Releases 7, 8 and 9

|  | Release 6 | Release 7 | Release 8 | Release 9 |
|---|---|---|---|---|
| 64QAM without MIMO | No | Yes | Yes | Yes |
| MIMO without 64QAM | No | Yes | Yes | Yes |
| 64QAM + MIMO | No | No | Yes | Yes |
| 64QAM + MIMO + Dual cell HSDPA | No | No | No | Yes |
| Peak data rate | 14 Mbps | 28 Mbps | 42 Mbps | 84 Mbps |

The above-mentioned solutions are called MIMO workaround solution and summarized in Table 15.1. The optimized MIMO transmission uses Secondary CPICH for antenna 2, Virtual Antenna Mapping and limited Precoding vector set. The benefit of the workaround solution is that non-MIMO UEs can co-exist on the same carrier as MIMO and that the powers from the two Node B power amplifiers are balanced and benefiting non-MIMO UEs too. Even with the workarounds HSDPA MIMO has still performance issues with code multiplexing when the precoding vectors are different. The activation of MIMO also decreases the cell capacity for non-MIMO UEs due to the additional interference caused by the S-CPICH transmission that cannot be utilized by non-MIMO UEs.

The combination of MIMO with other peak data rate features is shown in Table 15.2. MIMO in 3GPP Release 7 brings the peak data rate to 28 Mbps. The combination of 64QAM and MIMO comes in Release 8 pushing the peak rate to 42 Mbps. Release 9 further combines Dual cell HSDPA with MIMO enhancing the peak data rate to 84 Mbps.

64QAM and MIMO together improve the peak rate by 200% from 14 Mbps to 42 Mbps, and the dual cell HSDPA further doubles the peak rate to 84 Mbps. The average cell capacity improvement is far less than the peak rate improvement. The reason is that 64QAM and dual stream MIMO can only be utilized when signal-to-noise ratio is good enough. The capacity gain of these features is 10–50% depending on the network loading. With full loading the gain is typically below 10% both for 64QAM and with MIMO. The evolution of the peak and average rates are illustrated in Figure 15.24.

## 15.9 Transmit Diversity (TxAA)

The aim of the transmit diversity is to improve the link level performance for single antenna UEs by using two transmit antennas in Node B. HSDPA Release 5 specifications included transmit diversity using open loop STTD and closed loop mode 1. These transmit diversity options have practical problems: STTD performs poorly with equalizer terminals and closed loop mode 1 does not support continuous packet connectivity (CPC). As a side product of MIMO work, Release 9 specifications also bring improved transmit diversity, called TxAA. This solution is equal to MIMO single stream mode. TxAA solution can be used also for one antenna UEs while MIMO requires two antennas in UE. TxAA functionality is illustrated in Figure 15.25.

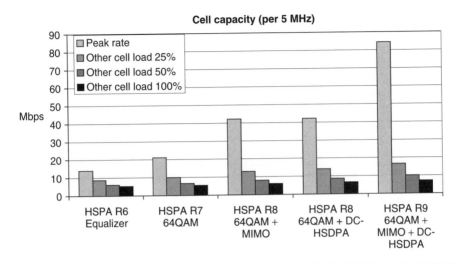

**Figure 15.24** Downlink peak rate and average cell rate evolution with HSDPA features

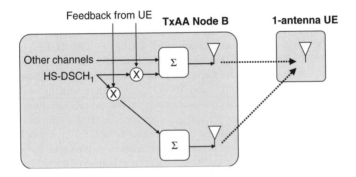

**Figure 15.25** TxAA functionality

## 15.10   Uplink 16QAM

HSUPA Release 6 uplink uses QPSK (dual-BPSK) modulation with multi-code transmission, providing 5.76 Mbps. The uplink data rate within 5 MHz can be increased by using higher order modulation or MIMO transmission. The challenge with uplink MIMO is that the UE needs to have two power amplifiers. Therefore, uplink single user MIMO is not part of HSPA evolution and not part of LTE in Release 8 either. The higher order 16QAM (or dual-4PAM) modulation together with multicode transmission was adopted as part of Release 7 for HSUPA, doubling the peak rate to 11.5 Mbps.

The higher order modulation improves the downlink spectral efficiency because the downlink has limited number of orthogonal resources. The same is not true for uplink since the HSUPA uplink is not orthogonal and there is practically an unlimited number of codes available in uplink. The highest uplink spectral efficiency can be achieved by using QPSK modulation only. In other words, HSUPA 16QAM is a peak data rate feature, not a capacity feature.

The multipath propagation affects high data rate performance. Therefore, the UE equalizer is used on HSDPA terminals. Also the Node B receiver can improve the uplink high bit rate HSUPA performance in multipath channels by using an equalizer. Another solution is to use four antenna receiver diversity in uplink.

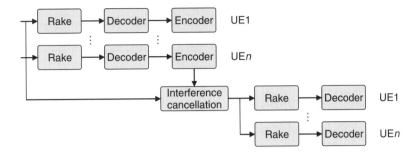

**Figure 15.26**   Parallel coded interference cancellation

High uplink data rates require also high $E_c/N_0$. The $E_c/N_0$ requirements in 3GPP are given for pre-defined HSUPA channel configurations called Fixed reference channels. Fixed reference channel 8 with 8 Mbps instantaneous data rate and 70% final throughput (5.6 Mbps) after HARQ retransmissions requires $E_c/N_0 = 12-16$ dB [8]. The corresponding uplink noise rise will also be similar $12-16$ dB impacting the coverage of other simultaneous users. It is therefore beneficial to utilize the uplink interference cancellation to subtract the high bit rate interference from other simultaneous users.

The Node B interference cancellation algorithms are not specified in 3GPP. An example algorithm can rely on the parallel coded interference cancellation as illustrated in Figure 15.26. Rake receiver is used as the initial stage for all UEs following by decoding. If the CRC check shows correct reception, the received signal can be cancelled from the other users in the cancellation stage. The decoding is repeated after the cancellation stage for those UEs where the reception failed after the initial Rake stage.

## 15.11   UE Categories

There are total of 28 UE categories for HSDPA and 9 for HSUPA. The most relevant single carrier UE categories for HSDPA are shown in Table 15.3 and for HSUPA in Table 15.4. The Release 7 and Release 8 HSDPA UE categories are defined both without any channel coding and with 5/6 coding, for example Category 13 obtains 17.6 Mbps with 5/6 coding while Category 14 gives 21.1 Mbps without any channel coding. The practical performance difference between those categories is mainly relevant

**Table 15.3**   HSDPA terminal categories

| Cat | Codes | Modulation | MIMO | Coding | Peak | 3GPP |
|-----|-------|------------|------|--------|------|------|
| 6 | 5 | 16QAM | – | 3/4 | 3.6 Mbps | Release 5 |
| 8 | 10 | 16QAM | – | 3/4 | 7.2 Mbps | Release 5 |
| 9 | 15 | 16QAM | – | 3/4 | 10.1 Mbps | Release 5 |
| 10 | 15 | 16QAM | – | 1/1 | 14.1 Mbps | Release 5 |
| 12 | 5 | QPSK | – | 3/4 | 1.8 Mbps | Release 5 |
| 13 | 15 | 64QAM | – | 5/6 | 17.6 Mbps | Release 7 |
| 14 | 15 | 64QAM | – | 1/1 | 21.1 Mbps | Release 7 |
| 15 | 15 | 16QAM | $2 \times 2$ | 5/6 | 23.4 Mbps | Release 7 |
| 16 | 15 | 16QAM | $2 \times 2$ | 1/1 | 28.0 Mbps | Release 7 |
| 17 | 15 | 64QAM or MIMO | | 5/6 | 23.4 Mbps | Release 7 |
| 18 | 15 | 64QAM or MIMO | | 1/1 | 28.0 Mbps | Release 7 |
| 19 | 15 | 64QAM | $2 \times 2$ | 5/6 | 35.3 Mbps | Release 8 |
| 20 | 15 | 64QAM | $2 \times 2$ | 1/1 | 42.2 Mbps | Release 8 |

**Table 15.4** HSUPA terminal categories

| Cat | TTI | Modulation | Coding | Peak | 3GPP |
|-----|------|------------|--------|-----------|-----------|
| 3 | 10 ms | QPSK | 3/4 | 1.4 Mbps | Release 6 |
| 5 | 10 ms | QPSK | 3/4 | 2.0 Mbps | Release 6 |
| 6 | 2 ms | QPSK | 1/1 | 5.7 Mbps | Release 6 |
| 7 | 2 ms | 16QAM | 1/1 | 11.5 Mbps | Release 7 |

in ideal channel conditions, but not in the practical macro cell field conditions. The peak rates in HSUPA categories in Release 7 and 8 are defined without any channel coding. HSUPA 16QAM gives the peak rate of 11.5 Mbps.

## 15.12   Layer 2 Optimization

WCDMA Release 99 specification was based on the packet retransmissions running from Radio Network Controller (RNC) to the UE on layer 2. The layer 2 Radio Link Control (RLC) packets had to be relatively small to avoid the retransmission of very large packets in case of transmission errors. Another reason for the relatively small RLC packet size was the need to provide sufficiently small step sizes to adjust the data rates for Release 99 channels. The RLC packet size in Release 99 is not only small, but it is also fixed for Acknowledged Mode Data and there are just a limited number of block sizes in Unacknowledged Mode Data. This limitation is due to fixed transport channel data rate options in Release 99.

The RLC payload size is fixed to 40 bytes in Release 99 for Acknowledged Mode Data. The same RLC solution is applied to HSDPA Release 5 and HSUPA Release 6 as well: the 40-byte packets are transmitted from RNC to base station in case of HSDPA. An additional configuration option to use the 80-byte RLC packet size was introduced in Release 5 to avoid extensive RLC protocol overhead, Layer 2 processing and RLC transmission window stalling. With the 2 ms TTI used with HSDPA, this leads to possible data rates being multiples of 160 kbps and 320 kbps respectively.

As the data rates are further increased in Release 7, increasing the RLC packet size even further would significantly impact on the granularity of the data rates available for HSDPA scheduling and the possible minimum data rates.

3GPP HSDPA and HSUPA allow the optimization of the Layer 2 operation since Layer 1 retransmissions are used and the probability of Layer 2 retransmissions is very low. Also, the Release 99 transport channel limitation does not apply to HSDPA/HSUPA since the Layer 2 block sizes are independent of the transport formats. Therefore, it is possible to use flexible and considerably larger RLC sizes and introduce segmentation to the MAC (Medium Access Control) layer in the base station.

This optimization is included for downlink in Release 7 and for uplink in Release 8 and it is called Flexible RLC and MAC segmentation solution. The RLC block size in flexible RLC solution can be as large as Internet Protocol (IP) packet, which is typically 1500 bytes for download. There is no need for packet segmentation in RNC. By introducing the segmentation to the MAC, the MAC can perform the segmentation of the large RLC PDU based on physical layer requirements when needed. The flexible RLC concept in downlink is illustrated in Figure 15.27.

The flexible RLC and MAC segmentation brings a number of benefits in terms of Layer 2 efficiency and in terms of peak bit rates:

- The relative layer 2 overhead is reduced. With the RLC header of 2 bytes, the RLC overhead is 5% in case of 40-byte RLC packet. When the RLC packet size increases to 1500 bytes, the RLC header

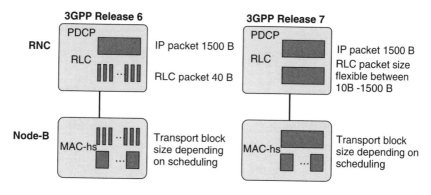

**Figure 15.27** Flexible RLC concept

overhead is reduced to below 0.2% and total L2 overhead to 1%. That reduction in the overhead can improve the effective application data throughput.

- The RLC block size can be flexibly selected according to the packet size of each application. That flexibility helps to avoid unnecessary padding which is no more needed in the flexible RLC solution. That is relevant especially for small IP packet sizes which are typical in VoIP or streaming applications.
- Less packet processing is required in RNC and in UE with octet aligned protocol header. The number of packets to be processed is reduced since the RLC packet size is increased and octet aligned protocol headers avoids bit shifting in high data rates connections. Both reduce the Layer 2 processing load and make the high bit rate implementation easier.
- Full flexibility and resolution of available data rates for the HSDPA scheduler.

## 15.13 Architecture Evolution

3GPP networks will increasingly be used for IP-based packet services. 3GPP Release 6 has four network elements in the user and control plane: base station, RNC (Radio Network Controller), SGSN (Serving GPRS Support Node) and GGSN (Gateway GPRS Support Node). The architecture in Release 8 LTE will have only two network elements: A base station in the radio network and an Access Gateway (a-GW) in the core network. The a-GW consists of control plane MME (Mobility management entity) and user plane SAE GW (System Architecture Evolution Gateway). The flat network architecture reduces the network latency and thus improves the overall performance of IP based services. The flat model also improves both user and control plane efficiency. The flat architecture is considered beneficial also for HSPA and it is specified as part of Release 7. The HSPA flat architecture in Release 7 and LTE flat architecture in Release 8 are exactly the same: the Node B responsible for the mobility management, ciphering, all retransmissions and header compression both in HSPA and in LTE. The architecture evolution in HSPA is designed to be backwards compatible: existing terminals can operate with the new architecture and the radio and core network functional split is not changed. The architecture evolution is illustrated in Figure 15.28.

Also the packet core network has flat architecture in Release 7. It is called the direct tunnel solution and allows the user plane to by-pass SGSN. With the flat architecture with all RNC functionality in the base station and using direct tunnel solution, only two nodes are needed for user data operation. This achieves flexible scalability and introduces the higher data rates with HSPA evolution with minimum impact on the other nodes in the network. This is important to achieve low cost per bit and enable

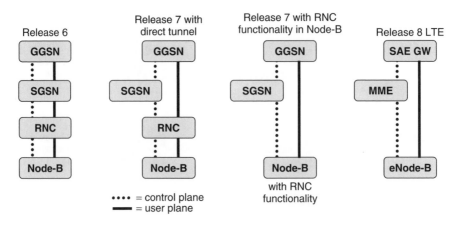

**Figure 15.28**  Evolution towards flat architecture

competitive flat rate data charging offerings. As the gateway in LTE has similar functionality to GGSN, it is foreseen to enable deployments of LTE and HSPA where both connect directly to the same core network element for user plane data handling directly from the base station.

3GPP Release 8 includes a new architecture to support small home Node Bs as well. The home Node Bs are called also as femto access points. The femtocells are covered in Chapter 19.

## 15.14   Conclusion

HSPA evolution has brought a number of important enhancements on top of HSPA Releases 5 and 6. HSPA enhancements provide major improvements to the end user performance in increasing peak bit rates, reducing mobile terminal power consumption and reducing latency. The peak bit rates can be tripled per 5 MHz carrier with 64QAM and 2x2 MIMO from 14 Mbps to 42 Mbps. These peak rate features bring some improvements to the spectral efficiency and cell capacity as well. The application performance is also improved by lower end-to-end round-trip time which is reduced below 30 ms with HSPA evolution.

The UE power consumption is considerably reduced in HSPA evolution due to discontinuous transmission and reception. The expected voice call talk time will improve by more than 50% and the usage time of bursty data applications increases even more. The practical talk times will improve further when the digital and RF technologies evolve and general power consumption is reduced. Fast dormancy feature can also reduce the power consumption for always-on applications.

HSPA evolution enhances basic voice service by mapping circuit switched (CS) voice on top of HSPA channels – a concept called CS voice-over HSPA. It is essentially a combination of Voice-over IP (VoIP) in the radio and CS voice in the core network. CS voice-over HSPA reduces UE power consumption and increases spectral efficiency compared to WCDMA voice while having no impact on the core network and end-to-end ecosystem. It will be easy to upgrade CS voice-over HSPA later to VoIP over HSPA since the radio solutions are very similar.

The set-up times are reduced by mapping also the RACH/FACH common channels on top of HSDPA/HSUPA. Actually all the services including signaling can be mapped on top of HSPA in HSPA evolution. There are only a few physical layer channels left from Release 99 specifications in Release 8 – otherwise, everything is running on top of HSPA. That explains also why the end user performance and the network efficiency have improved considerably compared to Release 99.

|  | | Peak rate | Average rate (capacity) | Cell edge rate | Latency gain | Talk time |
|---|---|---|---|---|---|---|
| **Downlink** | HSDPA 64QAM[1] | +50% | <10% | - | - | - |
| | HSDPA 2 × 2MIMO | +100% | <30% | <20% | - | - |
| | DC-HSDPA | +100% | +20-100% | +20-100% | - | - |
| **Uplink** | HSUPA 10 ms (2.0 Mbps)[2] | +600% | +20-100% | <100% | Gain 20 ms | - |
| | HSUPA 2 ms (5.8 Mbps) | +200% | <30% | - | Gain 15 ms | - |
| | HSUPA 16QAM | +100% | - | - | - | - |
| | Advanced Node B receiver | - | >30% | - | - | - |
| | DTX/DRX, Fast dormancy | - | - | - | - | > + 50% |
| | HS-FACH / HS-RACH | - | - | - | Setup time <0.1 s | - |
| | CS voice over HSPA | - | +80% (voice) | - | - | > + 50% |

[1]Baseline WCDMA Release 5 downlink 14.4 Mbps
[2]Baseline WCDMA Release 99 uplink 384 kbps

= clear gain >30%

= moderate gain <30%

**Figure 15.29**  Summary of HSPA evolution features and their benefits

3GPP Release 7 simplifies the network architecture. The number of network elements for the user plane can be reduced from four in Release 6 down to two in Release 7. The architecture of Release 7 is exactly the same as used in LTE in Release 8 which makes the network evolution from HSPA to LTE straightforward. Most of the 3GPP Release 7 and 8 enhancements are expected to be relatively simple upgrades to the HSPA networks as was the case with HSDPA and HSUPA in earlier Releases. The main features and their benefits in HSPA evolution are summarized in Figure 15.29.

# References

[1] 3GPP, Technical Specification 24.008 'Mobile Radio Interface Layer 3 Specification: Core Network Protocols', Version 8.3.0.
[2] Halonen, T., Romero, J. and Melero. J. *GSM, GPRS and EDGE Performance*, 2nd edition, New York: John Wiley & Sons, Ltd, 2003.
[3] Barreto, A., Garcia, L. and Souza, E. 'GERAN Evolution for Increased Speech Capacity', *Vehicular Technology Conference*, 2007. VTC2007-Spring, IEEE 65th, April 2007.
[4] 3GPP, 'Continuous Connectivity for Packet Data Users', 3GPP TR25.903V7.0.0, March 2007
[5] 3GPP, Technical Specification 25.308 'High Speed Downlink Packet Access (HSDPA); Overall Description; Stage 2', V.8.7.0.
[6] 3GPP, '64QAM for HSDPA', 3GPP R1-063335, November 2006.

[7]  3GPP, Technical Specifications 25.104 'Base Station (BS) Radio Transmission and Reception (FDD)',
     V.8.3.0.
[8]  3GPP, Technical Specifications 25.214 'Physical Layer Procedures (FDD)'.
[9]  Holma, H. and Toskala, A. (eds) *WCDMA for UMTS: HSPA Evolution and LTE*, 4th edition, New York:
     John Wiley & Sons, Ltd, 2007.
[10] Kurjenniemi, J., Nihtilä, T., Lampinen, M. and Ristaniemi, T. 'Performance of WCDMA HSDPA Network
     with Different Advanced Receiver Penetrations', *Wireless Personal Multimedia Communications (WPMC)*,
     Aalborg, Denmark, 17–22 September 2005.
[11] 3GPP, 'Further Discussion on Delay Enhancements in Rel7', 3GPP R2-061189, August 2006.
[12] Morais de Andrade, D., Klein, A., Holma, H., Viering, I., and Liebl, G. 'Performance Evaluation on
     Dual-Cell HSDPA Operation', submitted to IEEE ICC 2009.
[13] Holma, H., Kuusela, M., Malkamäki, E., Ranta-aho, K. and Tao C. 'VoIP over HSPA with 3GPP Release
     7', *PIMRC2006*, September 2006.
[14] 3GPP, 'HSDPA VoIP Capacity', 3GPP R1-062251, August 2006.

# 16

# HSPA Multicarrier Evolution

Harri Holma, Karri Ranta-aho and Antti Toskala

## 16.1    Introduction

WCDMA was specified with the chip rate of 3.84 Mcps and nominal carrier spacing of 5 MHz in Release 99. 3GPP releases included a number of evolution steps but all the enhancements were limited to the same 5 MHz bandwidth until Release 7. Release 8 brought dual cell HSDPA (DC-HSDPA) which enables a single UE to receive on two adjacent carriers. Release 9 further extends the dual cell approach to the uplink and combines the downlink DC-HSDPA with MIMO. Release 10 will specify support of three and four carriers in the downlink direction. The HSPA multicarrier evolution is summarized in Figure 16.1. The main benefits of multicarrier solutions are higher data rates and better trunking efficiency in multicarrier sites. The peak data rate evolution is shown in Table 16.1: four carriers and MIMO give a theoretical maximum rate of 168 Mbps. Multicarrier evolution co-exists with any legacy WCDMA and HSPA UEs on the same carrier, which makes the introduction of multicarrier HSPA a smooth process.

Release 9 also includes Multiband HSDPA where the two downlink carriers can be on different frequency bands, for example, one carrier on 900 MHz and another carrier on 2100 MHz. Release 10 will extend the downlink up to total four carriers on two bands. The uplink will remain on single band. The multiband evolution is illustrated in Figure 16.2.

This chapter presents in more detail the specifications and the benefits of multicarrier and multiband HSPA. The dual cell HSPA is based on the primary and secondary carrier. Just as in the single carrier operation, the primary carrier provides all the downlink physical channels for the UE both for downlink data transmission as well as the channels supporting the uplink data transmission: the High Speed Physical Downlink Shared Channel (HS-PDSCH), the corresponding High Speed Shared Control Channel (HS-SCCH), the Fractional Dedicated Physical Channel (DPCH) facilitating the uplink power control, the E-DCH Acknowledgement Indicator Channel (E-HICH), the E-DCH Absolute Grant Channel (E-AGCH) and the E-DCH Relative Grant Channel (E-RGCH). For Dual Cell HSDPA, the secondary carrier is used to transmit a second set of HS-PDSCHs and HS-SCCHs in downlink, and for Dual Cell HSUPA the secondary carrier also sends a duplicate set of downlink physical channels in support of the second uplink carrier. This leads to the fact that two uplink carriers can only be supported when two downlink carriers are also supported.

*WCDMA for UMTS: HSPA Evolution and LTE, Fifth Edition*    Edited by Harri Holma and Antti Toskala
© 2010 John Wiley & Sons, Ltd

**3GPP Release 7**
UE can receive and transmit on single 5 MHz carrier

**3GPP Releases 8-9**
UE can receive and transmit on two adjacent 5 MHz carriers

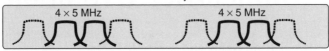

**3GPP Release 10**
UE can receive on four adjacent 5 MHz carrier

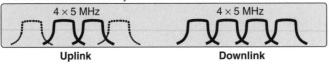

**Uplink**                                              **Downlink**

**Figure 16.1**   Overview of HSPA multicarrier evolution

**Table 16.1**   Downlink peak data rates with multicarrier HSPA

|            | Without MIMO | With MIMO |
|------------|--------------|-----------|
| 1 carrier  | 21 Mbps      | 42 Mbps   |
| 2 carriers | 42 Mbps      | 84 Mbps   |
| 3 carriers | 63 Mbps      | 126 Mbps  |
| 4 carriers | 84 Mbps      | 168 Mbps  |

For DC-HSDPA, the UE estimates Channel Quality Information (CQI) for both carriers to the link adaptation for both carriers separately. The UE also needs to provide HARQ acknowledgments in the uplink for both downlink carriers separately. All this uplink physical layer feedback is carried on a single High Speed Dedicated Physical Control Channel (HS-DPCCH) on the primary uplink frequency leading to the fact that DC-HSDPA can be supported with a single uplink carrier. There are two CQI values and two ACKs in uplink in case of DC-HSDPA and no MIMO. In the case of MIMO, four CQIs and ACKs are needed. With four carriers and with MIMO, the uplink feedback will be in total eight CQIs and ACKs.

The uplink E-DCH Dedicated Physical Control Channel (E-DPCCH) and Enhanced Dedicated Physical Data Channel (E-DPDCH) can be allocated on both carriers with DC-HSUPA in Release 9. The allocation of the physical channels is illustrated in Figure 16.3. In summary, all the physical channels can be carried on both frequencies except for HS-DPCCH which is only carried on the primary frequency. Any of the carriers can be primary and secondary.

The uplink power control runs independently for the different carriers because the fast fading can be different with a 5 MHz shift in the frequency domain, and the uplink power levels of the two carriers can also be different. The uplink Signal-to-Interference Ratio (SIR) is estimated by Node B from the uplink DPCCH separately for each carrier. The power control command is carried in F-DPCH on a corresponding downlink carrier. The uplink power control is illustrated in Figure 16.4. The secondary uplink frequency is faded and higher transmission power is used for the secondary frequency than for the primary frequency.

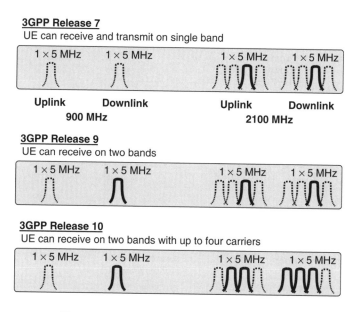

**3GPP Release 7**
UE can receive and transmit on single band

**3GPP Release 9**
UE can receive on two bands

**3GPP Release 10**
UE can receive on two bands with up to four carriers

**Figure 16.2** Overview of HSPA multiband evolution

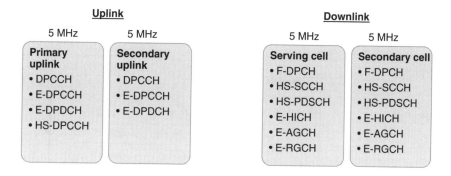

**Figure 16.3** Physical channel locations with DC-HSPA

All the other physical channels including the Primary Common Pilot Channel (P-CPICH), the Synchronization Channels (SCHs), the Acquisition Indicator Channel (AICH), the Paging Indicator Channel (PICH) and the Physical Broadcast Channel (P-BCH) can be carried on both frequencies allowing the legacy UEs to co-exist on the same carriers with the new multicarrier HSPA UEs. It is a major benefit for the multicarrier HSPA introduction that all legacy UEs can co-exist on any of the carriers where a multicarrier HSPA is used. It is also possible to operate DC-HSPA in such a way where the secondary carrier is dedicated for DC-HSPA use only. In that case the secondary carrier carries only P-CPICH for the channel estimation but no other common channels.

The packet scheduling is done jointly over the two or multiple carriers. The scheduler receives CQI reports from all downlink carriers. The scheduling decision is based on the number of input parameters including CQI, the amount of data in the buffer and the QoS priorities. The target of the scheduler is to maximize the user data rates and the system efficiency. It is possible to obtain gains from the frequency domain scheduling in the same way as in LTE.

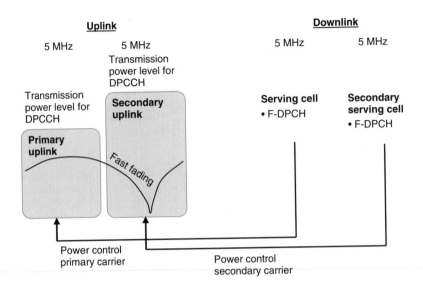

**Figure 16.4**   Uplink power control with DC-HSPA

The link adaptation and L1 retransmissions (HARQ) are run on different carriers separately. The retransmission uses the same frequency as the first transmission. DC-HSPA does not make any changes to RLC layer as the data stream is split between the two carriers in the MAC layer. The concept is illustrated in Figure 16.5.

The mobility is based on the primary frequency. That is, the mobility procedures are the same as with the single carrier HSPA.

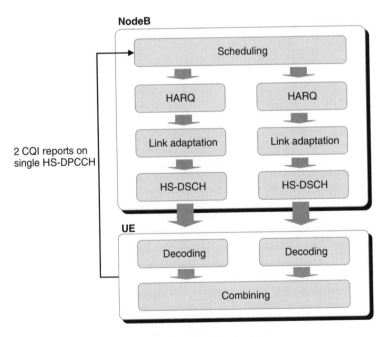

**Figure 16.5**   DC-HSDPA transmission concept

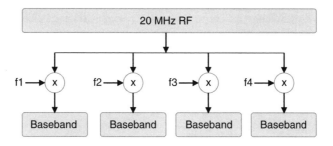

**Figure 16.6**  Multicarrier HSPA can utilize single RF chain

The multicarrier HSPA can be implemented with one wideband RF chain covering the whole 10–20 MHz channel. Each carrier is down-converted to baseband and each carrier can be processed in parallel. The high level structure is shown in Figure 16.6.

## 16.2  Dual Cell HSDPA in Release 8

Release 8 DC-HSDPA can double the user data rate if the number of users is low because a single user can utilize two parallel frequencies. When the system load increases, the probability decreases that a single user can get access to full capacity of both frequencies. But even at high load DC-HSDPA provides some capacity benefits compared to two single carriers. The principles of the DC-HSDPA gains are illustrated in Figure 16.7.

The capacity gain with DC-HSDPA is obtained because of the following features:

- *Frequency domain packet scheduling gain.* Since UE provides separate CQI reports on both carriers, the Node B packet scheduler can transmit the packets on the frequency that is not faded. The fast fading is heavily location-dependent and is uncorrelated when moving some tens of centimeters. Therefore, the fast fading is independent between two UEs. The principle of the frequency domain scheduling is shown in Figure 16.8. Also Long Term Evolution (LTE) uses frequency domain scheduling to obtain capacity gains. The frequency domain scheduling in LTE provides higher gains than in HSDPA since the LTE frequency resolution is 180 kHz while HSDPA is 5 MHz. The frequency domain scheduling gain in LTE has been shown to be 30–40% [1].

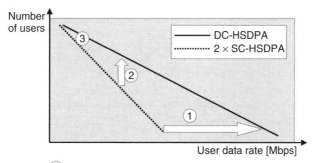

① = Double data rate at low number of users
② = Capacity gain with certain user data rate
③ = Slightly higher data rate at high number of users

**Figure 16.7**  Data rate gain of DC-HSDPA

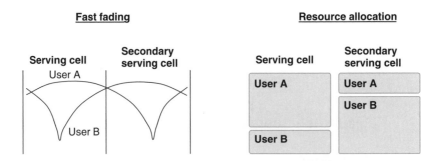

Figure 16.8   Frequency domain scheduling with DC-HSDPA

- *Statistical multiplexing or trunking gain.* The load can be balanced between two frequencies with 2 ms TTI resolution with DC-HSDPA while the load balancing with two SC-HSDPA requires slow inter-frequency handovers or redirections. Therefore, the load in practice in SC-HSDPA case is not ideally balanced.
- *Multiuser diversity gain.* HSDPA packet scheduling can utilize proportional fair algorithm in the time domain. Such a scheduling algorithm gives a higher gain when there is larger number of UEs to select from. DC-HSDPA now allows the users from two frequencies to be considered for the optimized scheduling.

The capacity gain results are shown in Figure 16.9. The simulations are run with Pedestrian A and Pedestrian B multipath profiles. The results compare the capacity of DC-HSDPA vs two SC-HSDPA

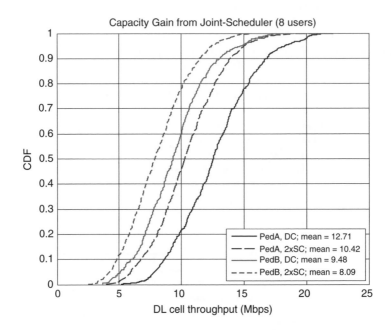

Figure 16.9   DC-HSDPA capacity gain with high load [2]

where the number of users is perfectly balanced between the carriers. The traffic model is a file download by total eight users. The results show that DC-HSDPA can improve the cell capacity by 20% compared to two SC-HSDPA carriers. In Pedestrian A channel, the two SC-HSDPA carriers together provide 10.4 Mbps while the DC-HSDPA solution offers 12.7 Mbps. The corresponding results in Pedestrian B channel are 8.1 Mbps and 9.5 Mbps. We can summarize that the data rate and capacity gain from DC-HSDPA vary between 20% at high load and 100% at low load.

## 16.3   Dual Cell HSUPA in Release 9

DC-HSUPA was included in Release 9 to boost the uplink user rates. There are a few differences between the dual cell benefits in uplink compared to downlink:

- UE transmission power is much lower than the Node B transmission power. The uplink data rate is more likely limited by the transmission power than the downlink data rate.
- Frequency selective packet scheduling is not as simple in the uplink as in the downlink because there is no similar fast CQI reporting as with HSDPA.

The following simulation results are obtained in two different cell sizes to estimate the impact of the UE power levels: Figure 16.10 shows the results in a small cell with 500 m inter-site distance (ISD) and Figure 16.11 shows them in a large cell with 1700 m inter-site distance. The number of users per MHz ($N$) is the same for the single carrier and DC-HSUPA case. Therefore, the DC-HSUPA simulation has in total $2 \times N$ users. The peak data rate in small cells doubles from 4 Mbps to 8 Mbps

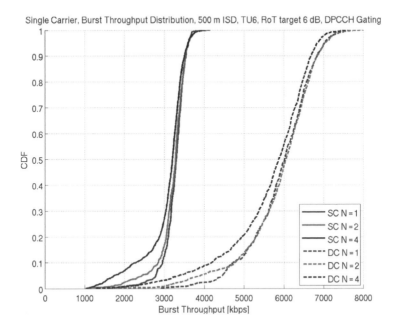

**Figure 16.10**   DC-HSUPA simulations in small cell [3]. *Note:* N is number of users per 5 MHz.

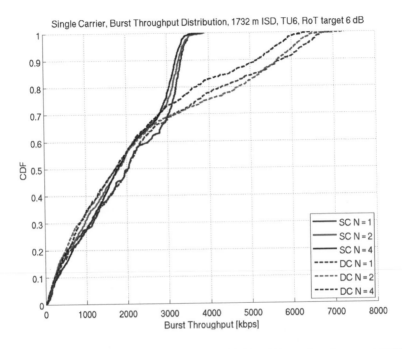

**Figure 16.11**   DC-HSUPA simulations in large cell [3]. *Note:* N is number of users per 5 MHz.

as expected. The median data rate improves from approximately 3.2 Mbps to 6 Mbps with 1–4 users per cell. The lowest 10% data rates are also improved from 3.0 Mbps to 4.7 Mbps with one user and from 2.3 Mbps to 4.2 Mbps with four users. In the large cells, the gain from DC-HSUPA is limited to those UEs that have enough transmission power to reach Node B with high data rates. In this case it is 30% of the UEs that benefit from DC-HSUPA. The median data rates are 1.8–2.0 Mbps both with and without DC-HSUPA in the large cells.

The DC-HSUPA results are summarized in Figure 16.12. The upper part shows the average data rates in small cells with 1–4 users and the lower part in large cells. The average data rates are nearly doubled in small cells. The data rates naturally decrease slightly when the number of users is increased. The average data rate gain in large cells is 20–40% with 1–4 users.

## 16.4   Dual Cell HSDPA with MIMO in Release 9

MIMO was defined in 3GPP Release 7 and DC-HSDPA in Release 8 and the combination of those two solutions comes in Release 9. MIMO can double the peak rate from 42 Mbps to 84 Mbps. We could say that only Release 9 brings the true MIMO gains. Release 7 MIMO did not double the data rate because the combination of MIMO + 64QAM was not defined. Release 8 MIMO did not double the data rate because the combination of MIMO + DC-HSDPA was not defined. Therefore, it is expected that wide MIMO deployments will not happen before Release 9.

Note that it is still possible to have MIMO UEs and DC-HSDPA UEs on the same carrier in Release 8 but both features cannot be used for a single UE in Release 8.

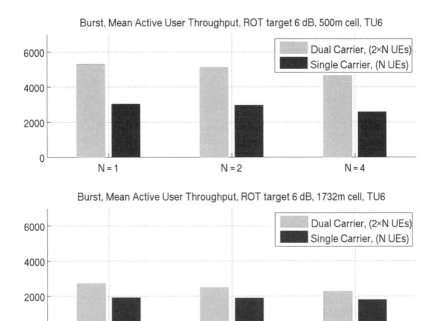

**Figure 16.12** DC-HSUPA data rate gains in small and in large cells [3]

## 16.5   Dual Band HSDPA in Release 9

It is not always possible to find enough spectrum space to deploy two or more HSPA carriers in one frequency band. Many operators have deployed HSPA carriers on two frequency bands. The aggregate bit rate could be increased if the UE were able to receive on carriers in separate bands at the same time. That kind of Dual Band HSPA is defined in Release 9. The band combinations are listed in Table 16.2. The European markets and some Asian markets can utilize configuration 1 with Band I (=2100 MHz) together with Band VIII (= 900 MHz band). The Americas market can use Band II (= 1900 MHz) and Band IV (= 1700/2100 MHz). The third possible configuration is Band I together with Band V (= 850 MHz) which is the deployment case in some Asian markets. The 3GPP Release 10 includes further band combinations between Bands I, II, IV, V and VIII.

The dual band HSDPA is relatively simple for the radio network if those two frequency variants are already deployed at the site. But the dual band HSDPA is more challenging for the UE implementation. The UEs typically support at least two HSPA frequencies today but those two frequencies are not

**Table 16.2** Dual band HSPA combinations in 3GPP Release 9

| Dual band configuration | Downlink band (MHz) | Uplink band (MHz) |
| --- | --- | --- |
| 1 | I and VIII (2110–2170 and 925–960) | I or VIII (1920–1980 or 880–915) |
| 2 | II and IV (1930–1990 and 2110–2155) | II or IV (1850–1910 or 1710–1755) |
| 3 | I and V (2110–2170 and 869–894) | I or V (1920–1980 or 824–849) |

**Figure 16.13**   Dual band HSDPA

used at the same time. In the case of dual band HSDPA, the UE must receive on two frequencies at the same time which increases the RF complexity in the UE as receivers on both bands must be able to operate simultaneously and regardless of which band the UE is transmitting in the uplink.

The lower frequency has better propagation which makes the coverage areas of the different bands different in size. The scheduling between the two bands can be done dynamically with 2-ms resolution based on CQI reports from UE. The CQI report directly takes into account the impact of noise and interference. The dual band case is illustrated in Figure 16.13.

The dual band HSDPA is defined for downlink only. The uplink transmission takes place on one frequency band only. RNC selects which of the two bands is used for the uplink transmission. There may be a need to make inter-frequency handover for the uplink carrier depending on the coverage: if the uplink is initially allocated to the higher frequency and UE runs out of coverage, the uplink transmission can be handed to the lower frequency.

LTE-Advanced in Release 10 will include similar multiband carrier aggregation features such as the HSDPA multiband feature. LTE Release 8 and Release 9 use continuous spectrum allocation.

## 16.6   Three and Four Carrier HSDPA in Release 10

Most countries have three or four operators at the 2100 MHz band which gives 15 or 20 MHz band per operator since the total 2100 MHz band is 60 MHz. Therefore, the spectrum allows in most cases three or four carriers to use HSPA. The 3C-HSDPA and 4C-HSDPA further improves the user data rates beyond DC-HSDPA and gives 5–10% capacity gain in the full buffer case. The data rate benefits of 4C-HSDPA are illustrated in Figure 16.14.

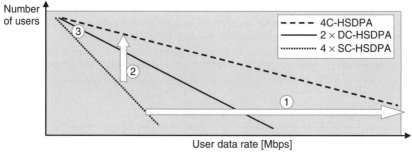

①  = Four times higher data rate at low number of users
②  = Capacity gain with certain user data rate
③  = Slightly higher data rate at high number of users

**Figure 16.14**   Data rate gain of 4C-HSDPA

4C-HSDPA uses 20 MHz RF in UE. 20 MHz bandwidth is mandatory for all LTE UEs. Similar RF receivers can now be used both for LTE and for 4C-HSDPA.

## 16.7  UE Categories

New UE categories are defined for DC-HSDPA and for DC-HSUPA. Release 8 included four categories for DC-HSDPA both with and without 64QAM and with 5/6 coding and without any coding. The peak rate without 64QAM is 28.0 Mbps and with MIMO 42.2 Mbps. Release 9 has a similar structure but all peak data rates are doubled with MIMO. Table 16.3 shows the DC-HSDPA UE categories and Table 16.4 shows the DC-HSUPA UE categories. The DC-HSDPA UEs can be implemented with either one or two antennas while the MIMO UEs must have two antennas. Therefore, the DC-HSDPA UE penetration will be higher than the MIMO UE penetration.

## 16.8  Conclusion

The main performance benefit from LTE compared to HSPA comes from wider bandwidth. Now HSPA+ can offer similar wideband performance benefits with dual cell and multicarrier evolution. Dual cell HSDPA was defined in Release 8 and dual cell HSUPA and the combination of MIMO and DC-HSDPA in Release 9. The work continues in Release 10 by defining three and four carriers for HSPA evolution bringing the theoretical peak rate beyond 150 Mbps.

Multicarrier HSPA can be deployed on the same frequencies with legacy WCDMA and HSPA devices, which makes the deployment easy. Therefore, multicarrier HSPA is the preferred technology for the existing HSPA frequencies while LTE is utilized on new frequency bands.

Many operators have three frequencies at the 2100 MHz band which enables 3-carrier HSPA and the peak rate beyond 100 Mbps with MIMO.

**Table 16.3**  UE categories for DC-HSDPA

| Cat | Codes | Modulation | MIMO | Coding | Peak | 3GPP |
|-----|-------|------------|------|--------|------|------|
| 21 | 15 | 16QAM | – | 5/6 | 23.4 Mbps | Release 8 |
| 22 | 15 | 16QAM | – | 1/1 | 28.0 Mbps | Release 8 |
| 23 | 15 | 64QAM | – | 5/6 | 35.3 Mbps | Release 8 |
| 24 | 15 | 64QAM | – | 1/1 | 42.2 Mbps | Release 8 |
| 25 | 15 | 16QAM | Yes | 5/6 | 46.8 Mbps | Release 9 |
| 26 | 15 | 16QAM | Yes | 1/1 | 56.0 Mbps | Release 9 |
| 27 | 15 | 64QAM | Yes | 5/6 | 70.6 Mbps | Release 9 |
| 28 | 15 | 64QAM | Yes | 1/1 | 84.4 Mbps | Release 9 |

**Table 16.4**  UE categories for DC-HSUPA

| Cat | Modulation | Coding | Peak | 3GPP |
|-----|------------|--------|------|------|
| 8 | QPSK | 1/1 | 11.50 Mbps | Release 9 |
| 9 | 16QAM | 1/1 | 23.0 Mbps | Release 9 |

# References

[1] Holma, H. and Toskala, A. *LTE for UMTS – OFDMA and SC-FDMA Based Radio Access*, Chichester: John Wiley, 2009.

[2] Morais de Andrade, D., Klein, A., Holma, H., Viering, I. and Liebl, G. 'Performance Evaluation on Dual-Cell HSDPA Operation', *IEEE Vehicular Technology Conference*, Fall 2009, Anchorage, Alaska, 2009.

[3] Repo, I., Aho, K., Hakola, S., Chapman, T. and Laakso F. 'Enhancing HSUPA System Level Performance with Dual Carrier Capability', submitted to IEEE ISWPC 2010.

# 17

# UTRAN Long-Term Evolution

Antti Toskala and Harri Holma

## 17.1 Introduction

This chapter presents the Universal Mobile Telecommunication Services (UMTS) Terrestrial Radio Access Network (UTRAN) Long-Term Evolution (LTE) work, which is defining a new packet-only wideband radio with flat architecture as part of the 3rd Generation Partnership Project (3GPP) radio technology family in addition to Global System for Mobile Communications (GSM)/General Packet Radio System (GPRS)/Enhanced Data rates for GSM Evolution and Wideband Code Division Multiple Access (WCDMA)/High-Speed Downlink Packet Access (HSDPA)/High-Speed Uplink Packet Access (HSUPA). This chapter covers the 3GPP schedule, background requirements and the technology principles around the UTRAN LTE physical layers, protocols and architecture as well as the outlook for the LTE-Advanced work in 3GPP. The first set of LTE Release 8 specifications was finalized at the end of 2007, but it took until March 2009 till a commercial baseline specification (from the LTE UE perspective) was achieved with all the corrections. The next Release, Release 9, was more or less completed by the end of 2009.The work on LTE-Advanced is aiming for 3GPP Release 10, with the first specs due towards the end of 2010 and finalization of the work in the first half of 2011.

3GPP started to investigate UTRAN LTE during 2004 by identifying the requirements from different players in the field. Following the workshop on the topic, it was agreed to start the feasibility study on the new packet-only radio system while the work on WCDMA evolution was continuing at full speed as well. The feasibility study was started in March 2005 and the key issues were to agree first the multiple access method to be used and the network architecture in terms of functional split between the radio access and core network.

The key requirements defined for the work [1] are:

- Packet-switched domain optimized.
- Server to User Equipment (UE) round-trip time below 30 ms and access delay below 300 ms.
- Peak rates uplink/downlink 50/100 Mbps.
- Good level of mobility and security ensured.

*WCDMA for UMTS: HSPA Evolution and LTE, Fifth Edition*   Edited by Harri Holma and Antti Toskala
© 2010 John Wiley & Sons, Ltd

- Improved terminal power efficiency.
- Frequency allocation flexibility with 1.4, 3, 5, 10, 15 and 20 MHz allocations; possibility to deploy adjacent to WCDMA, The actual specification phase adjusted the smallest bandwidths from 1,25 and 2,5 to be 1.4 MHz and 3.0 MHz. Especially the 1.4 MHz option was designed to suit a single cdma2000 carrier.
- Higher capacity compared with the Release 6 HSDPA/HSUPA reference case, in the downlink throughput three to four times and in the uplink two to three times reference scenario capacity.

## 17.2   Multiple Access and Architecture Decisions

From March 2005 until September 2006 3GPP conducted a feasibility study on the Evolved UTRAN (EUTRAN) technology alternatives and made a selection of the multiple access and basic radio access network architecture.

3GPP considered different multiple access options but came rather quickly to the conclusions that Orthogonal Frequency Division Multiple Access (OFDMA) is to be used in downlink direction and Single Carrier Frequency Division Multiple Access (SC-FDMA) is to be used in the downlink. The performance studies reflected in [2] showed that the requirements are achievable from the capacity sense as well.

In the previous round of the multiple access selection process, as covered in Chapter 4, WCDMA was selected as the multiple access method for UMTS in January 1998 based on the evaluation work done during 1997. An OFDMA proposal was already considered in the UMTS selection phase, but it was not mature enough at that point in time.

A number of things have changed since then to make OFDMA the preferred solution for the air interface, with the key developments being:

- *Larger bandwidth and bandwidth flexibility.* The UMTS spectrum allocation typically did not allow a larger carrier bandwidth than 5 MHz. WCDMA/High-Speed Packet Access (HSPA) shows attractive performance at 5 MHz when equalizer receivers are used. Also, the receiver complexity remains reasonably low. LTE work targets are for higher bit rates using larger bandwidth up to 20 MHz. OFDMA can provide benefits over a Code Division Multiple Access (CDMA)-based system when the bandwidth increases: the OFDMA signal remains orthogonal while CDMA performance suffers due to increased multipath components and the equalizer receiver gets more complex. Further, dealing with different bandwidths in the same system is more flexible with OFDMA.
- *Flat architecture.* In the architecture development, more intelligence is being added to the base station, similar to the trend set by HSUPA and HSDPA. The UMTS architecture initially was defined to be hierarchical, where the radio-related functionalities were located in the Radio Network Controller (RNC). In the flat architecture the radio-related functionalities are located in the base station. When the packet scheduling is located in the base station, fast packet scheduling can be applied, including frequency domain scheduling, as shown in Figure 17.1. The frequency domain scheduling is shown to improve the cell capacity up to 50% in the simulations. The frequency domain scheduling can be done in OFDMA but not in CDMA, where the interference is always spread over the whole carrier bandwidth.
- *Amplifier-friendly uplink solution with SC-FDMA.* One of the main challenges in OFDMA is the high peak-to-average radio of the transmitted signal, which requires linearity in the transmitter. The linear amplifiers have low efficiency; therefore, OFDMA is not an optimized solution for a mobile uplink where the target is to minimize the terminal power consumption. An LTE uplink uses SC-FDMA, which clearly enables better power-amplifier efficiency. SC-FDMA technology was not available when UMTS multiple access selection was done, but the first articles were just being published at the time, such as [3].

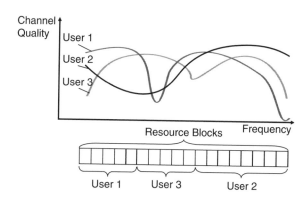

**Figure 17.1** Frequency domain scheduling

- *Simpler multi-antenna operation.* Higher bit rates can generally be obtained by using larger bandwidth and multiple antennas. Multiple input multiple output (MIMO) antenna technologies, emerging over the past few years, are required to achieve the LTE bit-rate targets. MIMO is simpler to implement with OFDMA than with CDMA.

3GPP decided to place the radio functionality fully in the base station, as shown in Figure 17.2. The new functionalities in the Base Transceiver Station (BTS) compared with HSDPA/HSUPA are now Radio Link Control (RLC) Layer, Radio Resource Control (RRC) and Packet Data Convergence Protocol (PDCP) functionalities. PDCP was first placed n the core network side but was later (early 2007) moved to EUTRAN (eNodeB). The architecture in Figure 17.2 shows the functional split between Radio Access and Core Network. While the only remaining element of the radio access network is eNodeB, more elements can be used on the core network side. The more comprehensive core network architecture is presented in Figure 17.3.

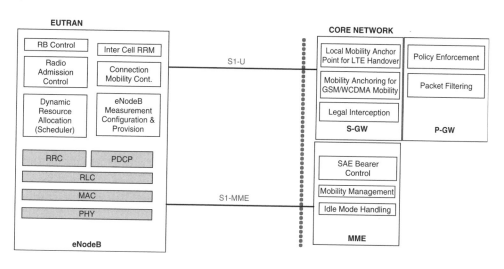

**Figure 17.2** Functional split between radio access and core network

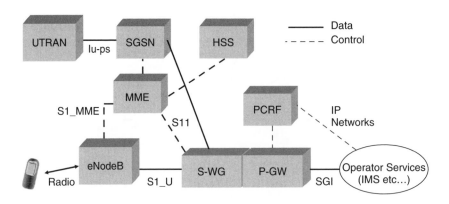

**Figure 17.3**   LTE/EPC architecture

From the radio access network point of view, the important trend was that there is no need for soft handover in the system, as was decided in 3GPP. This is the very same trend that has also been implemented with HSDPA, for example. In HSDPA, only physical layer control information maintains macro-diversity, user data do not. This was one of the enablers to place all radio functionality for the eNodeB, making the flat architecture support easier. As such, it would have been possible to support macro-diversity in a flat architecture as well, but this would have placed additional requirements for the transmissions links between the base stations.

## 17.3   LTE Impact on Network Architecture

On the core network side, additional elements, like registers, are needed as well for a fully functioning system. Figure 17.3 shows the more detailed description of the agreed 3GPP network architecture covering both radio and core network side. For the purposes of handover, there is also an interface between the eNodeBs, called X2. The interface between core and radio access networks is called S1, where S1 is defined in such a way that implementation in the core network side would be possible with having control- (S1_MME) and user-plane (S1_U) traffic processing in separate physical elements, as also presented in Figure 17.3. This facilitates good scalability as the increase in the data rate capabilities is not visible to the elements dealing only with control plane processing.

In the core network the term Evolved Packet Core (EPC) is often used to refer to the new core network to work together with LTE. On the EPC side the following entities have been defined:

- Serving Gateway (S-GW) and Packet Data Network Gateway (P-GW) for processing the user-plane data. These handle tasks related to the mobility management inside LTE, as well as between other 3GPP radio technologies. As shown in Figure 17.3, Serving GPRS Support Node for WCDMA (as well as for GSM) could connect to the gateways defined, thus handling the Gateway GPRS Support Node (GGSN) functionalities of the GSM/WCDMA network.
- Mobility Management Entity (MME) handles the control plane signaling, and especially mobility management and idle-mode handling. The S11 interface then connects MME to S-GW and P-GW if implemented in separate physical elements.
- Home Subscriber Server (HSS) presents the registers, covering functionalities such as the Home Location Register (HLR) and contains, for example, user-specific information on service priorities, data rates, etc.
- Policy and Charging Rules Function (PCRF) is related to quality of service policy as well as the charging policy applied.

The use of a flat network architecture means good scalability for increased data volumes with low dependency for the data volume itself, rather than adding network capacity mainly due to the increased number of users. This enables cost efficiency for both network rollout and capacity extension as the traffic increases.

## 17.4   LTE Multiple Access

As mentioned previously, the 3GPP multiple access is based on the use of SC-FDMA in the uplink direction and OFDMA in the downlink direction. While the basic principles of OFDMA are rather well known from, for example, wireless Local Area Network standards or digital TV standards, such as DVB-T/H, the use of SC-FDMA (with cyclic prefix) [3] represents a more recent technology not widely used in any of the existing systems. In the design, the physical layer parameter details have been chosen in such a way that implementing multimode GSM/WCDMA/LTE devices would be simpler, as well as facilitating the measurements to/from GSM/WCDMA for radio-based handovers to enable seamless mobility. The following sections will look at the SC-FDMA and OFDMA fundamentals and address the physical layer design details on LTE. The issue also worth noting is that the multiple selection is valid for both FDD and TDD modes of LTE, thus making it much simpler to make devices covering them both, as the receivers and transmitters can be the same on both modes with differences then mostly on the RF side.

### 17.4.1   OFDMA Principles

The principle of the OFDMA is based on the use of narrow, mutually orthogonal sub-carriers. In LTE, the sub-carrier spacing is typically 15 kHz regardless of the total transmission bandwidth. Different sub-carriers maintain orthogonality, as at the sampling instant of a single sub-carrier the other sub-carriers have a zero value, as shown in Figure 17.4. The actual transmission is then done by transmitting a signal after the Fast Fourier Transform (FFT) block, which is used to change between the time and frequency domain representation of the signal. The transmitted signal now has the following key properties:

- The symbol duration is clearly longer than, for example, with WCDMA the chip duration and also longer than the channel impulse response; thus, the channel impact is equal to a multiplication by a (complex-valued) scalar.

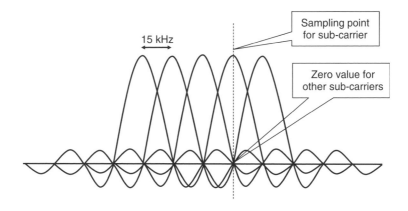

**Figure 17.4**   Adjacent sub-carrier with OFDMA

- There is no inter-symbol interference, as the transmitter uses a guard period (cyclic prefix) longer than the channel impulse response, which is ignored in the receiver and, thus, the effect of the previous symbol is not visible.
- The outcome of an FFT is thus a single signal which is basically a sum of sinusoids and having an amplitude variation that is larger the more sub-carriers have been used as an input to an FFT block.

This kind of signal is ideal from the receiver perspective as one does not need an equalizer but only needs to compensate the channel amplitude and phase impact on the different sub-carriers. On the receiver side one uses again the FFT to convert back from the frequency domain signal to the time domain representation of multiple sub-carriers, as shown in Figure 17.5.

The channel estimation is done based on the known data symbols that need to be placed periodically on parts of the sub-carriers. The equalizer in Figure 17.5 refers to the estimator to cancel out the complex-valued multiplication caused by the frequency-selective fading of the channel and does not present a great complexity. Also, the FFT or Inverse FFT (IFFT) operations are rather old numerical principles for which computationally efficient algorithms have long been developed. The fundamental functional of an FFT block is to transfer the time-domain signal into the frequency domain, representing basically the frequency components from which the time-domain signal is constructed. In the case of OFDMA the parallel inputs to an FFT could be considered as the frequency-domain components which are then converted to a single time-domain signal, thus carrying in a single long symbol typically up to 512, 1024 or 2048 modulated symbols. Use of FFT length of power of 2 is economical from a computational complexity point of view and, thus, is a generally applied principle. The IFFT/FFT operation is illustrated in Figure 17.6, where parallel sub-carriers with their own modulation and modulation order are going to the IFFT block, which is transforming the input to a single output signal and respectively again the receiver FFT block converts the signal back to parallel Quadrature Amplitude Modulation (QAM) symbols. The modulation of the neighbouring sub-carrier may also be different in terms of the modulation method being used. A practical transmitter needs additional blocks for the operation, such as a windowing operation to satisfy the spectrum mask. While the ideal OFDMA signal has nice spectral properties with steep sidelobes, the resulting non-idealities, and

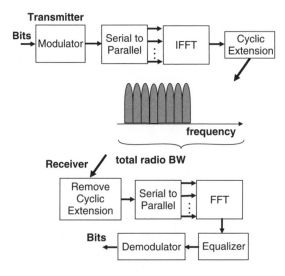

**Figure 17.5**   OFDMA transmitter and receiver diagrams

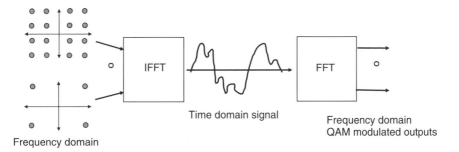

Frequency domain

Time domain signal

Frequency domain
QAM modulated outputs

**Figure 17.6** IFFT/FFT principle

especially clipping the peak amplitudes away, will raise the sidelobes; thus, similar function to the pulse shape filtering of WCDMA is needed.

The use of windowing corresponds to the multiplication of the transmitted signal with the particular shaped window. As shown previously, the practical transmitter, in addition to the windowing functionality, has a cyclic prefix to combat the inter-symbol interference. One could consider the guard interval as a kind of break in the transmission, but actually the waveform is continuous. The transmitter takes part of the waveform to be transmitted and copies part of that to the beginning of the symbol to be transmitted. This part just needs to exceed the channel impulse response duration, as shown in Figure 17.7, and then there is no inter-symbol interference. The receiver will ignore the particular cyclic prefix added. The receiver will actually see only a single symbol, as the channel impulse response is much smaller than the symbol duration, making the channel influence like a finite impulse response filter, and similar multipath components like with the Rake receiver in WCDMA cannot be seen.

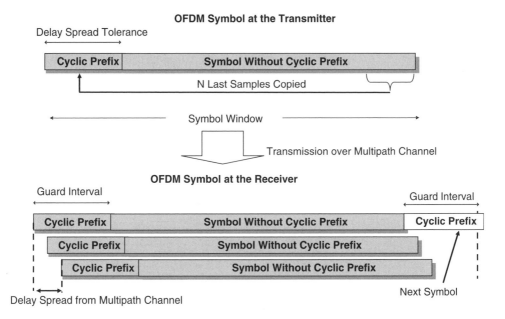

**Figure 17.7** Use of cyclic prefix for removing inter-symbol interference

The important element of OFDMA is that the transmission can be located in different places in the frequency domain, whereas with WCDMA, by definition, the transmission bandwidth was always independent of the information bandwidth. This allows use of the frequency-domain element in the scheduling, as shown in Figure 17.1.

## 17.4.2 SC-FDMA Principles

The SC-FDMA with cyclic prefix has a very simple transmitter structure in the basic form, just a QAM modulator coupled with the addition of the cyclic prefix, as shown in Figure 17.8. The signal bandwidth is thus dependent now on the momentary data rate of the modulator in use and only one (short) data symbol is being transmitted at a time.

The SC-FDMA with a cyclic prefix basically has the following benefits:

- Enabling the low-complexity equalizer receiver by eliminating inter-symbol interference with cyclic prefix.
- But allowing that with low peak to average (PAR) of the signal, as only one information bit is being transmitted at a time; thus, PAR is dominated by the modulation in use.
- Capability to reach a performance similar to OFDMA, assuming an equalizer is being used.

The practical transmitter is likely to take advantage of FFT/IFFT blocks as well placing the transmission in the correct position of the transmit spectrum in case of variable transmission bandwidth, as in LTE. While the maximum transmission bandwidth is up to 20 MHz, the minimum transmission bandwidth is down to 180 kHz, equal to the 12 times 15 kHz sub-carriers in the downlink direction or, rather, one resource block. The different transmitters (terminals) will use the FFT/IFFT pair to place otherwise equal bandwidth transmissions (bandwidth subject to base station scheduler allocations) in the different uplink frequency blocks by adjusting the 'sub-carrier' mapping between FFT and IFFT blocks, as shown in Figure 17.9. This is often referred to in the literature as frequency-domain generalization of the SC-FDMA transmission.

The receiver side is inherently more complicated, on the OFDMA side (for identical performance) as one needs to combat the multipath interference due to the short symbol duration with the use of an equalizer. The use of a cyclic prefix makes this simpler anyway, and as the equalizer is in the base station side, this does not represent any additional burden on the terminals.

The use of frequency-domain generation is very practical when the device needs to change the bandwidth and position of the transmission on the uplink frequency allocation. An example is shown in Figure 17.10, with users having identical transmission bandwidth but using the IFFT/FFT blocks

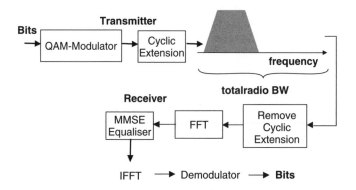

**Figure 17.8**  Simplified SC-FDMA transmitter and receiver chains

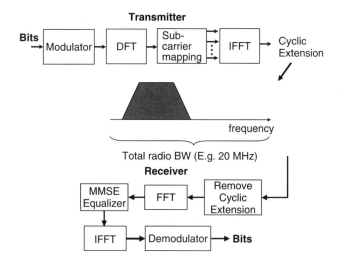

**Figure 17.9**   SC-FDMA with frequency-domain generation

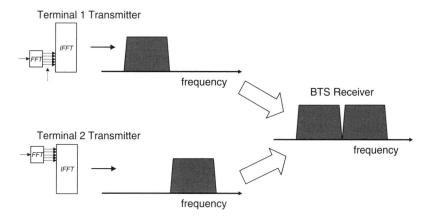

**Figure 17.10**   FDMA multiplexing of users with FFT/IFFT implemented in the transmitter

in the transmitters to place the resource blocks in the correct place. This is not expected to be in the system constant allocation, but as a response to the base station scheduler commands in the downlink. More details of the signaling are covered in the following section on the physical layer structure.

The fact of transmitting only a single symbol at a time ensures a low transmitter waveform, especially compared with the OFDMA case, as can be seen from Figure 17.11. This allows use of a similar amplifier as with the WCDMA case without having to use any over dimensioning or additional power back-off reaching maximum range and with low power consumption and good power conversion efficiency. The resulting PAR/CM impact on the amplifier is thus directly dependent on the modulation, whereas with the OFDMA case it is the number of sub-carriers that dominates the impact, as shown in Figure 17.11. Also pi/2-Binary Phase Shift Keying (BPSK) was originally considered in 3GPP, but as 3GPP performance requirements are such that the full (23 dBm) power level needs to be reached with Quadrature Phase Shift Keying (QPSK), then there are no extra benefits in the use of pi/2-BPSK; thus, this was eventually not included in the specifications.

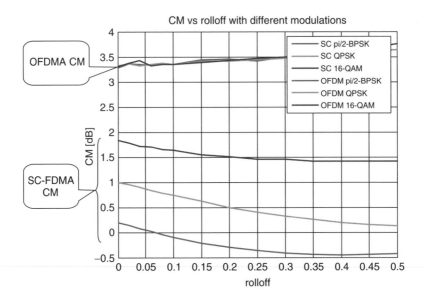

**Figure 17.11**  Cubic metric performance of SC-FDMA and OFDMA [5]

## 17.5   LTE Physical Layer Design and Parameters

The LTE physical layer, like the LTE in general, is designed for maximum efficiency of the packet-based transmission. For this reason there are only shared channels in the physical layer to enable dynamic resource utilization. This section will describe the general principles of the physical layer design and key parameters adopted in LTE.

The fundamental difference from WCDMA is now the use of different bandwidths, from 1.4 MHz up to 20 MHz. The parameters have been chosen so that the FFT lengths and sampling rates are easily obtained for all the operation modes as well as at the same time ensuring the easy implementation of dual-mode devices with a common clock reference. As can be seen from Table 17.1, the sampling rates can be derived easily from the WCDMA chip rate of 3.84 Mcps.

**Table 17.1**  Key parameters for different bandwidths

|  | 1.4 MHz | 3.0 MHz |  | 5 MHz | 10 MHz | 15 MHz | 20 MHz |
|---|---|---|---|---|---|---|---|
| Sub-frame (TTI) (ms) |  |  |  | 1 |  |  |  |
| Sub-carrier spacing (kHz) |  |  |  | 15 |  |  |  |
| Sampling (MHz) | 1.92 | 3.84 |  | 7.68 | 15.36 | 23.04 | 30.72 |
| FFT | 128 | 256 |  | 512 | 1024 | 1536 | 2048 |
| Sub-carriers | 72 + 1 | 180 + 1 |  | 300 + 1 | 600 + 1 | 900 + 1 | 1200 + 1 |
| Symbols per frame |  | 4 with short CP and 6 with long CP |  |  |  |  |  |
| Cyclic prefix |  | 5.21 µs with short CP and 16.67 µs with long CP |  |  |  |  |  |

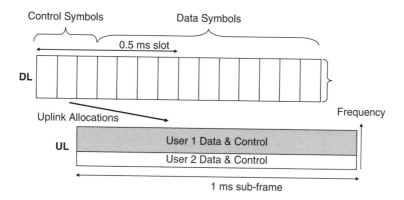

**Figure 17.12**   Downlink and uplink sub-frame structure

For the use of standalone carrier for broadcast purposes, the additional sub-carrier spacing of 7.5 kHz was considered, but was not included in the specification in Release 8 or 9, Release 9 allows Multimedia Broadcast Multicast Service (MBMS) use with shared carrier case based on the existing 15 kHz sub-carrier spacing and other L1 parameters.

Since there is no concept like Dedicated Channel (DCH) in WCDMA, all the resource allocations are short-term allocations, similar to the HSDPA and HSUPA operation. The downlink transmission also contains the control information on the uplink resources to be used, as shown in Figure 17.12. The first symbols, 1 to 3, depending on the need, are used for signaling purposes and rest of the symbols carry the actual data.

With the smallest bandwidth of 1.4 MHz, there can be 2 to 4 control symbols to ensure sufficient amount bits for the downlink control information in order to ensure good reception quality at the cell edge also.

Of the OFDMA symbols having a varying number of sub-carriers depending on the bandwidth, one allocates the resources as resource blocks of 180 kHz, resulting in 12 sub-carriers of 15 kHz each, as illustrated in Figure 17.13. Allocating the resources with such a granularity was necessary in order to limit the signaling overhead. Some of the sub-carriers on some of the symbols will need to carry reference data (pilot symbols) to facilitate the coherent channel estimation in the receiver. Similarly, in the uplink side, reference symbols also need to be embedded in the data stream to allow channel estimation at eNodeB.

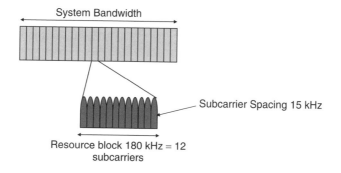

**Figure 17.13**   LTE physical layer resource block

Thus, the resulting data rate for a particular user will depend on:

- Number of resource blocks allocated
- Modulation applied
- Rate of the channel coding
- Whether MIMO is used or not
- Amount of overhead, including whether long or short cyclic prefix is used.

The following physical channels are available for user data:

- Physical Downlink Shared Channel (PDSCH) in the downlink direction. The channel supports QPSK, 16QAM and 64QAM modulations and may use up to 20 MHz bandwidth. The resulting peak data rate, especially with a two-antenna MIMO transmission, can reach up to 170 Mbps, depending on the level of channel coding and the resulting overhead for control. The channel coding for user data is based on the Release 99 WCDMA turbo encoding, but with a new turbo interleaver to facilitate more parallel operations for the actual implementation.
- Physical Uplink Shared Channel (PUSCH) in the uplink direction. The same set of modulations is supported, but use of 64QAM is optional for devices and depending on the device category as shown in Section 16.9. Also, multi-antenna uplink transmission is not specified in Release 8, but is going to be part of the LTE-Advanced Release 10 specifications. In the uplink direction, up to 20 MHz bandwidth may also be used, with the actual transmission bandwidth being a multiple of 180 kHz resource blocks, identical to downlink resource block bandwidth. The channel coding is the same as on the PDSCH. PUSCH may reach up to 60–80 Mbps depending on the modulation.

For signaling purposes and for a cell search operation, in addition, the following channels are specified, as presented in more detail in [4]:

- Physical Uplink Control Channel (PUCCH), covering uplink physical layer signaling needs.
- Physical Random Access Channel (PRACH)
- Physical Downlink Control Channel (PDCCH), covering the downlink physical layer signaling needs.
- Physical Broadcast Channel (PBCH) carries the master information block while actual system information blocks are on the PDSCH.
- Physical Control Format Indicator Channel (PCFICH), control the split between control and data resource allocation within the 1 ms sub-frame.
- Additionally, there are two types of signal defined: synchronization signal, to facilitate cell search (similarity to WCDMA Synchronization Channel), and reference signal for facilitating channel estimation and channel quality estimation taking care of the Common Pilot Channel and pilot symbol functionality in WCDMA.

The channels and structures defined earlier are valid for both LTE FDD and TDD mode of operation. The frame structure has slight differencies on a few aspects:

- The sub-frame when the transmission direction changes between downlink and uplink has a special structure as shown in Figure 17.14, having the Downlink Pilot Time Slots (DwPTS) and Uplink Pilot Time Slot (UpPTS) separated by the Guard Period to ensure time for the equipment to clear RF and change the transmission direction. The selected structure while minimizing FDD and TDD differences, it also faciliates the TD-SCDMA co-existence as discussed in Chapter 18.
- The location of synchronization signals is different than with FDD as the same locations can not be ensured to be always available in the downlink.
- The sounding reference signal (as covered in Section 17.6.2) can be sent additionally in the UpPTS as it cannot carry any data.

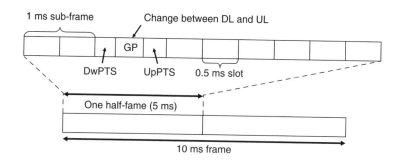

**Figure 17.14** LTE TDD mode frame structure

## 17.6 LTE Physical Layer Procedures

This section covers the key LTE procedures, starting from cell search and random access, data reception and transmission and Channel Quality Information (CQI) feedback and ending with power control and timing advance procedures.

### 17.6.1 Random Access

The same initial problem is valid with LTE as with WCDMA as well. The UE does not have accurate information as to what should be initial transmission power be in order the eNodeB to hear the random access transmission. Thus the similar preamble transmission as with WCDMA takes place with LTE, the UE will transmit the preamble and if not getting response, then sending the preamble again in the next allowed resource space. The key difference to the WCDMA operation, as covered in Chapter 6, is the steps after eNodeB hears the random access preamble. In the case of LTE, the random access channel does not contain any user data but the response from the eNodeB already gives coordinates where to place the following transmission using the PUSCH. Thus the RACH is only for transmission of the preamble and no actual user-specific data is carried in RACH since all user-specific data is on PUSCH. The LTE random access procedure is illustrated in Figure 17.15.

### 17.6.2 Data Reception and Transmission

The LTE physical layer is based on the dynamic nature of resource allocation, with both uplink and downlink resources allocated with 1 ms resolution in time domain and 180 kHz resolution in frequency domain. This is similar to the HSDPA operation where a 2 ms allocation period is being used, but in

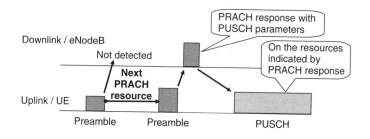

**Figure 17.15** LTE Random Access operation

the HSUPA case there is always some transmission on-going while with the LTE uplink the UE does not transmit anything if not allocated any resources.

The LTE physical layer operation for data reception has the following steps:

1. The scheduler in the eNodeB evaluates for different users, that have data in the buffer, what are the channel conditions, both in time and frequency domain, what are the retransmission needs etc. depending on the vendor specific implementation.
2. For a UE to be served during the next 1 ms TTI, the eNodeB identifies the necessary PDSCH parameters, how many sub-carriers (allocated in 12 sub-carrier physical resource blocks), which modulation can be used (QPSK, 16QAM or 64QAM) and whether MIMO transmission could be used from device category and channel quality point of view.
3. The eNodeB determines the number of devices to be scheduled in total during the 1 ms TTI and needed link adaptation for the signaling, and adjusts the signaling payload between 1 to 3 symbols per frame and sets the PCFICH value accordingly. Then the eNodeB selects a suitable PDCCH allocation that the device will decode (as there is finite amount of PDCCH cases each UE will try to decode) so that there is place for all devices to be scheduled. The PCFICH and PDCCH are in the beginning of the 1 ms TTI, as shown in Figure 17.16 and once the PDCCH allocation ends, then the rest of the symbols are filled with PDSCH data.
4. The UE decodes the set of possible PDCCHs in-line with the parameters defined, with the devices at the cell edge likely to get allocated PDCCH which has more resources used compared to the devices closer to the base station (or with better interference conditions in general). If the UE finds a suitable PDCCH (with CR check result correct after masking with the UE ID), it will receive the PDSCH based on the indication of the PDCCH.
5. The PDCCH parameters allow the device to combine the PDSCH data with correct HARQ process (there is also uplink data transmission related information on PDCCH as covered in the uplink transmission procedure).
6. Having decoded the PDSCH data (with possible combination of the data in the buffer), the UE will provide ACK/NACK for the packet after the 3 ms processing time (in the next available uplink slot in case of TDD).

The LTE physical layer Uplink Data transmission procedure has the following steps:

1. The eNodeB determines based on MAC layer reporting the transmission needs of the UE in the uplink direction.

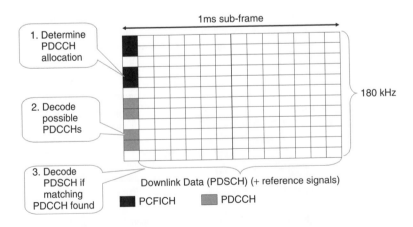

**Figure 17.16**  PCFICH and PDCCH positions within the sub-frame

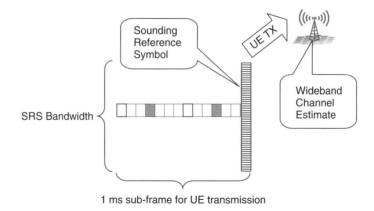

**Figure 17.17**   Sounding Reference Symbol transmission in the uplink direction

2. As the eNodeB needs information where in the frequency domain to place the UE transmission, the UE sends a Sounding Reference Symbol (SRS) which is basically a wideband transmission in the uplink direction where the UE has a reference signal sequence distributed in wider bandwidth, in a comb-like transmission as shown in the example in Figure 17.17. The SRS is sent with even power spectral density over the uplink transmission bandwidth (or part of it) and the eNodeB receiver will then measure the quality of the received SRS transmission at different parts of the uplink transmission bandwidth. The SRS bandwidth is a multiple of 4 PRB, thus as a minimum 720 kHz. Unlike in Figure 17.17, the SRS is not necessary overlapping the part of the spectrum used by the UE for actual data transmission in the corresponding sub-frame.

3. The scheduler will optimize the different transmissions and place the uplink transmission information for the PUSCH on the DPCCH space which the UE is decoding.

4. When the UE decodes the PDCCH with matching CRC, it will fill the n times 180 kHz allocation with the PUSCH and PUSCH transmission for the next 1 ms period.

5. The eNodeB will decode the PUCCH which has the necessary information (new data indicator) to ensure that combining is done with the correct data in the eNodeB buffer and otherwise for the data detection.

6. The eNodeB needs similarly to schedule the packet retransmissions in the uplink direction, as only PUCCH can be sent in a non-scheduled fashion in predetermined resources, for example, to request uplink transmission resources.

From the LTE uplink operation there is a fundamental difference to the HSUPA operation as there is no user data transmission taking place in the uplink unless permitted by the eNodeB. With HSUPA there is always some transmission on-going (excluding discontinuous slots or part of them with continuous packet connectivity or due to inter-frequency/inter-system measurements).

## 17.6.3   CQI Procedure

The CQI has basically the same interpretation as with HSDPA; it tells the eNodeB what is the data rate the UE believes it could receive successfully based on the current conditions experienced. The reported data rate, with an example table shown in Table 17.2, is based on the measurements of the signal strength and quality and other metrics, such as the UE receiver type, that can affect the situation on which the CQI is generated. The new element is to accommodate the frequency domain aspect of the scheduling. Thus the UE also states in which part of the spectrum the data reception would be most successful. This works especially with larger bandwidths, with the 1.4 MHz case there is

**Table 17.2**   LTE CQI table

| CQI value | Code rate × 1 024 | Modulation |
|-----------|-------------------|------------|
| 0         | N/A               | N/A        |
| 1         | 78                | QPSK       |
| 2         | 120               | QPSK       |
| 3         | 193               | QPSK       |
| 4         | 308               | QPSK       |
| 5         | 449               | QPSK       |
| 6         | 602               | QPSK       |
| 7         | 378               | 16-QAM     |
| 8         | 490               | 16-QAM     |
| 9         | 616               | 16-QAM     |
| 10        | 466               | 64-QAM     |
| 11        | 567               | 64-QAM     |
| 12        | 666               | 64-QAM     |
| 13        | 772               | 64-QAM     |
| 14        | 873               | 64-QAM     |
| 15        | 948               | 64-QAM     |

no need to send frequency selective CQI feedback as conditions do not vary that greatly over such a small bandwidth. Depending on the network parameterization, the UE is providing the feedback either for the predefined set of sub-bands or then the UE may select the best sub-bands to be reported. The added dimension, while allowing better performance compared to HSDPA, also increases the testing complexity for the CQI operation in the devices. The CQI feedback behaves as with HSDPA, while close to the base station with a good interference situation, a CQI value which indicates the use of 64QAM with high coding rate is likely, while at the cell edge the UE ends up reporting QPSK modulation with a lower code rate to have lot of protection from the channel coding to ensure successful data decoding.

## 17.6.4   Downlink Transmission Modes

The LTE downlink has the following seven transmission modes defined in Release 8:

1. Single-antenna port; port 0. This is the simplest mode of operation with no precoding and single antenna transmission only.
2. Transmit diversity. Transmit diversity with two or four antenna ports using space-frequency block coding, a similar approach to the open loop WCDMA transmit diversity described in Chapter 6.
3. Open-loop spatial multiplexing. This is an open loop mode with the possibility of rank adaptation (to select whether a single or multiple data stream is transmitted) based on the rank indication feedback. If rank = 1, transmit diversity is applied similar to transmission mode 2. With a higher rank, spatial multiplexing with up to four layers with large delay Cyclic Delay Diversity (CDD) is used.
4. Closed-loop spatial multiplexing. This is a spatial multiplexing mode with precoding feedback supporting dynamic rank adaptation.
5. Multi-user MIMO. Transmission mode for downlink Multi-User MIMO operation, where more than one user is sharing the same frequency and time resources.
6. Closed-loop Rank = 1 precoding. Closed loop precoding similar to transmission mode 5 without the possibility to do spatial multiplexing, i.e. the rank is fixed to one.

7. Single-antenna port; port 5. This mode can be used in a beamforming operation when UE specific reference signals are in use. This mode has a UE capability bit since mode 7 is not foreseen to be used in the first deployments in an FDD operation. In the case of a TDD operation this is expected to be used especially if deployed as the evolution solution to the TD-SCDMA. This is the only mode in Release 8 which uses the UE-specific reference signals while other transmission modes rely on the use of common reference signals.

In Release 9, there is additional mode, Mode 8 has been added. The purpose of the mode 8 is basically to combine the MIMO operation with the beamforming operation, thus enabling a dual stream operation with UE-specific reference signals. Thus combining the use of beamforming antenna array with a multi-stream transmission, including Multi-User MIMO operation.

## 17.6.5  Uplink Transmission Modes

LTE does not use multi-antenna transmission in the uplink direction, this will be available only from Release 10 onwards as part of the LTE-Advanced specification work. It is possible, however, to use multi-user MIMO (or virtual MIMO) in the uplink direction. This means that an eNodeB allocates users orthogonal references symbols and allocates them in the same resource space and then processes the transmission from two users like a MIMO transmission in order to separate the two independent data streams. In this case the uplink data rate of an individual UE is unchanged.

## 17.6.6  LTE Physical Layer Compared to WCDMA

The LTE physical layer, despite being based on different multiple access solution, is built on the experience of several WCDMA/HSPA Releases. The first LTE release covers from the start aspects known from particularly HSDPA and HSUPA, such as physical layer packet combining (HARQ), base station-based scheduling and link adaptation as well as support for multi-antenna downlink transmission. A comparison of the physical layer aspects in high level is presented in Table 17.3. In addition to the topics discussed earlier, the power control operation is not as critical in LTE due to the orthogonal uplink resources, thus a slow power control operation is used instead of a fast closed loop as in a WCDMA uplink.

## 17.7  LTE Protocols

The key trend with the LTE protocol design has been to put all radio-related functionalities on the BTS site, in the eNodeB. This follows the trend introduced in HSDPA and HSUPA, where Medium Access Control (MAC) layer functionality was added to the Node B. Now, further RLC functionality has been moved to the eNodeB, as well as the PDCP functionality with ciphering and header compression.

**Table 17.3**  LTE and WCDMA physical layer comparison

| Feature | LTE | HSUPA | HSDPA |
|---------|-----|-------|-------|
| Multiple access | OFDMA SC-FDMA | WCDMA | WCDMA |
| Bandwidth range | 1.4–20 MHz | 5–10 MHz | 5–10 MHz |
| Fast power control | No | Yes | No (associated DCH only) |
| Soft handover | No | Yes | No (associated DCH only) |
| Adaptive modulation | Yes | Yes | Yes |
| BTS-based scheduling | Time/Freq | Time/Code | Time/Code |
| Fast L1 HARQ | Yes | Yes | Yes |

*Note*: The physical layer specifications are 36.211 to 36.214; the latest versions are available from www.3gpp.org.

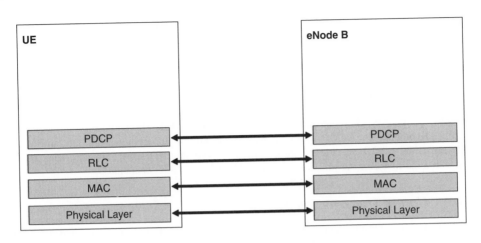

**Figure 17.18**  LTE user plane protocol stack

Thus, all the radio-related protocol layers previously in UTRAN divided between NodeB and RNC are now handled just between UE and eNodeB in LTE, with the control-plane protocol stack as shown in Figure 17.18.

In LTE, the MAC layer handles similar functionalities as with, for example, the HSDPA, including the following:

- Scheduling
- Priority handling
- Retransmissions
- Multiplexing of different logical channels on a single transport channel.

The RLC handles, as in WCDMA, the following functions:

- Retransmissions in case lower layer (MAC and L1) fails delivery, as in the case of acknowledged mode of operation of RLC in UTRAN side.
- Segmentation to fit for the protocol data units.
- Providing logical channels to higher layers.

For the PDCP the following is left, as indicated in Figure 17.19:

- Ciphering
- Header compression.

Also PDCP is located in the eNodeB, including the ciphering. This makes the functional grouping of LTE radio functionalities similar to the HSPA evolution architecture, as covered in Chapter 15. For the UE-specific control signaling, the PDCP also covers the security aspect, including ciphering and integrity check, thus unlike WCDMA, PDCP is part of the control plane protocol stack also.

In the control plane, the role of the RRC protocol (also in eNodeB) is similar to that on the UTRAN side. The RRC protocol configures the connection parameters, controls the terminal measurement reporting, carries the handover commands, etc. The same ASN.1 coding will be used for the LTE RRC specification as well, which allows extensions between different releases in a backwards-compatible way. The RRC protocol will contain less states than UTRAN; rather, only 'active' and 'idle' states

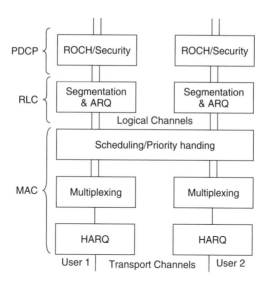

**Figure 17.19**   Distribution of functionality between MAC, RLC and PCCP layers

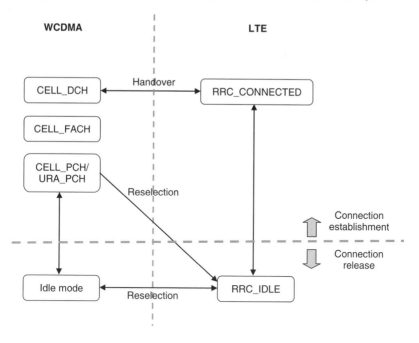

**Figure 17.20**   LTE and WCDMA RRC protocol states

are used in LTE due to the dynamic nature of the resource allocations. The two RRC states in LTE are named, and as shown in Figure 17.20 and described as follows:

• RRC_IDLE state, during which the device will monitor paging messages and use cell re-reselection for mobility. There is no RRC context being stored in any individual eNode B; the UE only has ID which identifies it in the tracking area.

- RRC_CONNECTED state, during which the UE location is known on the cell level and data can be transmitted and received. There is an existing RRC connection to an eNode B. Handovers controlled by the network are used for mobility. Handover is supported to both WCDMA and GSM to ensure service continuity (subject to data rate limitations obviously) when moving out of the LTE coverage area.

The control-plane functional split is shown in Figure 17.21, indicating that all radio-related control-plane signaling terminates in eNodeB, as well as showing the inter eNodeB interface, X2. The X2 interface in needed especially for inter eNodeB handover. Note that, as soft handover does not exist in LTE, there is no need to keep data running over the X2 interface continuously, but the interface may be used for some data forwarding in connection with the handover process. Non-Access Stratum signaling refers to the signaling between the core network and the UE.

There are further procedures than just handover (or reselection) between LTE and WCDMA. As the LTE is not connected to the CS core, there is additional means defined for when a device cannot obtain a desired service via the IMS connected to the LTE via EPC. Such an operation is referred to as the circuit switched fallback (CS fallback) handover. In the example for the mobile terminated case, the multimode UE gets the paging from MRC via the MME and following the location update to MCS, the UE does the handover to WCDMA and then performs paging response and CS call set-up via WCDMA network, as shown in Figure 17.22. CS fallback can use also GSM or cdma2000 network

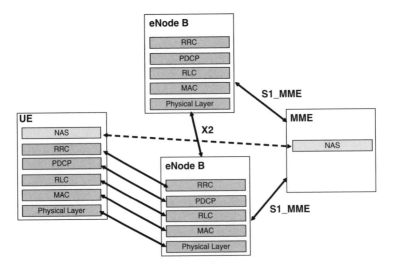

**Figure 17.21** LTE control-plane protocol stack

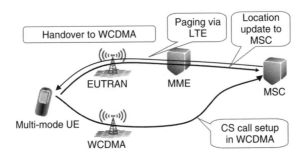

**Figure 17.22** CS fallback handover to WCDMA

depending on the multimode device capabilities. The use of CS fallback is an interim solution before full scale Voice Over IP (VoIP) implementation in LTE networks. Once the VoIP has been implemented, there is still a specific handover procedure, known as Single Radio Voice Call Continuity (SR-VCC) that allows handing over a VoIP call in LTE to GSM or WCDMA to be operated there as a CS voice call. In the case of WCDMA network, it is also possible to hand over the VoIP call to be continued as a VoIP call in WCDMA.

## 17.8 Performance

### 17.8.1 Peak Bit Rates

LTE provides high peak bit rates by using large bandwidth up to 20 MHz, high-order 64QAM and multistream MIMO transmission. The downlink peak bit rates can be calculated with Equation (17.1). QPSK modulation carries 2 bits per symbol, 16QAM 4 bits and 64QAM 6 bits. And $2 \times 2$ MIMO further doubles the peak bit rate. Therefore, QPSK 1/2 rate coding carries 1 bps/Hz, and 64QAM without any coding and with $2 \times 2$ MIMO carries 12 bps/Hz. The bandwidth is included in the calculation by taking the corresponding number of sub-carriers for each bandwidth option: 72 per 1.4 MHz and 180 per 3.0 MHz bandwidth. For the bandwidths 5 MHz, 10 MHz and 20 MHz, there are assumed 300, 600 and 1200 sub-carriers respectively. We assume 13 data symbols per 1 ms sub-frame. The achievable peak bit rates are shown in Table 17.4. The highest theoretical data rate is approximately 170 Mbps. If a $4 \times 4$ MIMO option is applied, then the theoretical peak data rate would double to 340 Mbps.

$$\text{Peak bit rate [Mbps]} = \frac{\text{bits}}{\text{Hz}} \times \text{Number of sub-carriers}$$

$$\times \frac{\text{number of symbols per sub-frame}}{1 \text{ ms}} \quad (17.1)$$

The uplink peak data rates are shown in Table 17.5: up to 86 Mbps with 64QAM and up to 57 Mbps with 16QAM. The peak rates are lower in uplink than in downlink since single-user MIMO is not specified in uplink. MIMO can be used in uplink as well to increase cell data rates, not single-user peak data rates. The LTE targets of 100 Mbps in downlink and 50 Mbps in uplink are clearly met.

### 17.8.2 Spectral Efficiency

The LTE spectral efficiency can be estimated in a number of different ways: using system-level simulation tools, using the Shannon formula combined with the macro-cellular interference distribution

**Table 17.4** Downlink peak bit rates

| Modulation coding | | Peak bit rate per sub-carrier/bandwidth combination | | | | |
|---|---|---|---|---|---|---|
| | | 72/1.4 MHz | 180/3.0 MHz | 300/5.0 MHz | 600/10 MHz | 1200/20 MHz |
| QPSK 1/2 | Single stream | 0.7 | 2.1 | 3.5 | 7.0 | 14.1 |
| 16QAM 1/2 | Single stream | 1.4 | 4.1 | 7.0 | 14.1 | 28.3 |
| 16QAM 3/4 | Single stream | 2.2 | 6.2 | 10.5 | 21.1 | 42.4 |
| 64QAM 3/4 | Single stream | 3.3 | 9.3 | 15.7 | 31.7 | 63.6 |
| 64QAM 4/4 | Single stream | 4.3 | 12.4 | 21.0 | 42.3 | 84.0 |
| 64QAM 3/4 | $2 \times 2$ MIMO | 6.6 | 18.9 | 31.7 | 64.3 | 129.1 |
| 64QAM 4/4 | $2 \times 2$ MIMO | 8.8 | 25.3 | 42.5 | 85.7 | 172.1 |

**Table 17.5** Uplink peak bit rates

| Modulation coding | | Peak bit rate per sub-carrier/bandwidth combination | | | | |
|---|---|---|---|---|---|---|
| | | 72/1.4 MHz | 180/3.0 MHz | 300/5.0 MHz | 600/10 MHz | 1200/20 MHz |
| QPSK 1/2 | Single stream | 0.7 | 2.0 | 3.5 | 7.1 | 14.3 |
| 16QAM 1/2 | Single stream | 1.4 | 4.0 | 6.9 | 14.1 | 28.5 |
| 16QAM 3/4 | Single stream | 2.2 | 6.0 | 10.4 | 21.2 | 42.8 |
| 16QAM 4/4 | Single stream | 2.9 | 8.1 | 13.8 | 28.2 | 57.0 |
| 64QAM 3/4 | Single stream | 3.2 | 9.1 | 15.6 | 31.8 | 64.2 |
| 64QAM 4/4 | Single stream | 4.3 | 12.1 | 20.7 | 42.3 | 85.5 |

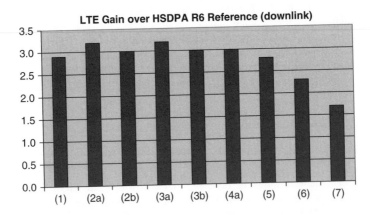

**Figure 17.23** Relative spectral efficiency of LTE compared with HSPA R6

or by estimating the expected relative capacity gain over HSPA radio. In this section we present the expected capacity gain over HSPA based on system simulations. The capacity estimates based on the Shannon formula can be found from [6].

The relative spectral efficiency of LTE Release 8 compared with HSPA Release 6 is presented in Figure 17.23. The different bars represent the simulated values from the different companies [2]. The gain of LTE is up to three times compared with HSPA Release 6, which was the original target of LTE.

Also, HSPA has further evolution in Release 7, improving the spectral efficiency beyond the Release 6 reference case used in 3GPP benchmarking. The main improvements are mobile equalizer receiver and 2 × 2 MIMO scheme. HSPA evolution is discussed in Chapter 15.

There are clear reasons why LTE improves the spectral efficiency compared with HSPA. LTE can exploit such capacity enhancement techniques that are not possible in HSPA or give only limited benefits in HSPA. The OFDMA modulation with frequency-domain scheduling keeps the users orthogonal in LTE. The CDMA transmission in HSPA causes some intra-cell interference in the multipath channel. Part of the intra-cell interference can be removed by the equalizer or by the interference cancellation, but there is still gain provided by OFDMA. The orthogonality is obtained because of long symbols and cyclic prefix in OFDMA. There is no interference between the consecutive symbols as long as the delay spread is shorter than the cyclic prefix. The typical delay spread is a few microseconds, whereas

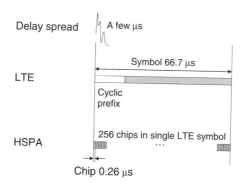

**Figure 17.24**  Principal of orthogonality

the cyclic prefix is clearly longer 5 or 17 μs in LTE. The HSPA chip is just 0.26 μs and the multipath propagation causes inter-chip interference. The HSPA chip and LTE symbol lengths are illustrated in Figure 17.24. On the other hand, the cyclic prefix takes some part of the capacity: 7.5% in the case of shorter cyclic prefix.

The fast fading is frequency dependent and the typical coherence bandwidth of the signal in macro-cells is in the order of 1 MHz. Within the LTE carrier bandwidth of up to 20 MHz there are some frequencies that are faded and some frequencies that are not faded. In the case of CDMA modulation in HSPA the signal is spread over the whole transmission bandwidth. The spreading gives frequency diversity gain by averaging over the different fading conditions. The spreading, however, is not the most efficient way of transmission. Ideally, the transmission should be done only using those frequencies that are not faded. This is possible in OFDMA in LTE. The transmission can be scheduled by resource blocks, each 180 kHz. The scheduling is based on the mobile channel quality reporting that indicates the most favorable resource blocks. Then the scheduler uses the most favorable resources not only in the time domain (as with HSPA), but also in the frequency domain, as illustrated in Figure 17.25.

The interference rejection combining or interference cancellation in OFDMA can operate per narrowband sub-carrier. Those algorithms tend to perform better with narrowband sub-carriers, compared with the wideband HSPA carrier.

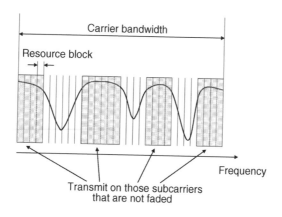

**Figure 17.25**  Principle of frequency domain scheduling

## 17.8.3   Link Budget and Coverage

The LTE link budget for uplink is presented in Table 17.6 for a 64 kbps data rate and two resource block allocation, giving 360 kHz transmission bandwidth. The terminal power is assumed 23 dBm and no body loss is included for the data connection. The base station receiver assumes a radio-frequency (RF) noise figure of 2.0 dB. The required signal-to-interference-and-noise ratio (SINR) value is taken from link simulations. The receiver sensitivity becomes 123.4 dBm without interference. We add 2.0 dB interference margin. It is assumed that the cable loss is compensated by the masthead amplifier (MHA), or, alternatively, the RF head is installed closed to the antenna. The three-sector macro-cell antenna gain is 18 dBi for 65-degree antennas. The maximum allowed path loss between mobile and base station antenna then becomes 162.4 dB.

The corresponding LTE downlink link budget is presented in Table 17.7 for a 1 Mbps data rate assuming antenna diversity in the terminal and 10 MHz bandwidth. The base station power is assumed 46 dBm. The mobile noise figure is assumed 7 dB. The required SINR value is taken from link simulations. The receiver sensitivity becomes −107.5 dBm without interference. We add 3.0 dB interference margin and include 1.0 dB for the control channel overhead. The maximum allowed path loss between base station and mobile antenna becomes 165.5 dB.

The maximum path loss values can be compared with the HSDPA/HSUPA link budgets in Chapter 12. The LTE link budgets show up to 3 dB higher values than HSDPA/HSUPA, which is mainly due to lower interference margins with orthogonal modulation.

The mapping of the path loss values to the cell range depends heavily on the system deployment assumption. We assume below maximum path loss of 163 dB in four different scenarios. Urban and suburban cases assume indoor coverage for mobile users, whereas rural assumes just outdoor coverage and rural fixed assumes directive antenna on rooftop for the fixed wireless case. The assumptions are presented in Table 17.8.

The corresponding cell ranges are shown in Figure 17.26. Note that the $y$-axis has a logarithmic scale. The cell range is shown for 900, 1800, 2100 and 2500 MHz frequency variants. The urban cell range varies from 0.7 to 1.5 km and suburban from 1.6 to 3.6 km. Such cell ranges are also typically found in existing GSM and UMTS networks. The rural case shows clearly higher cell ranges: 27 km

**Table 17.6**   Uplink link budget for 64 kbps with dual-antenna receiver base station

|   | Data rate (kbps) | 64 | |
|---|---|---|---|
| | *Transmitter – UE* | | |
| a | Max. TX power (dBm) | 23.0 | |
| b | TX antenna gain (dBi) | 0.0 | |
| c | Body loss (dB) | 0.0 | |
| d | EIRP (dBm) | **24.0** | = a + b + c |
| | *Receiver – Node B* | | |
| e | Node B noise figure (dB) | 2.0 | |
| f | Thermal noise (dBm) | −118.4 | = k (Boltzmann) × T (290 K) × B (360 kHz) |
| g | Receiver noise floor (dBm) | −116.4 | = e + f |
| h | SINR (dB) | −7.0 | From simulations |
| i | Receiver sensitivity (dBm) | −123.4 | = g + h |
| j | Interference margin (dB) | 2.0 | |
| k | Cable loss (dB) | 2.0 | |
| l | RX antenna gain (dBi) | 18.0 | |
| m | MHA gain (dB) | 2.0 | |
| | *Maximum path loss* | **162.4** | = d − i − j − k + l − m |

**Table 17.7**  Downlink link budget for 1 Mbps with dual-antenna receiver terminal

| | Data rate (Mbps) | 1 | |
|---|---|---|---|
| *Transmitter – Node B* | | | |
| a | HS-DSCH power (dBm) | 46.0 | |
| b | TX antenna gain (dBi) | 18.0 | |
| c | Cable loss (dB) | 2.0 | |
| d | EIRP (dBm) | **62.0** | $= a + b + c$ |
| *Receiver – UE* | | | |
| e | UE noise figure (dB) | 7.0 | |
| f | Thermal noise (dBm) | −104.5 | $= k$ (Boltzmann) $\times$ T (290 K) $\times$ B (9 MHz) |
| g | Receiver noise floor (dBm) | −97.5 | $= e + f$ |
| h | SINR (dB) | −10.0 | From simulations |
| i | Receiver sensitivity (dBm) | −107.5 | $= g + h$ |
| j | Interference margin (dB) | 3.0 | |
| k | Control channel overhead (dB) | 1.0 | |
| l | RX antenna gain (dBi) | 0.0 | |
| m | Body loss (dB) | 0.0 | |
| *Maximum path loss* | | **165.5** | $= d − i − j − k + l − m$ |

**Table 17.8**  Assumptions for cell range calculations

| Okumura–Hata parameter | Urban indoor | Suburban indoor | Rural outdoor | Rural outdoor fixed |
|---|---|---|---|---|
| Base station antenna height (m) | 30 | 50 | 80 | 80 |
| Mobile antenna height (m) | 1.5 | 1.5 | 1.5 | 5 |
| Mobile antenna gain (dBi) | 0.0 | 0.0 | 0.0 | 5.0 |
| Slow fading standard deviation (dB) | 8.0 | 8.0 | 8.0 | 8.0 |
| Location probability (%) | 95 | 95 | 95 | 95 |
| Correction factor (dB) | 0 | −5 | −15 | −15 |
| Indoor loss (dB) | 20 | 15 | 0 | 0 |
| Slow fading margin (dB) | 8.8 | 8.8 | 8.8 | 8.8 |

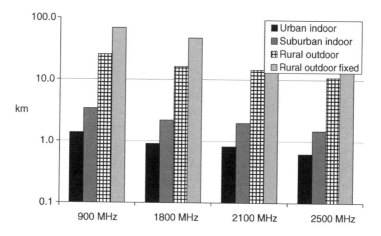

**Figure 17.26**  Cell range with the maximum path loss of 164 dB

for the outdoor mobile coverage and even beyond 70 km for the rural fixed installation. We need to note here that the Earth's curvature limits the maximum cell range to approximately 40 km with 80 m base station antenna height.

## 17.9    LTE Device Categories

The UE capability is based on five different categories, as shown in Table 17.9. These categories are valid for Releases 8 and 9, with some changes coming inevitably due to the LTE Advanced development as covered in Section 17.10 and also some discussion has been on-going in 3GPP whether the step between categories 1 and 2 is too big to be competitive with HSDPA device classes. Furthermore, support for RX diversity (and MIMO) with small devices with support for low frequency bands (below 1 GHz) has been also addressed, since the practical diversity antennas with a small UE form factor provide very limited gain at frequencies below 1 GHz.

## 17.10    LTE-Advanced Outlook

The work on LTE-Advanced is part of Release 10 work, with the work for Release 10 focusing on the following topics:

- Carrier aggregation work is defining the methods how to extend the LTE operation for bandwidths beyond 20 MHz, covering also the use of non-contiguous allocations, similar to the work done in Release 9 for dual-band HSDPA. The carrier bandwidth is kept at 20 MHz and larger bandwidths are implemented by aggregating the necessary number of 20 MHz (or smaller) carriers. This also allows Releases 8 and 9 devices to access the system when keeping the strcture for the 20 MHz (or below) carriers unchanged at least for those parts visible for the legacy devices. The carrier aggregation within a frequency band and between frequency bands is illustrated in Figure 17.27, with the European 800 MHz and 2600 MHz chosen as an example. The combination of frequency bands becomes important as typically an operator has at most 20 MHz on a given frequency band.
- Relay support, with relays being considered from the UE point of view as separate base stations. From the network perspective the relays are under the control of another eNodeB. The relays have either in-band or out-band backhaul connection to a 'regular' eNodeB (donor cell) including their own HARQ operation etc. Typical use is. for example, as illustrated in Figure 17.28, where an indoor UE has a weak link to the actual eNodeB while a Relay Node (RN) has a better link from the close proximity of the building to the UE and thus could use higher data rates to provide high

**Table 17.9**    LTE UE capabilities

| UE Category | 1 | 2 | 3 | 4 | 5 |
| --- | --- | --- | --- | --- | --- |
| Uplink Peak data rate | 5 Mbps | 25 Mbps | 50 Mbps | 50 Mbps | 75 Mbps |
| Downlink Peak data rate | 10 Mbps | 50 Mbps | 100 Mbps | 150 Mbps | 300 Mbps |
| Highest Downlink modulation | 64QAM | 64QAM | 64QAM | 64QAM | 64QAM |
| Highest Upnlink modulation | 16QAM | 16QAM | 16QAM | 16QAM | 64QAM |
| RX Diversity | Yes | Yes | Yes | Yes | Yes |
| Downlink MIMO support | Optional | Yes ($2 \times 2$) | Yes ($2 \times 2$) | Yes ($2 \times 2$) | Yes ($4 \times 4$) |
| Maximum RF bandwidth | 20 MHx | 20 MHx | 20 MHx | 20 MHx | 20 MHx |

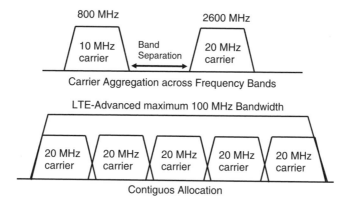

Figure 17.27   LTE Advanced carrier aggregation

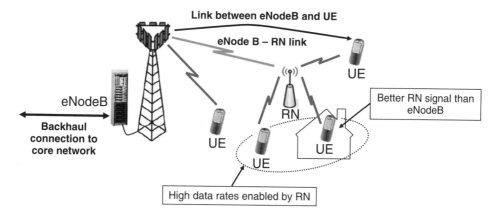

**Figure 17.28**   Relay Node use case with LTE-Advanced

data rate support to UE indoors. Some of the sub-frames are then configured so that the UE is not expecting any transmission so that the relay node can get the data from the eNodeB since obviously the relay cannot transmit and receive simultaneously on the same carrier.

- Uplink MIMO is introducing the UE multi-antenna, multi-stream transmission to enable higher data rates in the uplink direction.
- Downlink multi-antenna enhancements are enhancing the Releases 8 and 9 multi-antenna transmission capabilities, with the inclusion of topics like 8 antenna transmission support in the downlink direction for LTE-Advanced.

Additionally, many topics are being worked on for Release 10, including support for new frequency bands, Minimization of Drive Tests (MDT) as well as further enhancements for the Self Optimizing Networks (SON) functionalities. Especially the MDT and SON functionalities are intended to cut the network operation costs in reducing the need for manual tuning and testing activites in network optimization. Further information on LTE-Advanced activity can be found in [7] until the Release 10 specifications become available towards the end of 2010.

## 17.11    Conclusion

The 3GPP work on LTE aimed for specification availability for the end of 2007 and that was also reached with the actual protocol specification finalization (freeze) in March 2009. Now there is a new member in the 3GPP radio access family designed to fulfill the needs for high data rates and performance going beyond HSDPA and HSUPA evolution. The LTE design is such that integration with existing GSM and WCDMA deployments for seamless coverage offering is facilitated. The novel uplink technology ensures a power-efficient transmitter for the device transmission and maximizes the uplink range as well. The LTE uses a flat architecture and is optimized for packet-switched-only services. The LTE performance, together with a flat architecture, ensures low cost per bit for a competitive service offering for end users.

## References

[1] 3GPP technical Report TR 25.913, Version 2.1.0. 'Requirements for Evolved UTRA and UTRAN', June 2005.
[2] 3GPP Technical Report 25.814, Physical Layer Aspects of Evolved Universal Terrestrial Radio Access (UTRA), Version 7.1.0, September 2006.
[3] Czylwik, A. 'Comparison between Adaptive OFDM and Single Carrier Modulation with Frequency Domain Equalization', *IEEE Vehicular Technology Conference 1997, VTC-97*, Phoenix, pp. 863–869.
[4] Holma, H. and Toskala, A., *LTE for UMTS*, New York: John Wiley & Sons, Ltd, 2009.
[5] Toskala, A., Holma, H., Pajukoski, K. and Tiirola, E., 'Utran Long Term Evolution in 3GPP', *IEEE 17th International Symposium on Personal, Indoor and Mobile Radio Communications*, September 2006.
[6] Mogensen, P., Na, W., Kovacs, I., Frederiksen, F., Pokhariyal, A., Pedersen, K., Kolding, T., Hugl, K. and Kuusela, M., 'LTE Capacity Compared to the Shannon Bound', *VTC 2007*, Spring 2007.
[7] 3GPP Technical Report TR 36.912, Version 9.1.0, Feasibility Study for Further Advancements for E-UTRA (LTE-Advanced), December 2009.

# 18

# TD-SCDMA

Antti Toskala and Harri Holma

## 18.1   Introduction

From the start the use of the Time Division Duplex (TDD) operation was part of the Universal
Terrestrial Radio Access (UTRA) operation to deal with the unpaired spectrum allocations. In Release
99, the 3.84 Mcps TDD mode was included together with the first WCDMA release in the specifications. For the next release, Release 4, the 1.28 Mcps TDD mode was added, known more widely
as TD-SCDMA. The motivation for the adoption for a smaller chip rate was the expected equipment
complexity especially related to the use of adaptive antenna and joint processing solutions on the
network side and thus it was intended to achieve low enough sampling rates together with a low
enough chip rate. Later on still another TDD mode with 7.68 Mcps was added. While the 3.84 Mcps
and 7.68 Mcps TDD modes failed to achieve significant market deployment, the 1.28 Mcps TDD, or
TD-SCDMA, based on the Release 5 version with 3 carrier operation, is widely used and is partly
under construction in China. The commercial operation during 2009 is following a few years of a
test period and the issuing of 3G licenses in China in early 2009. TD-SCDMA is covered mainly in
Release 5 and later versions of the 3GPP 25.221 to 25.225 specifications [1–5]. Only the physical layer
specifications are specific to TD-SCDMA (or TDD modes in general), the higher layer specifications
are the same as in FDD. This chapter will focus in covering the TD-SCDMA operation. First, TDD
operation as a duplex method is introduced in Section 18.1 and Section 18.2 covers the radio access
network architecture differences between TD-SCDMA and WCDMA. The physical layer, including
the physical channels of TD-SCDMA, is introduced in Section 18.3 and Section 18.4 discusses the
achievable data rates and UE classes. Section 18.5 covers the TD-SCDMA physical layer procedures,
including the HSDPA and HSUPA operation principles with TD-SCDMA. Section 18.6 covers the
key interference issues, including also TD-SCDMA and TD-LTE (LTE TDD mode co-existence). The
background of UTRA TDD was covered in Chapter 4.

### 18.1.1   TDD

Three different duplex transmission methods are used in telecommunications: FDD, TDD and space
division duplex (SDD). The FDD method is the most common duplex method in cellular systems.

*WCDMA for UMTS: HSPA Evolution and LTE, Fifth Edition*   Edited by Harri Holma and Antti Toskala
© 2010 John Wiley & Sons, Ltd

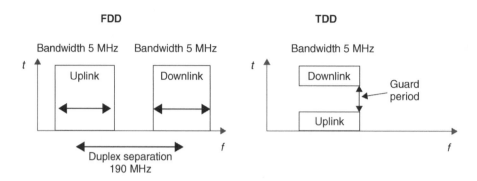

**Figure 18.1**   Principles of FDD and TDD operations

The FDD method is used, for example, in Global System for Mobile Communications (GSM) and with the Wideband Code Division Multiple Access (WCDMA) terminals system. The FDD method requires separate frequency bands for both uplink and downlink. The TDD method uses the same frequency band but alternates the transmission direction in time domain. TDD is used, for example, in Wireless LAN (WLAN) applications. or cordless telephone systems such as DECT. The SDD method is used, for example, in cellular backhaul operation with directional antennas in connection with microwave links. Also the use of beamforming techniques with FDD or TDD can be considered an SDD application as well. The fundamental principle of the FDD and TDD operation is shown in Figure 18.1, with the share of the TDD carrier in the time domain visible.

In both the FDD and TDD modes of operation, the terms uplink (or reverse link) and downlink (or forward link) are used to refer to the terminal or base station transmissions respectively. The TDD operation has the following characteristics which are not part of the FDD operation:

- *Discontinuous operation* of the power amplifier both in the terminal and the base station. As the same spectrum is shared, the transmitter side needs to be shut down while trying to receive in the same spectrum. To avoid interference leaking in the transmitter to the receiver side, one typically needs short guard times to allow the signal to 'die' from the transmitter RF parts before reception should occur.
- *Interference between the uplink and downlink.* As now the devices connected to the different base stations may end up transmitting while a nearby device may end up trying to receive on the same frequency, one can experience interference between the uplink and the downlink. Respectively the similar situation may arise between two base stations which are operating on the same carrier. Besides the same carrier, this may be an issue for other nearby carriers if the frequency separation is not large enough. For this reason, the practical TDD networks need to be operated with time synchronization between sites. The uplink/downlink interference is further analysed in Chapter 17.
- *Asymmetric spectrum use.* With a TDD system, one can adjust the division between uplink and downlink thus allocating more capacity to the transmission direction which needs more throughput. In practical networks this would need to be aligned between nearby cells due to the interference between the uplink and downlink.
- *No duplex spacing.* As the TDD system only uses a single frequency, there are no separate uplink and downlink frequencies with duplex spacing. Respectively in a TDD implementation one can proceed without the duplex filter.
- *Reciprocal channel.* While with FDD the fading in uplink and downlink directions is not correlated due to the frequency (duplex) separation, in the case of TDD there is more correlation due to the use of the same spectrum part for transmission and reception. This allows a more open loop type

of operation in many procedures such as power control or dealing with adaptive antenna solutions. However, if we aim to adjust the phase of the transmitted signal based on the received signal, it would require the same number of antenna branches on both transmitter and receiver and also needs some consideration for the RF chain calibration.

## 18.2   Differences in the Network-Level Architecture

TD-SCDMA differs from the FDD mode operation for those issues that arise from a different Layer 1. The protocols have been prepared following the principle that there are typically common messages, but the information elements then are specific to the mode (either FDD or TDD) being used. Use of the Iur interface is not needed for soft handover purposes with TDD, as only one Node B is transmitting for one user. If the TDD system uses relocation, then there is no need to transfer user data over the Iur interface in practice. Figure 18.2 illustrates the TDD key differences compared with the architecture in Chapter 5. The core network does not typically show any difference, except that from the achievable data rates. The highest data rate offered in the market with the TD-SCDMA devices at the moment (Spring 2010) seems to be 2.8 Mbps for the downlink direction. In practice, TDD networks are very few and none of the operators seem to running both at the moment in the same country. From the Radio Network Controller (RNC) point of view, the biggest issue to cope with is the Radio Resource Management (RRM), which in TDD is based on different measurements and resource allocation principles along with the use of Dynamic Channel Allocation (DCA).

## 18.3   TD-SCDMA Physical Layer

The TD-SCDMA physical layer is aligned with some parts (like channel coding) with WCDMA as covered in Chapter 6, but for the details of the physical channels in many cases only the name is the same. Also as the applied multiple access principle is a combination of TDMA and CDMA, a device chip set made for WCDMA is useless for TD-SCDMA consideration (and vice versa). The chip rate of 1.28 Mcps is thus harmonized, being 1/3 of the WCDMA chip 3.84 Mcps chip rate, and

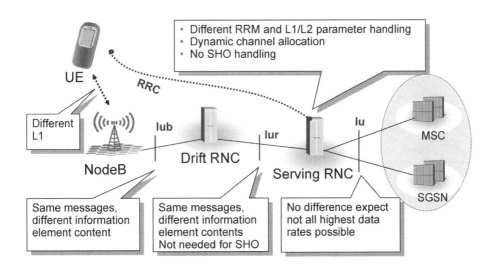

**Figure 18.2**   Difference between TD-SCDMA and WCDMA in the radio access network

in order to enable use of the same clocking reference for both, there would be need to be a dual mode (TD-SCDMA/WCDMA) device. There are also TD-SCDMA-specific procedures for the physical layer operation as described in Section 18.5.

## 18.3.1   Transport and Physical Channels

The following are the transport channels for TD-SCDMA in Release 5:

- Dedicated Channel (DCH)
- Broadcast Channel (BCH)
- Paging Channel (PCH)
- Forward Access Channel (FACH)
- Random Access Channel (RACH)
- High Speed Downlink Shared Channel (HS-DSCH)
- Downlink Shared Channel (DSCH)
- Uplink Shared Channel uplink (USCH).

The most notable change in Release 5, following the trials done based on Release 4 of TD-SCDMA, was the modification of the broadcast channel structure. To ensure operation at the cell edge as well, the original single carrier operation was modified to be the multi-frequency operation where three carriers are always used. The data can be on any of the carriers (but all channels for one user are only on one of the carriers) while the Broadcast Channel (BCH) is mapped into slot 0 in the primary frequency. So in this respect the data has basically reused 1 while the BCH (or the Physical Common Control Channel, PCCCH) is reusing 3 as illustrated in Figure 18.3.

Due to the late start of the commercial operation, the operation started with Release 5 devices and thus the downlink data was taken care of with the HSDPA features from the start. Thus in the description below also the HSDPA-related channels are covered.

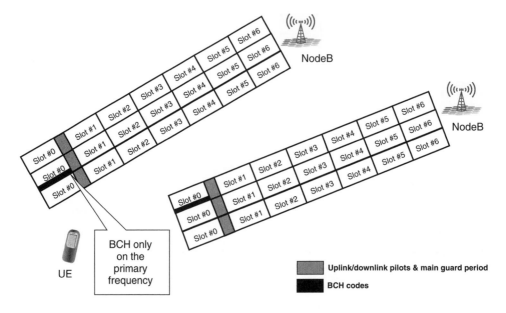

**Figure 18.3**   TD-SCDMA BCH operation in multi-frequency cells

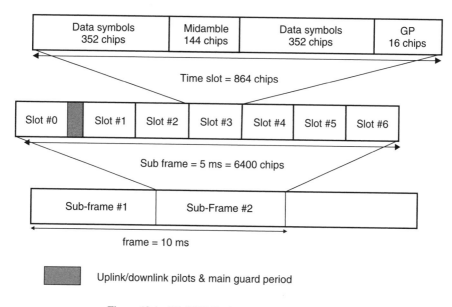

**Figure 18.4** TD-SCDMA time slot and frame structure

In Release 7 High Speed Uplink Packet Access (HSUPA) functionalities were added, again in the 3GPP specifications the term HSUPA is not used but the transport channel term is Enhanced DCH (E-DCH), as in WCDMA specifications.

The multiple access principle in TD-SCDMA allocates to a particular user and channel, a combination of codes in one or more time slots. Thus, a given physical layer (uncoded) data rate capability is constructed from the number of codes, the spreading factor, the time slots and the modulation method applied. The common pilot channel is not used as in the WCDMA, but the channel estimation in the downlink (and uplink) is done based on the midamble (training sequence). An example structure is shown in Figure 18.4. The midambles and the slot structure are addressed further in Section 18.3.3.

The TD-SCDMA physical channels are the following:

- Dedicated Physical Channel (DPCH), which carries the DCH. Similar to WCDMA, circuit-switched traffic is (before Release 8) normally carried on a DCH one, such as AMR speech calls or CS video calls.
- Primary Common Control Physical Channel (P-CCPCH) carries the BCH. P-CCPCH is always mapped onto two codes in slot 0 with spreading factor of 16. This enables the devices to find the channel after the cell search operation. As shown in Figure 18.3, only the primary frequency of multi-frequency operation carriers allows the P-CCPCH. This allows sufficient reliability also at the cell edge for the channel as the channel coding is not too strong.
- Secondary Common Control Physical Channel (S-CCPCH) carries the PCH and FACH. The S-CCPCH is also with spreading factor 16 but not with a fixed code allocation.
- Fast Physical Access Channel (FPACH) is used in TD-SCDMA to respond to the random access attempt (a bit similar to the AICH in WCDMA) but it provides more information than just a signature acknowledgment. The FPACH, with spreading factor 16, provides also timing and power level adjustment information for the device.
- Physical Random Access Channel (PRACH) carries the RACH. PRACH uses spreading factors 4, 8 or 16, depending on the system parameters. The PRACH uses a time shifter version of the same training sequence for all users.

- The synchronization channels are located in connection with the downlink to the uplink switching point after time slot 0, as shown in Figure 18.5. The downlink synchronization channel, DwPCH, is equal to the DwPTS and the uplink synchronization channel, UpPCH, is equal to the UpPTS in Figure 18.5.
- High Speed Physical Downlink Shared Channel (HS-PDSCH) carries the HS-DSCH and is dynamically mapped to one or more code channels, similar to the WCDMA HSDPA principle, as guided by the Shared Control Channel for HS-DSCCH (HS-SCCH). This allows dynamically frame by frame allocation of the physical channel resources in the downlink between the users that actually have transmission needs.
- High Speed Shared Control Channel (HS-SCCH) carries the same information as WCDMA, informing the receiver for example which codes to receive and which slot format (modulation and spreading factor) is being used. Also the momentary data rate (transport block size) as well as information whether the packet is a new packet for the HARQ process or not.
- Downlink Shared Channel (DSCH) is mapped onto the Physical Downlink Shared Channel (PDSCH) which is an older channel allowing also resource sharing but is not as advanced as HS-PDSCH. DSCH was originally developed to avoid slot shortage with packet data use but has in practice been replaced by the HS-DSCH use.
- Uplink Shared Channel uplink (USCH) is also a less dynamic channel (compared to the HSUPA operation of WCDMA in Release 6 or with TD-SCDMA in Release 7) for uplink resource sharing. USCH is mapped onto the Physical Uplink Shared Channel.
- Page Indicator Channel (PICH) carries the paging indicators. The key design principle is to allow devices to be first in only a short period of time to catch the paging indicator and then to allow the devices rapidly to decode the paging message to see if the message was intended for that particular device or not (since the paging indicators are corresponding to the paging group and not individual devices).
- Shared Information Channel for HS-DSCH (HS-SICH) carries the HSDPA operation-related control information in the uplink direction, including the Channel Quality Information (CQI).

In addition to the channel described, TD-SCDMA has been evolving beyond the first commercial release, Release 5. Release 7 was the next big step with the introduction of the High Speed Uplink Packet Access for TD-SCDMA as well. The principle follows those of WCDMA Release 6. The key principle has been with TD-SCDMA also to react faster to the bursty nature of the uplink transmission needs with faster resource allocation. Also the physical layer packet has been enabled and is available in the downlink direction.

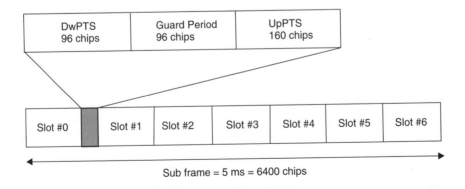

**Figure 18.5**  TD-SCDMA subframe structure

After the Release 5 version of the TD-SCDMA specifications, the following new physical channels were added:

- MBMS-related channels, to enable the similar MBMS functionality as with WCDMA as covered in Chapter 14.
- Physical Layer Common Control Channel (PLCCH) is used to carry power control and timing control information to multiple UEs. The idea behind is basically the same as with WCDMA with Fractional DPCH (see Chapter 15), to avoid allocation of a DCH in the downlink to transmit necessary power control information. Following the TD-SCDMA characteristics, now just the information also needed for all UEs, the timing control information, has been added, together with the power control information. Only spreading factor 16 is used and one slot format, no other information is carried on the PLCCH other than TPC and SS bits.
- E-DCH Physical Uplink Channel (E-PUCH) is applied to implement the HSUPA functionality with TD-SCDMA, as introduced in the Release 7 version of the 3GPP specifications. The E-PUCH operation also adopted the 16QAM modulation which was added in the WCDMA HSUPA operation in Release 7.
- The E-DCH Uplink Control Channel (E-UCCH) which is providing information in the uplink direction on the momentary data rate, transport block size and other necessary information to enable correct demodulation and packet combining in Node B.
- E-DCH Random Access Uplink Control Channel (E-RUCCH) carries uplink control information when E-PUCH resources are not available, to enable UE to enable the device to request uplink resources when transmission needs occur.
- E-DCH Absolute Grant Channel (E-AGCH) is carrying in the downlink direction the uplink (slot/code) allocation from Node B to devices using E-DCH.
- E-DCH Hybrid ARQ Acknowledgement Indicator Channel (E-HICH) follows the WCDMA principle in carrying feedback whether a packet has been decoded correctly or not by the BTS. Additionally, however, it has the functionality of sending the TPC and SS bits for the devices with non-scheduled packet transmission.

## 18.3.2    Modulation and Spreading

The modulation schemes with TD-SCDMA differ slightly from the WCDMA ones. For the Downlink Dedicated Physical Channel, there is in addition to the Quadrature Phase Shift Keying (QPSK) also 8PSK, which was added in the 1.28 Mcps TDD to enable the theoretical 2 Mbps peak rate to be reached. The spreading codes have the lengths 1, 2, 4, 8 or 16, but in the downlink only values 1 (no spreading) and 16 are used. After spreading there is chip-by-chip multiplication for scrambling. With the HS-PDSCH there is additionally 16QAM modulation (Release 5) and then also later 64QAM was added, but 8PSK is not used with HS-DPSCH.

With E-PUCH (HSUPA) operation in the uplink direction, QPSK and 16QAM modulation are used, 8PSK is not applied as the link adaptation steps are sufficient already without use of that with a combination of different modulation and coding rates.

## 18.3.3    Physical Channel Structures, Slot and Frame Format

As was shown in Figure 18.4, TD-SCDMA has 7 slots per 5 ms sub-frame and thus 14 slots within the 10 ms frame. This was due to the fact that numerology would not work even if using the same 15 slots as with WCDMA, the approach that was used with 3.84 Mcps and 7.68 Mcps TDD modes.

Within the sub-frame, the first slot is always allocated in the downlink direction and the second on the uplink direction. The rest of the slots would then be configured to be divided between uplink and downlink, depending on the expected transmission needs in the coverage area. When considering the multi-carrier transmission, all the carriers obviously need to have the same timing and uplink/downlink split to avoid interference between transmit direction (as discussed further in detail in Section 18.6).

The lack of continuous uplink allocation affects the link budget of TD-SCDMA and TDD systems in general. If using a similar power amplifier as WCDMA, there is less total output power available (peak power is the same but the aggregated energy during the frame is less) when compared to the continuous operation of the power amplifier. When compared with a WCDMA operation, the biggest difference is with small data rates as they have allocation typically only in one of the slots, and thus in such a case the average power is reduced by 10 x log(1/14), this roughly 11.5 dB reduction would correspond roughly to half the cell range, requiring four times as many base stations for the similar low data rate coverage. If the coverage target is a higher data rate, the difference reduces with more time slots activated in the uplink. One could of course consider using a higher output power amplifier, but then with the higher data rates, one would need to reduce the maximum transmit power, resulting in higher terminal power consumption. Thus dimensioning the power amplifier only based on the one slot operation is not seen as a practical approach. This fundamental difference explains the operator's roll-out strategy: if the FDD spectrum is available, it is better to use that first and then later in the capacity enhancement phase to consider the TDD option. Typically an operator also has only $1 \times 5$ MHz of unpaired spectrum (if considering the 1900–1920 MHz) compared to $2 \times 10$ MHz or $2 \times 15$ MHz for the paired spectrum.

### 18.3.3.1   Burst and Slot Structure

While 3.48 Mcps and 7.68 Mcps had multiple burst types, with TD-SCDMA there is only one burst format. As shown in Figure 18.6, there is always the constant length of 144 chips midamble (training sequence) and 352 chips data parts on both sides of the midamble. If additional physical layer control information is used in the slot type, it is placed on either side of the midamble. In addition, there is a 16 bit Guard Period (GP). The slot structure then depends additionally on modulation and the spreading factor, and then determines the actual data rate for a particular time slot. There are multiple slot structures due to the varying allocations:

- allocation of spreading factor and applied modulation;
- allocation of Transport Format Combination Indicator (TFCI), which indicates the parameter combination used to actually inform the receiver of the data rate and to enable explicit decoding of the data rate used by the transmitter;
- allocation of Transmission Power Control (TPC) bits for the purpose of the uplink power control. Even though TD-SCDMA has fewer users active simultaneously than WCDMA, it is still important to keep the power differences between the users small to limit the inter-user interference and inter-cell interference.
- use of the Synchronization Signal (SS) which is used to adjust the uplink timing of a particular time slot so that differences are aligned and retain as much power as possible from the orthogonality in the uplink direction.

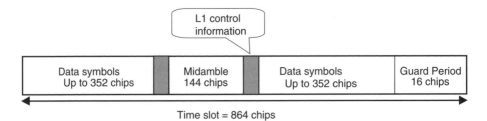

**Figure 18.6**   Place of the L1 control information in the slot

There are 24 time slot formats in the downlink and 69 in the uplink when considering only QPSK modulation. In the downlink, less time slot formats are needed as the spreading factor is always either 1 or 16, while in the uplink direction the intermediate steps are also possible to avoid unnecessary multi-code transmission.

As mentioned earlier, the midamble used is always 144 chips. Both the transmitter and receiver know which midamble belongs to which user and that information is used to facilitate channel estimation and joint detection (in the uplink) to separate different users. There is no further spreading or scrambling applied on top of the midamble, as is done for the data parts. The midambles used for different users are time shifted versions of the same basic midamble code. This allows one cyclic correlator to provide the channel estimates for all users. Interference between the cells (for channel estimation) is mitigated by having different basic midamble code users in neighboring cells.

With HS-PDSCH there are only four slot formats needed as the physical layer control information is carried on HS-SCCH. Thus, the only varying information is whether QPSK or 16QAM is used and whether the spreading factor is 1 or 16.

With E-PUCH there are again a lot of different slot formats, in Release 7 up to 69 different slot formats are defined to cover all possible combinations in terms of spreading factors, presence of different physical layer control fields and modulation applied.

### 18.3.3.2   PRACH

The Physical Random Access Channel (PRACH) can use three different spreading factors and thus three different uplink slot structures. The BCH will indicate which resources are available for the PRACH use, enabling scaling of the PRACH resources depending on the traffic needs.

### 18.3.3.3   SCH

In the 1.28 Mcps TDD, the downlink pilots, as indicated in Figure 18.5, contain the necessary synchronization information. The Downlink Pilot Channel (DwPCH) is transmitted in each 5 ms sub-frame over the whole coverage area. The pattern used on the 64 chips' information can have 32 different downlink synchronization codes.

### 18.3.3.4   Common Control Channels

Once synchronization has been achieved, the timing and coding of the Primary Common Control Physical Channel (P-CCPCH) are known. The BCH is mapped onto the P-CCPCH and it contains the necessary information to access the system, such as the PRACH parameters. As was mentioned earlier in Section 18.3.1, the P-CCPCH is transmitted only on the primary frequency.

The same principle is valid for the Secondary Common Control Physical Channel (S-CCPCH), which carries the Paging Channel (PCH) and the Forward Access Channel (FACH).

### 18.3.3.5   Shared Channels

While the TD-SCDMA specification contains the shared channels, the Downlink Shared Channel (DSCH) and the Uplink Shared Channel (USCH), the real interest and evolution are focused on the HSDPA channels (which are also deployed from the start of the network operation) and on the channels related to the HSUPA operation. While the DSCH and the USCH are like the DCH, they are allocated by a higher layer on a temporary basis to avoid resource shortages in the code and slot domains. Now with the introduction of HSDPA and later HSUPA channels, this brings a more dynamic allocation coupled with the link adaptation and fast feedback and retransmissions to increase not only resource allocation efficiency but also spectral efficiency of the operation.

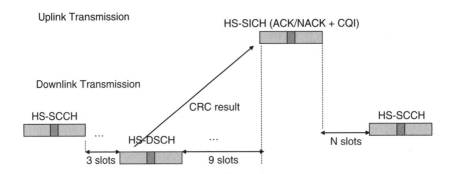

**Figure 18.7** HSDPA timing with TD-SCDMA

The HS-PDSCH uses those earlier mentioned four slots formats and thus in the case of spreading factor 16, either 88 or 176 uncoded bits are transmitted per code, depending on whether the QPSK or 16QAM modulation is in use. Adding more parallel codes, and using more slots, constructs the desired data rate. Another alternative to more parallel codes is to use spreading factor 1 (i.e. no spreading) which then permits 1408 or 2816 bits per slot depending on whether QPSK or 16QAM is in use. With Release 8, the use of 64QAM modulation was added (as well as the possibility of sending the TPC and SS bits on HS-PDSCH as well). With 64QAM, the data rate is not doubled any more than 16QAM, but for example, with a single code with spreading factor 16, the number of (uncoded) bits per slot is changed from 176 to 264.

The High Speed Shared Control Channel (HS-SCCH) is used to inform the UE of the data transmission on HS-PDSCH, giving the UE three slots time to decode the HS-SCCH and determine if HS-PDSCH slots are to be decoded or not, as shown in Figure 18.7.

With the HSUPA operation, the E-PUCH carries the data and the E-UCCH is used to carry information about the transport block size (including the modulation applied) as weak as the information of the HARQ process and re-transmission sequence number to avoid buffer corruption in case of a signaling error (and to indicate the redundancy version). The packet combining follows the same principle as with WCDMA as covered in Chapters 12 and 13 for HSDPA and HSUPA respectively.

### 18.3.3.6 TD-SCDMA Transmit Diversity

Similar to WCDMA, there are both open and closed loop transmit diversity methods defined for TD-SCDMA. The use of open loop transmit diversity is defined for most of the channels, while the closed loop transmit diversity is used for the DCH and HS-PDSCH and related control downlink channels. In Release 8, also the Multiple Input Multiple Output (MIMO) operation was introduced to the TD-.SCDMA specifications. The MIMO operation defined the support for the dual stream operation, when the UE would receive two independent data stream from different antennas, as introduced in Chapter 15 for the WCDMA operation.

## 18.4 TD-SCDMA Data Rates

The 1.28 Mcps TDD resulting data rate is around 8 kbps with one slot per sub-frame (two slots per 10 ms) and spreading factor 16 and the use of QPSK. The 69 different uplink formats with QPSK and the 24 different downlink formats can be used to build a particular data rate. The practical approach

**Figure 18.8**   Example TD-SCDMA terminal

is to use the DCH channels only for voice and then implement the data in the downlink direction with the HS-PDSCH channels. With the support of a single carrier operation, which at the end of 2009 seems to be the solution being used in devices, the downlink peak data rate supported by the TD-SCDMA devices is 2.8 Mbps. An example TD-SCDMA device, Nokia 6788, is shown in Figure 18.8 with HSDPA support.

With the multi-frequency operation (with three carriers) the Release 5 peak downlink data rate becomes (with HSDPA and 16QAM) 14 Mbps, which could be achieved by allocating 30 time slots across three carriers and not having channel coding overhead. This is similar to the WCDMA HSDPA peak data rate with 15 codes allocated and using 16QAM with such rate matching parameters that there is basically no overhead for channel coding. Both WCDMA and TD-SCDMA UE capability parameters are covered in [15].

## 18.5   TD-SCDMA Physical Layer Procedures

### 18.5.1   Power Control

In TD-SCDMA the power control is not in all cases as critical as in WCDMA, as there is also the TDMA dimension, which especially with high data rates can be used to keep uplink users orthogonal (if a device has not enough resources to fill the full slot). On many occasions, however, users have to share the same slots with the CDMA principle and then the power control becomes very important to ensure successful detection at the Node B receiver. In the theoretical case of ideal interference cancellation, the need for power control would be reduced, but practical implementation has limitations including a limited dynamic range for the Analog-to-Digital (AD) converters, thus power control also eases the receiver implementation. Both in the downlink and uplink directions, the use of power control allows minimization of the inter-cell interference. The power control parameters are shown in Table 18.1.

### 18.5.2   TD-SCDMA Receiver

In WCDMA, the basic receiver was a Rake receiver as introduced in Chapter 4. In TD-SCDMA the power control is less frequently compared to WCDMA and thus not that accurate, so advanced

**Table 18.1** Mcps TDD power control characteristics

| | Uplink | Downlink |
|---|---|---|
| Method | Initially open loop and then SIR based inner loop (for some control channel only open loop) | SIR based inner loop |
| Rate | Closed loop: 0–200 Hz | |
| | Open loop: variable delay depending on slot allocation | 0–200 Hz |
| Closed Loop Step sizes | 1, 2, 3 dB | 1, 2, 3 dB |

receiver technology is needed to separate the users from each other to avoid the near–far problem. Since the number of users within a time slot would be at most 16, the receiver complexity is simpler compared to a corresponding receiver on the WCDMA side. The use of smaller chip rate also reduced the needed processing power. In the uplink direction the receiver requirements are more demanding as there users have more power differences and each has a unique multipath propagation, while in the downlink direction more code orthogonality is preserved for the transmission coming from a single source. The use of interference cancellation or joint detection has been studied a lot in the literature and the algorithms are well known, with examples given in [6–14]. The receiver technologies use different approaches to tackle both the Inter-Symbol Interference (ISI) and interference between users, the latter known also as Multiple Access Interference (MAI). The performance requirements set in 3GPP for the physical layer operation were made assuming use of advanced receivers in Node B and UE.

## 18.5.3  Uplink Synchronization

In TD-SCDMA, the timing advance as with high chip rate modes is not used. Uplink synchronization is used to aim to reduce uplink interference. The principle is to have users in the uplink sharing the same scrambling code and to have uplink transmission partly orthogonal by coordinating the uplink TX timing with closed-loop timing control and small 1/5–1/8 chip resolution. For this purpose the slot structure contained the space for Synchronization Signals (SS) on different channels to enable provision of the uplink timing control, regardless whether dedicated or shared channels are being used. The timing control is a relative timing compared to the timing used previously, similar to closed loop power control. Different timing control step resolutions were allowed in order not to force the use of any particular sampling rate in the device implementation. With the use of 1.28 Mcps it is also a bit easier to achieve better uplink synchronization than with higher chip rates. Uplink synchronization was also considered for WCDMA but was eventually not adopted in 3GPP specifications.

## 18.5.4  Dynamic Channel Allocation

The TD-SCDMA operation, especially in the multi-frequency case, offers the possibility of the Dynamic Channel Allocation (DCA) in multiple domains as shown in Figure 18.9. This is important in dealing with the channel allocation as there is no soft handover available:

- In the frequency domain, the RNC can place the device on the carrier with least traffic and/or interference, as typically the 3 carrier deployment would be used. All the services of the UE need to be then on this single carrier.

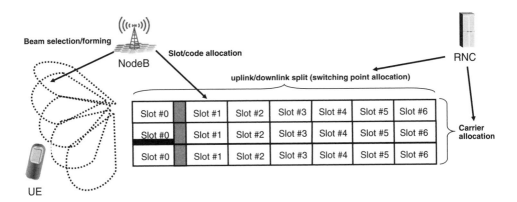

**Figure 18.9**  DCA functionality division between TD-SCDMA Node B and RNC

- In the time and code domain, which with HSDPA operation is part of the base station scheduler, for the DCH (or USCH/DSCH) the case is based on the slower RNC-based allocation. With the HSDPA operation, the PS data on HS-PDSCH is subject to link adaptation which makes it more robust to inter-cell interference as well, similar to the HSDPA operation with WCDMA, while avoiding sending data when the interference situation is really bad. Also with the introduction of HSUPA in Release 7, the Node B-based resource allocation is then enabled also in the uplink direction with TD-SCDMA.
- In the spacial domain, with the use of smart antennas (beamforming). In the TD-SCDMA network in China, beamforming antennas are actually in use, while with WCDMA no wide-scale deployments have been made.

## 18.5.5   Summary of the TD-SCDMA Physical Layer Operation

The physical layer differences between WCDMA and TD-SCDMA can be then summarized, as given in Table 18.2. The best alignment has been reached in the area of channel encoding where both use the same turbo encoding and the same rate of convolutional encoders, but due to the differences in design in slot structure and multiple access principles, the majority of the physical layer details are different. This makes reusing hardware between different systems not a trivial issue.

**Table 18.2**  Summary of the physical layer differences between TD-SCDMA and WCDMA

|  | TD-SCDMA | WCDMA |
| --- | --- | --- |
| Multiple access method | TDMA, CDMA (inherent FDMA) | CDMA (inherent FDMA) |
| Duplex method | TDD | FDD |
| Multirate concept | Multicode, multislot and orthogonal variable spreading factor (OVSF) | Multicode and OVSF |
| Detection | Coherent, based on midamble | Coherent, based on pilot symbols |
| Dedicated channel power control | Closed loop; rate $\leq 200\,\text{Hz}$ | Fast closed loop; rate $= 1500\,\text{Hz}$ |
| Intra-frequency handover | Hard handover | Soft handover |
| Channel allocation | Timeslot and code allocation | Code allocation |
| Intra-cell interference cancellation | Support for joint detection | Support for advanced receivers at base station |
| Spreading factors | 1–16 | 2–256 (512) |

## 18.6    TD-SCDMA Interference and Co-existence Considerations

The use of the same frequency for uplink and downlink also creates additional interference cases, especially the interference between base stations and the interference between the UEs, as indicated in Figure 18.10. The basic interference cases for the adjacent channel operation do exist with TD-SCDMA as well, similar to the WCDMA operation but the following section looks at the TDD specific cases that may occur within a TDD system and between FDD and TDD operations.

### 18.6.1    TDD–TDD Interference

Since both uplink and downlink share the same frequency in TDD, those two transmission directions can interfere with each other. By nature the TDD system is synchronous, and this kind of interference occurs if the base stations are not synchronized. It is also present if different asymmetry is used between the uplink and downlink in adjacent cells, even if the base stations are frame synchronized. Frame synchronization requires an accuracy of a few symbols, not an accuracy of chips. The guard period allows more tolerance in synchronization requirements. Figure 18.11 illustrates possible interference scenarios. The interference within the TDD band is analyzed with system simulations in [16]. Interference between uplink and downlink can also occur between adjacent carriers. Therefore, it can also take place between two operators.

In FDD operation, duplex separation prevents interference between uplink and downlink. The interference between a mobile and a base station is the same in both TDD and FDD operation and is not considered in this chapter.

#### 18.6.1.1    Mobile Station to Mobile Station Interference

Mobile-to-mobile interference occurs if the other mobile in Figure 18.11 is transmitting and the other one is receiving simultaneously in the same (or adjacent) frequency in adjacent cells. This type of interference is statistical because the locations of the mobiles cannot be controlled. Therefore, it cannot be avoided by network planning. Intra-operator mobile-to-mobile interference occurs especially at cell borders. Inter-operator interference between mobiles can occur anywhere where two operators' mobiles are close to each other and transmitting on fairly high power. Methods to counter mobile-to-mobile interference are:

- DCA and RRM
- Power control.

#### 18.6.1.2    Base Station-to-Base Station Interference

Base station-to-base station interference occurs if the TDD base station in Figure 18.11 is transmitting and the FDD base station is receiving in the same (or adjacent) frequency in adjacent cells. It

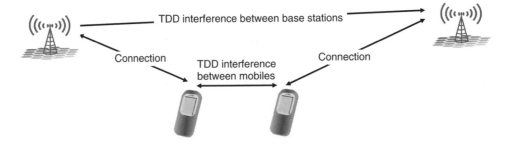

**Figure 18.10**    TDD specific interference cases

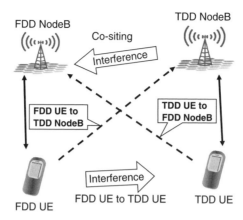

**Figure 18.11**  Interference cases between FDD and TDD operations

depends heavily on the path loss between the two base stations and, therefore, can be controlled by network planning.

Intra-operator interference between base stations depends on the base station locations. Interference between base stations can be especially strong if the path loss is low between the base stations. Such cases could occur, for example, in a macro-cell, if the base stations are located on masts above rooftops. The best way to avoid this interference is by careful planning to provide sufficient coupling loss between base stations.

The outage probabilities in [16] show that cooperation between TDD operators in network planning is required, or the networks need to be synchronized and the same asymmetry needs to be applied. Sharing base station sites between operators will be very problematic, if not impossible. The situation would change if operators had inter-network synchronization and identical uplink/downlink splits in their systems.

From the synchronization and coordination point of view, the higher the transmission power levels and the larger the intended coverage area, the more difficult will be the coordination for interference management. In particular, the locations of antennas of the macro-cell type tend to result in line-of-sight connections between base stations, causing strong interference. Operating TDD in indoor and micro/pico-cell environments will mean lower power levels and will reduce the problems illustrated.

## 18.6.2   TDD and FDD Co-existence

In Region 1 (including Europe) the FDD and TDD modes of operation have spectrum allocations that meet at the border at 1920 MHz; therefore TDD and FDD deployment cannot be considered independently, see Figure 18.12. DCA can be used to avoid TDD–TDD interference, but DCA is not effective between TDD and FDD, since FDD has continuous transmission and reception. The possible interference scenarios between TDD and FDD are summarized in Figure 18.11.

### 18.6.2.1   Co-Siting of UTRA FDD and TDD Base Stations

From the network deployment perspective, the co-siting of FDD and TDD base stations looks an interesting alternative. There are, however, problems due to the close proximity of the frequency bands. The lower TDD band, i.e. 1900–1920 MHz, is located adjacent to the FDD uplink band, i.e. 1920–1980 MHz. The resulting filtering requirements in TDD base stations are expected to be such that co-siting a TDD base station in the 1900–1920 MHz band with an FDD base station is not

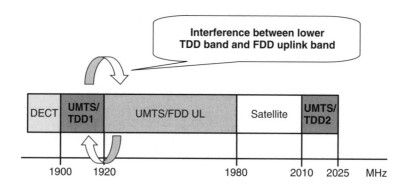

**Figure 18.12** Interference situation with the band edge at 1920 MHz in Region 1

considered technically and commercially a viable solution, with such an example studied in Chapter 14 for the use of TDD band for broadcast use.

The micro- and pico-cell environments change the situation, since the TDD base station power level will be reduced to as low as 24 dBm in small pico-cells and re-using the same antenna structure is not as critical as with the macro environment. On the other hand, the assumption of a decent antenna-to-antenna separation will not hold if antennas are shared between TDD and FDD systems, for example, when providing indoor coverage with shared distributed antenna systems for both FDD and TDD modes. Thus, the TDD system should create a separate cell layer. In the pico-cell TDD deployment scenario, the interference between modes is easier to manage with low RF powers and separate RF parts.

### 18.6.2.2   Interference from UTRA TDD Mobile to UTRA FDD Base Station

UTRA TDD mobiles can interfere with a UTRA FDD base station. This interference is basically the same as that from a UTRA FDD mobile to a UTRA FDD base station on the adjacent frequency. The interference between UTRA FDD carriers is presented in Section 8.5. There is, however, a difference between these two scenarios: in pure FDD interference there is always the corresponding downlink interference, while in interference from TDD to FDD there is no downlink interference. In FDD operation, the downlink interference will typically be the limiting factor; therefore, uplink interference will not occur. In the interference from a TDD mobile to an FDD base station, the downlink balancing does not exist as between FDD systems, since the interfering TDD mobile does not experience interference from UTRA FDD as was illustrated in Figure 18.11.

One way to avoid uplink interference problems is to make the base station receiver less sensitive on purpose, i.e. to desensitize the receiver. For small pico-cells indoors, base station sensitivity can be degraded without affecting cell size. Another solution is to place the FDD base stations so that the mobile cannot get very close to the base station antenna.

### 18.6.2.3   Interference from UTRA FDD Mobile to UTRA TDD Base Station

A UTRA FDD mobile operating in 1920–1980 MHz can interfere with the reception of a UTRA TDD base station operating in 1900–1920 MHz. Uplink reception may experience high interference, which is not possible in FDD-only operation. The inter-frequency and inter-system handovers alleviate the problem. The same solutions can be applied here as in Section 18.6.2.2.

### 18.6.2.4  Interference from UTRA FDD Mobile to UTRA TDD Mobile

A UTRA FDD mobile operating in 1920–1980 MHz can interfere with the reception of a UTRA TDD mobile operating in 1900–1920 MHz. It is not possible to use the solutions of Sections 18.6.1.2 because the locations of the mobiles cannot be controlled. One way to tackle the problem is to use downlink power control in TDD base stations to compensate for the interference from the FDD mobile. The other solution is inter-system/inter-frequency handover. This type of interference also depends on the transmission power of the FDD mobile. If the FDD mobile is not operating close to its maximum power, then the interference to TDD mobiles is reduced. The relative placement of UTRA base stations has an effect on the generated interference. Inter-system handover requires multimode FDD/TDD mobiles, and this cannot always be assumed.

### 18.6.2.5  Co-location of TD-SCDMA with LTE TDD

With the LTE FDD mode, the interference cases in general do not change from what has been considered for UTRA FDD. With the LTE TDD (also often referred as TD-LTE) mode of operation, there are specific considerations to enable co-location in close spectrums with TD-SCDMA. The LTE TDD mode frame structure otherwise follows that of LTE FDD, but the sub-frame (as covered in Chapter 16) for the TDD case has a specific structure when the transmission direction changes after the first sub-frame between uplink and downlink. The timing has been chosen so that with specific TD-SCDMA uplink/downlink splits the carrier timings can be organized so that there is no overlap between the uplink and downlink transmissions, as shown in Figure 18.13, and covered in more detail in [17].

## 18.6.3  Conclusions on TDD and TD-SCDMA Interference

Sections 18.6.1 and 18.6.2 considered those TDD interference issues that are different from FDD-only operation. The following conclusions emerge:

- Frame-level synchronization of each operator's TDD base stations is required and this is also applied in the commercial TD-SCDMA network in China.
- Frame-level synchronization of the base stations of different TDD operators is also highly recommended if the base stations are close to each other. As such this is not an issue as of today since there is only one TD-SCDMA network in commercial use.

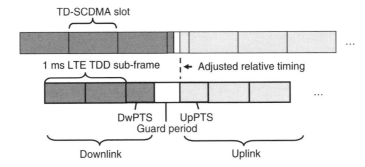

**Figure 18.13**   TD-SCDMA and LTE TDD mode co-location

- Cell-independent asymmetric capacity allocation between uplink and downlink is not feasible for each cell in the coverage area.
- DCA is needed to reduce the interference problems within the TDD band.
- Interference between the lower TDD band and the FDD uplink band can occur and cannot be avoided by DCA. At the moment the band in Region 1 is not too much in use, so the problem is not that severe. Also in China the TD-SCDMA deployment is not at this band (but at 2.1 GHz instead), so there the interference issue with WCDMA is not relevant at the moment.
- Inter-system and inter-frequency handovers provide a means of reducing and avoiding the interference.
- Co-siting of FDD and TDD macro-cell base stations is not feasible when expecting to use them at close proximity in frequency domain as well, and co-siting of pico base stations sets high requirements for TDD base station implementation.
- Co-existence of FDD and TDD can affect FDD uplink coverage area and TDD quality of service.
- With proper planning, the TDD operation can complement FDD as a capacity expansion.
- With TD-SCDMA the results would slightly vary when considering a single 1.28 Mcps carrier or whether one is considering deployment of the multi-frequency site with three carriers.

## 18.7    Conclusion and Future Outlook on TD-SCDMA

This chapter has covered the TDD mode of operation in general and especially the TD-SCDMA operation. The focus was on the physical layer issues, since the higher layer specifications are, to a large extent, the same as with WCDMA. In an actual implementation, the algorithms for both the receiver and RRM differ between TD-SCDMA and WCDMA, as the physical layers have different parameters to control. Especially in the TD-SCDMA base station, advanced receivers are needed, whereas for mobile stations the required receiver solution will depend on the details of performance requirements. With the development of technology also a WCDMA UE with HSDPA support would today have more advanced receivers than just the basic RAKE receiver, but mostly equalizer type of receivers to cope with the inter-symbol interference rather than inter-user interference.

From the service point of view, both UTRA TDD and FDD can provide both low and high data-rate services with similar qualities of service. The only exception for TD-SCDMA remains the highest data rates as a, after a certain point, the highest data rates are asymmetric and all the highest data rates are not available (or follow one or two 3GPP Releases later). The coverage of TD-SCDMA will be smaller for low and medium data-rate services than the comparable WCDMA service due to the TDMA duty cycle, but the difference is smaller when going to high data rates if several time slots can be allocated for the uplink direction.

Interference aspects for the TDD operation were analyzed and will need careful consideration for deployment. With proper planning, UTRA TDD can complement the UTRA FDD network, the biggest benefit being the separate frequency band that can be utilized only with the TDD operation.

During 2009, TD-SCDMA has been put to commercial use in China, following a trialling period over several years, and many of the major vendors have released TD-SCDMA-capable handsets. The higher chip rate modes have been used in some countries but that has not achieved major success in any of the markets, as also the big roll-out foreseen in Japan, as suggested in the 4th edition of this book, did not take place after all.

## References

[1] 3GPP Technical Specification 25.221 V8.6.0, Physical Channels and Mapping of Transport Channels onto Physical Channels (TDD).
[2] 3GPP Technical Specification 25.222 V8.4.0, Multiplexing and Channel Coding (TDD).
[3] 3GPP Technical Specification 25.223 V8.4.0, Spreading and Modulation (TDD).

[4] 3GPP Technical Specification 25.224 V8.3.0, Physical Layer Procedures (TDD).

[5] 3GPP Technical Specification 25.102 V8.1.0, UTRA (UE) TDD; Radio Transmission and Reception.

[6] Steiner, B. and Jung, P. 'Optimum and Suboptimum Channel Estimation for the Uplink of CDMA Mobile Radio Systems with Joint Detection', *European Transactions on Telecommunications and Related Techniques*, Vol. 5, 1994, pp. 39–50.

[7] Lupas, R. and Verdu, S. 'Near–far Resistance of Multiuser Detectors in Asynchronous Channels', *IEEE Transactions on Communications*, Vol. 38, 1990, pp. 496–508.

[8] Klein, A. 'Data Detection Algorithms Specially Designed for the Downlink of CDMA Mobile Radio Systems', *Proceedings of IEEE Vehicular Technology Conference*, Phoenix, AZ, 1997, pp. 203–207.

[9] Klein, A. and Baier, P.W. 'Linear Unbiased Data Estimation in Mobile Radio Systems Applying CDMA', *IEEE Journal on Selected Areas in Communications*, Vol. 11, 1993, pp. 1058–1066.

[10] Klein, A., Kaleh, G.K. and Baier, P.W., 'Zero Forcing and Minimum Mean Square-Error Equalization for Multiuser Detection in Code-Division Multiple-Access Channels', *IEEE Transactions on Vehicular Technology*, Vol. 45, 1996, pp. 276–287.

[11] Jung, P. and Blanz, J.J. 'Joint Detection with Coherent Receiver Antenna Diversity in CDMA Mobile Radio Systems', *IEEE Transactions on Vehicular Technology*, Vol. 44, 1995, pp. 76–88.

[12] Papathanassiou, A., Haardt, M., Furio, I. and Blanz, J.J. 'Multi-User Direction of Arrival and Channel Estimation for Time-Slotted CDMA with Joint Detection', *Proceedings of the 1997 13th International Conference on Digital Signal Processing*, Santorini, Greece, 1997, pp. 375–378.

[13] Varanasi, M.K. and Aazhang, B. 'Multistage Detection in Asynchronous Code-Division Multiple-Access Communications', *IEEE Transactions on Communications*, Vol. 38, 1990, pp. 509–519.

[14] Väätäjä, H., Juntti, M. and Kuosmanen, P. 'Performance of Multiuser Detection in TD-CDMA Uplink', *EUSIPCO-2000*, Tampere, Finland, 5–8 September 2000.

[15] 3GPP Techical Specification, TS 25.306, V6.9.0, UE Radio Access Capabilities.

[16] Holma, H., Povey, G. and Toskala, A. 'Evaluation of Interference Between Uplink and Downlink in UTRA TDD', *VTC'99/Fall*, Amsterdam, 1999, pp. 2616–2620.

[17] Holma, H. and Toskala, A., *LTE for UMTS*, Chichester: John Wiley & Sons, Ltd, 2009.

# 19

# Home Node B and Femtocells

Troels Kolding, Hanns-Jürgen Schwarzbauer, Johanna Pekonen, Karol Drazynski, Jacek Gora, Maciej Pakulski, Patryk Pisowacki, Harri Holma and Antti Toskala

## 19.1 Introduction

Now that the radio spectrum is being fully utilized and as the spectral efficiency of WCDMA/HSPA is very mature, it is a challenge how to improve capacity and coverage to manage the next 5–10 years predicted data growth using the existing installed base station sites. Pushing the spectral efficiency further gets more difficult and opting to use smaller cells will eventually be a necessity to permit more capacity and higher data rates for the end users.

With a potentially high penetration loss from outdoor to indoor on the order of tens of dBs, there is an excellent opportunity to build very small and isolated indoor cells which have very high signal quality, few users per site, and which only interfere marginally with the wide area network outside. As shown in Figure 19.1, there are many product options to provide indoor coverage, such as active or passive distributed antenna systems (DAS), Pico base stations utilizing the normal 3GPP architecture and Femto base stations that use the special architecture discussed in this chapter. The differentiating factors are required user capacity, needed quality of service and importance of coverage. For the zero-touch and plug & play market use cases, the dedicated 3GPP radio technology is the Home Node B (HNB) which belongs to the Femto segment in Figure 19.1. 'Home' relates to usage in a customer's home, e.g. also denoted customer premise equipment (CPE), but this term is rather limited compared to the discussed uses for the radio technology which encompasses also small to medium enterprise (SME) and hot-spot applications as can be seen in Figure 19.1.

The intended business drivers for the traditional home Node B (HNB) relate to different aspects such as:

1. better indoor coverage in homes;
2. lower cost of data delivery using low cost HNBs and relying on end users' own DSL lines;
3. fast enabling of new end user services in homes with a true mobility component;
4. stimulus of extended data usage also in 3GPP wide area;
5. reduction of churn and increased customer loyalty.

*WCDMA for UMTS: HSPA Evolution and LTE, Fifth Edition*   Edited by Harri Holma and Antti Toskala
© 2010 John Wiley & Sons, Ltd

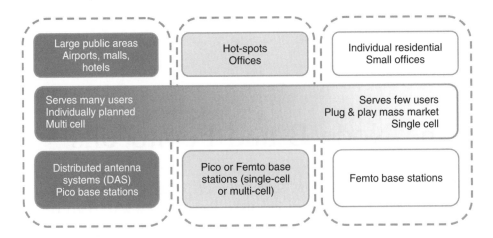

**Figure 19.1**  Examples of indoor solutions and their target application areas

Operating in licensed bands, the HNB offers an unparalleled backwards compatibility with all devices that operate in a wide area and the offload can be done in a true seamless fashion via handovers re-using the same set of 3GPP radio principles. For a study of business drivers, the reader is referred to e.g. [1, 2].

The HNB is fundamentally a new small base station in the sense that it has a downlink receiver for own transmission band (e.g. mini-'UE') built-in which enables it to measure and assess the radio conditions in its intended transmission bands. As this can be done prior to registration (e.g. when it actually starts transmitting), it opens up new possibilities for both low-cost self-optimization and for sending radio measurements to a centralized node for more centralized deployment optimization.

Further, and similar to Wi-Fi access points, the HNB specifications allow either the operator or the hosting party (e.g. the end user hosting the HNB and who has the contract with the operator) to define strict access control only for certain subscribers. An HNB can belong to one of three access classes, as will be described further in Section 19.5:

1. *Closed*, meaning that only the UE that belongs to the configured closed subscriber group (CSG) can access it;
2. *Open*, meaning that all UEs of the configured operator can access it;
3. *Hybrid*, meaning that all UEs of the configured operator can access it but members of the CSG may receive preferential treatment.

As a terminology issue in this chapter, the term *femtocell* denotes a small cell, typically an indoor one with a radius in the order of up to several tens of meters. It is the cell hosted by an HNB. When we talk about the specific 3GPP radio technology for the femtocell including the specifications side, the term *HNB* is used. When touching upon the topic of architecture for the HNB, the term *3G Femto* denotes an HNB which uses the 3GPP UMTS Femto architecture to connect to the core network which is discussed in Section 19.4.

Although this chapter specifically addresses the 3G HNB, most of the information is relevant to both 3G and LTE. In 3GPP, a large amount of the home base station work for the two radio technologies is done jointly although some differences apply, related to e.g. architectural aspects. Some of the key differences are briefly outlined throughout the chapter. The LTE version of the femto access point is the *Home eNode B (HeNB)*.

Femto is a complete new way of building the mobile network. It covers a number of new aspects including interference control, architecture, mobility, security and access control, which are all different from the macro cell networks. Therefore, this chapter gives more detailed description about the femto-related solutions and their backgrounds compared to other chapters in the book.

## 19.2    Home Node B Specification Work

The 3GPP specified the WCDMA HNB system as part of its Release 8 work, with enhancements in later releases. The specifications ensure the possibility of using femto solutions in a real multivendor environment with open interfaces between the network elements, and as fully integrated with other 3GPP-based cellular systems. The 3GPP Release 8 specifications contain the baseline functionality for residential HNB deployment and the basic HNB architecture definition. Release 9 specifications include add-ons to that basic functionality such as inbound mobility and open and hybrid access mode for the HNB. During Release 10 specification work, further enhancements and solution optimizations were considered which are particularly aimed at business use. As an indicative roadmap of HNB specific features that are discussed as part of 3GPP work, see Table 19.1. Even if Releases 8 to 10 bring several enhancements to the HNB operation, it is still possible to use legacy UEs prior to Release 8 to connect to the HNB.

As an HNB belongs to the customer premise equipment (CPE) category, 3GPP builds upon the foundation of the Broadband Forum for device management (CPE WAN management protocol TR-069) who in parallel have developed the TR-196 Femto Access Point service data model which carries the specific HNB configuration parameters. IETF protocols for IP security are also adopted.

In addition to 3GPP activity, the Femto Forum is a separate body outside of 3GPP working with both technical as well as business aspects of the femto deployments [1] and creating inputs to 3GPP. Many companies in the Femto Forum also participate in 3GPP.

**Table 19.1**    Main HNB-related features supported in different 3GPP releases

| Features supported for 3GPP Release 99-7 UEs | 3GPP Release 8 | 3GPP Release 9 | Potential topics for 3GPP Release 10 |
| --- | --- | --- | --- |
| Closed Subscriber Group (CSG) concept via HNB-GW access control Hand-out active mode mobility Idle mode incoming and outgoing mobility | Architecture aspects: Functional split for CN, HNB-GW, HNB APs U-Plane handling C-Plane handling Closed Subscriber Group (CSG) concept and Idle mode mobility Hand-out active mode mobility CSG User Authentication including backwards compatibility for pre-Rel8 UE 3G HNB RF requirements | Mobility topics: Hand-in scenario Handovers between H(e)NBs Open access mode Hybrid access mode HNB security aspects HNB OAM support Introduction of Operator controlled CSG List Study Item on HNB and macro BTS interference management | IMS interworking Features for Enterprise environment Local IP Access and Internet Offload support for HNB system incl. security Optimization of CSG to CSG mobility Optimized HNB – macro cell interference management Remote access feature |

## 19.3    Technical Challenges of Uncoordinated Mass Deployment

The HNB specifications in Releases 8 and 9 are a good starting point but need to mature further to permit a fully flexible zero-touch error-resistant deployment on a very large scale. Operators will have to take on huge responsibility in deploying HNBs so that they fulfil the regulatory requirements and may need to allocate a dedicated HNB spectrum to avoid interference between macro and femto layers. In this section, some of the key technical challenges related to the HNB deployment are discussed in overview. The main issues are summarized in Figure 19.2.

A key issue relates to interference. In a traditional network roll-out each base station's impact on the network is carefully considered to ensure good capacity and coverage everywhere. With uncoordinated placements of small base stations that may not be accessible by all users (e.g. the closed CSG type), wide area coverage may receive detrimental and uncontrolled effects from the femtocell layer, and vice versa, the femtocell layer may offer an unsatisfactory end user experience due to interference from the macro layer or even from within its own layer in a densely deployed scenario. Interference issues and mitigation methods are discussed in detail in Section 19.7.

As the HNB is envisioned to use a shared backhaul connection to the Internet, the cellular operator may be unable to guarantee a satisfactory end-to-end service experience for the end user. For some services such as cellular voice, end users are accustomed to very high quality levels and this requirement will be expected in the HNB domain as well. Further, commercially available DSL data rates may in many cases be limiting compared to peak data rates offered by a 3G/HSPA cellular operator in either uplink and downlink directions or both. This raises a dilemma of service prioritization related to the customers' own backhaul for which there may be many interested service providers. Commercially available home routers today provide prioritized traffic shaping to ensure that e.g. voice continues smoothly even in the presence of heavy data download activity. If a femtocell is added to the home network, the HNB can provide its own local prioritization of the air interface capacity related to a stable backhaul capacity, but overall prioritization needs to be conducted compared to other running services such as IPTV, gaming, communication services, etc. This calls for packet classification and priority rules for cellular-related traffic in the residential gateway. On top, special service level agreements may be needed between the cellular operator or the customer on one hand, and the fixed network operator on the other, to reserve sufficient backhaul bandwidth for the HNB.

The original network design for 3G cellular system mobility is based on the assumption of a reasonable amount of well-defined neighbors. In the mass deployment of uncoordinated cells with demands for good seamless handover functionality, a new range of roaming and handover challenges arise. Some relate to maintaining the lists of dynamic handover neighbors while others relate to handling cell identification ambiguity. Finally, the capacity for a femtocell may be limited; either by hardware limitations in the HNB or due to backhaul limitations, and ensuring robust handovers requires either a well-informed or a very conservative approach. The aspects of mobility are covered in detail in Section 19.5.

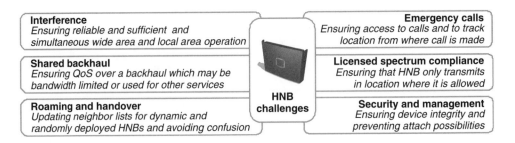

**Figure 19.2**    The key challenges for zero-touch and uncoordinated mass deployment of HNBs

The deployment of HNBs in an operator's licensed spectrum (including the geographical area) is constricted by the requirements of regulatory compliance. These requirements must be facilitated both by the HNB equipment itself but also in its integration into the core network. As an example key issue, the HNB deployment must enable dispatching and tracking of emergency calls. An HNB operating in the closed mode, allowing only configured members to make use of it, has to meet the regulatory requirement of allowing other subscribers to establish emergency calls. National laws furthermore require that the information about the actual geographical location of the UE is communicated to the emergency central which, due to the limited femtocell coverage, is basically the location of the HNB the UE is attached to.

Another regulatory issue is that of licensed spectrum compliance, e.g. to ensure that the radio interface is operated only within spectrum ranges and geographical area(s) according to the license of the responsible operator. As an HNB may easily be moved to another country and plugged into any available Internet router, this is a key concern related to uncoordinated mass deployment.

Generally, the main implementation challenge is to create a zero-touch system to update the legacy core network about new femtocells and initiate location changes for these femtocells automatically. Evidently, obtaining knowledge of the geographical position of an HNB forms the basis of ensuring regulatory compliance. The use of a Global Navigation Satellite System (GNSS), such as a Global Positioning System (GPS), is facilitated but although it is by far the most accurate method, its use may be limited by the indoor deployment of the HNBs. Further, the HNB can perform scans of its radio environment to extract information about the surrounding macro base stations and operators in the area to determine if it is located in a permitted location. Based on network planning data, the management system is able to calculate the geographical position with reasonable accuracy that is comparable to determining the UE position in the macro network. However, as one reason for the HNB is to provide coverage where there is none, this is not always an acceptable solution either. Further, as the HNB is located behind an Internet router, the publicly assigned IP address can in some cases be used to determine the location at e.g. city/street/house level but is not fully reliable in many cases. Finally, DSL-line ID-based location verification is an accurate option but calls for a major integration of HNB management and the DSL management systems.

Finally, any cellular device deployed by an end user constitutes generally a set of security and management challenges. Current 3GPP standards on management are based on the requirement that an HNB is allowed to switch on its radio transmitter only after the geographical location has been verified and the configuration parameters have been sent down to the HNB. An HNB therefore needs to have basic knowledge about how to establish the initial contact with the relevant administration system of the mobile network operator. A related security challenge is to check the integrity of the HNB device and either allow or block access to the operator's core network. This includes verification that the software has not been tampered with. Also integrity of the communication between the HNB and the network elements of the HNB architecture needs to be guaranteed. Confidentiality may further need separate handling depending on e.g. operator policy and country-specific settings. On top of this, the management system needs to ensure robust and scalable communication flows between potentially millions of HNBs that may be switched off at the subscriber's discretion. More information about the involved architecture and protocols are described in the following section.

## 19.4   Home Node B Architecture

A key point in architecture design has been to ensure scalability regarding a potential large volume of connected HNBs. For a typical macro cell deployment, a single macro site may easily support several thousands of households within its coverage range. Given that some percentage of these households adopt an HNB, it is clear that the number of femtocells can easily exceed the number of macro cells by large orders of magnitude. Hence, in terms of reference architecture, a new concentrator network element, the Home Node B Gateway (HNB-GW), has been introduced. A new interface between

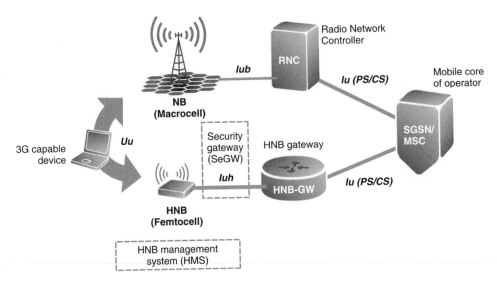

**Figure 19.3**   3GPP Iu-based HNB architecture [3]

HNBs and the HNB-GWs, called the Iuh, has been defined which re-uses and extends the existing Iu protocol. The Iuh is an open interface and allows interoperability in multi-vendor networks. The HNB architecture reference model is shown in Figure 19.3 [3].

The HNB network architecture consists of the HNB serving the femtocell, the HNB gateway (HNB-GW), a security gateway (SeGW), and an HNB management system (HMS). The HNB-GW acts as the concentrator of potentially a large number of HNB connections towards the core network. In this respect, this is very similar to the function of the RNC in the traditional 3G architecture and the function of the HNB-GW in the HNB architecture. In addition to that, each HNB-GW handles HNB and UE registration over the HNB-specific interface, the Iuh.

The HMS assists the HNB in the HNB-GW discovery procedure, performs the HNB location verification, and further configures the accepted HNBs with appropriate operational parameters. The HMS additionally provides the HNB with information about the serving HMS and SeGW, if these are different from the initial ones.

The SeGW provides a secure connection between the HNB and the HNB-GW. As HNBs use commercial Internet connections to connect to the operator core, there is no inherent security embedded and therefore additional measures have been included in the HNB architecture as well as in the protocols. The SeGW is situated on the border between the trusted operator's network and the unsecured public domain. The SeGW facilitates an encrypted communication channel to assure the integrity of the data exchanged between the user and the network and it participates in the authentication of the HNB. The SeGW is a logical entity and can be integrated into the HNB architecture in a flexible way, e.g. there can be a single SeGW per multiple HNB-GWs, or a dedicated SeGW per backhaul network operator, etc.

## 19.4.1   *Home Node B Protocols and Procedures for Network Interfaces*

When defining the HNB architecture and protocols, 3GPP reused as many protocols known from standard UTRAN (e.g. RANAP, GTP-U) as possible. Furthermore, due to its relationship with the IP-based Internet access, 3GPP have also adopted well-known protocols from other standardization bodies such as the real time protocol (RTP) and the real time control protocol (RTCP) of the IETF.

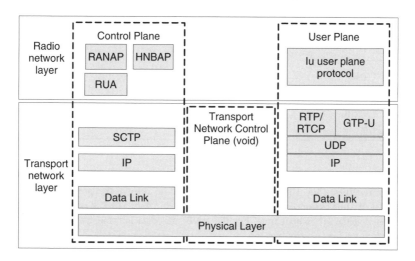

**Figure 19.4** Iuh protocol structure (CS and PS) as of Release 8. In Release 9, also protocols SABP and CS-Mux are supported on top of RUA and UDP respectively [6, 7]

In addition to that, certain new protocols and procedures were introduced for the control plane. The Home Node B Application Part (HNBAP) protocol [7] is responsible for the registration procedures of the UE and the HNB itself, while the RANAP User Adaption (RUA) protocol [6] is used to convey RANAP messages with additional information. Figure 19.4 shows the user and the control plane stacks for the Iuh interface. Although all communication is established over the IP, the security tunnels have been omitted in Figure 19.4 for simplicity.

For the circuit switched (CS) user plane, the AMR voice frames are transported using the RTP protocol while the optional RTCP protocol is used for quality measurements. For the Iuh packet switched (PS) user plane traffic, the GTP-U protocol is used as in legacy UTRAN; see Chapter 5 for more information. The voice service will take nearly 100 kbps in the HNB transport, so the relative overhead is large for low data rate services but it is typically not a problem because the number of simultaneous voice users is low.

The HNB architecture calls for new functional procedures to take care of certain tasks specific to managing the HNB environment:

- Since the communication from the HNB to the operator's core network (CN) goes through the public unsecured Internet, the IPSec tunnelling based on IETF protocol is used.
- The provisioning of configuration parameters to the HNB has been adapted from DSL world and is based on the TR-069 protocol from the Broadband Forum [4], although extended with an HNB specific data model, TR-196 [5], as well as some security extensions profiled in [12].
- As mentioned in Section 19.3, the HNB location verification may be done using a number of measuring techniques by signaling provided by TR-069.
- Verification of a UE's right to access to a given HNB requires a check of whether the UE belongs to the HNB's closed subscriber group (CSG); either by means of an HNB-specific CSG check procedure or by looking at the access control list (ACL) for legacy UE; the latter is located in the HNB-GW and optionally in the HNB and is based on IMSI numbers.

The main procedures required to bring an HNB into service are:

- *HNB boot and discovery/registration*. Once powered up or reset to factory defaults, the HNB performs an autonomous device integrity validation to ensure that the device hasn't been tampered

**Table 19.2**   Information assumed available upon boot at first power-up of HNB device

| In HNB | In public IP network | In HMS and possibly SeGW |
| --- | --- | --- |
| URL and root certificate of the operator's SeGW<br>URL of the operator's initial HMS<br>Own device certificate containing unique HNB identity | DNS resource record to resolve the URL of the operator's SeGW | Root certificate for the HNB device certificate<br>HNB identity |

with. It then receives a local IP address and uses DNS to resolve the preconfigured URL of the operator's SeGW. It is assumed that the information listed in Table 19.2 is available in the respective network nodes at the HNB boot time. For the HNB, this data is typically pre-provisioned by the HNB manufacturer for security reasons.

- *HNB registration using HNBAP*. Once the previous steps are completed, the HNB is now ready to notify the HNB-GW that it is available at a particular IP address. As part of this procedure, HNB provides an HNB-GW with the parameters and identity it has and then activates the radio transmission only following the positive completion of the procedure with the HNB-GW.

Once the HNB is operational, UE may now attempt to connect. The HNBAP protocol is used to register each HNB-connected UE (HUE) at both the HNB and the HNB-GW when entering the femtocell in idle mode. HNB forwards the UE identification data to the HNB-GW where the UE access control procedure takes place in case of a non-CSG capable UE (pre-Release 8 UEs). For a CSG-capable UE, the core network performs a check of the CSG member status. There are special rules for handling an emergency call (which is basically always allowed) while for other services only UE with proper rights will be served (others need to be redirected).

## 19.4.2   Femtocell Indication on a Terminal Display

It may be beneficial to indicate on the terminal display that the terminal is camping on a femtocell. The display is relevant especially if the femtocell is subject to different pricing than macro cells. The display may also be useful if there are problems to identify if the problem is associated with the femto connection or with the macro cell connection. There are a few options available on how to indicate the femtocell presence on the display.

- *Option 1: USIM: PLMN/LA (Public Land Mobile Network/Location Area) range in USIM*: The femtocells are using a specific PLMN or LA range, which will be mapped to the femto indication by USIM. It is possible to reconfigure the USIM information over the air.
- *Option 2: NITZ*: The core network sends the PLMN name in NITZ (Network Identity and Time Zone). NITZ is a mechanism for providing the local time and date, as well as network provider identity information to the terminal. NITZ has been part of the 3GPP standard since Release 96. NITZ is primarily used to automatically update the system clock of the terminal, but it can also be used to provide network identity. In case of femtocells, the network identity would give the femto indication.
- *Option 3: Terminal memory*: The PLMN name is pre-stored in the terminal memory during manufacturing or the pre-stored list of PLMN names is updated in maintenance.

# 19.5   Closed Subscriber Group

There are three fundamentally different HNB classes, also known as cell access modes, envisioned to match requirements from different use cases:

- *Closed Subscriber Group (CSG) HNB*: This is meant for the business or home application, where the customer of the HNB service wants to restrict its usage to own demands, e.g. the cell is not part of the public coverage for the operator. The cell in closed access mode is accessible in normal service state only for the members of the CSG of that cell.
- *Hybrid HNB*: In this case, part of the capacity is 'reserved' for the UE belonging to the configured CSG, but a part of the capacity is left open for more public usage. This solution benefits the operator as part of the interference problem is alleviated (as described in Section 19.7) and this may in return lead to a lower HNB operation cost for the end user (or even revenue generated from sharing backhaul and radio capacity). The CSG membership can be also used to differentiate in the service offer between the subscribers.
- *Open HNB*: The last category is fully open to all subscribers and its use is thus envisioned for hot-spot applications as part of the managed wide area network. From the UE's perspective, the open HNB cell is like a normal macro cell.

In the following section, the procedure for managing the CSG list is detailed as well as the access control system applicable to both CSG capable and legacy UE.

## 19.5.1   Closed Subscriber Group Management

The CSG which is associated with a femtocell is configured during the set-up of the HNB and is denoted by its CSG Identity (CSG ID). The network operator allocates CSG ID values throughout the network and multiple HNB cells can be part of the same closed subscriber group, e.g. they will in this case have the same CSG ID configured. The HNB cell operating in closed access or in hybrid mode always has a CSG ID allocated. The open HNB does not belong to any dedicated closed subscriber group and therefore does not have any CSG ID allocated. As part of the HNB registration procedure the HNB informs the HNB-GW about the CSG ID.

The list of closed subscriber groups to which an end user belongs is maintained as part of the subscription information. The CSG ID list is stored on the USIM card/UE and in the operator's core network in the HLR/HSS as part of the subscription data. The CSG ID list of a subscriber received in the NAS may be divided into two parts: (1) the allowed CSG list, which is managed by the HNB system operator/end user and/or by the operator; and (2) the operator CSG list, which is controlled only by the operator. The list available to the UE in the access stratum is a combination of those lists and is called the CSG-Whitelist. As mentioned earlier, the CSG concept is not supported by pre-Release 8 UE and in this case the CSG subscription information is stored in the HNB-GW instead. There may also be UEs of later Release without CSG support because the CSG concept is optional.

An update of the CSG ID list in the UE is done either with an application tool (e.g. via a web interface or an application) or with the help of manual CSG selection, which is initiated by the end user.

## 19.5.2   Closed Subscriber Group Access Control

CSG access control is always done when the UE is performing a location update towards a CSG cell. During CSG access control the CSG subscription information of the UE is checked against the CSG ID of the serving cell, to figure out whether the UE has the right to use the CSG cell in the normal service state. Otherwise, the access request is rejected. Emergency calls are naturally allowed, regardless of the CSG subscription.

In the case of the hybrid cell, the CSG subscription information is checked in the same way as in the case of CSG access control to identify if the UE is a member of the hybrid cell. The membership information may be used for admission control or to offer dedicated service for the members in the hybrid cell. Different from the CSG access control in CSG cell, the access request of a non-member is not rejected in the hybrid cell case.

## 19.6   Home Node B-Related Mobility

In this section, the mobility procedures specific to HNB cells are listed. To be complete, different HNB types as well as different handover scenarios are summarized in Table 19.3. In the following subsections the special HNB cases are considered. Also related signaling examples for the handovers are presented.

WCDMA and HSUPA use soft handover in the traditional macro cellular network architecture to support mobility and interference control. The soft handover between different femtocells or between femtocells and macro cells is not supported but instead hard handover is always used.

### 19.6.1   Idle Mode Mobility

As shown in Table 19.3 only the CSG and hybrid HNB have special procedures compared to normal macro cell operation covered in Chapter 7. The CSG HNB cell indicates via the BCCH the CSG ID of the closed subscriber group it belongs to and with a separate parameter that it is a closed access mode HNB cell. However, since there may be many cells in an area, it may be too tedious for a UE to extract this information for all cells. Hence, the network may reserve a certain range of primary scrambling code (PSC) values for CSG cells. This range is broadcast by all CSG HNBs of the network. Also the macro cells may broadcast that PSC range as optional information. The UE, when receiving the PSC range for CSG cells, can consider it to be valid for 24 hours.

Based on this CSG-specific PSC range, the UE will know which measured neighbor cells are CSG cells without the need to extract the specific CSG information (e.g. reading the SIB) of all the measured neighbor cells. This information will help the idle mode mobility procedures in two ways:

1. The UE without a CSG subscription can ignore any CSG cells in its cell ranking, because normal access would not be allowed for such an UE anyway.

**Table 19.3**   Mobility/relocation cases and different HNB classes

| Handover/HNB type | | CSG | Hybrid | Open |
|---|---|---|---|---|
| Idle mode | | Special methods and handling for pre-Rel8 UE and Rel8+ UE, see Section 19.6.1. | Special methods and handling for pre-Rel8 UE, Rel8 UE, and Rel9+ UE, see Section 19.6.1. | Same as macro only case, see Chapter 7. |
| Active mode | Outbound | Same as macro only case, see Chapter 7. Neighbor cell identification discussed in Section 19.6.2. | | |
| | Inbound | Special methods and handling for pre-Rel8 UE, Rel8 UE, and Rel9+ UE, see Section 19.6.3. | Special methods and handling for pre-Rel8 UE, Rel8 UE, and Rel9+ UE, see Section 19.6.3. | Same as macro only case, see Chapter 7. |
| | Inter-HNB | No specific aspects, see Section 19.6.4. | No specific aspects, see Section 19.6.4 | Same as macro only case, see Chapter 7. |

2. The UE with a CSG subscription can utilize the PSC range information when searching for allowed CSG cells, because it will help to reduce the number of potential cell candidates, from which the BCCH SIBs need to be decoded during the CSG cell search.

If one or more carrier frequencies are used for dedicated CSG deployment, that is broadcast on the macro cells and the CSG cells. This information can be used by the UE to avoid unnecessary measurements on that frequency even though the normal cell measurement rules would require measurements of the frequency carrier.

The legacy UE (pre-Release 8 UE) is not aware of any CSG-specific information broadcast on the BCCH of the CSG cell or macro cell. Therefore the UE will treat the CSG cell as a normal macro cell. If the cell appears to be the best cell based on the cell ranking criteria, the UE will try to access the CSG cell.

The HNB-GW, and optionally HNB, is responsible for the CSG access check for legacy UE, because it knows whether the UE is CSG capable (UE release number and CSG capability). The operator may configure the cell reselection parameters, such as Qoffset and Qhyst, to bias the reselection of CSG cells.

The hybrid access mode for the HNB cell was introduced only as part of 3GPP Release 9 specification work. Therefore the pre-Release 9 UE, even with CSG capability, will regard the hybrid cell as a normal cell for access and cannot identify a hybrid cell as a member hybrid cell even though the cell's CSG identity would be in the UE's Allowed CSG list. The CSG UEs, from Rel-9 onwards, are able to recognize whether the cell is a CSG or a hybrid cell based on an indication broadcast on the BCCH.

The hybrid HNB cell appears to the non-member UE as a normal cell. This will require that the PSCs of the hybrid cells are broadcast as part of the neighbor cell list on BCCH.

## 19.6.2   Outbound Relocations

Generally speaking, the relocation procedures from an HNB cell (regardless of its cell access mode) to a macro cell is based on the same procedures as the inter-RNC relocation methods described in Chapter 7. During the self-configuration phase, the HNB cell identifies the neighboring macro cells by conducting measurements with its downlink receiver, see recommended HNB measurements in Section 19.7.2. This is one of the key autonomous functionalities for zero-touch deployment to maintain good mobility features.

The macro neighbor cell list is defined in the HNB based on these measurements and possibly with the help of configuration information received via the HMS where further location-based information can be used to optimize the handover neighbor list generation. The macro cell neighboring list is indicated to the HNB-connected UE as in a macro cell case.

## 19.6.3   Inbound Relocations

Macro cell handover decisions are based on measurement report messages from the UE containing received signal power as well as the target cells' primary scrambling code, the PSC. These measuring reports are configured by measurement control messages and/or system information. However, there are two main reasons why macro inbound mobility procedures are not fully applicable to HNBs:

- As will be detailed in Section 19.6.8, there can be a very large amount of HNBs under the coverage of one macro station which causes confusion in the sense that the macro base station may have multiple HNBs within its coverage range re-using the same PSC, which is used to define target cells in normal handover procedures.
- As a new problem for macro base stations, only a UE with a matching CSG subscription can be handed over to CSG HNBs. In other words, inbound mobility to CSG/hybrid cells requires target

cell identification and access rights verification (by the access control procedure) in addition to normal procedures.

In order to satisfy those additional HNBs requirements, certain procedures had to be added to the inbound mobility procedure. That is why, similar to idle mode, the UE for intra-frequency relocation can be configured with a PSC range (corresponding to the CSG HNBs PSC range) for which special attention is required. For example, to alleviate the risk of confusion, the UE should get from the BCCH channel the Cell Global Identity (CGI) which is used to resolve cell identity ambiguity rather than just using the PSC. The CGI reported to the network, in combination with the PSC, allows the network to effectively mitigate the problem of correct target cell addressing although acquiring CGI values calls for additional measurement complexity for the UE.

In order to avoid handover attempts to CSG cells which are not permitted, the UE has to verify its access rights on the measured CSG/hybrid cell *prior* to reporting it as a handover target. This process is called a preliminary access check. This is the same process as described for the idle mode in Section 19.6.1, where the CSG-ID broadcast on the BCCH channel is matched against the CSG-Whitelist. The network typically initiates handover preparations to CSG cells for which the UE has verified its subscription.

As mentioned above for intra-frequency inbound relocation to CSG cells, the network can configure a range of PSCs for which the UE should report other relevant handover preparation information in addition to the result of the (preliminary) access check. However, inter-frequency inbound mobility requires BCCH channel acquisition on a different frequency than the serving cell frequency for which a measurement configuration is needed. The UE should notify the network by sending a proximity indication about the necessity to request a measurement configuration on the indicated frequency/RAT. For this reason, the UE has to be aware of its location and available CSG HNBs in its vicinity. This information should be made available to the UE from an implementation specific autonomous search function based on location fingerprint information. It is also possible for the UE to perform inter-frequency measurements autonomously during idle times as long as the ongoing serving cell data reception is not interrupted.

After checking access rights and providing additional information to the network, the rest of the inbound procedure proceeds as usual and as described in Chapter 7.

The inbound mobility in the case of uncoordinated deployment of hybrid cells will require additional reporting from the UE in a similar manner as for CSG cells, in order to identify the correct target hybrid HNB cell. The network may reserve for the hybrid HNB cells a dedicated range of PSC values to avoid any PSC confusion situation with the macro neighbor cell reporting. However, contrary to the case of CSG cells, such a hybrid PSC range is not broadcast in BCCH. Additionally, with the help of a dedicated PSC range, the serving cell is able to identify the neighbors for which the additional reporting from the UE side is necessary to get the target cell identity. The consequences are that the pre-Rrelease 9 UE cannot be supported for inbound mobility to the hybrid cell if there is a risk of PSC confusion (i.e. more than one hybrid cell is using the same PSC value within the macro cell coverage) or if the cell identities of the surrounding neighboring hybrid HNB cells are not known in advance, e.g. based on some network internal configuration information.

If the hybrid HNB cells are deployed in a coordinated manner, the issues with PSC confusion or with cell identities not being known can be avoided and the normal inter-RNC relocation procedure can be supported with legacy UE.

## 19.6.4 Relocations between HNB Cells

The relocations between HNB cells do not invoke any additional procedures on the radio interface apart from those already described for different HNB types. The handover cases are dealt with using the inbound and outbound mobility procedures as covered in the following sections.

## 19.6.5  Paging Optimization

The UE is paged in case of incoming calls in all cells part of the same location area. One location area may consist of multiple cells, e.g. both macro and HNB cells. The CSG HNB cells that reside in the same location can belong to different CSGs. When a UE is camping on an allowed CSG HNB cell, it will be paged via all cells part of the current location of the UE. With a high HNB density, it may be of great importance to optimize the paging procedure in such a manner that the paging message for the UE is broadcast only via cells that are likely to be the current cell of the UE, and this is achieved by sending the paging message only to macro cells and allowed CSG (or open) HNB cells within the location, based on the information in the HNB-GW of CSG IDs of the different cells in the location area and the CSG ID list of the subscriber. While the HNB-GW can access the CSG ID list, this information is not available generally on the core network side in 3GPP architecture and thus such an optimization on paging needs to be done in HNB-GW side.

## 19.6.6  Home Node B to Macro Handover

The procedure for an outbound handover from an HNB to the macro network is basically equivalent to the inter-RNC handover when no Iu-r interface exists. As the UE is handed over to a macro cell, no CSG-related access control takes place. The HNB-GW acts as an RNC towards the core network.

## 19.6.7  Macro to Home Node B Handover

The procedure defined for inbound handover from a macro cell to an HNB is also applicable to other relocation scenarios in which an HNB is the target. During the handover procedure the UE CSG support capability, the UE CSG membership status, the target HNB access type, and the target HNB CSG ID are retrieved in order to provide appropriate handling of the UE request. For example, the CSG UEs should have preferential handling over the non-CSG UEs when attaching to its own hybrid cell.

The handover procedure is illustrated in Figure 19.5 and starts with the source RNC taking the decision to relocate the UE to an HNB. This can be preceded by a dedicated specific measurement

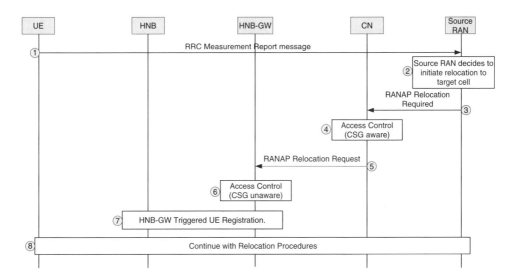

**Figure 19.5**   General message flow for inbound handover to HNB

procedure (performed by a CSG-capable UE) aimed at identifying CSG-IDs of the detected HNBs. The source RNC identifies the target RNC (in this case the appropriate HNB-GW) and sends the RANAP Relocation Required message to the core network in step 3. If the UE is CSG-capable, the CSG-related data is included in that message, so that the core network can perform the access control procedure in step 4 and send a Relocation Required message in step 5 if allowed access. Alternatively, if the UE is not CSG-capable, the core network does not perform access control and sends the Relocation Required message to the HNB GW. In this case the target HNB-GW instead performs access control based on the access control list (ACL) (step 6). In both cases of CSG-capable and non-CSG-capable UE, the HNB-GW or the target HNB verifies if the requested CSG-ID matches the actual CSG-ID of the target cell. In step 7 the HNB-GW triggers an implicit UE registration procedure towards the HNB via the RUA protocol and it allocates the Context-ID for the UE.

## 19.6.8  Home Node B Cell Identification Ambiguity

Uncoordinated mass deployment of many HNBs creates two different challenges related to cell identification ambiguity:

1. *PSC collision*: When there is more than one cell with overlapping coverage area and with same PSC value. This means that UE sees two or more HNBs with same PSC value.
2. *PSC confusion*: When there is more than one neighboring cell with the same PSC value and the serving cell does not know which one the UE is measuring and reporting.

PSC collision should not appear more frequently than in the macro case, because the most typical collision cases should be avoidable with the help of proper PSC range allocation during the HNB cell set-up.

As briefly discussed in previous sections, PSC confusions constitute one of the fundamental problems of uncoordinated deployment of femtocells. Due to a limited set of possible values, a certain PSC needs to be re-used across a network. In a planned deployment, the operator can ensure that a source cell never gets confused as to which target cell is identified with a certain PSC value in a handover measurement. With uncoordinated deployment this is no longer guaranteed. For example, if 100 PSC values are reserved for CSG cells, then the confusion happens if there are more than 100 HNB cells located within the coverage area of a single macro base station; no matter how well the PSC values are distributed among the HNBs.

To be completely certain about a measured target cell in the inbound mobility case, the macro RNC can instruct the UE to measure and report its Cell Global Identity (CGI) and further report, based on the CSG-specific information, if the UE is CSG member. The source cell can rely on the UE only triggering handover requests to allowed CSG cells. Although the UE will still be required to measure the CSG ID to check that admission is OK, the measurement does not need to be transmitted over the network if there is no risk of confusion. The network will control the extra reporting to minimize additional network complexity and high UE measuring load.

It may be attractive in general to ensure that PSC confusion is already minimized in the HNB configuration process. In general, to avoid reporting GCI measurements in the network the following conditions must be fulfilled:

- The macro cell can assume that there is no risk of PSC confusion, i.e. only one neighbor with a certain PSC value exists.
- The macro cell can store and use the information about the GCI (and optionally the CSG ID) of a certain HNB if it is once reported by a UE. This is an RNC implementation issue.
- The macro cell should be also aware of the allowed CSG IDs for a UE in order to avoid handover attempts to unauthorized CSG HNB cells.

As described in Section 19.1, the booting HNB measures its surroundings including which macro cells it sees (e.g. GCI) and what is the associated path loss to the macro cell. This measurement is reported to the management system prior to registration finalization with the HNB-GW, e.g. prior to allowing the HNB to start transmitting. Following such an approach, the management system can keep track of which HNBs are present within a certain macro cell's coverage range and reduce confusion by actively taking part of the PSC configuration process for the HNB prior to registration. Further, the management system may also indicate confusion status to the macro cell layer to minimize problems when GCI measurements are requested in the network.

### 19.6.9  Summary of Home Node B-Related Mobility

The mobility solutions with HNBs are summarized below:

- Pre-Release 9 UEs support outgoing handovers from HNB to macro cells.
- Pre-Release 9 UEs use idle mode selection to HNB in the case of uncoordinated HNB deployment.
- Release 9 UEs can support incoming handovers from the macro to the HNB.
- The incoming handovers to the dedicated HNB frequency need compressed mode measurements. The measurement triggering needs proximity information about the UE location.
- Handovers between two HNBs are supported.
- Soft handover is not possible between macro cells and HNB, and not between two HNBs.
- Closed subscriber groups (CSG) can be used to control access to HNBs both in idle mode reselections and in handovers. The CSG control for the pre-Release 8 UEs is located in HNB-GW. The CSG control for Release 8 UE is located in the core network based on the information provided by the UE.

## 19.7  Home Node B Deployment and Interference Mitigation

### 19.7.1  Home Node B Radio Frequency Aspects

To enable a low cost implementation and uncoordinated deployment of HNB products, it was found that the RF requirements defined for normal UTRA base station classes were not directly re-usable and thus a new Home Base Station class was introduced [8]. On the user equipment side, there were no HNB-specific RF requirements introduced in order to ensure maximum backwards compatibility and mobility. It is possible to use older (all the way to Release 99) UEs with HNBs even if performance enhancements are included in Release 8 and 9 specifications. The enhancements are naturally not available for the legacy UEs.

In general, it is desirable to relax the HNB hardware and radio frequency (RF) requirements to allow more compact and low-cost designs. However, relaxing RF requirements can have detrimental effects that would prevent co-channel macro and femto deployment as well as dense HNB deployments. As an example, it is not desirable to relax the HNB receiver performance as this will call for increased transmission power of any UE connected to the HNB and thus create more interference from the femtocell to the surrounding macro cells or femtocells.

As its most characteristic requirement, the maximum allowed transmission power of the Home Base Station is set to 20 dBm (17 dBm with MIMO per transmission branch) [10, 11]. In practice, and using an adaptive maximum power-setting algorithm, the allowed transmit power is typically lower. However, no algorithm is specified in 3GPP as of Release 9 to protect macro cells on the same frequency as the HNB and in principle the maximum power setting can be freely used, although this is not desirable. One exception, however, which is specified in [11] is a requirement related to protection of a macro UE which is located in the vicinity of the HNB but which is connected to a macro cell on the adjacent carrier. In this case, the Home Base Station is specified to limit its maximum transmission power according to the measured interference level on the adjacent channels licensed to other operators. This requirement is only applicable to the Home Base Station class and was needed in order to provide protection between adjacent bands of different operators where no other preventive mechanisms can be expected.

In order to facilitate lower cost design, the frequency error requirement has been relaxed, compared to larger base station classes, to 250 parts per billion under the assumption that the maximum speed of a Home NB connected UE is 30 km/h. For reference, the wide area base station requirement is five times tougher at 50 parts per billion. Besides based on the assumption of lower supported mobility, the testing environments for the Home Base Station are also based on channel profiles with less channel dispersion (e.g. Ped-A channel profile).

## 19.7.2   Recommended 3G Home Node B Measurements

The specifications as of Release 9 do not mandate femtocell behavior in terms of interference management but provide some guidelines in [8]. Full specification would require detailed test cases and methods which have been omitted so far. This means in practice that operators are left with the challenging task of ensuring robust operation of its network even in a dense and uncoordinated 3G HNB environment. In the following section, various possible methods for automating the interference management process are discussed. The baseline is an assumption that the 3G HNB has the capability to measure its surroundings and provide meaningful information to the management system. Such recommended measurements are summarized in Table 19.4.

**Table 19.4**   Recommended measurements to be collected and used by HNB and its management system

| Purpose | Measurements (source) | Surrounding macro | Surrounding HNB |
|---|---|---|---|
| Identify neighbor base station type | LAC, RAC (HNB DL Rx) | Yes | Yes |
| Identify detailed operator and cell IDs | PLMN ID, Cell ID, GCI/PSC, CSG ID (HNB DL Rx) | Yes | Yes |
| Calculate path loss to neighbor to adjust generated interference in co-channel DL | P-CPICH Tx power, co-channel CPICH RSCP (HNB DL Rx) | Yes | Yes |
| Calculate path loss to neighbour to adjust generated interference in adjacent channel DL | P-CPICH Tx power, adjacent-channel CPICH RSCP (HNB DL Rx) | Yes | Yes |
| Calculate path loss from neighbor to connected HUE to adjust generated interference in co-channel UL | P-CPICH Tx power, co-channel CPICH RSCP (HUE) | Yes | Yes, except P-CPICH Tx power |
| Calculate path loss from neighbor to connected HUE to adjust generated interference in adjacent-channel UL | P-CPICH Tx power, adjacent-channel CPICH RSCP (HUE) | Yes | Yes, except P-CPICH Tx power |
| Calculate co-channel intercell interference towards HUEs | Co-channel carrier RSSI, CPICH $E_c/N_0$ (HNB DL Rx, HUE) | Yes, total interference | |
| Calculate adjacent-channel intercell interference towards HUEs | Adjacent channel carrier RSSI, CPICH $E_c/N_0$ (HNB DL Rx, HUE) | Yes, total interference | |
| Calculate UL interference at HNB | RTWP (HNB PHY) | Yes, total interference | |

To assess its surrounding environment, the measurements shown in Table 19.4 are generally recommended to be done and reported before a 3G HNB is registered with the gateway and is allowed to start transmitting. As shown in Table 19.4, the measurements can be obtained using the built-in HNB downlink receiver (e.g. prior to registration or during run-time if such an interaction is facilitated between the management system and the HNB) or use measurements obtained from any Home Node B connected UE (HUE) (e.g. obtained during run-time). These measurements are then reported to the management. The operation and management system may then provide reasonable settings for key parameters such as the maximum transmission power, etc. For more details related to measurement definitions, see [8].

The actual deployment of HNBs in a given network depends on many factors including:

1. the use case (e.g. coverage or capacity driven by operational needs or special services driven by user needs),
2. to what extent the location of HNBs is coordinated, how much spectrum is available to the cellular operator and what is the load from the wide area,
3. whether HNB supports only a closed subscriber group or is fully or partially open;
4. product availability and maturity of the underlying radio standard.

The deployment of HNBs will most likely be a slower, gradual process due to the technical challenges and the need to demonstrate capability of the measures taken to mitigate them. Figure 19.6 presents one possible scenario from trial to mass deployment. In the first commercial step 2, HNBs will most likely be deployed in areas of restricted macro cell coverage or where for some reason this coverage cannot be provided. Later on, as the users' demand for high throughput increases, the level of HNB penetration will gradually increase, up to the point where a significant part of the operator's total traffic will be carried by the femtocell layer. The HNB deployment itself, somehow controlled by the operators at the beginning, has to eventually become fully uncoordinated to provide sufficient flexibility for end users.

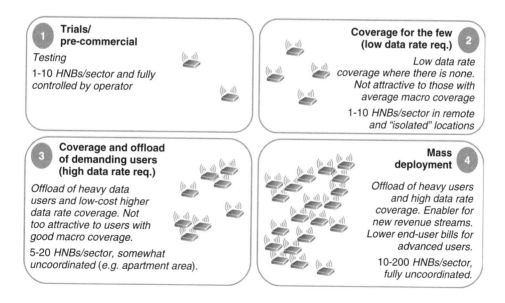

**Figure 19.6**   A possible scenario of mass deployment of femtocells

From a requirement point of view, the HNB deployment and configuration need to facilitate the safe operation of HNBs in a spectrum shared with a macro. Initially, e.g. step 2 in Figure 19.6, there will be too few femtocells deployed to always justify dedicating a spectrum for their operation. Finally, in the latest mass deployment stage where also traffic requirements in the wide area will grow similarly, it may again not be possible to reserve a certain spectrum for the femtocell layer. In this section, different aspects of femtocell interference are presented.

### 19.7.3  Home Node B Interference Considerations

For an example of interference mechanisms related to the deployment of HNBs and macro Node Bs (MNBs) in the same geographical region, see Figure 19.7. The arrows in Figure 19.7 represent the six-way undesired interference paths; namely the femto to macro, the macro to femto, and the femto to femto interference paths in both the uplink and downlink directions. The affected victim is indicated in Figure 19.7 as well. As can be seen, a key issue is the amount of isolation that the femtocell has regarding its surroundings. Envisioning that the femtocell, including where the HUE is located, is isolated by walls with high penetration loss, the interference paths are significantly reduced and co-existence becomes more or less straightforward. However, very high penetration losses cannot always be assumed due to different construction materials, open windows, location of multiple HNBs within a small apartment space, etc. The different interference mechanisms are thoroughly discussed in Table 19.5 as well as the qualitative measures that can be conducted to control the impact [8, 9]. On top of the interference scenarios depicted in Figure 19.7, additional cases are considered in 3GPP, e.g. for co-existence with other systems deployed in the femtocell coverage range (e.g. DECT) or if a UE is located very close to the HNB, see e.g. [8].

For the deployment of HNBs in a dedicated frequency band, the interference paths macro to femto and femto to macro are effectively mitigated benefiting both the macro-connected UE (MUE) and the HNB-connected UE (HUE). There is an exception if HNBs are deployed in a carrier adjacent to a macro carrier, due to potential adjacent channel leakage from the HNB to any macro-connected UE which may be located close to the HNB. The difference in levels of received power from macro

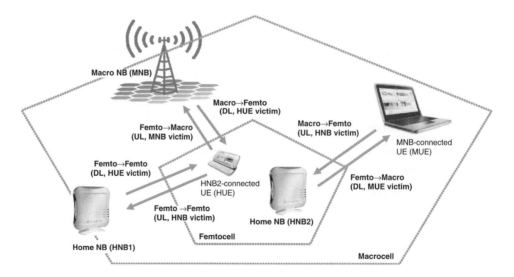

**Figure 19.7**  Illustration of the interference mechanisms in a co-located femtocell and macro cell network

**Table 19.5** Interference cases, their impact on system performance, and key remedies

| Interference case | Downlink impact | Possible remedies | Uplink impact | Possible remedies |
|---|---|---|---|---|
| Femto→Macro | **High**<br><br>Can reduce macro coverage significantly | Adaptive control of the maximum HNB transmit power<br>Use of open or hybrid type HNBs<br>Relying on inter-carrier handovers for MUE in the vicinity of the femtocell | **Medium**<br><br>Pathloss to HNB is normally low calling for low Tx powers unless very high data rates. However, fast power control can take very high dynamic values | Adaptive control of the maximum HUE transmit power and allowed noise rise<br>Adaptive control of the Maximum HNB transmit power also helps since cell selection then effectively prevents HUE to be too far from HNB center |
| Macro→Femto | **Medium**<br><br>Presence of co-channel MNB will significantly limit SINR in up to 25% of deployed HNBs | Adaptive control of the maximum transmit power to ensure minimum coverage of HNB (e.g. balancing HNB transmit power with received macro NB power levels) | **Medium**<br><br>E.g. when MUE is located inside femtocell or unshielded femtocell is close to MUE at cell edge | MUE inter-frequency handover to femtocell-free neighbor carrier if close to CSG HNB<br>Use of open/hybrid femtocells.<br>Different cell selection criteria (e.g. priority to HNBs) |
| Femto→Femto | **High**<br><br>Will be limiting SINR in densely deployed scenario with limited penetration losses | Frequency planning for HNBs (frequency reuse) or automatic frequency selection from given set.<br>Adjust outbound handover parameters<br>Use of hybrid/open femtocells | **Low**<br><br>Typically dominated by downlink performance issues unless cell selection criterion emphasizes uplink | Adjust handover parameters to make earlier use of MNB<br>Adaptive control of the maximum HUE transmit power |

*Note*: MUE denotes a UE connected to a macrocell while a HUE denotes a UE connected to a femtocell

cells and femtocells at the macro-connected UE antenna can in this case be as high as 50 dB [9]. Such a difference is enough to cause significant performance degradation for the macro-connected UE. Between different operators, the maximum HNB transmit power is therefore specified to consider also the received macro interference level in the adjacent carriers in the case of a different operator, see Section 19.7.1. However, such a mechanism is also recommended even in the case where adjacent carriers are used by the same operator.

For uncoordinated but rather dense deployment of femtocells in a dedicated carrier, the effective coverage range for a certain data rate may be rather limited. The only very effective method that an operator has in this scenario is to create some frequency flexibility for the HNBs, either by allowing them to auto-select a carrier from a set of available carrier frequencies or by randomly distributing HNBs over different frequency carriers. The dilemma here is that for soft frequency reuse methods to work, the operator needs to release multiple frequency carriers for femtocell usage, preferably 2–3. Many operators have problems allocating even one frequency to HNB, and it is practically impossible to allocate more frequencies to HNB in most operator cases.

In the case of a co-channel deployment, all the interference paths need to be considered and as a further aspect, the access mode strategy of HNBs is an important factor. CSG cells are merely interference generators to the UE that cannot access them. Opened or hybrid configurations ease this effect unless available backhaul capacity or other hardware limitations prevent further access. Each cell not accessible to a UE generates from its point of view a hole in the network coverage. The area affected can to some extent be controlled by regulating the femtocell maximum transmit power. This method, however, does not eliminate the problem completely, and by lowering the HNB transmit power the benefits of having a femtocell are minimized. Further, some mitigating effects can be achieved by adjusting handover and cell selection criteria, e.g. ensuring sooner or later inbound or outbound handovers to scale the effective cell ranges. However, the only certain way to provide full macro coverage is to reserve at least one carrier frequency which is free of CSG type femtocells, e.g. an escape carrier for macro users.

In Table 19.5, the six general interference paths are summarized including a short summary of the impact extent and type as well as some available options to mitigate the effect in the deployment. As can be seen from Table 19.5, in particular the impact of the femtocell layer on the macro cell layer needs to be considered. If the operator is unable to separate femtocells into their own frequency carrier, adaptive control of the maximum HNB and HNB-connected UE transmit powers constitute key methods. Such methods will be discussed in more detail in this section.

## 19.7.4  Adaptive Control of Home Node B Transmit Powers

One of the most promising and important methods for controlling the femto to macro downlink interference while simultaneously ensuring a minimum coverage area for the femtocell is to adjust the maximum allowed HNB transmit power based on measurements of the nearest macro Node B. A further target for the power algorithm (together with cell selection methods based on downlink SINR) is to limit the HNB coverage within exterior walls to minimize probability of a HUE connection to HNB at the same time as having a line of sight to a macro Node B without exterior wall isolation. For example, the downlink transmit power algorithm also provides mitigation of the uplink interference path indirectly.

One algorithm which is assumed throughout this section is based on the equation:

$$P_{HNB,TX} = \max\{\min\{P_{\max}, \alpha \cdot P_{MNB} + \beta\}, P_{\min}\} \qquad (19.1)$$

The parameters $P_{\max}$, $\alpha$ and $\beta$ can be signaled to the HNBs by the management system or be pre-configured (e.g. be operator- or vendor-specific). It is advisable that the operator maintains control over the power settings of HNBs operating in their own frequency band. We here assume also a $P_{\min}$ setting to ensure a minimum coverage range. Here we utilize the value of 0 dBm which provides a fair overall trade-off between femtocell and macro cell performance. As the algorithm is not standardized, its use must be agreed between vendor and operator. The value for $P_{max}$ can be set either to the maximum specified value for the home base station to allow maximum femtocell range or set more conservatively. The $\beta$-value denotes an offset over the received macro power and is effectively adjusting the cell size of the femtocell. Same goes for $\alpha$.

The power received from the strongest macro Node B at the HNB location can be measured by the HNB itself as described in Section 19.1. The algorithm can be further enhanced to take into account the presence of nearby HNBs and other optimization factors. The effect of the discussed power control algorithm on the macro-connected UE (MUE) performance is shown in Figure 19.8. Figure 19.8 assumes a deployment of respectively 12 and 100 HNBs in the dense urban scenario (apartment building structure) and it is assumed that most of the macro cell users (80%) on the same carrier are located in the building [10]. It is clear that for the simulated case, the available macro cell SINR for the macro UE is improved significantly by restricting the power of the HNB. It is also clear that this comes as a penalty on the HNB side where the probability of reaching a certain SINR now degrades. It is, however, also clear that even if one has power control mechanisms (e.g. the 100 HNB case), this is not sufficient to guarantee good indoor coverage for macro cell UE in all cases. Figure 19.8 shows Es/N0 probability in macro cells and femtocells. Es/N0 is defined as the narrowband signal-to-noise ratio on HS-DSCH.

Hence, it is generally recommended to keep at least a single carrier free from CSG type HNBs to allow macro users to 'escape' via inter-frequency handover. For hybrid or open HNB, this conclusion may apply as well, although it depends on the predicted backhaul and air interface capacity.

A similar possibility of managing the interference scenario for uplink exists by adjusting the maximum allowed transmission power for UE connected to the HNB (HUE). The HUE power regulation can be done with a fixed maximum value or dynamically. The dynamic approach is more effective but also more demanding in terms of required measurements and signaling [8]. The dynamic power control involves estimating the amount of interference (or effective noise rise) caused by a UE on the surrounding macro cells. This estimation can be done using UE measurements of pathloss towards considered base stations. Knowing which settings will not cause the noise rise at nearby stations to exceed an acceptable level, the serving HNB can inform the UE about the exact settings it can use.

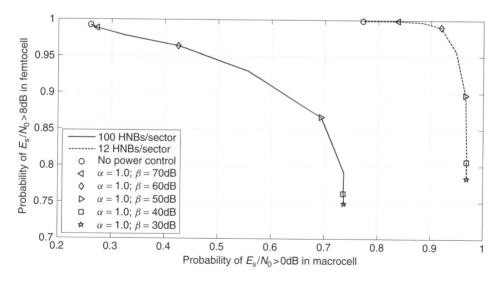

**Figure 19.8** Impact of the HNB adaptive power control on UE downlink performance. *Note:* $E_s/N_0$ relates to the HS-DSCH assuming that only one multi-code is transmitted and no orthogonality loss.

## 19.7.5 Femtocell Interference Simulations

The most basic factors for all studies on femtocells are: (1) the level of cell isolation such as, e.g. wall penetration losses as well as (2) the user distribution criterion. Unrealistic assumptions for these factors lead to over-optimistic or pessimistic conclusions. The values of penetration losses proposed for 3GPP studies are most often in the range of 10–20 dB for indoor-outdoor and 5 dB for room-to-room walls, see e.g. [10]. Those values provide quite good isolation between HNB and MNB or HNBs located in different buildings. However, in case of dense deployments, where many HNBs are located within a small area (e.g. apartment building), this is no longer the case. In Figure 19.9, such an example is shown in the form of a map where SINR availability (equivalent to HS-DSCH $E_c/N_0$, perfect orthogonality and only one multi-code transmitted) towards the strongest base station (either HNB or macro Node B) is indicated. In the example, a macro base station layer is also simulated on same frequency with the closest macro base station placed at grid point (0.0)m. Other macro base stations on a hexagonal grid are also simulated but not shown in Figure 19.9.

Distribution of users within the simulated environment is also an important factor in cases where statistical user distributions (instead of maps) are used. Placing UE in close proximity to an HNB which is not accessible (e.g. a CSG type HNB or an over-loaded hybrid/open) poses a threat in both downlink and uplink directions. In suburban or rural area deployment this may not be a serious issue, but for dense scenarios it becomes critical. As for dense urban simulations, the probability that a macro-connected UE is in the same building as a femtocell is specified as 80% in e.g. [10], the interferences between HNB and macro Node B are strong. The user distribution also plays an important role in how much offload effect is produced by the installed femto layer.

As the case of dense urban HNB deployment depicted in Figure 19.9 nicely reflects, the interference challenges for a mass-deployed HNB system (e.g. for apartments) in this case are the general simulation case considered throughout this section. For more information related to detailed simulation assumptions, the reader is referred to the detailed descriptions in [10]. In the simulations the following aspects are generally assumed:

- SINR is used as an abbreviation and denotes the $E_c/N_0$ available for the HS-DSCH. In the assessment no loss of orthogonality is considered and it is the $E_c/N_0$ achieved provided that only a single multi-code is transmitted with the total available HS-DSCH power. There is an upper-bound value of 22 dB due to an assumed level for error vector magnitude (EVM).

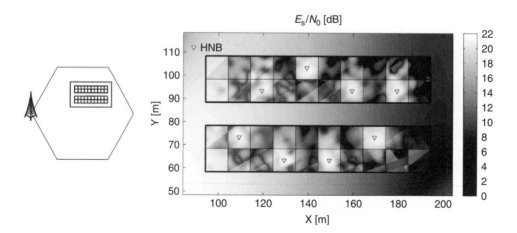

**Figure 19.9** SINR (HS-DSCH $E_s/N_0$) availability with co-channel dense-urban HNB deployment

- For downlink performance, we generally utilize the SINR value but in several cases we also conduct a mapping to user throughput utilizing HSDPA link adaptation methods and assuming that 15 codes and full UE data rate capability are available. We assume generally peak throughput values in our considerations. When considering performance with multiple active users in macro Node B or HNB, we assume equal round-robin type sharing of the resources.
- For uplink performance impact, we consider mainly the noise rise measured at the macro cell layer which indicates the reverse impact compared to the available noise rise target.
- When discussing offload of macro cells, we assume that all users in the network demand the same service level (256 kbps downlink and 64 kbps uplink) and look at the reduced load in the macro cell network, given the presence of HNBs.
- To look at different penetration levels, we generally consider the number of HNBs per macro Node B sector (12–100). The HNBs are randomly placed in the two building blocks. Three floors are simulated in accordance with [10].

In the following two sections, illustrative simulation results are presented for the case of dedicated channel and co-channel deployment of femtocells and macro cells.

### 19.7.5.1   Dedicated Channel Deployment of Femtocells

In deployments in suburban or rural areas, with high cell-to-cell isolation, there are practically very few detrimental interference effects limiting femtocell performance in a dedicated frequency band. However, with dense deployments of femtocells, the femto-to-femto interference becomes a seriously limiting performance factor, as is well known from other local area wireless standards such as Wi-Fi. Internal walls separating rooms in large apartment buildings may often not provide sufficient shielding, which results in a significant SINR reduction.

In Figure 19.10, simulations are presented for a different number of HNBs deployed per macro cell sector. The dense urban scenario discussed in the previous section is considered. It is clear from the results that dedicated channel deployment of femtocells provides the best overall end user throughput when comparing it to the co-channel deployment case which is included here for reference, e.g. macro cell interference will dominate the available SINR in many cases. For the co-channel deployment this is further visible from the fact that the available SINR does not depend significantly on how many HNBs are deployed in the sector. For dedicated channel deployment, the sensitivity to femto-to-femto interference is thus larger and as 100 HNBs are deployed per sector, the total received interference becomes approximately the same for the co-channel and the dedicated channel (e.g. the macro effect becomes negligible).

Figure 19.10 clearly indicates that using more carriers for HNB deployment, hence reducing the number of HNBs per sector, is an effective solution to further enhance the SINR performance in dedicated channel deployment. In order for this solution to work properly, the assignment of HNBs to specific carriers would have to be done by the operator (coordinated deployment) or using some autonomous techniques. For example, the HNB can be set to auto-select its best frequency from a set of possible carriers. The problem is that this approach creates a bandwidth-hungry femtocell deployment for the operator overall, e.g. less carriers can be dedicated to macro cell usage. In a co-channel deployment, the results in Figure 19.10 indicate that there is still some, although heavily reduced, potential gain in using less frequency re-use among the HNBs.

For the uplink direction, a dedicated channel deployment also helps the situation. As a macro cell operates on a different carrier, now there is no longer the threat of strong MUE to HNB interference. Hence, the HUE to HNB interference becomes the dominating effect for overall uplink performance in the femtocells. Generally, this effect is not as serious as the one caused by a macro-connected UE in co-channel deployment but it may be more likely, particularly for dense HNB deployments. Again, the cell range should be effectively limited by careful transmit power control in the downlink and the

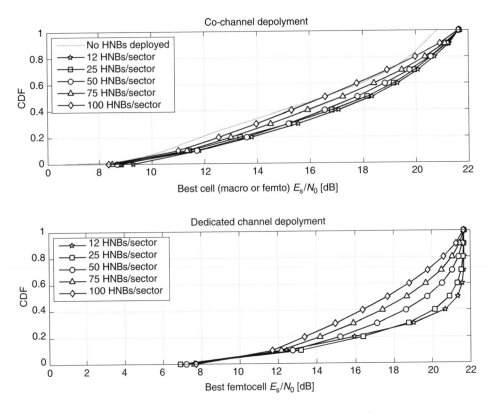

**Figure 19.10** $E_s/N_0$ distribution for HUE with dense urban deployment

good setting of handover parameters that a HUE further away from its own HNB may switch to a MNB. Other effective methods to control this interference include the adaptive HUE power limitation and use of open or hybrid femtocell configurations. Distributing femtocells over several frequency carriers also provides very significant improvements in this case.

### 19.7.5.2 Co-channel Deployment of Femtocells and Macro cells

Unless HNBs are deployed for basic coverage in areas where macro coverage is poor, the macro-to-femto interference generally dominates the HNB performance in the co-channel deployment case. To illustrate this aspect, consider Figure 19.11 where the benefit or drawback of using the HNB over the macro Node B in a given location is considered. The location of the femtocells is randomized uniformly over the macro coverage area and we use the urban type assumption where we have good femtocell-macro cell isolation and are not limited by femto-to-femto interference. The benefit or drawback is marked as a 'change' in throughput, e.g. positive if the user experience on the HNB is better than using the nearest MNB and negative if it is worse. As mentioned earlier we are assuming use of the power setting algorithm described in Section 19.7.4. Hence, a minimum femtocell coverage area is ensured since if a large amount of macro cell interference is detected, the HNB is allowed to increase its own transmission power. However, as maximum HNB power is limited, it is not always possible to ensure optimal conditions.

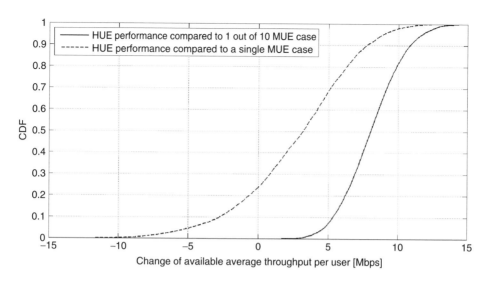

**Figure 19.11** Gains in home Node B UE performance from using HNB over nearest macro Node B

In the simulated result for a single femtocell per building in Figure 19.11, it is evident that for about 25% of the cell area (e.g. closest to the macro Node B) the femtocell cannot provide a higher SINR (or throughput) than what is available in the same location from the macro cell from outdoors. However, as is also shown in Figure 19.11, the available macro cell throughput often needs to be shared with many other users in the macro cell. Here, the example of 10 macro users is given, assuming a round-robin sharing of resources. For this example, the end user performance in the HNB is still consistently higher due to the lower load in the smaller cell.

Also the macro-to-femto uplink interference is important for the overall femtocell performance. As mentioned earlier, the CSG access mode poses an interference issue also in the uplink direction. A transmitting macro-connected UE positioned close to an HNB, but not connected to it (e.g. due to CSG), generates additional noise rise that cannot be managed directly by the affected HNB. This is especially visible at the edge of the macro cell. A macro-connected UE there uses a high transmit power in order to overcome pathloss to the macro Node B. In such a case, a nearby HNB observes an increased level of interference. For the macro-connected UE to HNB interference, the handover of the macro-connected UE to another carrier is the most promising solution and may be triggered in many cases since the downlink performance of the macro-connected UE is heavily limited by the presence of the CSG HNB.

The opposite interference path, e.g. the femto-to-macro uplink interference contribution, is considered next. This is the case where the HUE is the aggressor and the MNB is the victim. As discussed in Section 19.7.4, this effect is reduced by adaptively controlling the HNB downlink transmit power so that there is seldom direct line of sight between the HUE and the macro Node B. However, in the mass deployment case, the cumulative effect may still become significant. The simulated noise rise observed at a MNB due to an active HUE is shown in Figure 19.12. We consider the same dense urban deployment scenario as before and each deployed HNB has an active HUE placed uniformly within its cell range. The noise rise is assessed versus different power levels for the UE. For the simulation it is assumed that all HUEs transmit with the same power and simultaneously, e.g. we simulate the absolute worst-case scenario. As was indicated in Table 19.5 and is clear in Figure 19.12, an effective solution for the HUE to avoid macro Node B interference is dynamic power restriction for HUE. Compared to a total typical noise rise target for a macro Node B around 6 dB, it is clear

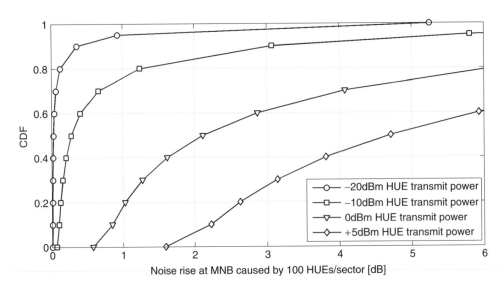

**Figure 19.12**  Noise rise at macro Node B generated by simultaneously active HUE

that some restrictions would be preferred unless the operator can rely on large trunking efficiency that many HUEs are seldom active at the same time or that required uplink data rates call for consistently lower HUE transmit powers. At −20 dBm HUE power capping, only a small fraction of macro Node Bs experience a noise rise effect larger than 1 dB. As the number then reflects an offload of 100 UEs in this particular case, the overall benefit for the macro layer (e.g. offload) should be intact.

## 19.7.6  Network Planning Aspects

As a core deployment issue it is important to understand the impact of which deployment strategy is used and what end user data rates can be promised. First, some basic link budget calculations are given in order to understand basic coverage aspects. Next, available user throughputs are studied for the dense urban and the suburban deployment cases.

### 19.7.6.1  Femtocell Range

This section gives estimates about the typical femtocell range considering the basic coverage that can be provided in an interference-free environment, e.g. dedicated channel and single HNB deployment case. The maximum femtocell radius is typically downlink limited due to the lower power level of the HNB. The maximum path loss calculation for an HSPA-capable femtocell is shown in Table 19.6. For reference, the path loss for WLAN 802.11g is also shown. The femtocell path loss is 115 dB assuming output power of 15 dBm and UE sensitivity of −100 dBm for 1 Mbps data rate. The WLAN maximum path loss is 110 dB. The used frequencies are also similar to HSPA using 2.1 GHz in most markets and WLAN using an unlicensed band at 2.4 GHz. Therefore, the corresponding cell range for a femtocell is expected to be somewhat longer than the WLAN cell range.

In the calculations, the indoor propagation model from [10] is utilized and the path loss as a function of cell range at 2 GHz is shown in Figure 19.13. It is assumed that the walls are of light construction material and that each wall adds a 5 dB attenuation. A 10 dB fading margin is considered which makes

**Table 19.6** Path loss calculation for HSPA femtocell and WLAN access point

|  | HSPA femtocell | WLAN 802.11g |
|---|---|---|
| Access point transmit power | 15 dBm | 20 dBm |
| Terminal sensitivity (1 Mbps) | −100 dBm (RSCP −110 dBm) | −90 dBm |
| Max path loss | 115 dB | 110 dB |
| Spectrum downlink | 2110–2170 MHz | 2400–2483 MHz |

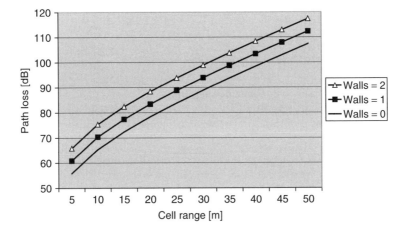

**Figure 19.13** Path loss as a function of cell range and number of walls according to [10]

the maximum path loss equal to 105 dB corresponding to a cell range of 35 to 45 meters with up to 2 walls between HNB and UE.

The maximum cell range as a function of HNB power is shown in Figure 19.14. To achieve a 20 meter cell range with 1–2 walls, −5–0 dBm HNB power is needed. Therefore, the femto power levels can be quite low covering a typical apartment or a small house.

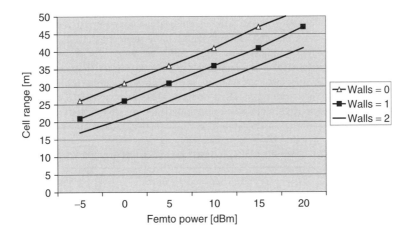

**Figure 19.14** Femto cell range as a function of femto power – no interference margin

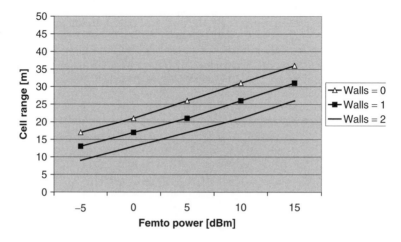

**Figure 19.15**  Femto cell range as a function of femto power – 10 dB interference margin

Figure 19.15 illustrates the femtocell range with an assumed 10 dB interference margin. That margin corresponds to co-channel interference from macro cells or from other femtocells. If we now want to provide a 20 meter cell range, the HNB transmission power needs to be 5–10 dBm. These calculations illustrate the benefit of using adaptive power settings in HNB depending on the interference environment.

### 19.7.6.2  Home Node B User Throughput

Simulation results have been conducted for an HNB density of 100 HNBs per macro sector. As the distribution algorithm is set so that all UE connects to the closest HNB, it corresponds to the case where HNBs are either assumed to be hybrid or open (or UE is part of all HNBs' CSG ID) or where it is a CSG cell but handover to a macro carrier is conducted whenever the own-CSG cell is offering too low a performance. The cumulative density functions (CDFs) of the available maximum user throughput rates (e.g. peak data rates) are plotted in Figure 19.16. In Figure 19.16, the performance of femtocell users (HUE) is compared for the two cases of co-channel versus dedicated channel deployment. The uppermost figure is for a dense urban environment with larger multi-floor apartment buildings whereas the suburban case illustrates more scattered HNB deployments protected by high penetration losses (e.g. single houses, single HNB per house, see e.g. [10]). The available user throughput values measured over the femtocell coverage zones are summarized in Table 19.7. The dedicated channel gives only moderate gains in data rates in dense urban areas. The reason is that most of the interference comes from other HNBs because of the very high HNB density. In suburban areas the dedicated frequency gives more gain because the main source of interference is macro cells.

**Table 19.7**  Resulting downlink femtocell data rates, 5 MHz deployment, HSPA, 100 HNBs/sector

| Environment | Dense urban deployment | | Suburban deployment | |
|---|---|---|---|---|
| Deployment | Co-channel | Dedicated channel | Co-channel | Dedicated channel |
| Minimum user throughput (10% CDF) | 2.2 Mbps | 3.3 Mbps | 3.0 Mbps | 13.5 Mbps |
| Typical user throughput (50% CDF) | 5.5 Mbps | 7.5 Mbps | 7.3 Mbps | 15 Mbps |

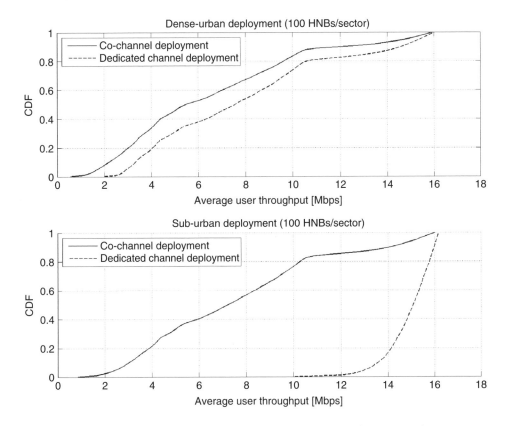

**Figure 19.16**   Available user performance with different HNB deployment scenarios

### 19.7.6.3   Macro Cell Offload

As mentioned, macro cell offloading via HNB deployment is one of the key use cases. Offload is possible for all types of HNBs, although open and hybrid types of course offer a larger offload effect in practice. When measuring the offload effect in the following we are assuming that all users in the network have a certain data rate requirement and we then look at how the required MNB transmission power (to fulfil the requirement) is reduced as HNBs are added to the sector. Due to the very aggressive simulation assumptions in [10] that an HNB always has a user present in its vicinity, the numbers are generally very optimistic. The femtocells have been simulated for the densely deployed scenario where there is high 80% probability of having users indoor where HNBs are deployed. In Figure 19.17, the results are shown in the sense of resulting downlink and uplink load factors (in percentage compared to full transmission power or noise rise) versus how many HNBs are deployed per sector. The users requiring offload have a downlink throughput of 256 kbit/s and an uplink throughput of 64 kbit/s. To take into account the situation of both open and CSG cells, simulations have been conducted where femtocells admit any nearby UE or only specific UE.

As can be seen for the HNBs configured as open, there is a significant offload in the cell as more HNBs are added (and a higher fraction of the active users will utilize those HNBs). As a large percentage of the users in the cell are located indoors, the presence of open HNBs adds significant improvements for the end users in many cases. Hence, the offload is significant and as 12 HNBs are added per sector, only a half load (measured in average required transmission power compared

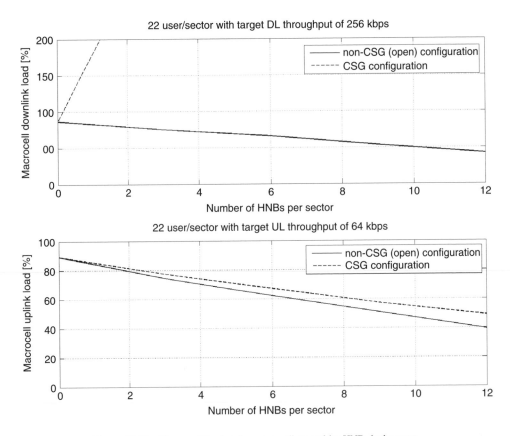

**Figure 19.17**   Change of load at the macrocell caused by HNB deployment

to maximum transmission power) remains in the macro cell as can be seen in the uppermost plot in Figure 19.17. A key here is that the chosen simulation scenario forces the HNB to be very active due to the biased distribution of users inside the buildings adopted in 3GPP. For the CSG case, this in turn creates detrimental effects immediately as the first added CSG HNB makes the macro-delivered SINR inside the same building so bad that the targeted downlink throughput can no longer be met from the macro cell without doubling the transmit power. This shows clearly that an operator should keep at least one 'escape' carrier free of CSG-type HNBs.

For the uplink case, there is a similar significant benefit as the HUE does not have to compensate for the indoor–outdoor path loss to get coverage via the HNB. Similar trends are observed here, although the difference between having an open versus a CSG configuration is less pronounced. The main reason is that HUE will use very low transmission powers to achieve a 64 kbit/s uplink throughput and hence only creates limited noise rise measured at the MNB. Hence, overall there is still offload capability for the uplink scenario deploying CSG HNBs.

## 19.7.7   Summary of Home Node B Frequency Usage

The frequency usage with HNBs is summarized below:

- HNBs can re-use the same frequency as macro cells in the case where the macro signal is weak and HNB is targeted to improve the coverage.

- HNBs benefit from dedicated frequency in the case of a strong macro signal in order to provide good data rates
- Operators have practical problems allocating a dedicated frequency for HNBs in high traffic areas when they have typically in total only two to four frequencies available. The challenges get even larger with Dual cell HSDPA (DC-HSDPA) that uses two adjacent frequencies to provide higher data rates.
- Closed Subscriber Group (CSG) may cause interference issues for macro cell UEs if the UE is not part of CSG. Therefore, at least one clear macro frequency is needed in case of CSG usage to allow UEs to escape the interference with inter-frequency handover.

## 19.8    Home Node B Evolution

As has been discussed throughout the chapter there is still more HNB-specific items that will be considered for standardization in future 3GPP releases. Some items that have already been discussed for 3GPP Release 10 were listed in Table 19.1. Key issues relate to further use cases of HNB technology, particularly optimization for business. In particular, the inter-CSG HNB mobility will be further optimized. Further, to ensure a high scope of applicability in the home and enterprise environment, there will be an increased focus on standardization solutions for local IP access, local breakout for efficient offload of the macro core network, and remote access solutions where end users may access their home LAN from wide area 3G devices. End-to-end QoS solutions will also be further discussed to improve the service offering over shared backhaul.

As first commercial HNB deployments are being made, it will also be considered if interference management needs to be taken to the next stage in 3GPP or if good vendor solutions can provide sufficient robustness. This trend is already seen in standardization work for the LTE Home eNode B (HeNB) where standardized interference mitigation and coordination are considered. In general, it is likely that standardization for WCDMA and LTE in this area may see further alignment as many technology aspects and challenges are essentially the same.

HNB evolution needs to consider the impact of the legacy UEs: what new Release 10 HNB features with UE impacts will be acceptable when we would need to simultaneously consider pre-Release 8, Release 8, Release 9 and Release 10 UEs which end up working differently.

## 19.9    Conclusion

Femtocell (Home Node B, HNB) represents a new approach to improving the indoor coverage and to pushing the network capacity. 3GPP specifications support femto deployment with enhancements in Release 8, 9 and 10 standards, but femtocells can be utilized by legacy pre-Release 8 UEs as well. The new features include femto architecture where RNC functionalities are mostly embedded in femto access points and the access points are aggregated via a femto gateway (HNB-GW) towards the core network. 3GPP specifications also cover inbound and outbound mobility from the macro cells to femtos and vice versa, interference control, security and closed subscriber groups.

Femtocells can be deployed on the same frequency as macro cells if femtocells are mainly targeted to improve indoor coverage in those areas with weak macro signal. If the macro signal is strong, it will be beneficial to use a dedicated frequency for femtocells to avoid interference between macro and femtocells and to offer higher data rates with femtocells. The interference between femtocells may also affect the user data rates in cases of dense femto deployment in apartment buildings. It is not simple to allocate a dedicated frequency to the femtocells because the total number of frequencies per operator is typically only two to four. The challenge gets even larger with Dual Cell HSDPA (DC-HSDPA) that uses two frequencies in parallel.

Commercial femto deployments started in 2009 by some operators mainly as the fill-in solution to improve home coverage. In the long term, femtocells can also be used to offload high traffic from the macro network. Femtocells change also the operator's business assumptions because the access point is typically purchased and installed by the end user and the transmission and the electricity are paid by the end user.

# References

[1] Femto Forum, Signals Research Group LLC, 'Femto Forum Femtocell Business Case Whitepaper', http://femtoforum.org, June 2009.

[2] Saunders, S., Giustina, A., Carlaw, S., Siegberg, R. and Bhat, R.B. *Femtocells: Opportunities and Challenges for Business and Technology*, New York: John Wiley & Sons, Ltd, 2009.

[3] 3GPP TS 25.467 v9.1.0, 'UTRAN Architecture for 3G Home Node B (HNB)'

[4] Broadband Forum, 'TR-069 CPE WAN Management Protocol', http://www.broadband-forum.org, May 2004.

[5] Broadband Forum, 'TR-196 Femto Access Point Data Model', http://www.broadband-forum.org, April 2009.

[6] 3GPP TS 25.468 v9.0.0 'UTRAN Iuh Interface RANAP User Adaption (RUA) Signalling'.

[7] 3GPP TS 25.469 v9.0.0 'UTRAN Iuh Interface Home Node B Application Part (HNBAP) Signalling'.

[8] 3GPP TR 25.967 v9.0.0 'Home Node B Radio Frequency (RF) Requirements (FDD)'

[9] Gora, J. and Kolding, T.E., 'Deployment Aspects of 3G Femtocells', the 20th Personal, Indoor and Mobile Radio Communications Symposium (PIMRC), Tokyo, Japan, September 2009.

[10] 3GPP RAN4, 'Simulation Assumptions and Parameters for FDD HeNB RF Requirements', R4-092042, May 2009.

[11] 3GPP TS 25-104 v9.2.0 'Base Station (BS) Radio Transmission and Reception (FDD)'.

[12] 3GPP TS 33.320 v9.0.0. 'Security of Home Node B (HNB)/Home Evolved Node B (HeNB)'.

# 20

# Terminal RF and Baseband Design Challenges

Laurent Noël, Dominique Brunel, Antti Toskala and Harri Holma

## 20.1   Introduction

This chapter presents an overview of the Universal Mobile Telecommunication Services (UMTS) Terrestrial Radio Access Network (UTRAN) design challenges imposed by some 3rd Generation Partnership Project (3GPP) test cases, with a focus on radio-frequency (RF), talk-time improvements due to Continuous Packet Connectivity (CPC), and also partly covering the impacts on baseband (BB). Section 20.2 presents design challenges in the WCDMA transmitter chain. Section 20.3 similarly covers the receive path, talk-time savings are presented in Section 20.4 and Section 20.5 opens a discussion related to the design of next-generation 'world-wide' phones which have to support multiple frequency bands as well as radio systems other than WCDMA.

3GPP has produced new Releases recently at roughly two-year intervals (Release 5 first version March 2002, Release 6 first version December 2004 and Release 7 December 2007) accelerating with Release 8 (December 2008) and Release 9 (early 2010), and thus is delivering plenty of new features for designers to cope with when designing new Release capable devices. The design cycle of a mobile phone takes on average 2 to 3 years, from first system requirements documents, to Integrated Circuit (IC) tape-outs, integration of hardware and software, hardware re-spins, to conformance tests, to handset available off-the-shelf. As the hardware (HW) development is to start in the first phase, having a Release with the physical layer as stable as early as possible is a key requirement for a fast time to market. The first Release 99 compliant handsets appeared in March 2003, but were built with ICs that were originally designed in the early stages of the Release. The challenge in early generations consisted in managing to fit all the circuitry into a reasonable volume and into a single two-sided printed circuit board (PCB). The radios were based on a discrete heterodyne architecture, and baseband sections were mostly an assembly of discrete components, such as Analogue to Digital Converters (ADCs), dedicated DSP chip, memory, etc. The next step consisted in reducing considerably the handset volume while still using two-sided boards. Large-scale integration CMOS technology allowed rapid development of single chip baseband with embedded ADC/DACs and the heterodyne radio was shrunk into highly integrated ICs. Recent advances in RF integration via the use of CMOS technology

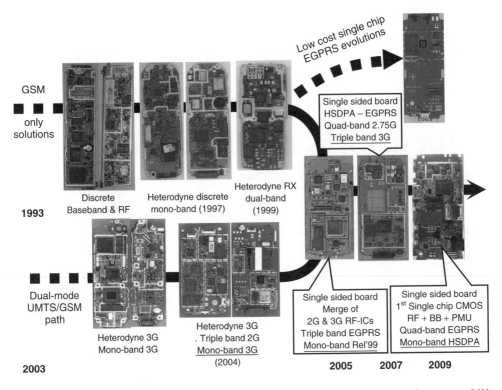

**Figure 20.1** GSM versus dual-mode handset evolution steps. Dual-sided GSM PCB picture is courtesy of [1]. The WCDMA dual-mode handset PCB pictures are reproduced by permission of [2] & [3]

has led to the emergence of feature rich, and yet reasonably low cost single-chip RF-BB and sometimes Power Management Unit (PMU) devices. Single chip solutions allow a significant reduction of the mobile phone hardware circuitry since only four ICs are now needed: an external SDRAM-DDR memory (referred to as 'DDR' throughout this chapter), a power amplifier (PA), a front-end antenna switch/filter module and the single-chip IC. The different steps are illustrated in Figure 20.1.

In their early days the first 3G handsets were heavily criticized on the market availability and on the limited model variety, and in many cases for a shorter talk time and standby time when compared to e.g. GSM devices. Between 2003 and now, the average talk time in 3G connected mode, normalised to a 900 mAH battery capacity, has nearly tripled, from 100 to approximately 270 minutes. Today, despite significant improvements in 3G handset technology, their talk-time has not quite reached the same level as their GSM counterparts. The 3G talk-time improvement brought by the introduction of CPC within the 3GPP standard is covered in Section 20.4.

Heterodyne architectures present the advantage of providing the best trade-off between achieving fast time to market while optimizing RF performance, at the expense of high power consumption and poor cost efficiency. It is no surprise to see that several millions of dual mode handsets in the 2005–2006 timeframe and a few remaining models in 2008 were still using this architecture [4, 5]. Yet, it is the direct conversion architecture (DCA) that acted as a key technology enabler to higher levels of integration. For a long time, the DCA was not put into production due to well-known DC-offsets/self-mixing, sensitivity to in- phase (I) / quadrature (Q) mismatches and flicker noise problems. The use of fully differential architectures, adequate manufacturing process and design models made IC production a reality around 1995 in 2G. Numerous inherent advantages make this technology essential in dual

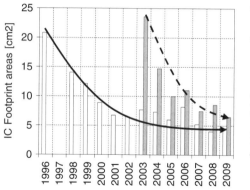
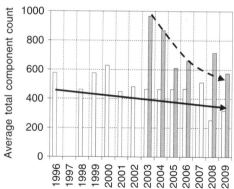

**Figure 20.2**  GSM versus dual-mode average IC footprint area and component count versus years. IC footprint area is defined as the total assembly area expressed in square centimeters occupied by the packaged and surface-mounted devices for ICs with eight pins or more. Mono-mode GSM handset: white bars, Dual mode 2G/3G handsets: grey bars

mode handsets: substantial cost and PCB footprint area is saved by removing intermediate frequency filters of the heterodyne architecture. Frequency planning is considerably simplified in DCAs, with only one voltage controlled oscillator (VCO) and phase-locked loop (PLL) required, thereby not only contributing to a reduction in power consumption, but also in providing a flexible solution to the design of multiple band RF subsystems. Finally, with channel filtering implemented at baseband, DCAs offer an elegant solution to the design of a single chip multi-mode receiver by enabling reconfigurability of the active low pass filter selectivity.

This handset evolution is best illustrated by Figure 20.2 which plots component footprint area and total component count vs. year. The graph was generated with data extracted from the analysis of 65 GSM and 38 GSM/WCDMA handsets [6]. While the total number of components in GSM stayed almost constant from 1996 to 2005, the IC area reduced considerably from 14 cm$^2$ (1998) to approximately 6 cm$^2$ (2006), representing a reduction of 57%. Over the last three years, the emergence of low cost solutions based upon single-chip CMOS RF-BB-PMU solutions have marked a trend in lowering the footprint area even further down to 4.5 cm$^2$. This graph must be compared with the simultaneous huge increase in mobile phone features. In 1998, handsets were using black and white LCD screens, without any multi-media capabilities. This is a considerable difference to the latest feature rich devices, now even including a built-in RGB video projector!

Starting with a nearly identical level of complexity in 2003, it only took UMTS handsets 3 years to reach the level of complexity that took nearly 8 years to mature in 2G. The slight increase in 2006–2008 3G handsets is partly due to the emergence of the first multiple band 3G devices, together with additional features such as support for WLAN or GPS.

## 20.2  Transmitter Chain System Design Challenges

### 20.2.1  The Adjacent Channel Leakage Ratio/Power Consumption Trade-Off

The adjacent channel leakage ratio (ACLR) is defined as the mean power the UE is allowed to transmit in the adjacent channel ($\pm 5$ MHz offset) and the alternate neighboring channels ($\pm 10$ MHz). The requirement is expressed as a ratio relative to the power the UE transmits on its assigned channel,

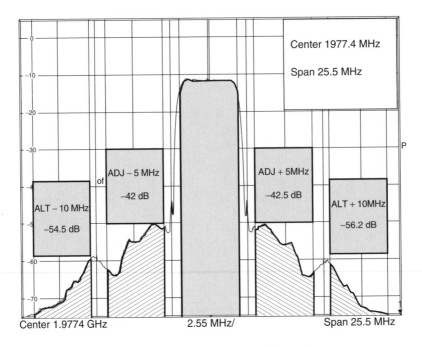

**Figure 20.3** Example of ACLR in a WCDMA terminal at maximum output power

and should not exceed −33 dB and −43 dB at ±5 and ±10 MHz distances respectively. This test represents UEs located at the edge of their own cell, thereby transmitting close to their maximum output power capability, while being located close to a Node B operating on an adjacent channel. ACLR is an important parameter which ensures that adjacent cells do not suffer from capacity loss due to UE noise emission leakage. The UE ACLR is therefore not only specified at maximum output power, but also across its entire output power dynamic range. An example of ACLR for a commercial power amplifier is shown in Figure 20.3. All powers are measured in 3.84 MHz bandwidth via a RRC filter.

With I/Q DAC capable of achieving better than −53 dB at ±5 MHz offsets, the ACLR is mostly dictated by the performances of the RF transmit modulator and power amplifier (PA). Since both AM/AM and AM/PM conversion contribute to ACLR, the antenna performance is not the simple sum of each block's contribution, as explained further in the following sub-section. For example, it is possible, via an AM/PM phase rotation in the RF modulator, to obtain a better ACLR performance at the PA output than at the RF modulator output if the PA distortions compensate for the AM/PM distortions in the modulation conversion.

### 20.2.1.1 ACLR Mechanisms: AM-AM and AM-PM

The term AM−AM is used to represent a change in the amplitude modulation (also called envelope) present on the output signal caused by the amplitude modulation of the signal present at the input of the non-linear device. Respectively the term AM-PM refers to the phase modulation caused by the amplitude modulation.

As the mean input power of such an AM stimulus is increased, the power amplifier (PA) is progressively driven outside its linear region to reach the saturation region – see regions I to III of Figure 20.4(A). The 1 dB compression point ($CP_1$) is most often used to define this transition. It corresponds to the power level (input or output referred) at which the amplifier gain has deviated from

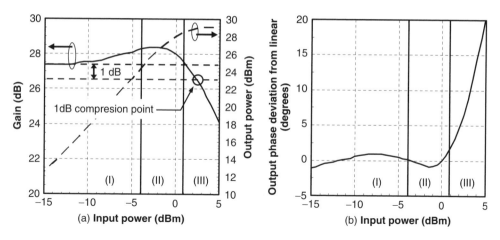

**Figure 20.4** Linear (I), saturation (II) and compression (III) regions in power amplifiers, (A): AM–AM, gain and output power vs. input power - (B): AM–PM

its linear value by 1 dB. Increasing further the input power level leads to clipping of the waveform. During the clipping process, ACL growth occurs as a consequence of odd order non-linearities. To illustrate this, the effect of compression in a device exhibiting only 3rd order distortion is modelled in Figure 20.5(A).

Injecting an AM stimulus composed of two closely spaced tones at frequency $f_1$ and $f_2$, results in an envelope peak to average power ratio (PAR) of 3 dB (assuming two tones of equal power) as shown in Figure 20.5(A). In the compression process, two ACL products are generated at the respective frequencies of $2f_1 - f_2$ and $2f_2 - f_1$. Figure 20.5(B) shows that the optimum power added efficiency (PAE $\approx$ 40% at 1 dBm input power) of the PA is obtained for mean input power close to the PA CP$_1$. PAE is derived by calculating the output power of the PA minus the input power divided

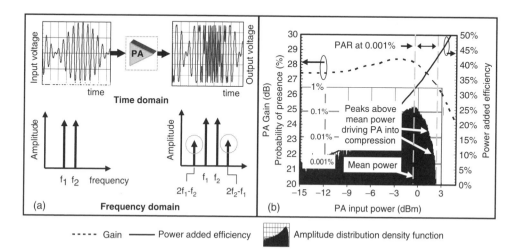

**Figure 20.5** Example of ACLR due to AM–AM in a PA modeled as $V_{out}(t) = 12V_{in}(t) - 4V_{in}^3(t)$. (A) Impact of a two-tone stimulus. (B) Release 99 reference channel 12.2 kbit/s ($\beta_c = \beta_d$) statistical analysis vs. PA gain vs. PA input power vs. PA power added efficiency

by the DC power consumption. Estimating PAR in WCDMA is more difficult than in the two tone case because the input signal appears as random chips in the time domain. The use of statistical tools such as the amplitude probability distribution (APD) overlaid in Figure 20.5(B) provides a valuable insight into the reference measurement channel (RMC) 12.2 PAR with equal amplitudes of the uplink DPDCH ($\beta_d$) and DPCCH ($\beta_c$). It becomes apparent that for a given mean input power close to the PA $CP_1$, the higher the PAR, the higher the probability of a peak driving the PA into compression, hence resulting in an increase of the adjacent channel leakage. Therefore, modulation schemes minimizing the PAR help operating the PA at its optimum PAE.

In WCDMA, the dual channel QPSK (also known in several sources as Hybrid-PSK (HPSK)) with complex valued scrambling operation, as described in Chapter 6, results in avoiding zero crossings of the constellation diagram. Under Release 99 test cases, HPSK manages to maintain the PAR nearly constant, ranging from 3.0 dB to 3.1 dB at 0.1% independently of all $\beta_c/\beta_d$ combinations.

Similarly high input powers will cause phase modulation (AM-PM) in a PA, thus the phase is rotated and therefore the resulting output spectrum is that of a phase modulated carrier, thereby also contributing to ACL products. Figure 20.4(B) was plotted with a CW tone (constant envelope) and shows that the phase is maintained nearly constant as long as the PA is driven in its linear region (region I). In this example, driving the PA into compression (region III) induces a phase rotation at a rate of 4 degree per dB. Note that typically, AM-PM occurs before AM−AM. This property has a significant impact in HSDPA applications, see Section 20.2.

### 20.2.1.2   The Maximum Output Power/ACLR Trade-Off

The 3GPP standard has two practical power classes: class 4 is defined for UEs with 21 dBm mean output power +2/−2 dB, while class 3 UEs transmit 24 dBm within +1/−3 dB. In HSDPA, mobiles must cope with a new uplink control channel, called the 'High-Speed Dedicated Physical Control Channel' (HS-DPCCH) as covered in Chapter 12. Depending on the relative amplitude of the uplink DPDCH $\beta_d$, DPCCH $\beta_c$ and HS-DPCCH $\beta_{hs}$, the increase in PAR at 0.1% probability ranges from 0.6 dB to 1 dB relative to the Release 99 reference 12.2 kbps test case. Through the AM−AM mechanism, the impact of the PAR increase onto ACLR can be solved in different ways:

- *Linearization techniques:* this concept relies on the use of less linear and therefore less expensive and more power-efficient PAs. The PA distortions are compensated using either feed forward or digital pre-distortion architectures. Despite being used in base-station equipment, the cost and complexity of these techniques have so far prevented them from being implemented in handsets.
- *Increase PA linearity:* this is equivalent to shifting the 1 dB compression point to higher input powers.
- *Output power back-off:* this technique consists in reducing the mean input power to avoid clipping.

It was decided to facilitate this latter technique in the 3GPP specifications by allowing a reduced maximum output power for each UE class. This has the advantage of allowing existing Release 99 PA designs to be re-used in HSPA applications and on the other hand as the situation occurs only when the uplink DPDCH data rate is below 64 kbps or so depending on the gain factor settings. However, backing-off the PA, or increasing its linearity, comes at the expense of reduced power efficiency, and therefore a reduced talk time.

In 3GPP discussions it was acknowledged that pure PAR is not the best metric to estimate the actual amplifier impact and use of tabular approach would have been too complicated as number of combinations was increasing for Release 6. Since ACLR at $\pm 5$ MHz is dominated by third order distortion products, the cubic metric (CM) was instead introduced in Release 6 as a way to generalize the amount of PA back-off allowed to fulfill the ACLR requirements. The reference CM value is set to 1 for $\beta_c/\beta_d = 12/15$, $\beta_{hs}/\beta_c = 24/15$. For all other combinations, the CM is defined with rounding to the highest number with 0.5 dB granularity using:

$$CM = CEIL \left\{ \frac{20\log_{10}((v\_norm^3)_{rms}) - 20\log_{10}((v\_norm\_ref)_{rms})}{k}, 0.5 \right\},$$

where $k$ is 1.85 for signals using the lower half of the channelization code tree, and 1.56 otherwise, v_norm is the normalized voltage waveform of the input signal, and v_norm_ref is the normalized voltage waveform of the reference signal (12.2 kbps AMR Speech). The maximum CM value is 3.5 and the associated maximum power back-off is set to CM − 1 (dB). This is used with uplink 16QAM modulation in Release 7 as well, where the maximum CM value is around 3 dB if not all gain factor combinations are allowed. The exact definition for the use of CM with 16QAM has been available since December 2007 and in later versions of the Release 7 specifications.

### 20.2.1.3  The ACLR/Talk-Time Trade-Off

In a class 3 handset, Figure 20.6(B) shows that the PA in its high power (HP)/high linearity mode remains the largest consumer of battery resources at full output power (54% share). We have seen that this is the price to pay to fulfill the ACLR requirements. However in the field, the UE spends very little time transmitting at maximum output power. Figure 20.6(A) shows a typical UE transmit power distribution collected from drive tests in a commercial UMTS band I urban environment [7]. The mean output power is close to −1 dBm.

In Figure 20.6(A), the dashed lines show that the PA in its High Power (HP) mode, consumes a constant 70 to 80 mA, whether the UE is transmitting 0 dBm or −50 dBm, thereby contributing to nearly $^1/_3$ of the entire power consumption (Figure 20.6(B) at −10 dBm output power). This represents a severe drainage on the battery capacity. To overcome this problem, one of the following PA control schemes may be used:

- *PA bias control.* This solution allows changing the PA quiescent current in either two (as shown in Figure 20.6(A)) or three steps. The current consumption is drastically reduced down to 20–30 mA

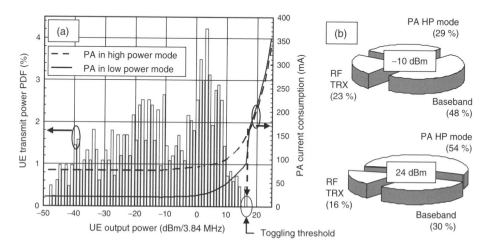

**Figure 20.6**  (A): UE transmit power density function vs. PA current consumption & PA power mode. (B) RF baseband power consumption distribution at −10 and 24 dBm output power

(plain lines). Latest PA technology can push this quiescent consumption to less than 10 mA, thereby further enhancing battery life. In the urban profile of Figure 20.6(A), toggling the PA at 12 dBm output power results in saving nearly 35 minutes of talk time using a 900 mAH battery.

- *PA power supply control.* This scheme uses a DC-DC converter to switch the PA power supply between two levels. This solution comes at the expense of an increase in PCB area and cost due to the required large choke inductor. Further readings on power consumption analysis and PA power consumption optimization schemes can be found in Section 20.4.

- Either control scheme calls for a careful selection of the toggling threshold. On one hand, the higher the power at which the PA is maintained in its low bias mode, the better the talk time improvement. On the other hand, PAs exhibit poorer ACLR performance in their low power (LP) mode. Therefore, the PA must be toggled at powers low enough to guarantee a good margin on the ACLR performance.

## 20.2.2   Phase Discontinuity

There is always a price to pay when implementing talk time enhancement techniques. PA toggling introduces two major impairments: a sudden gain variation ($\Delta G$) and a phase discontinuity ($\Delta\varphi$) occur in this process as can be seen in Figure 20.7(A). Both phenomenon cause problems: $\Delta G$ exceeds the 1 dB TPC step size, and therefore this step must either be calibrated in mass production, or measured on-board using an accurate power detector.

In reality, the problem is more severe as $\Delta G$ actually varies with temperature, carrier frequency, and battery supply voltage. In this case, if a power detector is used, then it must also be temperature and frequency compensated. If not, these variations require extra mass production calibration time.

To meet the 1 dB $\pm$ 0.5 dB accuracy of the TPC test pattern, the RF transmit modulator must also simultaneously adjust its output power to compensate for $\Delta G$. Consequently, the phase rotation shown in Figure 20.7(B) results from both PA and RF transceiver contributions. Too large a phase discontinuity severely affects the base-station detection performance. As a consequence, Release 5 contains the phase discontinuity requirements summarized in Table 20.1. Not only must the UE meet the $\pm$ 0.5 dB TPC accuracy and the ACLR requirements, but it must also ensure that phase jumps never

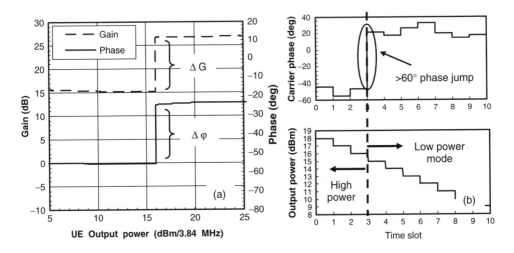

**Figure 20.7**   Example of PA phase and gain jump vs. output power (A), vs. time slot (B)

**Table 20.1** Phase discontinuity requirements

| Phase discontinuity $\Delta\theta$ in degrees | Maximum allowed rate of occurrence in Hz |
| --- | --- |
| $\Delta\theta \leq 30$ | 1500 |
| $30 < \Delta\theta \leq 60$ | 300 |
| $\Delta\theta > 60$ | 0 |

exceed 60 degrees at the Release 99 slot boundaries. Discontinuities greater than $30°$ are allowed, but hysteresis prevents this from occurring again within the next 5 consecutive slots (cf. Table 20.1). In the design one could use e.g. hysteresis for PA bias toggling or make use a baseband phase pre-distorter to compensate for the RF phase jump.

It was discovered that AM-PM also plays an important role in UE phase shifts, with the most demanding case being when the HS-DPCCH changing from an off state to a fully on state, resulting in a sudden 7 dB step in total uplink output power. This power rise could easily trigger the transition from low power to high power mode, resulting in a sudden phase jump in the middle of a Release 99 slot. In addition, when driven close to its maximum output power, a PA could easily exhibit a 4 degree phase rotation per dB step (see Figure 20.4(B)), which would translate into a sudden 28 degree phase jump! A 30 degree phase rotation in the HS-DPCCH results in a 0.4 or 0.2 dB loss in AWGN and VA30 detection performance respectively [8]. With 30 degrees set as the maximum value never to be exceeded in Release 6, a digital baseband phase pre-distorters will be required to compensate for RF imperfections.

## 20.3   Receiver Chain Design Challenges

Of the received test cases of the 3GPP standard, the reference sensitivity presented in Section 20.3.1 is the most demanding, setting not only tough performance requirements on the receive section, but also setting strict control of the transmitter to receiver chain isolation. In a WCDMA hand-set, this task is accomplished by a key component: the duplexer. The adjacent channel selectivity (ACS) test case covered in Section 20.3.2 sets a design trade-off between cost and complexity of the RF – baseband sections.

In the Release 99, RF receiver-related performance tests use the 12.2 kbps reference measurement channel (RMC), except for the maximum input power test case where the interference is generated with an orthogonal channel noise simulator (OCNS). The pass/fail criterion is based on a target bit error rate (BER) which must be better than $10^{-3}$ at a given downlink wanted channel power ($I_{or}$), and a dedicated physical channel energy per chip (DPCH_$E_c$) to $I_{or}$ power ratio [9]. Yet, the 3GPP baseband related performance pass/fail criterion is based upon a block error rate (BLER) $\leq 10^{-2}$ for DPCH_$E_c/I_{oc} = -17.6$ dB in the presence of OCNS, and not a target BER in a 12.2 kbps RMC. Link level simulations published in [10] show that a BLER of $10^{-2}$ and a BER of $10^{-3}$ are reached at nearly equal DPCH_$E_c$ to noise ratios, respectively at DPCH_$E_c/I_{oc} = -19.6$ dB, and $-19.8$ dB. Under the static, single path AWGN test conditions, the OCNS can be considered as orthogonal noise. Therefore these results can be applied to the 12.2 kbps RMC test case used in RF conformance. Throughout this section, $N_0$ denotes the power spectral density (PSD) of a white noise source. The model used to derive the NF budget throughout the next sections assumes a required BB DPCH_$E_c/N_0$ to $-19.5$ dB to account for a 0.3 dB penalty due to both baseband imperfections and the impact the RF I/Q receiver chain impairments. Measurements on recent commercial UEs, as shown in Figure 20.8, indicate that this model represents a worst case since measured DPCH_$E_c/N_0$ values range between $-19.6$ and $-20.1$ dB, even in presence of their respective RF-IC receiver chain impairments.

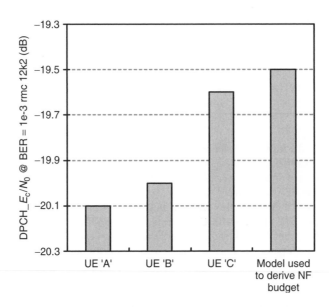

**Figure 20.8** Measured DPCH_$E_c/N_0$ at $I_{or} > -60$ dBm, required to meet BER $\leq 10^{-3}$ for RMC 12k2 (in the sense of the RF reference-sensitivity test case) in recent WCDMA handsets vs. model used for NF budget estimation. SNR is varied using an AWGN signal generator

## 20.3.1   UE Reference Sensitivity System Requirements

The receiver sensitivity test emulates the case of a UE located at the edge of a cell. Due to the full duplex nature of WCDMA, as the UE travels away from the Node B, not only does it receive a lower signal power for which it must maintain BER $\leq 10^{-3}$, but it must also increase its transmitted output power, to eventually reach its maximum output power capability. Figure 20.9 illustrates the test set-up seen from a UE RF sub-system point of view. We will see that the UE reference sensitivity depends on the UE Noise Figure (NF) budget, which in turn is a function of three variables:

$$\text{UE NF budget (dB)} = \text{NF}_{\text{BB intrinsic}} \text{ (dB)} - D_{\text{TX noise}} \text{ (dB)} - D_{\text{TX leakage}} \text{ (dB)},$$

where $\text{NF}_{\text{BB intrinsic}}$ is the noise figure derived with the transmitter chain 'virtually switched-off', derived from the intrinsic baseband DPCH_$E_c/N_0$ performance, $D_{\text{TX noise}}$ is the allowed desensitization due to TX noise leakage, and $D_{\text{TX leakage}}$ is the allowed desensitization due to the mixer second order distortion products caused by the leakage of the transmit carrier through the duplexer. In either case, the net result is a degradation of the SNR: the second order inter-modulation (IMD$_2$) products fall in-band, while the leakage of TX chain noise is directly added to the intrinsic LNA referred noise floor and therefore also degrades the $I_{or}$ SNR. In the following sections we show that keeping both aspects of desensitization under control used to come at the expense of using two external band-pass filters for each UMTS band of operation. Recent advances in DCA transceivers and in duplexer technology have made either transmit and/or receive chain SAW-less subsystems a reality in mass production today. Section 20.4 shows that the removal of these filters is the key to achieving a low cost multiple band dual mode solution. For the sake of simplicity, secondary desensitization factors due to LNA gain cross-compression and reciprocal mixing of the transmitter carrier leakage with the receiver LO-phase noise are neglected.

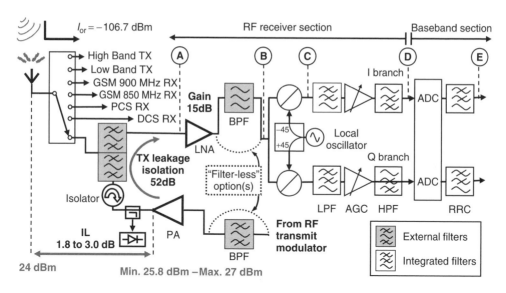

**Figure 20.9** Reference sensitivity from a UE perspective (BB: baseband; PA: power amplifier; BPF: band-pass filter; LPF: low-pass filter; HPF: high-pass filter; LNA: low-noise amplifier; ADC: analogue to digital converter; RRC: root-raised cosine filter). The coupler/detector combination is optional – refer to Section 20.4.3 for more details on PA control schemes

### 20.3.1.1  Baseband Modem Performance Intrinsic Noise Figure Requirements

In Band I devices, the injected composite input power at the UE antenna port is $-106.7$ dBm, in which the DPCH_$E_c/I_{or}$ is set to $-10.3$ dB, i.e. at $-117$ dBm. Assuming a DPCH_$E_c/N_0$ ratio of $-19.5$ dB at BER of $10^{-3}$, the maximum acceptable noise level referred to the antenna connector must be less than $-97.5$ dBm. At room temperature, the thermal noise power level in a 3.84 MHz bandwidth is $-108.1$ dBm. Consequently, the maximum $NF_{BB\ intrinsic}$ is 10.6 dB as shown in Figure 20.10. With reference to Figure 20.8, it is worth noting that the $NF_{BB\ intrinsic}$ budget in state of the art 3G phones ranges between 10.7 and 11.2 dB in Band I.

Using this NF budget as a baseline for Band I operation, Table 20.2 shows that the reference sensitivity requirements across 3GPP bands set a NF budget ranging from 10.6 to 13.6 dB. The budget is relaxed in bands where the duplex distance and short duplex gaps place tougher design constraints on the duplexer than in band I. This is the case, for example, in Band II where the short 80 MHz duplex distance is a challenge for duplexer manufacturers, resulting in higher insertion losses (typically 3 dB in Band II, compared with only 2 dB max in Band I), and therefore a higher NF is tolerated.

In early generations of dual-mode Enhanced General Packet Radio Service (EGPRS)/WCDMA handsets, only one 3G band was supported. Today, the de-facto standard is at least dual band 3G support in addition to the quad band for EGPRS, with a variety of combinations depending on the targeted world region. The most common combinations are Band I and VIII, II and V, while mono-band products come either as Band I or Band IV. The next generations of low cost, large volume 3G entry phones must support at least dual band and in certain markets triple Band 3G operations, while in the high-end segment the number of bands can be as high as 4 simultaneous bands. In the context of the low cost, large volume market segment, the use of a single antenna is key to maintaining the manufacturing cost of the UE at a reasonable level. This calls for a single pole (SP) – multiple throw

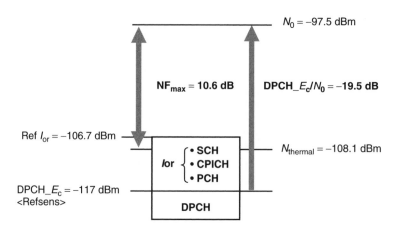

**Figure 20.10**  WCDMA band I reference sensitivity test case scenario (NF: Noise Figure; $N_{thermal}$: thermal noise power in 3.84 MHz bandwidth at 25 °C)

**Table 20.2**   3GPP bands, co-located standards and associated reference sensitivity

| 3GPP | Uplink band | Downlink band | Duplex distance | Region | Reference sensitivity | Co-located standards |
|------|-------------|---------------|-----------------|--------|-----------------------|----------------------|
| I    | 1920–1980   | 2110–2170     | 190 | WW | −117 | None |
| II   | 1850–1910   | 1930–1990     | 80  | US | −115 | GSM, CDMA |
| III  | 1710–1785   | 1805–1880     | 95  | EU | −114 | GSM |
| IV   | 1710–1755   | 2110–2155     | 400 | US | −117 | CDMA |
| V    | 824–849     | 869–894       | 45  | US | −115 | GSM, CDMA, AMPS, NADC |
| VI   | 830–840     | 875–885       | 45  | Japan | −117 | PDC |
| VII  | 2500–2570   | 2620–2690     | 120 | WW | −115 | WiMAX |
| VIII | 880–915     | 925–960       | 45  | EU | −114 | GSM |
| IX   | 1750–1785   | 1845–1880     | 95  | Japan | −116 | PDC |
| X*   | 1710–1785   | 2110–2170     | 400 | US | −117 | CDMA |
| XI   | 1427.9–1452.9 | 1475.9–1500.9 | 48 | Japan | −117 | |
| XII  | 698–716     | 728–746       | 30  | US | −114 | |
| XIII | 777–787     | 746–756       | 31  | US | −114 | |
| XIV  | 788–798     | 758–768       | 30  | US | −114 | |
| XIX  | 830–845     | 875–890       | 45  |    | −117 | |
| XXI  | 1447.9–1462.9 | 1495.9–1510.9 | 48 | Japan | −117 | |

antenna switch as shown in Figures 20.36 and 20.37 of Section 20.5, where it can be seen that the switch complexity varies from SP9T to SP7T depending on whether co-banding is implemented or not. Refer to Section 20.5 for more details on the trade-off between cost and performance in multiple band handsets. With antenna switch insertion losses (IL) ranging from 0.8 to 1.2 dB, and duplexer IL ranging from 1.8 to 2.1 dB in band I [11] it can be seen that the 10.6 dB NF band I budget gives plenty of headroom for RF IC designers to pass conformance test. As a consequence, most 3G phones today pass the Band I reference sensitivity test with at least a 3 dB margin [12].

### 20.3.1.2 Impact of TX Noise Leakage into the RX Band

The noise leakage budget is evaluated in five elementary steps as shown in Figure 20.11. We assume that the PCB leakage is negligible compared to the duplexer TX to RX isolation*. Let's evaluate the noise leakage budget for a 5 dB NF, assuming a maximum desensitization $D_{TX\ noise}$ of 0.5 dB (steps 1 and 2 of Figure 20.11).

To meet this target the TX noise leakage power received at monitor 'A' (see Figure 20.9) must be set to a level 9.1 dB below the intrinsic noise floor of our receiver, i.e. at $-112.1$ dBm, which expressed in Power Spectral Density (PSD) translates into $-177.9$ dBm/Hz (steps 2 and 3). This is below thermal noise and is hence a rather unusual concept. Assuming a worst case 43 dB duplexer noise isolation, the maximum acceptable noise level as measured at the duplexer TX port is $-134.4$ dBm (step 4), a level achieved by most commercial PAs [13, 14]. This requirement becomes a real challenge once referred to the WCDMA transmitter output port, with a maximum noise PSD of $-166$ dBm/Hz (step 5)! Figure 20.12(A) extends the previous derivation over a wider range of UE NF and shows that, with this PA/duplexer combination, it is impossible to avoid at least 0.25 dB desensitization in a filter-less solution. A generalization of this example is presented in Figure 20.12(B) which shows an interesting dilemma: to meet a given $D_{TX\ noise}$ budget, the better the receiver NF, the tougher the transmit modulator noise requirements. Alternatively, since all worst case conditions will seldom simultaneously occur in a single handset, one may want to consider allowing enough desensitization to just pass the reference sensitivity, i.e. with 0 dB margin. Using these assumptions, the TX noise requirements could be relaxed to $-150$ dBm/Hz depending on the NF.

Until the beginning of 2009, RF-IC transmitters which went into mass production were far from reaching the noise requirements for a filter-less solution. Thus the use of an external band-pass filter was essential in practical implementations. Recent press Releases [15] announced the availability of the first transmit path filter-less RF solutions. Considering the time delay between the Release of a new platform and their first appearance in commercial products, it is believed that first TX Saw-less commercial phones should appear in early 2010.

### 20.3.1.3 Impact of TX Leakage and Mixer IIp2 Budget on Reference Sensitivity

In Direct Conversion Receivers (DCRs), self-mixing occurs and is known as one of the mechanisms generating $IMD_2$ products [16, 17]. Self-mixing is due to finite isolation between the RF and the LO port of the down-conversion mixer, as illustrated in Figure 20.13. Under these conditions, the mixer behavior may be approximated as that of a squarer. In the 3GPP standard, three types of blockers will cause SNR degradation via second order distortions: out-of band sinusoidal (CW) blockers, in-band CW blockers, and the mobile's own transmitter leakage. The squaring of CW blockers is not generally a problem in WCDMA since the IMD2 product is a simple DC term which can be easily rejected using either High Pass Filters (HPF), or AC coupling capacitors. In the case of AM blockers such as the transmit leakage, the baseband $IMD_2$ products are rather more troublesome: their amplitudes vary in time and their bandwidth is twice that of the signal source, thereby directly overlapping the spectrum of the wanted channel. This can be clearly seen in Figure 20.13 (right) which shows the spectrum as observed at the I/Q output of the mixer. The strong DC component is rejected by the RX HPF. However, the receive channel LPF cannot avoid SNR degradation. To overcome this problem, one can either increase the mixer linearity, but this usually results in an increase in current consumption.

---

*Passive filter manufacturers have made significant efforts in shrinking the packages of duplexers for UMTS. Duplexer package size has shrunk from $5 \times 7$ mm in 2003, to $3 \times 2.5$ mm in 2009, and with future package developments down to $2 \times 2.5$ mm. With so small footprints, maintaining a high duplexer isolation from the TX port to the RX port now also implies a well controlled PCB layout design so that leakage is not dominated by PCB technology.

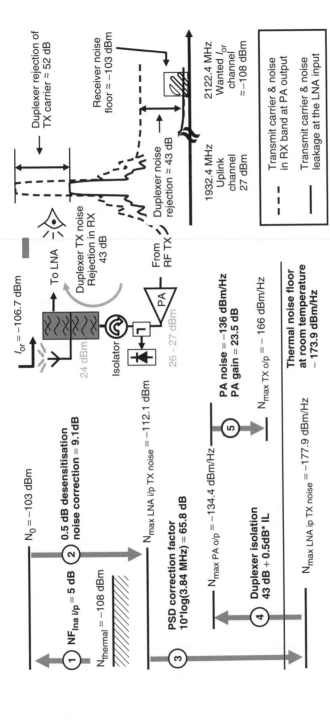

**Figure 20.11** Transmitter noise leakage and impact on RF modulator output noise - PSD = power spectral density – PA = power amplifier. Assumptions: 0.5 dB extra insertion loss due to isolator, coupler and PCB losses, 43 dB duplexer isolation, 23.5 dB PA gain, PA noise emission of −136 dBm/Hz [13]

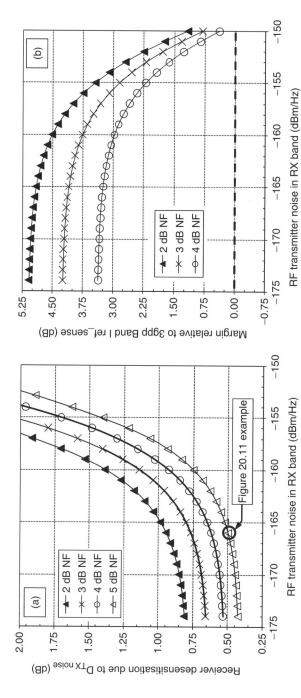

**Figure 20.12** A): WCDMA TX modulator noise impact on receiver desensitization - (B): Margin to the band I 3GPP reference sensitivity level in a class III example. *Plots generated using worst case assumptions: 43 dB duplexer isolation, 23.5 dB PA gain, 23.2 dBm antenna output power, maximum PA noise of −136 dBm/Hz [13], 3.2 dB PA to antenna insertion losses, and 2.2 dB antenna to LNA input pin insertion loss*

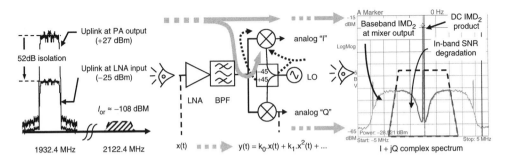

**Figure 20.13** Self mixing in direct conversion receivers – Left: TX leakage at LNA input - Right: I/Q spectrum observed at mixer output; dashed lines represent the wanted channel

Alternatively, the blocker can be rejected with the use of an external band-pass filter, but this comes at the expense of increased cost.

In a filter-less architecture, the mobile's own TX leakage places the most stringent requirements on the mixer IIp2 which must receive a weak input signal ($\approx -90$ dBm), in the presence of a TX leakage mean power of approximately* $-10.5$ dBm! Considering the rather large number of variables required to derive the mixer IIp2 requirements**, the IIp2 requirements can only be derived accurately using link-level simulations. The method presented in [18] is used in Equation (20.1) as a 'hand computation' to evaluate an order of magnitude of the LNA input referred two tones (CW) IIp2:

$$IIp_2 = 2P_{\text{TX leakage}} - P_{\text{Max IMD2}} - 3.01\,\text{dB} - L_{\text{HPF}} - L_{\text{LPF}} + K_{\text{mod}}, \qquad (20.1)$$

where $P_{\text{TX leakage}}$ is the transmitter leakage at the LNA input port, $P_{\text{Max IMD2}}$ is the maximum tolerable IMD$_2$ product for a given $D_{\text{TX leakage}}$, $L_{\text{HPF}}$ is the rejection of the DC product in the HPF, $L_{\text{LPF}}$ is a correction factor to estimate the integrated IMD2 product falling in the RRC LPF bandwidth, and $K_{\text{mod}}$ is a correction factor which is crest factor dependent, and is equal to 1.07 [18, 19] to reflect the RMC 12.2 TX leakage PAR. Using previously listed assumptions, an IIp2 of 52 dBm is required, which translates to >70 dBm IIp2 at the mixer input. This figure represented a tough design challenge for RF-IC designers until 2007–2008. The use of an inter-stage filter was therefore needed. In the above example, a filter offering 35 dB of rejection [20] reduces dramatically the requirement to less than 10 dBm. This is because a 1 dB reduction in transmit leakage reduces IIp2 by 2 dB. Thus, calibrating the maximum output power of a handset to reach 22 dBm instead of 24 dBm can also help. Careful design of fully differential structures that reduce mixer imbalances and therefore increase the IIp2, in either passive or active mixer topologies, as well as the integration of either automatic, or mass production-assisted internal IIp2 calibration circuitry, has brought to the market a new breed of DCRs which offer >52 dBm IIp2 [21, 22]. We will see in Section 20.5 that this perspective paves the way for low-cost multi-band 3G radio solutions.

---

*Transmit leakage mean input power at mixer input $\approx$ PA output power (+27 dBm) – isolator/coupler losses (0.5 dB) – duplexer isolation (52 dB) + LNA gain (15 dB) = $-10.5$ dBm.

**Required variables to estimate mixer IIp2 requirements in Equation1: UE power class (assume 24 dBm antenna power), duplexer TX-RX transmitter carrier isolation (minimum 52 dB), LNA gain (15 dB), LNA referred NF (let's assume 4 dB), tolerable desensitization $D_{\text{TX leakage}}$ (let's say 0.2 dB), PA to antenna insertion losses (2.8 dB worst case), combined analog LPF and baseband RRC filter rejection (assume 3 dB), HPF rejection of DC terms, and finally DC to IMD$_2$ power ratio relationship (assume 9 dB).

## 20.3.2   Inter-Operator Interference

This section presents the design trade-off between rejecting sufficiently the adjacent carriers due to other operators and avoiding the degradation of the modem BB BER performance. The two types of adjacent interferer defined in the 3GPP Release 99 [9] are illustrated in Figure 20.14:

- The adjacent channel selectivity (ACS) defines the minimum UE tolerance of the presence of a WCDMA modulated carrier located at $+/-5$ MHz offset from its assigned carrier frequency.
- The narrowband blocking test uses a GMSK modulated carrier and aims at ensuring WCDMA handsets can co-exist in frequency bands where both GSM and WCDMA operate. This situation is foreseen in bands II, III, IV, V, VIII and X. In bands using the 200 kHz raster, without the use of the 100 kHz offset, the GMSK blocker power is set to $-56$ dBm and located at 2.8 MHz offset (e.g. Bands III and VIII), while in other bands the spacing is tighter, at 2.7 MHz, with a relaxed power of $-57$ dBm. These offsets allow the introduction of WCDMA carriers in the middle of a 5 MHz block of EGPRS spectrum (minimum spacing $= 5/2 + 0.2$ MHz).

In an ideal world, all WCDMA Node Bs would be co-located and the base-station PAs would be so linear that the output spectrum would be perfectly rectangular (top of Figure 20.14). In such a case, all carriers would suffer from an identical path loss and, therefore, reach the UE antenna with a power ratio identical to that set at the Node B. As this is not the reality, the ACS test case I requests the UE to maintain a BER $\leq 10^{-3}$ (RMC 12.2 kbps) in the presence of either interferers of Figure 20.14, while transmitting at 20 dBm in class III operation. A poor ACS performance may lead to dropped calls in certain areas of the cells, also called 'dead zones', the worst case being the proximity of an adjacent carrier generated by a Node B located at the edge of the assigned UE cell. An evaluation of the size of dead zones can be found in Chapter 8.

The handset performance in presence of ACI sets a design trade-off between the receiver channel filter rejection, the ADC resolution, the AGC loop accuracy and the LNA linearity. Assuming an LNA linear enough in first approximation, the rejection of these interferers is achieved by the DCR analog

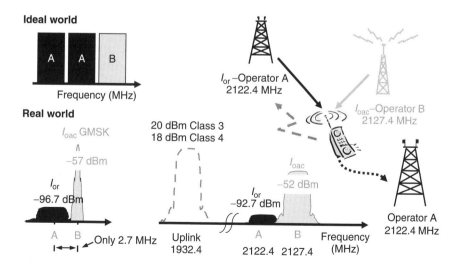

**Figure 20.14**   Adjacent channel interference (ACI) from a handset perspective

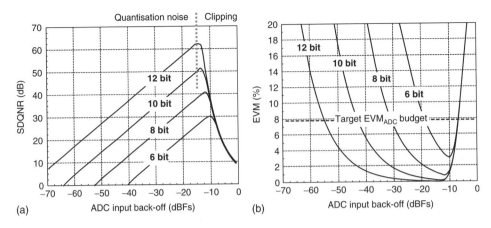

**Figure 20.15** (A): SDQNR (in dB) at ADC output against ADC resolution and ADC input back-off (BO) or ADC clipping ratio. (B): Corresponding ADC output EVM performance assuming AWGN noise to EVM relationship. The dashed horizontal sets a target ADC$_{EVM}$ budget of 7.8% for the sake of illustrating the design trade-off discussion

LPF and the digital baseband RRC filter, as shown in Figure 20.9 (monitor D and E). In between, the ADC is the key block which influences both RF and baseband architectural choices. The ADC input power level must be set to maximize the ADC output SNR. The automatic gain control loop (AGC) operates to adjust the analogue gain so that the total input power is optimally set. For the sake of simplicity, in the following discussion, we estimate the minimum required ADC resolution by assessing the ADC EVM budget (denoted EVM$_{ADC}$) based upon the pseudo quantization noise model. The graph plotted in Figure 20.15(A) shows the optimum Signal to Distortion Quantization Noise Ratio (SDQNR) at the ADC output is met for ADC clipping ratios, also denoted 'ADC back-off' (ADC BO), ranging from $-12$ to $-14$ dBFs. The resulting EVM$_{ADC}$ is plotted in Figure 20.15(B).

An example of EVM$_{ADC}$ budget of 7.8% (Figure 20.15(B)) is introduced to illustrate the design trade-off discussion. Let's assume a UE targeting HSDPA 16 QAM categories, say, category 8 for e.g. The UE EVM budget is estimated from the required SNR to achieve the highest 16QAM throughput. From [23], less than 5% throughput loss is met if the total composite EVM is less than 12.5%. Assuming that each EVM impairment is AWGN-like, we take an example where the Node B EVM (EVM$_{Node\ B}$) is equal to 5%, the duplexer EVM (EVM$_{dux}$) at band edge of a Band VIII device is 2.5% (worst case), and the RX RF-IC EVM (EVM$_{RF-IC}$) performance is equal to 8%. This sets an EVM$_{ADC}$ budget of:

$$EVM_{ADC} = \sqrt{(12.5)^2 - [EVM_{Node\ B}^2 + EVM_{dux}^2 + EVM_{RF-IC}^2]} = 7.8\,\% \qquad (20.2)$$

In the following, we use two extreme cases to illustrate the design trade-off in designing channel filtering for a given EVM constraint:

- Ideal DCR with infinite adjacent carrier rejection. This situation is depicted in Figure 20.16(A), where the optimal ADC input back-off is always met within the accuracy of the AGC loop. In Figure 20.16, a real-life histogram of ADC BO recorded over the duration of a voice call in the case 1–3 km/h fading model is overlaid to illustrate that at any moment in time, the actual ADC BO deviates with a 6 dB spread from the target ADC BO. This is due to the slow update rate of the loop (10 ms update rate). A faster update rate would result in less spread, and therefore would reduce ADC dynamic range (DR) requirements. Applying this experimental histogram recording, the wanted

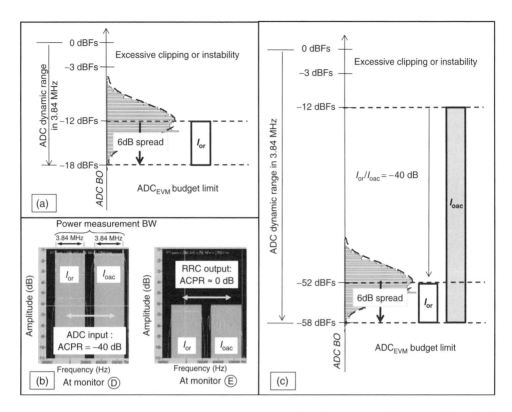

**Figure 20.16** (a) ADC dynamic range estimation in the case of an ideal "DCR" with infinite ACPR (b) FFT at ADC input and RRC digital filter output in the case of an "all-pass" DCR. Plot generated with WinIQSim™ [24], (c) ADC dynamic range estimation in the case of an "all-pass" DCR. The ADC BO histogram shows a recorded distribution captured over a 10 minute long WCDMA BLER measurement performed in a fading test case-1, 3 km/h. The AGC loop updates of the analog gain at 10 ms intervals, resulting in a 6 dB spread of actual back-off compared to the target back-off

carrier will never experience an ADC BO lower than $-18$ dBFs. Reverting to Figure 20.15(B), the minimum ADC resolution to meet $<7.8\%$ EVM$_{ADC}$ is 6 bit. From a baseband perspective, this solution is ideal: the lower the number of quantization bits, the smaller the number of multiply & additions and therefore the lower the power consumption. However, in this case the high ACPR requirement calls for sharp channel filters. The sharper the filter, the higher the in-band distortions, the greater the ICI, and therefore the poorer the BER vs. DPCH_$E_c/N_0$ performance.

- DCR behaving as an all-pass filter, performing simple, distortion free, down-conversion. Figure 20.16(B) was generated with the commercially available product WinIQSIM™ [24] and shows that a 20 tap long digital RRC filter can easily reject the ACS by 40 dB. Hence, very little analogue channel filtering is required since the role of this block can be restricted to prevent SNR degradation through aliasing during the conversion process. This has the advantage of minimizing RF-IC EVM and therefore the BER degradations due to ICI. However, in this case, the AGC now effectively reduces the analog gain so that the total power $I_{or} + I_{oac}$ meets the optimum back-off. Since $I_{or}$ is 40 dB below the total ADC input power, this is equivalent to setting the $I_{or}$ back-off, on average, to $-52$ dBFs (see Figure 20.16(C)). With our ADC BO histogram example, the minimum BO experienced by the wanted carrier is now $-58$ dBFs. From Figure 20.15(B), the ADC effective in-band resolution can be estimated to approximately 12.6 bits. A few years ago, this figure would not have been practical using conventional pipeline ADCs because of the associated high cost and high power consumption. The situation has recently changed with the widespread use of sigma-delta ADC converters which can deliver up to 12 bit in-band effective resolution at less than 10 mW power consumption per ADC [25, 26].

A trade-off between these two extremes is necessary to reach an optimum solution. The use of high resolution $\Sigma\Delta 12$ ADC technology for WCDMA reduces considerably the complexity of the analogue channel filter. With 12 bit effective resolution, Figure 20.15(B) shows that our EVM$_{ADC}$ example budget sets a minimum wanted carrier BO of approximately $-52$ dBFs. Compared to the 'all-pass' DCR budget (Figure 20.16(C)), this translates into an ACPR requirement of only 6 dB channel attenuation. Thus it can be seen that with today's technology, the design of the analogue channel filter can be considerably relaxed. The resulting modern CMOS radio can now rely on digital filters to perform the ACPR requirement with a very low penalty on EVM, and yet with low power consumption.

Finally, as far LNA linearity is concerned, as $I_{or}$ is increased, some handset performance is limited by the LNA ACL performance. For example, current commercial handsets pass the ACS test with 15 to 20 dB margin at $I_{or} = -92.7$ dBm, i.e. $I_{oac}$ ranging from $-37$ dBm to $-32$ dBm. This is very close to the UE maximum input power. At $I_{or} = -65.7$ dBm, the handset must be able to cope with an adjacent carrier power of $-25$ dBm (also referred to ACS test case II). This sets an LNA design challenge to ensure that:

- The gain is sufficiently dropped to relax the mixer linearity requirements.
- The gain is not too low to prevent the LNA NF from failing to meet the HSDPA SNR requirements (see Section 20.3.3).
- Preferably, the LNA gain toggling avoids generating phase discontinuities which would impact BB modem HSDPA demodulation.

Yet the design must also maintain a high LNA linearity to prevent ACL products from falling directly in-band, thereby degrading the wanted channel SNR.

## 20.3.3  Impact of RF Impairments on HSDPA System Performance

From an RF perspective, the use of low spreading factor (SF) 16, and 16 QAM in HSDPA sets a tighter I/Q impairment budget than in Release 99 QPSK modulation. This is due to the fact that in a 16QAM constellation diagram, the symbol decision area is smaller than in a classical QPSK constellation as shown in Chapter 12. Hence, a given RMS error vector magnitude (EVM) that does not cause symbol errors in QPSK may do so in 16QAM. In HSDPA, the throughput is adjusted based upon CQI feedback from the UE, with a linear function between SNR and CQI. The higher the SNR, the lower the effective coding rate applied to the physical channel. As opposed to Rel99, the turbo encoder/decoder settings are maintained unchanged, and instead, variable puncturing together with incremental redundancy is used to adjust the throughput. In this respect, the introduction of HSDPA sets two new RF requirements:

- At low $I_{or}$, the UE RF receive chain must guarantee a sufficiently high SNR to deliver maximum throughput. For the highest categories, an SNR of 15 dB is required [27]. As far as the RF section is concerned, this translates into tailoring the distribution of gain stages, including the LNA gain steps, so that the AGC lookup table delivers the highest SNR over the greatest range of $I_{or}$,
- At high $I_{or}$, i.e., under high $I_{or}/I_{oc}$ conditions, the UE RF receive chain must guarantee a low EVM performance. The EVM requirement is defined by the acceptable throughput loss for a given modulation and coding rate scheme. A good reference for UE design is the EVM requirement

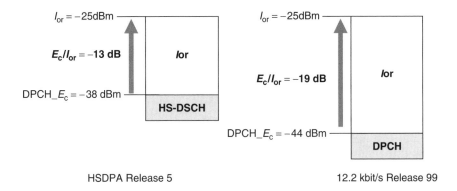

**Figure 20.17**   Maximum input power test case in Release 99 and Release 5 handsets

defined in [23] which sets a maximum of 12.5%. In conformance test, the Node B EVM is usually much better than this value. Yet, to ensure good field performance, this budget must be shared between the UE and the Node B. Obviously the lower the RF-IC RX chain EVM, the better, and with the arguments presented in earlier sections, today an $EVM_{RF-IC} < 5\%$ is easily achievable [28].

In 3GPP [9], a specific HSDPA RF conformance test case has been defined to check the impact of RF EVM onto throughput performance at maximum input power. This test represents a UE located close to the base station. The test cases for both Release 99 and Release 5 are shown in Figure 20.17. The total received power level is set to $-25\,\mathrm{dBm}$ in both cases. The quality metrics in HSDPA require a minimum throughput of 700 Kbit/s to be maintained using four codes and transmission in every third TTI. The $DPCH\_E_c/I_{or}$ ratio is set to $-19\,\mathrm{dB}$ for Release 99 and $-13\,\mathrm{dB}$ for the HS_DSCH. This test not only checks signal integrity, but also ensures that LNA and mixer linearity have been designed with sufficient margins to avoid non-linear distortions of the wanted signal.

## 20.4   Improving Talk-Time with DTX/DRX

Despite the impressive progress made in the design of dual-mode GSM-WCDMA system solutions (Section 20.1), even the best WCDMA handset today barely matches the talk-time (TT) of a GSM phone. This is partly explained by the full duplex nature of FDD-WCDMA, in which the entire cellular circuitry must be maintained powered in the cell-DCH state. Even with the anticipated further shrinking of CMOS processes, the discussion in Section 20.4.1 shows that the maturity of future WCDMA handsets are on the brink of reaching a plateau. This will take WCDMA TT to a level matching at best that of a modern GSM chipset. Therefore, only a key change to the access technology can help commercial products make a significant step in improving 3G phones TT. The examples presented show that the power consumption of next generation WCDMA handsets is dominated by the RF subsystem, the rest of the UE, including baseband modem, being a minor contributor. In that respect, by allowing the gating of the RF subsystem, both concepts of Discontinuous Transmissions (DTX) and Discontinuous Reception (DRX) introduced in the Continuous Packet Connectivity (CPC) work item of 3GPP Release 7 (cf. Chapter 15), appear as a very promising 3GPP solution to increasing the UE battery lifetime.

Since talk-time improvement in CPC relies upon a gating concept, a model of the UE consumption vs. UE output power is required for each key battery resource consumer. This is the objective of Sections 20.4.2 to 20.4.4, in which two models are defined: a model of the best UE at the time of writing, (denoted '2009 UE') and a theoretical model of a next generation phone (denoted '2011 UE') derived by interpolating the power consumption of the 2009 UE key blocks. Battery life savings are

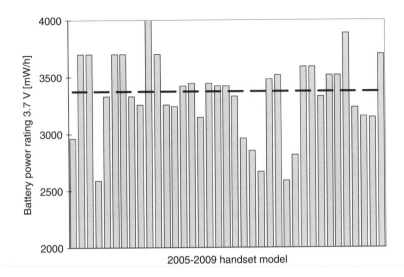

**Figure 20.18** 2005–2009 dual mode handset battery survey. PDA and smartphones excluded from selected UEs. Average rating = 3360 mW/h under 3.7 V, i.e. ≃ 900 mAH

estimated in Section 20.4.5 by applying a best case DTX/DRX gating pattern to both UE models. Each talk-time estimation uses a reference battery of 900 mAH applied to the GSM Association (GSMA) [29] guidelines. The battery capacity is defined as the average battery capacity surveyed over 41 'small form factor' dual mode UEs released over the past three years as shown in Figure 20.18.

In the following sections, the authors have conscientiously and carefully tried to estimate and forecast each contributor's power consumption that prevailed at the time of publication. However, the presented material may change as new data become available and the reader is encouraged to consult a variety of sources. For this reason, the presented estimations should be used as a tool to help better understand the technology relative trends and should not be considered as absolute performance indicators.

## 20.4.1 Talk-Time Benchmark of Recent WCDMA Handsets

Benchmark measurements are often obsolete at the time of publication because of the two to three year lead time required to produce a mobile phone. However, they provide very useful trend indicators. This section focuses on experimental assessments of various WCDMA UEs because power consumption breakdown analyses are rarely available to the public, mostly for confidentiality reasons. For the sake of simplicity, the experimental results of the best UE are split into four key contributors in Sections 20.4.2, 20.4.3 and 20.4.4:

- Radio receiver chain
- Radio transmitter chain
- Power amplifier
- A global category entitled 'rest of the UE' which encompasses Power Management Unit (voltage regulators, both linear and switched mode DC/DC), BB-IC modem, voice activity related circuitry and external DDR memory accesses.

The experimental setup uses a universal radio-communication tester (a CMU200 from R&S™) as a Node B emulator, and an RF co-axial connection to the UE antenna port. Power consumption is monitored according to the guidelines of [29] in both a voice and a reference measurement channel (RMC) 12k2 connection. [29] defines the mobile phone power consumption test methods. Throughout the rest of this section, this measurement technique is referred to as 'DG09'. Five representative UEs have been selected to reflect the evolution of WCDMA handset power consumption with a focus to measure the performance of the cellular core circuitry. Each UE is measured inside a dark faraday cage, which ensures that adaptive backlight schemes (if any) dim the backlight to its minimum value to prevent excessive power consumption during voice calls. PDA and smart phones are excluded from the analysis due to their large LCD screens and associated backlighting circuitry.

The UEs' main features are:

- class 3, small form factor designs,
- manufactured by the top 4 market leaders,
- built with chipset belonging to the 3 top market leaders,
- commercially released from Q1-2007 to Q3 2009, thereby representing a snapshot of the state of the art performance in the 2005–2007 timeframe.

Figure 20.19 shows the RMC 12k2 measured power consumption against the UE transmitted output power. Apart from UE 'E' which is a mono-band, band IV product, all other UEs are tested in band I.

It is interesting to note that power consumption is nearly constant across the output power range of −50 dBm to approximately −10 dBm. Furthermore, sudden small power consumption steps occur at −10 dBm for UE 'C' and 'D', +10 dBm and +13 dBm for UEs 'B' and 'E' respectively. These

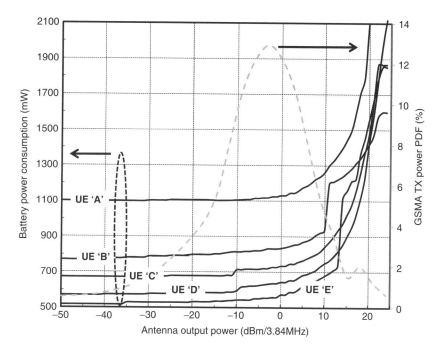

**Figure 20.19** Measured WCDMA handset power consumption vs. output power in a static RMC 12k2 test case. The GSMA UE transmit power probability density function overlaid in dashed lines, secondary *y*-axis

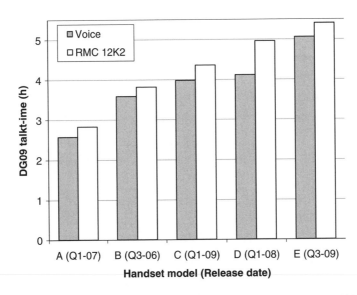

**Figure 20.20** UE equivalent talk-time integrated over the DG09 PDF and normalized to our 900 mAH reference battery. Grey: voice, white: RMC 12k2

steps are signs of different PA power management schemes: UE 'C' and 'D' use a closed loop PA bias and power supply control scheme, UE 'B' uses an hybrid scheme where both power supply and PA mode switching is used, while in UE 'E', the PA is directly connected to the battery supply, and uses gain/mode switching only.

The measurements presented in Figure 20.20 indicate that from UE 'A', released in Q1-2007, to UE 'E' released in Q3-2009, the battery life has nearly doubled, from 2.5h to approx 4.5h respectively. Also note the significant difference from UE to UE between a voice TT and an RMC 12k2 TT. This is partly due to each manufacturer's choice of LCD and backlight management strategies. For example, in UE 'C', both LCD and backlight dim quickly but the LCD stays 'on' permanently over the whole duration of the call. In UE 'D', the LCD remains in the 'on' state, but backlight circuitry is dimmed dynamically based upon the feedback of an ambient light sensor resulting in higher voice call power consumption. UE 'E' uses a more radical approach where both backlight and LCD appear switched off after a given delay, thereby minimizing the extra power consumption during voice calls.

We choose UE 'E' as the baseline to build the '2009 UE' power consumption model. A maturity trend is then estimated for each key contributor in order to derive a snapshot of what could be expected in 2 to 3 years time. The resulting UE power consumption model, called the '2011 UE' model, is used to compute battery savings in an HSPA phone using DTX/DRX.

## 20.4.2 Trend in RF-IC Power Consumption and Model

The power consumption breakdown analysis is best assessed at −50 dBm since all RF circuitry power consumption is at its minimum. Figure 20.21 shows that UE 'E' consumes 514 mW at that power level.

### 20.4.2.1 Radio Receiver Consumption Model

With a high level of integration and confidentiality associated with sophisticated RF ICs, estimating either RF-RX or RF-TX power consumption is not a simple exercise. We use a visual inspection of

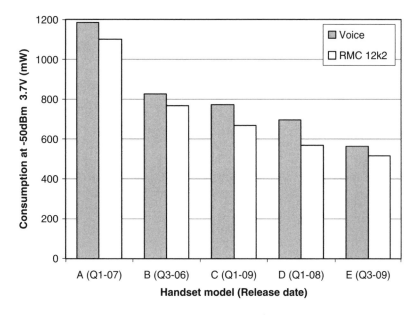

**Figure 20.21**   UE power consumption at $-50\,\text{dBm}$ output power: Grey: voice, white: RMC 12k2

the IC markings of the tested UEs to identify the chipmakers. We then assume the recent RF-IC design publications of these companies provide a reliable source of information on power consumption.

A survey of IEEE publications is presented in Figure 20.22. Consumptions are scaled to a 3.7 V battery, using either a linear voltage regulator or a DC/DC converter with 87% efficiency supply. The most

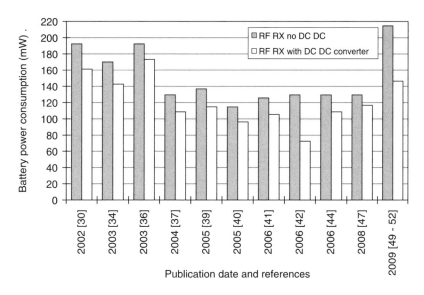

**Figure 20.22**   RF receiver power consumption survey published in IEEE, normalized to a 3.7 V battery. Grey: linear regulator, white: DC/DC converter with 88% efficiency

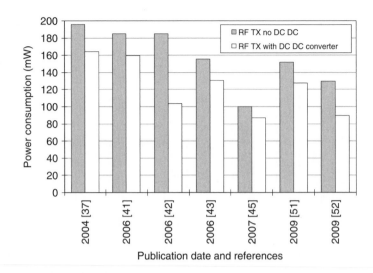

**Figure 20.23** IEEE survey of RF-IC transmitter power consumption at their minimum rated transmit power, normalized to a 3.7 V battery. Grey: linear regulator, white: DC/DC converter with 88% efficiency. 18 mW is added to [52] reflect our assumption of a linear regulator supply to the PLL synthesiser (approx 4.8 mA/2.7 V)

recent publications indicate that RF-RX power consumptions are converging towards 137 mW (37 mA under 3.7 V). Based on visual inspections of the IC markings in UE 'E', we use data presented in [49] and [52] to model the '2009 UE' RF receiver, i.e. at approximately 148 mW (40 mA under 3.7 V).

We assume that by 2011, RF IC solutions will mostly be designed in CMOS. As stated in Section 20.5, this choice of process not only reduces the cost of manufacturing, but also allows shifting most of the Bi-CMOS analog circuitry, such as channel filters, into the digital domain. In turn, this shifts current consumption onto a lower supply domain that modern CMOS processes offer. In conjunction with the use of high efficiency DC/DC converters, the net result is a battery power consumption gain. We use a conservative 2011 model, where the receiver chain consumption is set to 35 mA/3.7 V, i.e. about 130 mW. This power consumption is assumed constant across the whole receiver dynamic range.

In the longer term, it is likely that WCDMA transceivers will benefit from the natural LTE HW reconfigurability requirements. It is therefore anticipated that WCDMA RF RX chain will be able to adapt its power consumption according to the air interface context. For example, a different configuration could be used in HSDPA mode than in WCDMA mode. It would also make sense to use adaptive IP2 as a function of the UE own TX power since UEs spend very little time transmitting at their maximum output power. A further few extra mA could be saved with such schemes, especially at low output powers.

### 20.4.2.2 Radio Transmitter Consumption Model

The survey of RF transmitter IC design shown in Figure 20.23 indicates a decrease in power consumption over time, but the trend is less obvious than for the RX chains.

Power consumption extraction is less accurate in TX chains mainly because, as opposed to RX chains, the power consumption depends on the transmitted output power. The '2009 UE' TX power consumption model is based on [49] and [52]. The data in [52] is normalized to our reference battery, as shown in Figure 20.24.

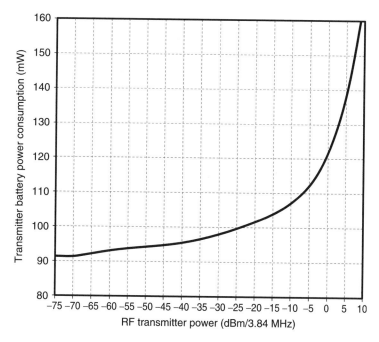

**Figure 20.24** RF TX IC power consumption against transmit IC output power from [52]. Assumes 88% DC/DC efficiency, 3.7 V battery voltage and an extra 4.8 mA at 2.7 V for PLL-synthesizer (18 mW) which we assume to be supplied via a linear voltage regulator

Careful PCB inspection of UE 'E' and probing of the 3G PA control pins estimate the following front-end parameters when the UE is transmitting −50 dBm:

- PA to antenna port insertion losses of ≃ 2.5 dB
- A PA gain of approximately 16 dB,

Thus, the RF TX IC output power is roughly −63.5 dBm. With reference to Figure 20.24, this translates into a 92 mW power consumption. This model is considered as the benchmark for our '2011 UE'.

## 20.4.3 Power Amplifier Control Schemes and Power Consumption Model

The PA power consumption/ACLR design trade-off has been the subject of many studies, not only in 3GPP contributions, but also in commercial platforms (Section 20.2.1). To date, two main optimization schemes are used in production:

- A switched mode scheme which can be used in either an open loop or a closed loop fashion, with two or three power mode settings pre-defined by the PA manufacturer. Variants include PA directly supplied from the battery, or one PA supply voltage step supplied by a programmable DC/DC converter. This scheme can be clearly seen in UE 'B' and 'E' in Figure 20.19.
- A variable bias scheme operated in a closed loop configuration, where continuous PA bias and PA power supply adjustments are achieved via use of auxiliary DACs and a DC/DC converter. This is the case of UE 'C' and 'D'.

### 20.4.3.1 Variable Bias and Supply Scheme

The principle of operation of the scheme is simple, but its implementation far more challenging. The scheme consists in tuning the PA bias and supply voltage to optimize the power consumption/ACLR trade-off across a range of output powers over which PA contributions impact the UE TT performance. In practice, the scheme requires the use of an RF power detector, a directional coupler, and an auxiliary ADC to sample the PA output power. The closed loop algorithm compares the computed RF rms power with a target value and makes small TX chain gain corrections until the target is reached. Since the 3GPP transmit power control (TPC) dynamic range ($-50$ to 24 dBm) exceeds most cost effective power detectors, the loop can only be closed over a limited output power range. For example, in handsets 'C' and 'D', a sudden consumption increase occurs at Pout $\geq -10$ dBm, implying that the loop operates over nearly 34 dB dynamic range. Figure 20.25 (left) illustrates the overall loop architecture.

The associated PA gain variations of Figure 20.25 (right) require continuous, real-time adjustments via either the RF TX VGA, or via digital and/or analog I/Q fine tuning. Such loops offer numerous advantages:

- Relaxes both the RF transmitter and the PA gain stability requirements, since they are detected and corrected automatically.
- Allows fine tuning of both PA supply and bias voltage to deliver optimum ACLR/talk-time trade-off over the most interesting range of UE output power.
- Eases the replacement of a given PA model with another.
- Delivers accurate power control performance independently of PA gain variations.

Yet, this scheme comes at the expense of a higher bill of material (BOM) and a higher SW complexity than the switched mode scheme.

### 20.4.3.2 Switched Mode Control Scheme

Mode switched PAs are controlled via a dedicated HW pin, which sets the PA into two or three biasing states, each optimized for a given range of output power. For example, a low, mid and high power mode is made available in [53]. Certain PAs also support scaling of their power supply below a certain threshold to provide further savings [14]. The concept is shown in Figure 20.26.

Beyond the low BOM inherent to this control scheme, this solution remains attractive today. It delivers competitive ACLR performance while offering near identical or better TT performance than a closed-loop operation. However, there are several challenges associated with the scheme as discussed in Section 20.2.1. In particular, when PA supply switching is used, the associated gain transient duration can easily excess the duration of a WCDMA timeslot. Specific correction look-up tables are required, via a mix of mass production calibration, and proprietary compensation techniques.

### 20.4.3.3 PA Power Consumption Model

Measurements performed on evaluation boards presented in Figure 20.27 show that the PA operated in a closed loop fashion delivers superior ACLR performance. However, there is a slight power consumption advantage to switched mode and switched supply PA, which consumes nearly 10 mA less current at $-50$ dBm. Integrating both profiles over the DG09 PDF, the switched mode PA saves 20 mW which translates in 20 minute extra battery life.

In the context of a widespread use of low cost, large volume solutions, the switched mode scheme presents a significant cost/performance trade-off advantage. Thus, our 2011 PA model relies on the associated transfer function plotted in Figure 20.27.

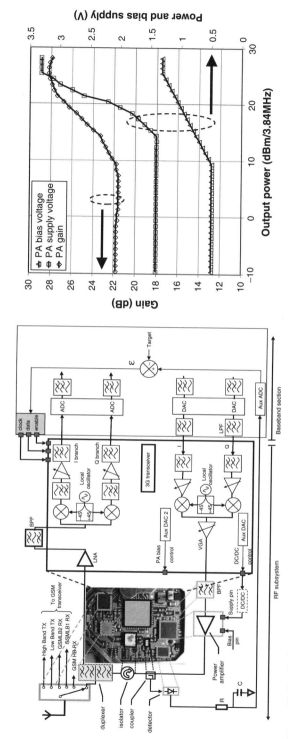

**Figure 20.25** Left: example of a closed loop HW implementation similar to that observed in UE "C" and "D" PCB [7]. Right: PA gain measurements under an example bias and supply profile

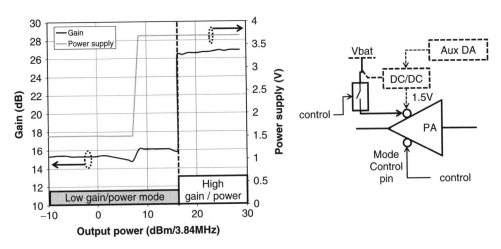

**Figure 20.26** Mode and supply switching concept showing gain steps due to supply switching and mode control. Gain on primary left *y*-axis, power supply on right *y*-axis

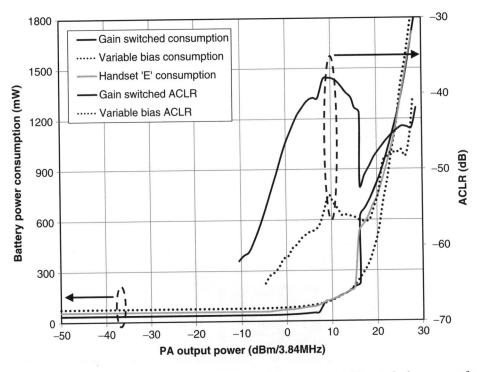

**Figure 20.27** Comparison of ACLR vs. power consumption performance of both PA control schemes vs. performance of PA in handset "E". Handset "E" PA data is extracted from the manufacturer's datasheet. Primary *y*-axis: power consumption, secondary *y*-axis: ACLR

## 20.4.4  UE Power Consumption Models

### 20.4.4.1  Estimating the 'Rest of the UE 'E' 'Power Consumption

The 'rest of the UE' power consumption depends on many variables: modem architecture (Rake, G-Rake or equalizer), power supply distribution and adaptive voltage and frequency scaling strategies, proprietary software and algorithms activity etc.

Restricting the number of contributors that fall under the term 'Rest of the UE' to only two groups simplifies the analysis: a baseband category includes digital baseband (DBB), analog baseband (ABB) as well as audio and clock generation units, and another category regroups the contributions of both power management unit (PMU) and external DDR memory. Both LCD and backlight are ignored. In UE 'E', the rest of the phone power consumption is estimated by performing a simple subtraction from the data collected at −50 dBm as shown in Table 20.3 and illustrated in Figure 20.28. It is worth noting that the RF subsystem contributes to 58% of the total UE power consumption.

### 20.4.4.2  Estimating the 'Rest of UE' Model

The 2011 model assumes a shrinking of DBB and ABB into the next CMOS process node. A common rule of thumb states that the power consumption is reduced by 30 to 40% when changing nodes. We assume that by the 2011–2012 timeframe the 'rest of UE' consumption should approach 130 mW (35 mA) including analog audio and DDR activity. The assumption is based upon a CMOS 40 nm solution with the use of low voltage supply rails (1 V or below), and extensive use of dynamic voltage and frequency scaling according to CPU/DSP activity. The use of low digital supply voltage

**Table 20.3**  UE 'E' power consumption breakdown at −50 dBm in an RMC12k2

| Contributor | Power consumption |
| --- | --- |
| Total UE power consumption | 514 mW (139 mA @ 3.7 V) |
| PA (cf. Figure 20.27) from PA vendor datasheet | 59 mW (16 mA @ 3.7 V) |
| RF receiver chain (cf. Section 20.4.2.1), | 148 mW (40 mA @ 3.7 V) |
| RF transmitter chain (cf. Section 20.4.2.2.) | 92 mW (25 mA @ 3.7 V) |
| Rest of the UE contribution | 514 mW − (59 + 148 + 92) = 215 mW (≃ 60 mA/3.7 V) |

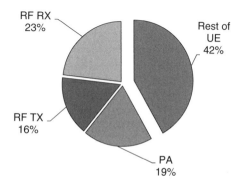

**Figure 20.28**  UE 'E' power consumption breakdown estimate at −50 dBm, rmc 12k2

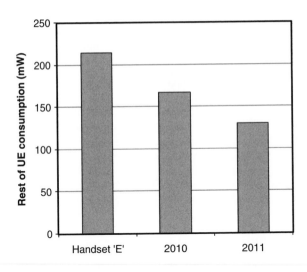

**Figure 20.29** "rest of the UE" power consumption trend. For illustration purposes

rails benefits from high DC-DC efficiency and allows greater battery life savings. We anticipate the emergence of intermediate solutions in 2010 at about 160 mW as shown in Figure 20.29.

### 20.4.4.3 Estimating Voice Activity Related Power Consumption

From the list of tested handsets, UE 'E' is the best device to evaluate the extra power consumption due to voice activity. This is because the phone's LCD and backlight circuitry appears completely switched off after a few seconds delay during voice calls. With reference to Figure 20.21, this is estimated at $\simeq 567 - 514 = 53$ mW (14 mA).

The resulting UE models are summarized in Figure 20.30. The UE 'E' model matches with a 53 mW offset fairly accurately the experimental measurements reported in Figure 20.19. In the following we assume that the PMU/DDR contributions account for up to 11% of the rest of the UE consumption.

### 20.4.4.4 Talk-time Evolutions for 2011 WCDMA

Table 20.4 summarizes the power consumption of each UE model integrated over the DG09 profile. The right hand-side column shows the scaling factors used to derive battery life savings in the HSPA handset. We have deliberately adopted a conservative approach by assuming that only 20% of the baseband contribution is gated during DTX/DRX.

The WCDMA 2011 power consumption is clearly dominated by the RF subsystem group (Figure 20.31) which now accounts for up to 72% of the total consumption. By allowing gating of the RF circuitry, one can expect that the use of DTX/DRX will provide optimum battery life savings. Thanks to the power savings assumed through CMOS shrinking, the 2011 handset has benefited from a significant reduction of the BB power consumption. Looking at the resulting pie chart of Figure 20.31, the BB now becomes a minor contributor to the total consumption. Consequently, shrinking even further would only bring minimal improvements to the total consumption. Similar comments can be made to the RF IC components. We can therefore consider that WCDMA technology will reach a power consumption plateau once this split becomes a commercial reality.

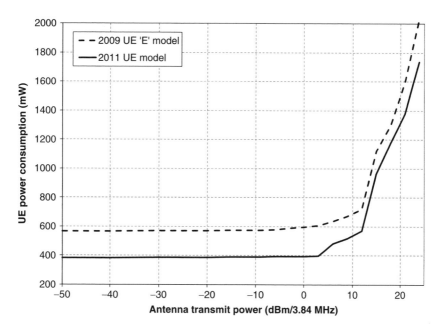

**Figure 20.30** UE power consumption model vs. transmit output power: dashed line UE 'E' model, plain lines: 2011 UE model

**Table 20.4** DG09 WCDMA power consumption split using our 2009 and 2011 UE models. Power consumption scaling factors due to HSPA DTX/DRX are listed in right-hand side column

|  |  | 2009 WCDMA UE 'E' | 2011 WCDMA model | 2013 impact of HSPA DTX/DRX operation |
|---|---|---|---|---|
| Power Amplifier | (mW) | 122 | 93 | × TX activity |
| RF transmit IC | (mW) | 103 | 103 | × TX activity |
| RF receive IC | (mW) | 148 | 130 | × RX activity |
| Baseband (DBB, ABB, audio & clocks) | (mW) | 239 | 130 | Assume 80% fixed – 20% L1 activity dependent. Variable part is multiplied by the average of RX and TX activity. |
| PMU-DDR | (mW) | 29 | 14 | Fixed consumption |
| Total | (mW) | 641 | 456 |  |
| Equivalent talk-time | (h) | 5.1 | 7.3 |  |

## 20.4.5  Talk-Time Improvements in Circuit Switched Voice over HSPA with DTX/DRX

### 20.4.5.1  L1 Activity Estimation Assumptions

Three scenarios are used to derive the TX and RX activity factors of Table 20.4:

1. UL voice, DL silence insertion descriptor (SID) frames – with 45% probability of occurrence,
2. DL voice, UL SID – with 45% probability of occurrence,
3. DL and UL SID – with 10% probability of occurrence.

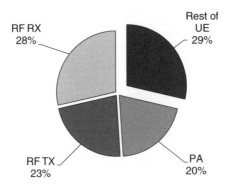

**Figure 20.31**   WCDMA 2011 power consumption split over DG09 PDF

In the following, the activity factors are estimated and graphically illustrated with a set of timing diagrams. In each scenario, the estimations are based upon a near 'best case' set of L1 assumptions and represent the best theoretical activity factors that could be expected in a HSPA DTX/DRX handset. In particular, we assume:

- Nearly ideal timing alignment between UL and DL frames.
- The downlink E-HICH ACK delay jitter of 9.7 to 12.6 time slot is taken into account in each timing diagram.
- 2 pre-amble and 1 post-amble slots are assumed during uplink DPCCH gated transmissions.
- During DRX, the reception gaps of the F-DPCH follow the uplink transmission pattern. This is because the Node B needs to compute UE TPC commands based upon initial reception of uplink pilot information.
- To avoid excessive overloading of the timing diagrams, the reception pattern of E-AGCH (or E-RGCH) overlaps that of the E-HICH reception. We have therefore removed this activity from the timing diagrams.

We also assume the following HW implementation of a specific set of parameters:

- DL AGC requires 1 slot to converge upon each RX chain wake-up.
- Channel estimation requires approximately an average of 4 slots every 40 ms.
- TX chain settling time is dominated by the RF LO-PLL, at a worst case 1/3 of a slot.

### Uplink Voice, Downlink SID L1 Activity

Figure 20.32 and Table 20.5 show one possible example of both uplink and downlink activity factors during an uplink VoIP transmission, while downlink frames are SID. It is interesting to note that the RX activity factor remains fairly high (45%). This is due to the fact that RX activity is tied to the transmission pattern of HSUPA packets for which reception of E-HICH (and E-AGCH or E-RGCH) is required.

### Downlink Voice, Uplink SID L1 Activity

With 26% TX and approximately 51% RX activity, the scenario illustrated in Figure 20.33 and summarized in Table 20.6 results in a near identical activity factor than uplink voice. Timing offsets have been adjusted in order to maximize the overlapping of both TX and RX activities.

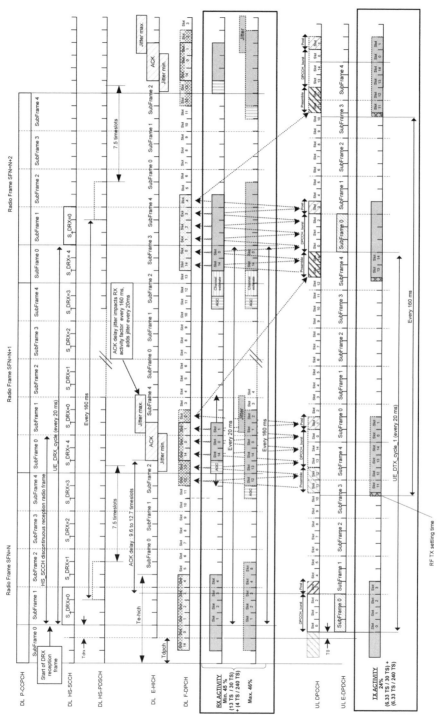

**Figure 20.32** Uplink voice, downlink SID timing diagram example

**Table 20.5**   TX and RX activity factor summary in uplink voice, downlink SID

|  | Uplink activity factor |
| --- | --- |
| First transmission | 2 ms of TX + postamble & preamble<br>= 4 ms every 20 ms = 20% |
| Retransmission (10%) | 2% |
| Downlink ACK/NACK | 6 slots = 4 ms every 160 ms = 2.5% |
| TX PLL settling time | 0.2 ms per 20 ms + 0.2 per 160 ms = 1.1% |
| Total TX activity | 26% |
|  | Downlink activity factor |
| DRX cycle | 6 slots per 20 ms = 20% |
| Downlink reception due to uplink<br>  ACK/NACK | 3 slots per 10 ms = 10% |
| SID related activity | Min: 3 slots per 160 ms = 1.2%<br>Max: 5 slots per 160 ms = 2% |
| AGC and channel estimation activity | 4 slots per 20 ms = 13.3% |
| Total RX activity | 45 to 46% |

***Uplink and Downlink SID L1 Activity***
Activity during both uplink and downlink SID frames is summarized in Figure 20.34 and Table 20.7.

#### 20.4.5.2   Talk-time Estimation in HSPA

The HSPA activity factors without packet bundling are summarized in Table 20.8.

Further savings can be reached using two packet bundling as shown in Table 20.9.

Talk-times are computed over the standard DG09 profile for WCDMA. In HSPA mode, DTX or DPCCH uplink gating significantly reduces the uplink interference. To account for the lower cell noise rise in DTX, the HSPA talk time is computed by integrating the UE gated models integrated over a shifted DG09 PDF profile. We assume that the UE average TX power is reduced by 30%. The instantaneous UE TX peak power is adjusted accordingly so that the new target average output power is met.

Figure 20.35 compares the associated talk times for WCDMA and HSPA both for 2009 and 2011 technologies. The talk time with HSPA can be even up to 15 hours if LCD and backlight is not considered and up to 12 hours with 2011 technology if LCD power consumption is included, representing 100% and 80% battery savings respectively. Taking LCD activity into account, the talk-time gain of HSPA compared to WCDMA is 65% to 80% for 2009 and 2011 technologies. As predicted, the savings are greater in the 2011 model than in 2009 UE, because of the higher RF circuitry contribution (cf. Figure 20.31). The neighbor cell measurement and reporting activity is not considered in these calculations. On the other hand, the use of packet bundling can further improve the talk time.

## 20.5   Multi-Mode/Band Challenges

### 20.5.1   From Mono-Mode/Mono-Band to Multi-Mode/Multi-Band and Diversity

From 1992 to 2002, the mobile phone industry increased its service to the user through small incremental steps building on the widespread GSM 2G standard. It started with voice only in the European 900 MHz band and later expanded to the 1800 MHz band. Data calls with GPRS and increased data

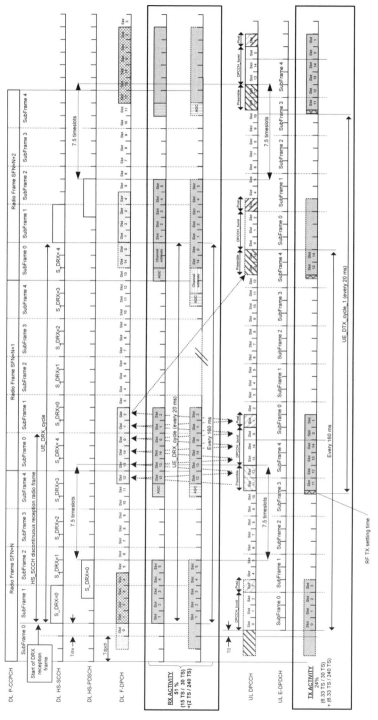

**Figure 20.33** Downlink voice, uplink SID timing diagram example

**Table 20.6**   TX and RX activity factor summary in downlink voice, uplink SID

| Uplink activity factor | |
| --- | --- |
| First transmission | 2 ms of TX + postamble & preamble |
| | = 4 ms every 20 ms = 20% |
| Retransmission (10%) | 2% |
| Downlink ACK/NACK | 6 slots = 4 ms every 160 ms = 2.5% |
| TX PLL settling time | 0.2 ms per 20 ms + 0.2 per 160 ms = 1.1% |
| Total TX activity | 26% |
| Downlink activity factor | |
| DRX cycle | 11 slots per 20 ms = 36.6% |
| | + retransmissions 1% = 37.6% |
| Downlink reception due to uplink transmission | 2 slots per 160 ms = 0.8% |
| AGC & channel estimation activity | 4 slots per 20 ms = 13.3% |
| Total RX activity | 51.7% |

rate with the more complex EDGE modulation were then introduced, similarly world-wide roaming based on 3 or 4 bands is now supported. All these new features were implemented at almost no extra cost, size nor power consumption as RF transmit architectures evolved towards solutions avoiding any external filtering thus saving two to four SAW filters depending on the number of bands supported.

More recently with the introduction of 3G, the radio designer has been faced with a rapid evolution of 3GPP standardization and an explosion of bands to be supported (Table 20.2). Additionally, 3G terminals must in most cases support 2G (GSM) standard and sometimes the 2G and 3G bands coincide which creates further complexity.

For further increased data rates, 2 antenna Multiple Input Multiple Output (MIMO) technology was introduced in Release 7, as covered in Chapter 15. Fortunately for the handset talk-time these MIMO schemes only require duplication of the receiver path, similar to the RX diversity in the receiver. However, MIMO still adds to the complexity of the radio design, especially in the front end of the RF system and the antenna design, particularly when multi-mode and multi-band constraints are considered.

## 20.5.2   New Requirements Due to Co-existence

The explosion of additional features in the devices results in the need for many wireless sub-systems to co-exist in the same handset. Co-existence itself is not new, as cellular 2G and Bluetooth have been found in phones for years. However, more recently the number of possible radio combinations in a phone has increased significantly with the addition of GPS, Mobile TV, WiFi, FM RX and soon WUSB, FM TX, ZigBee and Galileo. To add further complexity, some of these standards may support MIMO schemes and operate in different frequency bands depending on the region. A summary can be found in Table 20.10. Today's most complex handsets can require more than 10 antennas and cover more than 10 frequency bands.

The former combination of 2G and Bluetooth was relatively easy to handle for several reasons:

- The low transmitted power of 0 dBm for Bluetooth creates negligible interference for the cellular receiver.
- The relatively relaxed Bluetooth sensitivity makes it difficult for the cellular transmitter noise to desensitize the Bluetooth receiver, especially in the earpiece case where the link is quite robust.

- The relatively high separation between the 1 GHz or 2 GHz cellular bands and the 2.4 GHz Bluetooth band allows easy filtering of Bluetooth transmitted noise and of cellular transmitter leakage to the Bluetooth receiver.
- Both systems are TDMA and hence the probability of collisions is relatively low.

The new radios being added to cellular solutions pose more significant issues as most of them require very good sensitivity (GPS, WLAN, Mobile TV). Also, in some cases they have very low band separation: Bluetooth and WiFi are in the same band, GPS is very close to DCS TX band, Mobile TV

**Table 20.7**  TX and RX activity factor summary during uplink and downlink SID

| Uplink activity factor | |
| --- | --- |
| First transmission | 2 ms tx + pre/postambles |
| | = 4 ms per 160 ms = 2.5% |
| Retransmission (10%) | 0.25% |
| Downlink ACK/NACK | 6 slots = 4 ms every 160 ms = 2.5% |
| CQI | Aligned with transmission |
| TX PLL settling time | 0.2 ms for every transmission = 0.4 ms |
| | per 160 ms = 0.25% |
| Total TX activity | 5.5% |
| Downlink activity factor | |
| DRX cycle | 3 slots per 20 ms = 10% |
| SID related activity | Min: 10 slots per 160 ms = 4.1% |
| | Max: 12 slots per 160 ms = 5% |
| AGC and channel estimation activity | 3 slots per 20 ms + 1 slot per 160 ms = 10.4% |
| Total RX activity | 24.5 to 25.5% |

**Table 20.8**  Overall summary of TX and RX activity factors without packet bundling

| HSPA | Uplink | Downlink |
| --- | --- | --- |
| Uplink voice | 26% | 46% |
| Downlink voice | 26% | 52% |
| SID frames | 5.5% | 25.5% |
| Average activity | 24% | 47% |

**Table 20.9**  Overall summary of TX and RX activity factors with two packet bundling

| HSPA | Uplink | Downlink |
| --- | --- | --- |
| Uplink voice | 14.2% | 40.3% |
| Downlink voice | 15.1% | 26.7% |
| Average activity | 14% | 33% |

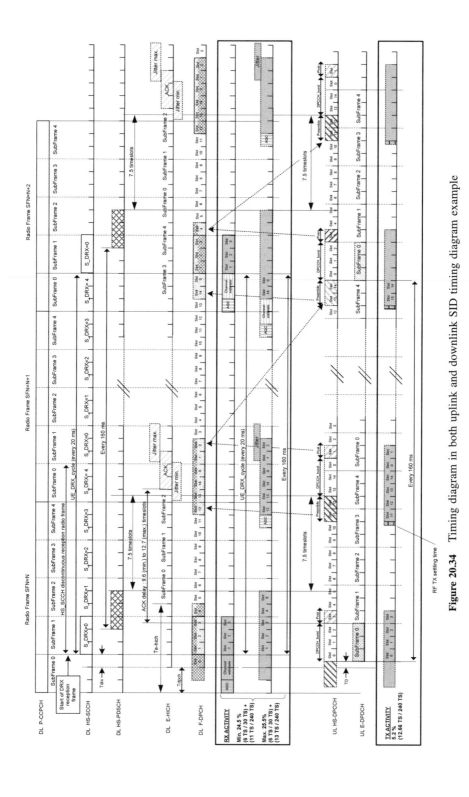

**Figure 20.34**  Timing diagram in both uplink and downlink SID timing diagram example

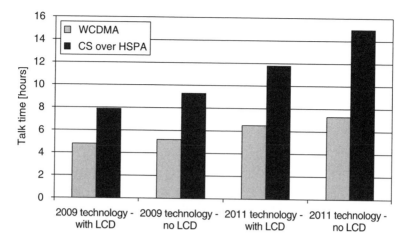

**Figure 20.35**   WCDMA vs. HSPA absolute talk time evolutions

**Table 20.10**   Set of possible wireless standard combination in future handsets

| Service | Standard | Number of bands | Modes | AntennaDiversity |
|---------|----------|-----------------|-------|------------------|
| Cellular | **GSM** | 4 | GMSK, GPRS, EDGE | Release 7 (RX) |
|  | **WCDMA** | 10 | R99, HSDPA, HSUPA | HSDPA (RX) |
|  | WiMax | 3 | OFDM | Depends on profile |
| WLAN | **WiFi** | 2 | 802.11a,b,g,n | 11n (RX&TX) |
| WPAN | **Bluetooth** | 1 | GFSK, 8PSK | no |
|  | UWB | 7 | OFDM | no |
| Localization | **GPS** | 1 | CDMA | no |
|  | Gallileo | 1 | CDMA | no |
| TVoM | **DVB-H&T** | 3 | OFDM | Play & record? |
|  | DMB | 2 | OFDM | Play & record? |
| Radio | **FM/RDS** | 1 | FM | no |
|  | FM TX | 1 | FM | no |
|  | AM | 1 | AM | no |
|  | DAB | 1 | OFDM | no |
|  | DRM | 1 | OFDM | no |

*Note:* Standards in bold are co-existence use-cases that can be found in today's products

being very close to GSM TX band. In the case of WLAN, the high transmit power and a relaxed spectral mask creates a source of strong interference and elevated noise into the frequency bands where other systems operate.

Simplifying the analysis to consider only one-to-one interaction and only antennas coupling, provides a valuable insight into the complexity of co-existence. The first order interactions between a first radio, A, and a second one, B, are listed below:

- Sharing the same band, for example with Bluetooth and WLAN, with 'air' time sharing.
- Transmitter noise of A falling into the receive band of B: This is the case when the bands are adjacent.

- AM content of transmit modulation of radio A being demodulated by receiver B through its second order non-linear response. This is particularly of importance for direct conversion receivers.
- Harmonics of transmitter A falling into radio B receive band.
- Spurious responses of receiver B falling into radio A transmit frequency. This can be the case for DVB-H receiver having spurious response in cellular and WLAN transmit bands, especially at harmonics of the receiver's local oscillator.
- Receiver B blocked by transmitter A leakage due to insufficient IP3 or input compression point.

In most of the cases, specific filtering or increased linearity may need to be used to solve the problem. The situation is complicated enough when only one to one interactions are considered, but the situation is even more complicated when three radios are considered. If in addition these different subsystems are integrated into the same shielded cavity and in some cases on the same chip, further interaction mechanism can occur such as coupling of the different local oscillators, clocks. A totally new level of problematic is created if radio elements from different vendors are integrated in the same device or with the possible use of memory cards with their own clock having undefined (from the 3GPP radio perspective) clock frequency and radio interference tolerance.

### 20.5.3   Front End Integration Strategies and Design Trends

To date, the dual-mode cellular solutions simply place a 2G radio next to a 3G radio, and in some cases the two modes don't even share a common antenna. Integration effort is underway, and dual-mode base band processors can be found. More recently, multi-band dual-mode RF transceivers have been available, but they are still two separate 2G and 3G transceivers in the same die or package. Supporting a 2G band calls for only one RX band select filter (usually a SAW filter) thanks to the success of filter-less architecture, such as the offset loop transmitter introduced in 1995. The full duplex nature of 3G results in the need for four filters to support a single 3G band (two in the duplexer filter, one between the LNA and Mixer to relax the front-end linearity and one between the transmitter and the PA to meet transmitted noise requirements in the receive band). Considering there are 10 bands in 3G, with four of them common to 2G and 3G, and that every region or operator requires a different subset of them, there is much pressure to simplify the banding options. A comparison between a quad-band 2G-only solution and a quad-band 2G plus triple-band 3G dual-mode solution block diagrams representative of today's solution illustrates the additional complexity in Figure 20.36 The cost and size of the 12 extra filters more than offset the cost of an extra RF-IC in the system.

### 20.5.4   Impact on Today's Architectures

The new requirements related to co-existence, together with the explosion of bands to be supported in cellular, call for architecture changes to reduce the terminal size and cost. The main objective is to eliminate the need for external filters. Effective reuse of hardware for 2G and 3G is also essential. Reconfigurable radios are attractive for multi-mode cellular capability, as they facilitate extensive sharing of hardware, as shown in Figure 20.37. However, they must provide low-noise transmitters and a receiver with high blocking tolerance. These architectures should also encompass standards such as WLAN and WiMAX, and they should also be compatible with 3G Long-Term Evolution covered in Chapter 17. If these architectural steps can be taken, then the multi-mode multi-band block diagram of Figure 20.36 is simplified to that shown in Figure 20.37 resulting in the elimination of eight external filters. This solution offers simplified support for multiple banding options, allowing up to eight 3G bands.

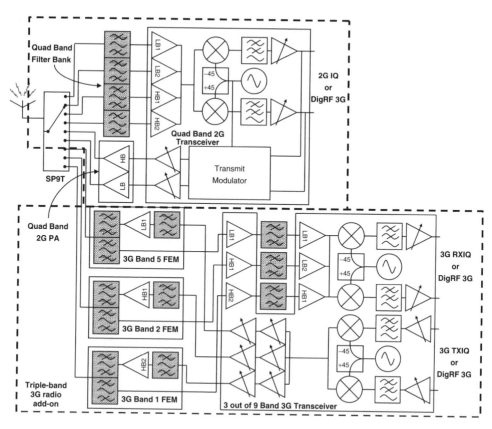

**Figure 20.36**  Dual mode Quad Band 2G triple band 3G radio block diagram. *Legend:* DigRF = Digital RF interface. FEM = Front End Module. LB = low band, HB = high band

Beyond cost and size optimization, the drive for a higher level of integration continues. Single-chip cellular phones are a reality in 2G today if the PA, RF front end and peripherals are ignored. This trend will certainly be followed for higher-feature dual-mode phones where the 2/3G solution plus extra solutions like GPS, Bluetooth or FM-radio will be integrated in a single system-on-chip (SoC). These more complex chips incorporating digital base-band functions will have to be processed in the latest deep sub-micrometer standard digital CMOS technology available. Integration of analogue and RF functions' deep sub-micrometer CMOS technology is challenging due to the wide process spread, device mismatches, leakage and noisy environment. There are thus advantages in looking to advanced architectures, particularly those that reduce the analogue circuitry. For the receiver, early digitization using a wideband high dynamic range Sigma–Delta ADC is attractive. The generation of a clean transmit signal by means of new modulator architectures with extensive digital pre-distortion also requires high dynamic range wideband DACs. Use of calibration and digital processing is also necessary to allow fast porting of analogue sections to the next CMOS node.

These efforts in developing more robust receivers, cleaner transmitters, and radios that can be easily reconfigured (with increased signal processing and performance in the digital domain) are paving the way to competitive Software-Defined Radios, which, as of today, cannot be considered for consumer applications.

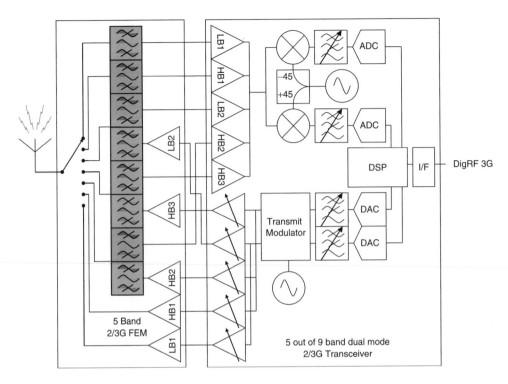

**Figure 20.37** Optimized 5 band, dual-mode 2/3G radio block diagram with SAW-less transceiver architecture. *Legend:* I/F = interface. DSP = digital signal processor

## 20.6   Conclusion

This chapter presented an overview of the design challenges for 3G RF and baseband design. From the early days of power hungry and bulky 3G devices, the market has reached the 500 million WCDMA subscribers milestone. The future WCDMA device will provide competitive talk and stand-by time compared to GSM only devices, while HSPA Release 7 and 8 handsets will further improve the battery life time with discontinuous transmission and reception for voice and for data connections. The trend is towards highly integrated architectures that should cover efficiently numerous frequency bands while still supporting several non-cellular features such as GPS etc. The use of multiple radio technologies in a single device will add further challenges to the device design to ensure proper operation with multiple radios active.

## References

[1] Dual sided GSM PCB picture, courtesy of Rolf Dieter Zimmermann, NXP Semiconductors Nuremberg internal 'museum', Germany.
[2] WCDMA dual-mode handset PCB pictures reproduced with authorization from Portelligent teardown reports, www.portelligent.com.
[3] Some WCDMA handset PCB pictures reproduced with permission from ST Ericsson.
[4] Portelligent, 'Sony Ericsson Walkman 900i', teardown report #118.20-060116-SW
[5] Portelligent, 'Sony Ericsson K530i' Report #11000-090114-TCe.
[6] Data extracted from several individual teardown reports produced by Portelligent.

[7] Moss, J. Holma, H., Toskala, A., Ahonen, T.T., Wiffen, N. and Noël, L. 'WCDMA system and operation', the University of Oxford, Department for Continuing Education, http://www.conted.ox.ac.uk/cpd/electronics/courses/UMTS.asp.

[8] 3GPP R4-060431, 'HS-DPCCH Phase Discontinuity Simulation Results', Ericsson, May 8-12, 2006.

[9] 3GPP TS 25.101V3.19.0 (2006-12), User Equipment (UE) Radio Transmission and Reception (FDD) (Release 1999), www.3GPP.org

[10] TSGR4#8(99) 99689, 3GPP TSG RAN WG4 #8 'Simulation Results for UE Downlink Performance Requirements', Motorola, Sophia Antipolis, France, 26–29 October 1999.

[11] Murata SAW duplexer for WCDMA band I (DPX), SAYFP1G95AA0B00 Datasheet, http://search.murata.co.jp/Ceramy/image/img/PDF/ENG/SAYFP1G95AA0B00.pdf.

[12] Holma, H. and Toskala, A. (eds), *LTE for UMTS-OFDMA and SC-FDMA Based Radio Access*, 2009, New York: John Wiley & Sons Ltd, Chapter 11.

[13] Avago Technologies, 'ACPM-7381-TR1UMTS2100 $4 \times 4$ Power Amplifier (1920-1980MHz)', http://www.avagotech.com/docs/AV02-0646EN.

[14] Anadigics 'AWU6601 HELP3™ Band 1/WCDMA/TD-SCDMA 3.4V/28.25 dBm Linear PA Module -Data Sheet - Rev 2.1', http://www.anadigics.com/products/handsets_datacards/wcdma_hspa_power_amplifiers/awu6601.

[15] Fujitsu Microelectronics Now Shipping Industry's First 3G SAW-less Transceiver: a New Multimode, Multiband UMTS/GPRS/EDGE Solution, 14 September 2009 press release, http://www.fujitsu.com/us/news/pr/fma_20090914.html.

[16] Laursen, S., 'Second Order Distortion in CMOS Direct Conversion Receivers for GSM', *Proc. Eur. Solid State Circuits Conf. (ESSCIRC)*, Sept. 1999, pp. 342–345.

[17] Manstretta, D., Brandolini, M. and Svelto, F. 'Second-Order Intermodulation Mechanisms in CMOS Downconverters', *IEEE Journal of Solid-State Circuits*, Vol. 38, 2003, pp 394–406.

[18] Iversen, C. R. 'A UTRA/FDD Receiver Architecture and LNA in CMOS Technology', PhD thesis, RF Integrated Systems and Circuits Group, Aalborg University, Denmark, November 2001.

[19] Jussila, J. 'Analog Baseband Circuits for WCDMA Direct-Conversion Receivers', Dissertation for the degree of Doctor of Science in Technology, Department of Electrical and Communications Engineering, Helsinki University of Technology, (Espoo, Finland), June 2003.

[20] Murata WCDMA receive band I SAW filter, SAFEB2G14FB0F00, www.murata.com.

[21] Kaczman, D. et al. 'A Single–Chip 10-Band WCDMA/HSDPA 4-Band GSM/EDGE SAW-less CMOS Receiver with DigRF 3G Interface and +90 dBm Iip2', *IEEE Journal of Solid-State Circuits*, Vol. 44, March 2009.

[22] Dufrêne, K. et al. 'Digital Adaptive Iip2 Calibration Scheme for CMOS Downconversion Mixers', *IEEE Journal of Solid-State Circuits*, Vol. 43, no. 11, November 2008.

[23] 3GPP TS 25.104, www.3GPP.org.

[24] Rohde and Schwarz Software WinIQSIM™ for Calculating I/Q Signals for Modulation Generator AMIQ, www.rsd.de.

[25] Olujide, A. and Demosthenous, A. 'Constant-Resistance CMOS Input Sampling Switch for GSM/WCDMA High Dynamic Range $\Sigma\Delta$ Modulators', *IEEE Transactions on Circuits and Systems: Regular Papers*, Vol. 55, No. 10, November 2008.

[26] Fujimoto, Y. et al. 'A 100MS/s 4MHz Bandwidth 70dB SNR $\Sigma\Delta$ ADC in 90nm CMOS', *IEEE Journal of Solid-State Circuits*, Vol. 44, no. 6, June 2009.

[27] R1-02-0675, 'Revised CQI Proposal', Motorola, Ericsson, Paris, France, April 9–12, 2002.

[28] Tenbroek, B. et al. 'Single-Chip Tri-Band WCDMA/HSDPA Transceiver without External SAW Filters and with Integrated TX Power Control', ISSCC 2008/SESSION 10/CELLULAR TRANSCEIVERS/10.2.

[29] GSM Association, DG09. v5.1 - Battery Life Measurement Technique, September 2009, http://gsmworld.com/newsroom/document-library/technical_documents.htm.

[30] Brunel, D. et al. 'A Highly Integrated 0.25um BiCMOS Chipset for 3G UMTS/CDMA Handset RF Sub-system', 2002 IEEE Radio Frequency Integrated Circuits Symposium.

[31] Aparin, V. et al. 'A Highly-Integrated Tri-Band/Quad-Mode SiGe BiCMOS RF-to-Baseband Receiver for Wireless CDMA/WCDMA/AMPS Applications with GPS Capability', ISSCC 2002/SESSION 14/CELLULAR RF WIRELESS/14.3.

[32] Ryynänen, J. et al. 'A Single-Chip Multimode Receiver for GSM900, DCS1800, PCS1900, and WCDMA', *IEEE Journal of Solid-State Circuits*, Vol. 38, No. 4, April 2003.

[33] Rogin, J. et al. 'A 1.5-V 45-mW Direct-Conversion WCDMA Receiver IC in 0.13- m CMOS', *IEEE Journal of Solid-State Circuits*, Vol. 38, No. 12, December 2003.

[34] Gharpurey, R. et al. 'A Direct-Conversion Receiver for the 3G WCDMA Standard', *IEEE Journal of Solid-State Circuits*, Vol. 38, No. 3, March 2003.

[35] Reynolds, S. K. et al. 'A Direct-Conversion Receiver IC for WCDMA Mobile Systems', *IEEE Journal of Solid-State Circuits*, Vol. 38, No. 9, September 2003.

[36] Yoshida, H. et al. 'Fully Differential Direct Conversion Receiver for WCDMA Using an Active Harmonic Mixer', 2003 IEEE Radio Frequency Integrated Circuits Symposium.

[37] Thomann, W. et al. ' A Single-chip 75-GHz/0.35-pm SiGe BiCMOS WCDMA Homodyne Transceiver for UMTS Mobiles', 2004 IEEE Radio Frequency Integrated Circuits Symposium.

[38] Gatta, F. et al. 'A Fully Integrated 0.18- m CMOS Direct Conversion Receiver Front-End with On-Chip LO for UMTS', *IEEE Journal of Solid-State Circuits*, Vol. 39, No. 1, January 2004.

[39] 'Fully-Integrated WCDMA SiGeC BiCMOS transceiver',Bruno Pellat et al. Proceedings of ESSCIRC, Grenoble, France, 2005, Paper 9.F.3.

[40] Tamura, M., Nakayama, T., Hino, Y. et al. 'A Fully Integrated Inter-Stage-Bandpass-Filter-Less Direct-Conversion Receiver for WCDMA, *IEEE RFIC, Symp*., June 2005, pp. 269–272.

[41] Kaczman, D. L. et al. 'A Single-Chip Tri-Band (2100, 1900, 850/800MHz) WCDMA/HSDPA Cellular Transceiver', *IEEE Journal of Solid-State Circuits*, Vol. 41, No. 5, May 2006, pp. 1122–1132.

[42] Tomiyama, H., Nishi, C., Ozawa, N. et al. 'A Low Voltage (1.8 V) Operation Triple Band WCDMA Transceiver IC', *IEEE RFIC Symp. Dig, Papers*, June 2006, pp. 165–168.

[43] Villain, F. and Burg, O. 'Fully Integrated Multi-Band WCDMA Transmitter with Minimum Carrier Leakage and Optimized Power Consumption Mode', Philips Semiconductors, Solid-State Circuits Conference, 2006. ESSCIRC 2006, *Proceedings of the 32nd European Conference*, Sept. 2006, pp. 267–270.

[44] Koller, R., Rühlicke, T., Pimingsdorfer, D. and Adler, B. 'A Single-chip 0.13 $\mu$m CMOS UMTS WCDMA Multi-band Transceiver,' IEEE RFIC Symp., June 2006. pp. 187–190.

[45] Jones, C. et al. 'Direct-Conversion WCDMA Transmitter with – 163 dBc/Hz Noise at 190MHz Offset', ISSCC 2007/Session 19/Cellular and Multi-Mode Transceivers/19.1.

[46] Eloranta, P. et al. 'A Multimode Transmitter in 0.13m CMOS Using Direct-Digital RF Modulator', *IEEE Journal of Solid-State Circuits*, Vol. 42, No. 12, December 2007.

[47] Tenbroekl, B. et al. 'Single-Chip Tri-Band WCDMA/HSDPA Transceiver without External SAW Filters and with Integrated TX Power Control', ISCC 2008, session 10, cellular transceiver 10.2, pp. 202–203.

[48] Mirzaei, A. and Darabi, H. 'A Low-Power WCDMA Transmitter with an Integrated Notch Filter', *IEEE Journal of Solid-State Circuits*, Vol. 43, No. 12, December 2008, pp. 2868–2881.

[49] Hadjichristos, A. et al. 'Single-Chip RF CMOS UMTS/EGSM Transceiver with Integrated Receive Diversity and GPS', 8–12 Feb. 2009, pp. 118–119, 119a.

[50] Gaborieau, O. et al. 'A SAW-Less Multi-Band WEDGE Receiver', ISSCC 2009/SESSION 6/CELLULAR AND TUNER/6.2, pp. 144–166.

[51] Sowlati, T. et al. 'Single-chip Multiband WCDMA/HSDPA/HSUPA/EGPRS Transceiver with Diversity Receiver and 3G DigRF Interface without SAW Filters in Transmit RX 3G Path', *Solid-State Circuits Conference - Digest of Technical Papers*, 2009, ISSCC 2009. IEEE International, 8–12 Feb. 2009, pp. 116–117, 117a.

[52] Cassia, M. et al. 'A Low-power CMOS SAW-Less Quad Band WCDMA/HSPA/HSPA+/1X/EGPRS Transmitter', *IEEE Journal of Solid-State Circuits*, Vol. 44, No. 7, July 2009.

[53] Avago Technologies 'ACPM-7382 UMTS Band1 (1920-1980MHz) 4 × 4 Power Amplifier Module', http://www.avagotech.com/docs/AV02-1890EN.

# Index